IMPA Monographs

This series, jointly established by IMPA and Springer, publishes advanced monographs giving authoritative accounts of current research in any field of mathematics, with emphasis on those fields that are closer to the areas currently supported at IMPA. The series gives well-written presentations of the "state-of-the-art" in fields of mathematical research and pointers to future directions of research.

Edson de Faria • Pablo Guarino

Dynamics of Circle Mappings

Second Edition

Springer

Edson de Faria
Instituto de Matemática e Estatística
Universidade de São Paulo
São Paulo, São Paulo, Brazil

Pablo Guarino
Instituto de Matemática e Estatística
Universidade Federal Fluminense
Niterói, Rio de Janeiro, Brazil

IMPA Monographs
ISBN 978-3-031-67494-5 ISBN 978-3-031-67495-2 (eBook)
https://doi.org/10.1007/978-3-031-67495-2

Mathematics Subject Classification: 37E10, 37C40, 37E20, 37F25, 37F31, 37F50

This Springer imprint is published by the registered company Springer Nature Switzerland AG
The registered company address is: Gewerbestrasse 11, 6330 Cham, Switzerland

À memória de meus pais, Jacyra e Dorival (EdF).

Aos meus filhos, Guga e Caio (PG).

Preface

One-dimensional Dynamics is a rich and beautiful subject, and the most authoritative work entirely dedicated to it still is, unquestionably, the book written by de Melo and van Strien (1993). Thus the reader may ask: why bother writing another book about this subject?

It is a fair question. The main reason is that much has happened since 1993: more than half of the present book's contents deals with recent developments in the area. Moreover, rather than aiming at being comprehensive, our book delves deeper into a specific topic in One-dimensional Dynamics, namely, the dynamics of *invertible circle maps*. Let us say a few words explaining how this topic fits into the general framework of the modern theory of Dynamical Systems.

One of the major general goals in the area of Dynamical Systems is to solve the *smooth classification problem*: given two smooth dynamical systems which are topologically equivalent, when are they smoothly equivalent? In somewhat vague terms, this problem is tantamount to understanding the fine-scale geometric properties of such systems.

In such general setting, and particularly in higher dimensions, the above classification problem seems rather daunting (perhaps even hopeless). Hence one should first attempt to understand low-dimensional systems. At least at an intuitive level, the problem should be much simpler for one-dimensional systems; after all, in dimension one the linear order structure and "lack of ambient space" should impose severe restrictions on the possible geometries of such systems, thereby facilitating their smooth classification. However, even here the problem turns out to be rather subtle. A basic distinction that must be made in the one-dimensional context is between *invertible* dynamics—to wit, homeomorphisms of the circle—and *non-invertible* dynamics, such as the dynamics of unimodal or multimodal maps of the interval (or the circle).

In this book—written for a series of lectures delivered by both authors at the *33rd Brazilian Mathematics Colloquium*—we deal with *invertible* dynamical systems on the circle, concentrating on two major classes: global diffeomorphisms and smooth homeomorphisms with critical points. In the case of smooth diffeomorphisms of the circle, deep results have been obtained from the mid to late seventies onwards, starting with M. Herman's thesis and culminating with the work of J.-C. Yoccoz, with important contributions by Y. Katznelson and D. Ornstein, among others. After describing those results, we will focus on the case of smooth homeomorphisms with critical points, a topic to which both authors have dedicated several years of research. In this context, the notions of *renormalization*, *rigidity*, and *universality* play a decisive role, and have been widely studied in the last 30 years.

The material in this book is divided into four parts. In the first part we study rigid rotations and then circle homeomorphisms, introducing the notion of *rotation number*, a dynamical invariant introduced by Poincaré at the end of the nineteenth century. We also describe some connections between dynamical properties of the rotation number with the theory of continued fractions. In the second part we study circle diffeomorphisms, presenting some classical results due to Denjoy and discussing some of the main ideas in the Arnold–Herman–Yoccoz theory. We present the subject by developing it from its basic principles in a self-contained way. In particular, together, these two initial parts can be used in a first graduate-level course on one-dimensional dynamics. The book contains around 140 exercises, varying widely in their level of difficulty; these should help the students enhance their understanding of the subject.

The third part of this book introduces *multicritical circle maps*, which are smooth homeomorphisms of the circle with a finite number of critical points, an important and active topic in the area of one-dimensional dynamics. The fourth and last part of this book is devoted to renormalization theory, focusing on the analysis of the fine geometric structure of orbits of multicritical circle maps, as well as on certain complex-analytic aspects of the subject. We will describe in these final chapters several important results by K. Khanin, M. Martens, C. McMullen, W. de Melo, D. Sullivan, A. Teplinsky and M. Yampolsky, among others. We would like to remark that, since these ideas are quite deep, the narrative in this final part is by necessity very sketchy.

Throughout the book, we provide, for the most part, complete proofs of several fundamental results in circle dynamics, such as the Poincaré classification, Denjoy's classical results and constructions, Arnold's conjugacy theorem for analytic circle diffeomorphisms with Diophantine rotation number (we also describe his counterexamples to linearizability), a conjugacy theorem for finitely smooth diffeomorphisms with Diophantine rotation number, Yoccoz's theorem on minimality of multicritical circle maps, the *real bounds*, quasisymmetric rigidity, the fact that exponential convergence of renormalization implies smooth rigidity, Lipschitz continuity of the renormalization operator (for maps with a single critical point), and the *complex bounds*. We also survey, skipping many details, the proof of the exponential convergence of renormalization for critical circle maps, both in the analytic and the smooth case. The book closes with a list of open questions and three appendices: the first describing some aspects of the ergodic theory of continued fractions, the second presenting a proof of a linearization theorem for finitely smooth diffeomorphisms with Diophantine rotation number, and the third discussing ergodic properties of a certain skew product over the Gauss map.

The present book is primarily aimed at graduate students and young researchers working in Dynamical Systems, but we hope it will have something to offer to other mathematicians interested in the subject. As prerequisites, it assumes that the reader is familiar with the contents of a standard graduate course in Real Analysis (including Metric Spaces, Measure Theory and basic Functional Analysis) in addition to some notions of Ergodic Theory and Dynamical Systems. In Chaps. 4, 11, 13, and 14, basic knowledge of Complex Analysis is needed as well.

Due to limitations of time and space, many interesting topics of circle dynamics have been left out of this book. These include interval exchange transformations, maps with break points, maps with flat spots, mode locking universality, dynamics of endomorphisms (including the notion of rotation set), thermodynamic formalism, random dynamical systems, and groups acting on the circle, among others.

In recent years, we have benefited from conversations with many friends and colleagues, among them Peter Hazard, Mikhail Lyubich, Marco Martens, Bruno Nussenzveig, Sebastian van Strien, Dennis Sullivan, Charles Tresser, Björn Winckler, Misha Yampolsky, and most notably Welington de Melo. Several parts of this book have been inspired by these interactions.

We are also grateful to two anonymous referees whose keen comments and suggestions have led to a substantial improvement of our book over the original colloquium lecture notes.

Finally, we would like to thank the organizers of the 33rd Brazilian Mathematics Colloquium for the opportunity to present the course on which this book is based. Special thanks go to Paulo Ney de Souza for his extremely professional editorial help. Readers are encouraged to send comments and suggestions as well as corrections to our institutional addresses.

São Paulo, Brazil Edson de Faria
Niterói, Brazil Pablo Guarino

Contents

Part I Basic Theory

1 Rotations ... 3
1.1 Topology and Combinatorics of Rotations ... 3
1.1.1 A Dichotomy ... 4
1.1.2 Sequence of Closest Returns ... 6
1.2 Rotations and Continued Fractions ... 7
1.2.1 Basic Theory of Continued Fractions ... 7
1.2.2 Best Approximations ... 12
1.3 Weyl's Equidistribution Theorem ... 14
1.3.1 Equidistribution ... 14
1.3.2 A Simple Application ... 17
1.4 Ergodicity of Irrational Rotations ... 18
Exercises ... 20

2 Homeomorphisms of the Circle ... 25
2.1 Translation and Rotation Numbers ... 25
2.1.1 The Classical Definition ... 26
2.1.2 The Order Definition ... 27
2.1.3 The Measure-Theoretic Definition ... 28
2.1.4 Properties of the Rotation Number ... 28
2.2 Topological Dynamics of Homeomorphisms ... 31
2.2.1 Rational Rotation Number ... 31
2.2.2 Irrational Rotation Number ... 32
2.3 Invariant Measures and Semi-Conjugacies ... 35
Exercises ... 37

Part II Diffeomorphisms

3 Diffeomorphisms: Denjoy Theory ... 43
3.1 The Naive Distortion Lemma ... 43
3.2 Denjoy's Theorem ... 45

3.2.1 The C^2 Version 45
3.2.2 The Bounded Variation Version 46
3.3 Denjoy's Examples 49
3.3.1 The Basic Construction 49
3.3.2 Moduli of Continuity 55
3.3.3 Further Results 57
3.4 Ergodic Properties 60
3.4.1 Ergodicity with Respect to Lebesgue Measure 60
3.4.2 Zero Lyapunov Exponents 62
3.4.3 Further Ergodicity Results 63
Exercises 68

4 Smooth Conjugacies to Rotations 73
4.1 Herman's Invariants 74
4.2 Small Denominators: Arnold's Theorem 77
4.2.1 The Linearized Equation 79
4.2.2 Non-linear Estimates 82
4.2.3 Proof of Arnold's Theorem 86
4.3 Counterexamples to Linearizability 89
4.3.1 One-Parameter Families 89
4.3.2 Residual Sets of Non-linearizable Parameters 91
4.3.3 Singular Measures and Conjugacies 93
4.4 Further Local Theory: The Brjuno Condition 97
4.5 Global Theory: Herman–Yoccoz Results and Beyond 99
Exercises 100

Part III Multicritical Circle Maps

5 Cross-Ratios and Distortion Tools 105
5.1 Cross-Ratios 105
5.2 The Schwarzian 106
5.2.1 Definition 106
5.2.2 Koebe's Nonlinearity Principle 108
5.2.3 The Minimum Principle 110
5.3 Distortion and Cross-Ratio Distortion 110
5.3.1 Koebe's Distortion Principle 110
5.3.2 Distortion and the Schwarzian 111
5.3.3 Behavior Near Critical Points 114
5.4 The Cross-Ratio Inequality 116
5.5 A Cancellation Lemma 118
Exercises 122

6 Topological Classification and the Real Bounds 125
6.1 Definition and Examples of Multicritical Circle Maps 125
6.1.1 Blaschke Products 126
6.1.2 The Arnold Family 128

6.2 Topological Classification ... 129
6.2.1 Dynamically Symmetric Intervals ... 130
6.2.2 Proof of Yoccoz's Theorem ... 134
6.3 Real a Priori Bounds ... 135
6.3.1 Dynamical Partitions ... 135
6.3.2 The Real Bounds ... 138
6.3.3 On the Notion of Comparability ... 143
6.4 First Consequences ... 144
6.4.1 C^1 Bounds ... 144
6.4.2 Sums of Polar Ratios ... 147
6.5 A Negative Schwarzian Property ... 148
6.6 Beau Bounds ... 150
Exercises ... 153

7 Quasisymmetric Rigidity ... 157
7.1 Quasisymmetry and Fine Grids ... 157
7.1.1 A Criterion for Quasisymmetry ... 159
7.1.2 A Criterion for Smoothness ... 162
7.2 Quasisymmetric Conjugacies ... 164
7.3 Almost Parabolic Maps ... 166
7.3.1 Yoccoz's Inequality ... 167
7.3.2 Balanced Decompositions ... 170
7.4 Quasisymmetric Rigidity ... 173
7.4.1 More on the Geometry of Dynamical Partitions ... 174
7.4.2 Building a Suitable Fine Grid ... 177
7.4.3 The Punchline ... 184
Exercises ... 184

8 Ergodic Aspects ... 187
8.1 The Integrability of log Df ... 187
8.2 No Invariant σ-Finite Measures ... 190
8.2.1 The Katznelson Criterion ... 191
8.2.2 Proof of Theorem 8.1 ... 194
8.2.3 Negative Schwarzian Redux ... 204
8.3 Lyapunov Exponents ... 208
8.3.1 The Collet–Eckmann Condition ... 209
8.3.2 The Key Step ... 209
8.4 Further Ergodic Properties ... 217
8.5 Hausdorff Dimension ... 218
Exercises ... 219

9 Orbit Flexibility ... 223
9.1 Geometric Equivalence of Orbits ... 223
9.1.1 Orbit-Flexibility ... 224
9.1.2 Statement for Unicritical Maps ... 225
9.1.3 Statements for Multicritical Maps ... 226

9.1.4 Centralizers ... 227
9.1.5 Unbounded Geometry ... 228
9.2 Renormalization Trails and Ancestors ... 229
9.3 The Skew Product ... 231
9.3.1 The Fiber Maps ... 232
9.3.2 The Skew Product ... 233
9.4 Proof of Theorem 9.6 ... 233
9.5 Even-Type Rotation Numbers ... 236
9.6 Proofs of Theorems 9.1 and 9.2 ... 239
9.7 The C^∞ Realization Lemma ... 242
9.7.1 Admissible Pairs ... 242
Exercises ... 244

Part IV Renormalization Theory

10 Smooth Rigidity and Renormalization ... 249
10.1 Smooth Rigidity ... 250
10.2 Renormalization of Commuting Pairs ... 253
10.3 A Fundamental Principle ... 258
10.3.1 Main Theorem ... 259
10.3.2 Comparing Orbits of Two Almost Parabolic Maps ... 260
10.3.3 Proof of Theorem 10.4 ... 263
10.4 The C^m-Approximation Lemma ... 270
10.5 Counterexamples to $C^{1+\alpha}$ Rigidity ... 273
10.5.1 Saddle-Node Surgery ... 274
10.5.2 The Counterexamples ... 276
Exercises ... 277

11 Quasiconformal Deformations ... 279
11.1 Quasiconformal Homeomorphisms ... 280
11.1.1 The Geometric Definition ... 280
11.1.2 The Analytic Definition ... 281
11.1.3 Measurable Riemann Mapping Theorem ... 282
11.2 A Simple Dynamical Application ... 283
11.3 Holomorphic Approximation Lemma ... 285
Exercises ... 288

12 Lipschitz Estimates for Renormalization ... 293
12.1 Lipschitz Estimates for Controlled Commuting Pairs ... 293
12.2 Standard Families ... 295
12.2.1 Glueing Procedure and Translations ... 296
12.2.2 Standard Families of Commuting Pairs ... 297
12.2.3 Renormalization of Standard Families ... 298
12.3 Orbit Deformations ... 301
12.4 Composition ... 316
12.5 Order ... 322

12.6 Synchronization 324
12.7 Lipschitz Estimate 327
Exercises 327

13 Exponential Convergence: The Smooth Case 329
13.1 The Shadowing Property 330
13.1.1 Extended Lifts of Critical Circle Maps 332
13.1.2 Almost Schwarz Inclusion 335
13.1.3 A Bidimensional Glueing Procedure 338
13.1.4 Main Perturbation 345
13.1.5 The Shadowing Sequence 353
13.2 Bounding the C^{r-1} Metric 355
13.3 Proof of the Exponential Convergence 356
13.4 The Attractor of Renormalization 357
Exercises 358

14 Renormalization: Holomorphic Methods 361
14.1 Sullivan's Program 361
14.2 Holomorphic Commuting Pairs 365
14.3 Pull-Back Argument 367
14.4 Existence and Limit-Set qc-Rigidity 367
14.4.1 Construction of Examples 367
14.5 Complex Bounds 373
14.6 McMullen's Dynamic Inflexibility Theorem 384
14.7 Proof of Exponential Convergence 386
14.8 Hyperbolicity of Renormalization 387
14.8.1 Cylindrical and Parabolic Renormalizations 388
Exercises 394

Epilogue 397

A Ergodic Theory of Continued Fractions 401
A.1 Expansions as Itineraries 401
A.2 The Gauss Measure and Almost Surely Properties 402
A.3 Diophantine Approximations Revisited 407
Exercises 409

B Cohomological Equations and Smooth Conjugacies 411
B.1 The Cohomological Equation 411
B.2 Solving the Cohomological Equation 412
B.3 Distortion Estimates 413
B.3.1 Distortion Estimates in $C^{1+\mathrm{BV}}$ 414
B.3.2 Distortion Estimates in $C^{2+\alpha}$ 416
B.4 C^1 Exponential Convergence 420
B.4.1 Further Distortion Estimates 422
B.5 The Diophantine Condition 423
B.5.1 Proof of Proposition B.3 424

B.6 Proof of Theorem 4.11 427
B.6.1 Hölder Continuity of the Invariant Density 429

C A Skew Product Over the Gauss Map 435
C.1 An Absolutely Continuous Invariant Measure 436
C.2 Markov Property 439
C.2.1 A Countable Markov Partition 439
C.3 Ergodicity 441
C.3.1 Bounding Jacobian Distortion 441
C.3.2 A Lebesgue Density Argument 444
C.3.3 End of Proof 446

References 446

Index 455

List of Symbols

$[M, T]$	cross-ratio of intervals M, T, page 106
$\lfloor x \rfloor$	greatest integer $\leq x$, page 4
$\{x\}$	fractional part of x, page 4
$\|\cdot\|_{C^k}$	C^k norm, page 74
$\|\cdot\|_{\mathrm{BV}}$	bounded variation norm, page 46
$\|\cdot\|_{\sigma}$	C^0 norm of functions defined on S_σ, page 81
$\mathbb{A}$	special class of rotation numbers, page 259
$B_f(\cdot,\cdot)$	bi-Schwarzian of f, page 113
$\mathrm{BV}(\boldsymbol{S}^1)$	space of functions of bounded variation on the circle, page 46
$C^{1+\mathrm{BV}}$	space of maps f with $\log Df \in \mathrm{BV}(\boldsymbol{S}^1)$, page 49
$\mathbb{C}(I)$	the plane slit along two rays, page 335
$\mathrm{CrD}(f; M, T)$	cross-ratio distortion of f on intervals M, T, page 111
$\boldsymbol{D}$	class of all standard C^1 Denjoy examples, page 55
$D^0(\boldsymbol{S}^1)$	space of lifts of elements of $\mathrm{Diff}^0_+(\boldsymbol{S}^1)$, page 29
$\mathrm{Diff}^{1+Z}(\boldsymbol{S}^1)$	space of circle diffeomorphisms with $\log Df \in Z$, page 49
$\mathrm{Diff}^0_+(\boldsymbol{S}^1)$	space of orientation-preserving homeomorphisms of $\boldsymbol{S}^1$, page 29
$\dim_H(\cdot)$	Hausdorff dimension, page 60
$D_\theta(a, b)$	Poincaré disk, page 335
$\mathscr{E}$	Epstein class, page 391
$\mathbb{E}_\infty$	set of irrationals $\rho = [a_1, a_2, \cdots] \in (0, 1)$ with each a_n even and $a_n \to \infty$, page 226
$G(\cdot)$	Gauss map, page 11
$\mathscr{H}_r(\cdot)$	Herman's C^r invariant, page 74
$I_n(x)$	n-th renormalization interval of x, page 135
m	normalized Lebesgue measure on the unit circle, page 36
$\mathscr{N} f$	nonlinearity of f, page 43
$\mathscr{P}_n(x)$	n-th dynamical partition associated with x, page 135
$\mathscr{P}(\boldsymbol{S}^1)$	space of Borel probability measures on $\boldsymbol{S}^1$, page 64
R_α	counterclockwise rotation by angle $2\pi\alpha$, page 4
$\mathscr{R}(\cdot)$	renormalization operator, page 257

$\mathrm{SL}(2,\mathbb{R})$	special linear group in dimension 2, page 8
$\boldsymbol{S}^1=\mathbb{R}/\mathbb{Z}$	unit circle or one-dimensional torus, page 3
Sf	Schwarzian derivative of f, page 107
S_σ	the strip $\{z\in\mathbb{C}:\lvert\mathrm{Im}(z)\rvert<\sigma\}$, page 79
$\mathrm{Var}(\varphi)$	total variation of φ, page 46
$\boldsymbol{W}^0$	class of C^0 Denjoy examples, page 55
$\boldsymbol{W}^1$	class of C^1 Denjoy examples, page 55
Z	Zygmund class, page 48
$Z_0(f)$	centralizer of f in $\mathrm{Diff}_+^0(\boldsymbol{S}^1)$, page 37
$\alpha_T(x)$	α-limit set of x under T, page 31
χ_A	characteristic or indicator function of A, page 15
$\chi_f(x)$	Lyapunov exponent of f at x, page 63
μ_f	Beltrami coefficient of a QC homeomorphism f, page 282
ν	Gauss measure, page 7
$\omega_T(x)$	ω-limit set of x under T, page 31
$\Omega(T)$	non-wandering set of T, page 31
$\rho(f)$	rotation number of f, page 26
$\rho_T(\cdot)$	Poincaré density of interval T, page 106
$\tau(F)$	translation number of F, page 26
$\widehat{\varphi}(n)$	n-th Fourier coefficient of φ, page 81
$\zeta=(\eta,\xi)$	critical commuting pair, page 253

Part I
Basic Theory

Chapter 1
Rotations

This opening chapter is devoted to the simplest dynamical systems on the circle that are not entirely trivial: the *rigid rotations*. Under a rigid rotation, all orbits look exactly the same. There are only two possible behaviours for such orbits. Either they are all dense on the circle, or else they are all periodic with the same period. This dichotomy can be read off from the angle by which points on the circle are rotated. The ratio of this angle to a full turn is called the *rotation number*. If the rotation number of a rigid rotation is rational, then all orbits are periodic. If the rotation number is irrational, then all orbits are dense. Moreover, the way the points of an orbit deploy themselves on the circle can be read off from the continued fraction development of the rotation number. Due to this connection with continued fractions, it is fair to say that the dynamical study of rotations was started by the ancient Greeks.

1.1 Topology and Combinatorics of Rotations

The dynamical systems we wish to study have as their phase space the *unit circle*, denoted S^1 in this book, which can be defined in at least two ways. One way is to regard it as the affine one-dimensional manifold $\mathbb{R}/\mathbb{Z}$, also called the *one-dimensional torus*. Another way is to regard it as the boundary of the unit disk in the complex plane, namely $\partial\mathbb{D} = \{z \in \mathbb{C} : |z| = 1\}$. These two representations of the unit circle are equivalent, the equivalence being induced by the *exponential covering map* $\exp : \mathbb{R} \to \partial\mathbb{D}$ given by $\exp(t) = e^{2\pi i t}$. Both representations make it clear that S^1 is also a topological group, the group operation being addition modulo 1 in the first representation and complex multiplication in the second. The reader should keep in mind the equivalence between these two representations. In most of what we do in this book, we use the additive representation, but will switch to the multiplicative representation whenever convenient.

E. de Faria, P. Guarino, *Dynamics of Circle Mappings*, IMPA Monographs,
https://doi.org/10.1007/978-3-031-67495-2_1

1.1.1 A Dichotomy

Given a real number α, let us denote by $R_\alpha : S^1 \to S^1$ the *counterclockwise rotation* of the unit circle by an angle equal to $2\pi\alpha$. This map is given by $R_\alpha(x) = x + \alpha \pmod 1$ in additive notation, or equivalently by $R_\alpha(z) = e^{2\pi i\alpha}z$ in multiplicative notation. We are interested here in the orbit structure of rotations, both from the topological and metric viewpoints. When we speak of S^1 as a metric space, we always take the distance between two points x, y to be the one induced from the real line by the exponential covering map, i.e., $d(x, y) = \min\{|u - v| : \exp(u) = x\,,\ \exp(v) = y\}$. The group of orientation-preserving isometries of S^1 under this metric is precisely the group of rotations.

From the topological viewpoint, the dynamical behavior of rotations is very simple. There is a dichotomy, according to whether α is rational or irrational.

(1) If α is *rational*, say $\alpha = p/q$ in irreducible form, then every $x \in S^1$ is a periodic point with period q. In other words, we have

$$R_\alpha^q(x) = x + q\alpha = x + p = x \pmod 1\ ,$$

for all $x \in S^1$.

(2) If α is *irrational*, then every orbit $\mathcal{O}^+(x) = \{R_\alpha^n(x) :\ \ n \geq 0\}$ is *dense* in S^1. This follows from Lemma 1.1, stated and proved below (see also Proposition 1.1).

The result alluded to above is a classical one, discovered by Dirichlet in 1842. Its proof uses the well-known *pigeonhole principle*.[1] We need some notation. Given any real number x, we denote by $\lfloor x \rfloor$ the greatest integer less or equal to x, and by $\{x\}$ the *fractional part* of x, i.e. $\{x\} = x - \lfloor x \rfloor$.

Lemma 1.1 *If α is an irrational number, then*

(i) For each positive integer Q there exist an integer p and a positive integer q with $q \leq Q$ such that

$$|q\alpha - p| < \frac{1}{Q}\ . \tag{1.1}$$

(ii) There exist infinitely many rational numbers p/q such that

$$\left|\alpha - \frac{p}{q}\right| < \frac{1}{q^2}\ . \tag{1.2}$$

[1] Also known as the *box counting principle*, it simply states that if $N + 1$ objects are placed in N boxes, then at least one box will contain at least two objects.

Proof Partition the interval $[0, 1)$ into Q sub-intervals of equal length, namely

$$\Delta_{j,Q} = \left[\frac{j-1}{Q}, \frac{j}{Q}\right), \quad j = 1, 2, \ldots, Q .$$

These are our boxes. Then consider the $Q + 1$ numbers

$$0, \ \{\alpha\}, \{2\alpha\}, \ \ldots, \{Q\alpha\} \ \in [0, 1) .$$

These are all distinct, because α is irrational. By the pigeonhole principle, there exist $j \in \{1, 2, \ldots, Q\}$ and $n_1, n_2 \in \{0, 1, \ldots, Q\}$ distinct such that both $\{n_1\alpha\} \in \Delta_{j,Q}$ and $\{n_2\alpha\} \in \Delta_{j,Q}$ hold. Writing $m_1 = \lfloor n_1\alpha \rfloor$ and $m_2 = \lfloor n_2\alpha \rfloor$, we see that

$$|(n_1 - n_2)\alpha - (m_1 - m_2)| = |\{n_1\alpha\} - \{n_2\alpha\}| \leq \left|\Delta_{j,Q}\right| = \frac{1}{Q} .$$

Hence, taking $p = m_1 - m_2$ and $q = n_1 - n_2$, we deduce (1.1). Equality in (1.1) cannot happen, because α is irrational. This proves (i), and (ii) is a direct consequence of (i). □

For our next lemma, we shall use the following simple property of fractional parts: if x, y are real numbers with $\{x\} + \{y\} < 1$, then $\{x + y\} = \{x\} + \{y\}$.

Lemma 1.2 *If α is irrational, then the sequence $\alpha_n = \{n\alpha\}$, $n \in \mathbb{N}$, is dense in* $[0, 1]$.

Proof Given $\varepsilon > 0$, let N be a positive integer such that $1/N < \varepsilon$. By Lemma 1.1, there exists a positive integer n with $1 \leq n \leq N$ such that $\alpha_n < 1/N$. Since $\alpha_n > 0$ is irrational, there exists a (unique) $k \in \mathbb{N}$ such that $k\alpha_n < 1 < (k+1)\alpha_n$. Therefore the points $j\alpha_n$ with $j = 1, 2, \ldots, k$ are *ε-dense* in $[0, 1]$. But by the simple property given before the statement of this lemma, we have

$$j\alpha_n = j\{n\alpha\} = \{jn\alpha\} = \alpha_{jn} \quad , j = 1, 2, \ldots, k .$$

In other words, we have proved that the set $\{\alpha_n, \alpha_{2n}, \ldots, \alpha_{kn}\}$ is ε-dense in $[0, 1]$. Since ε is arbitrary, it follows that (α_n) is dense in $[0, 1]$. □

As a corollary, we have the following result.

Proposition 1.1 *If α is irrational, then for all $x \in \mathbf{S}^1$ the positive orbit $\mathcal{O}^+(x) = \{R_\alpha^n(x) : n \in \mathbb{N}\}$ of x under the rotation R_α is dense in $\mathbf{S}^1$.*

Proof Note that

$$R_\alpha^n(x) = R_\alpha^n(R_x(0)) = R_x(R_\alpha^n(0)) .$$

Since R_x is an isometry of the unit circle, it follows that the sequence $(R_\alpha^n(x))_{n\geq 0}$ is dense if and only if the sequence $(R_\alpha^n(0))_{n\geq 0}$ is dense. But $R_\alpha^n(0) = \{n\alpha\}$, and this last sequence is dense by Lemma 1.2. □

This proposition justifies the dichotomy we stated at the beginning of this section.

1.1.2 Sequence of Closest Returns

Let $R_\alpha : \mathbf{S}^1 \to \mathbf{S}^1$ be the rotation of angle $2\pi\alpha$ on the unit circle, where α is *irrational*. We define a sequence of positive integers (q_n) recursively as follows. Let $q_0 = 1$, and for each $n > 0$, let

$$q_n = \min\left\{i > q_{n-1} : d(x, R^i(x)) < d(x, R^{q_{n-1}}(x))\right\} .$$

Here, x is any point on the circle. It does not matter which x we choose in this definition, because R_α is an isometry. The positive integer q_n is called the *n-th closest return time* of the orbit of (any) x. The meaning is clear: each iterate $R_\alpha^{q_n}(x)$ is closest to x than any previous iterate $R_\alpha^i(x)$, $1 \leq i < q_n$.

It so happens that *consecutive* closest returns to x occur on *opposite sides* of x. Being "on opposite sides of x" might seem somewhat ambiguous (we are on a circle, after all!), but the ambiguity disappears if we remove $R_\alpha^{q_0}(x) = R_\alpha(x)$ from $\mathbf{S}^1$: it is then legitimate to speak of opposite sides of x in the arc $\mathbf{S}^1 \setminus \{R_\alpha(x)\}$. The precise statement is as follows.

Lemma 1.3 *Let $J_n(x) \subset \mathbf{S}^1$ be the interval of endpoints $R_\alpha^{q_{n-1}}(x)$ and $R_\alpha^{q_n}(x)$ that contains x, and let $J_n'(x) \subset J_n(x)$ be the interval of endpoints x and $R_\alpha^{q_{n-1}}(x)$. Then $R_\alpha^{q_{n+1}}(x) \in J_n'(x)$.*

This lemma, in turn, is a consequence of the following result.

Lemma 1.4 *Let $I_n(x) \subset J_n(x)$ be the interval with endpoints x and $R_\alpha^{q_n}(x)$. Then the intervals $R_\alpha^j(I_n(x))$, with $j \in \{0, 1, 2, \ldots, q_{n+1} - 1\}$, are pairwise disjoint.*

Proof Let $0 \leq i < j \leq q_{n+1} - 1$ be such that $R_\alpha^i(I_n(x)) \cap R_\alpha^j(I_n(x)) \neq \emptyset$. Then $k = j - i$ satisfies $I_n(x) \cap R_\alpha^k(I_n(x)) \neq \emptyset$, and obviously $0 < k < q_{n+1}$. Recall that R_α is an isometry, so $|R_\alpha^k(I_n(x))| = |I_n(x)|$. Since R_α is orientation-preserving, we see that either $x \in R_\alpha^k(I_n(x))$ or $R_\alpha^k(x) \in I_n(x)$ (see Fig. 1.1). In either case, we have

$$d(x, R_\alpha^k(x)) \leq |I_n(x)| = d(x, R_\alpha^{q_{n+1}}(x)) .$$

But this can only happen if $k \geq q_{n+1}$, which is certainly not the case. □

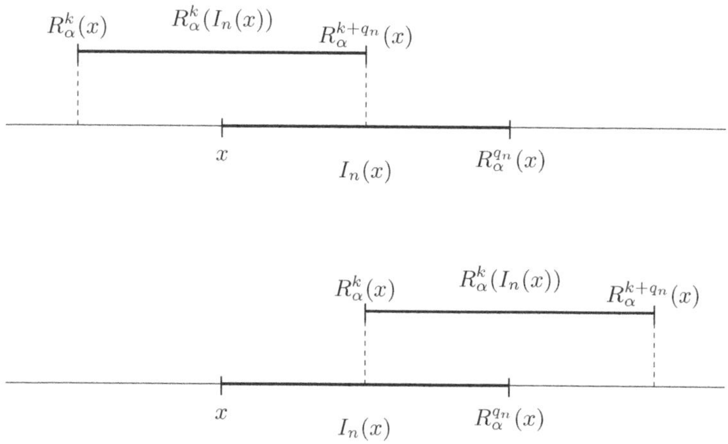

Fig. 1.1 Two possibilities

Remark 1.1 The intervals $I_n(x)$ defined above are called the *closest return intervals* associated to the point x. We sometimes omit the point x and write I_n instead of $I_n(x)$

1.2 Rotations and Continued Fractions

In this section we look at rotations from an arithmetic viewpoint. Given $R_\alpha : S^1 \to S^1$, we introduce the *continued fraction* development of α and show that the denominators of the sequence of *best rational approximations* to α (obtained by truncating the continued fraction expansion of α) are precisely the closest return times for R_α introduced in Sect. 1.1.

1.2.1 Basic Theory of Continued Fractions

Let us consider the group G of 2×2 real matrices with determinant ± 1. This group acts on the extended real line $\widehat{\mathbb{R}} = \mathbb{R} \cup \{\infty\}$ (the one-point compactification of $\mathbb{R}$) as the group $\mathscr{M}(\mathbb{R})$ of *real fractional linear* (or *Möbius*) transformations. More precisely, to each matrix

$$A = \begin{pmatrix} a & b \\ c & d \end{pmatrix} \in G$$

we associate a Möbius transformation T_A given by

$$T_A(\zeta) = \frac{a\zeta + b}{c\zeta + d} .$$

Matrix multiplication in G corresponds to composition of maps in $\mathcal{M}(\mathbb{R})$, in other words, if $A, B \in G$ then $T_{AB} = T_A \circ T_B$. Thus we have a homomorphism $G \to \mathcal{M}(\mathbb{R})$, and it is easy to check that such homomorphism is surjective and that its kernel is $\{\pm I\}$. In particular, $\mathcal{M}(\mathbb{R}) = G/\{\pm I\}$[2]

We consider certain special elements of G. Given $x \in \mathbb{R}$, let

$$\sigma_x = \begin{pmatrix} x & 1 \\ 1 & 0 \end{pmatrix} .$$

The associated Möbius transformation T_{σ_x} is given by

$$T_{\sigma_x}(\zeta) = x + \frac{1}{\zeta} .$$

Now, given any sequence $x_0, x_1, \ldots, x_n, \ldots$ of real numbers, we associate to it the sequence of matrices A_n given by

$$A_n = \sigma_{x_0}\sigma_{x_1} \cdots \sigma_{x_n} = \begin{pmatrix} x_0 & 1 \\ 1 & 0 \end{pmatrix} \begin{pmatrix} x_1 & 1 \\ 1 & 0 \end{pmatrix} \cdots \begin{pmatrix} x_n & 1 \\ 1 & 0 \end{pmatrix} .$$

The sequence of corresponding Möbius transformations T_{A_n} is therefore given by

$$\begin{aligned} T_{A_n}(\zeta) &= T_{\sigma_{x_0}} \circ T_{\sigma_{x_1}} \circ \cdots \circ T_{\sigma_{x_n}}(\zeta) \\ &= x_0 + \cfrac{1}{x_1 + \cfrac{1}{x_2 + \cfrac{1}{\cdots + \cfrac{1}{x_n + \cfrac{1}{\zeta}}}}} . \end{aligned} \tag{1.3}$$

The entries of the matrices A_n can be determined by recurrence. Writing

$$A_n = \begin{pmatrix} p_n & p_{n-1} \\ q_n & q_{n-1} \end{pmatrix}$$

[2] Through this identification, $\mathcal{M}(\mathbb{R})$ contains a copy of $\mathrm{PSL}(2, \mathbb{R}) = \mathrm{SL}(2, \mathbb{R})/\{\pm I\}$ (the *projective* special linear group) as a subgroup of index 2.

and taking into account that $A_{n+1} = A_n \sigma_{x_{n+1}}$, we see that

$$p_{n+1} = x_{n+1} p_n + p_{n-1} \tag{1.4}$$

$$q_{n+1} = x_{n+1} q_n + q_{n-1} \, . \tag{1.5}$$

We also have $p_0 = x_0$, $p_1 = x_0 x_1 + 1$ and $q_0 = 1, q_1 = x_1$. It readily follows from these facts that p_n and q_n are given by polynomials of degree $n + 1$ and n, respectively, in the variables $x_0, x_1, \ldots, x_n$. Moreover, since

$$\det A_n = \prod_{j=0}^{n} \det \sigma_{x_j} = (-1)^{n+1} \, ,$$

we see that

$$p_n q_{n-1} - p_{n-1} q_n = (-1)^{n+1} \, , \quad \forall \, n \geq 1 \, . \tag{1.6}$$

Let us now specialize our discussion to the case when each x_n is an integer, say $x_n = a_n \in \mathbb{Z}$. We assume also that all a_n's are *positive*, with the possible exception of a_0. For later reference, let us repeat here the defining recurrence relations for the p_n's and q_n's in this case:

$$p_{n+1} = a_{n+1} p_n + p_{n-1} \tag{1.7}$$

$$q_{n+1} = a_{n+1} q_n + q_{n-1} \, . \tag{1.8}$$

Since $q_0 = 1$ and $q_1 = a_1 \geq 1$ and $a_n \geq 1$ for all $n \geq 1$, we deduce from (1.7) that $q_{n+1} \geq q_n + q_{n-1}$. This tells us that the sequence (q_n) grows at least as fast as the Fibonacci sequence: by an easy inductive argument, it follows from this last inequality that

$$q_n \geq \frac{1}{\sqrt{5}} \left(\frac{1 + \sqrt{5}}{2} \right)^n \, .$$

Presently, what is more important for our purposes is that $q_n \to \infty$ exponentially fast as $n \to \infty$. Dividing both sides of (1.6) by $q_{n-1} q_n$, we have

$$\frac{p_n}{q_n} - \frac{p_{n-1}}{q_{n-1}} = \frac{(-1)^{n+1}}{q_{n-1} q_n} \tag{1.9}$$

This shows that the sequence of rational numbers p_n / q_n is a Cauchy sequence (recall here that $q_n \to \infty$ exponentially fast), and therefore the limit

$$\alpha = \lim_{n \to \infty} \frac{p_n}{q_n} \tag{1.10}$$

exists. This number must be irrational (why?). We stress that the rational numbers p_n/q_n are already in irreducible form, for (1.6) implies that $\gcd(p_n, q_n) = 1$. These rational approximations to α are called the *convergents* of α, while the coefficients a_n are called the *partial quotients* of α.

Next, we note that (1.9) implies, by a simple telescoping trick, the relation

$$\frac{p_n}{q_n} = a_0 + \sum_{j=0}^{n-1} \frac{(-1)^{j+1}}{q_j q_{j+1}} . \tag{1.11}$$

Letting $n \to \infty$ here yields α as the sum of an infinite convergent series, namely

$$\alpha = a_0 + \sum_{j=0}^{\infty} \frac{(-1)^{j+1}}{q_j q_{j+1}} . \tag{1.12}$$

These facts tell us that the convergents p_n/q_n *alternate* around α (their limit). In fact, again using that (q_n) is an increasing sequence, we have

$$\frac{p_0}{q_0} < \frac{p_2}{q_2} < \cdots < \frac{p_{2n}}{q_{2n}} < \cdots < \alpha < \cdots < \frac{p_{2n+1}}{q_{2n+1}} < \cdots < \frac{p_3}{q_3} < \frac{p_1}{q_1} . \tag{1.13}$$

Furthermore, if we subtract (1.11) from (1.12), we get

$$\alpha - \frac{p_n}{q_n} = \sum_{j=n}^{\infty} \frac{(-1)^{j+1}}{q_j q_{j+1}} ,$$

and from this it follows that

$$\left| \alpha - \frac{p_n}{q_n} \right| < \frac{1}{q_n q_{n+1}} < \frac{1}{q_n^2} ,$$

for all $n \geq 0$.

Summarizing, we have proved one half of the following result.

Theorem 1.1 *Given a sequence of integers $a_0, a_1, \ldots, a_n, \ldots$ with $a_n \geq 1$ for all $n \geq 1$, there exists a unique irrational number α with the following properties.*

(i) Writing, for each $n \geq 0$

$$\frac{p_n}{q_n} = a_0 + \cfrac{1}{a_1 + \cfrac{1}{a_2 + \cfrac{1}{\cdots + \cfrac{1}{a_n}}}}$$

as an irreducible fraction, we have

$$\left|\alpha - \frac{p_n}{q_n}\right| < \frac{1}{q_n q_{n+1}} \quad \textit{for all} \quad n \geq 0 \,.$$

(ii) The convergents p_n/q_n *alternate around* α*, and their limit is* α*.*

Conversely, given an irrational number α*, there exists a unique sequence of integers* $a_0, a_1, \ldots, a_n, \ldots$ *with* $a_n \geq 1$ *for all* $n \geq 1$ *such that*

$$\alpha = \lim_{n\to\infty} a_0 + \cfrac{1}{a_1 + \cfrac{1}{a_2 + \cfrac{1}{\cdots + \cfrac{1}{a_n}}}} \,. \tag{1.14}$$

Proof By now, only the converse statement at the end requires proof. Given the number α, define $a_0 = \lfloor \alpha \rfloor$ and let $\alpha_0 = \alpha$. Next, define

$$\alpha_1 = \frac{1}{\alpha_0 - a_0} \quad \text{and} \quad a_1 = \lfloor \alpha_1 \rfloor \,.$$

Note that α_1 is a well-defined, positive irrational number, and that a_1 is a positive integer. Now proceed inductively in this fashion: having defined the positive irrational number α_n and the positive integer a_n, let

$$\alpha_{n+1} = \frac{1}{\alpha_n - a_n} \quad \text{and} \quad a_{n+1} = \lfloor \alpha_{n+1} \rfloor \,.$$

This produces the desired sequence of integers. We leave to the reader the task of proving that, indeed, (1.14) is satisfied. □

We close this section with the following remark. Let $G : [0, 1] \to [0, 1]$ be defined by $G(0) = 0$ and

$$G(\alpha) = \left\{\frac{1}{\alpha}\right\} , \quad \text{for all } \alpha \neq 0 \,.$$

This is the so-called *Gauss map*, which is extremely useful in the study of continued fractions. Also, let $a_1 : [0, 1] \to \mathbb{Z}^+$ be given by

$$a_1(\alpha) = \left\lfloor \frac{1}{\alpha} \right\rfloor$$

This is, of course, the first (non-zero) partial quotient of $\alpha \in (0, 1]$. Then, if $a_n(\alpha)$ denotes the n-th partial quotient of α, we have

$$a_{n+1} = a_1 \circ G^n(\alpha)\ , \quad \text{for all } n \geq 0\ . \tag{1.15}$$

Thus, there is an intimate relationship between the continued-fraction development of a real number in $[0, 1]$ and the dynamics of the Gauss map. In particular, many interesting statistical properties of the partial quotients can be derived from the *ergodic theory* of the Gauss map. This will be fully explained in Appendix A.

Remark 1.2 We warn the reader that, for notational convenience, later in this book we will write a_0 for the first partial quotient of a number $\alpha \in (0, 1]$, instead of a_1. Thus, the indices in the sequence of partial quotients will all be shifted by 1.

1.2.2 Best Approximations

The convergents of an irrational number α are the best rational approximations to α, in a sense that is made precise in the following result.

Theorem 1.2 *If p_n/q_n denotes the n-th convergent of the irrational number α, for $n = 0, 1, \ldots$, then*

(i) For all $n \geq 0$, we have

$$\frac{1}{q_n(q_n + q_{n+1})} < \left|\alpha - \frac{p_n}{q_n}\right| < \frac{1}{q_n q_{n+1}}\ . \tag{1.16}$$

(ii) We have

$$|q_0\alpha - p_0| > |q_1\alpha - p_1| > \cdots > |q_n\alpha - p_n| > \cdots\ .$$

(iii) If p, q are non-zero integers such that

$$|q\alpha - p| < |q_n\alpha - p_n|\ . \tag{1.17}$$

for some $n \geq 0$, then $q \geq q_{n+1}$.

Proof The right-most inequality in (1.16) was proved in Theorem 1.1. Since the convergent p_{n+2}/q_{n+2} lies between α and the convergent p_n/q_n, we have

$$\left|\alpha - \frac{p_n}{q_n}\right| > \left|\frac{p_{n+2}}{q_{n+2}} - \frac{p_n}{q_n}\right|\ . \tag{1.18}$$

The recurrence relations defining p_n, q_n easily imply the identity

$$p_{n+2}q_n - p_nq_{n+2} = (-1)^n a_{n+2} .$$

Using this identity in (1.18), and taking into account that $a_{n+2} \geq 1$, we get

$$\left|\alpha - \frac{p_n}{q_n}\right| > \frac{a_{n+2}}{q_nq_{n+2}} = \frac{a_{n+2}}{q_n(a_{n+2}q_{n+1} + q_n)} \geq \frac{1}{q_n(q_{n+1} + q_n)} ,$$

and this establishes the left-most inequality in (1.16). Hence, (i) is established.

In order to prove (ii), we multiply the inequalities in (1.16) by q_n, obtaining

$$\frac{1}{q_n + q_{n+1}} < |q_n\alpha - p_n| < \frac{1}{q_{n+1}} .$$

These inequalities are valid for all $n \geq 0$. But if $n \geq 1$ then $q_{n+1} \geq q_n + q_{n-1}$, and therefore

$$|q_n\alpha - p_n| < \frac{1}{q_n + q_{n-1}} < |q_{n-1}\alpha - p_{n-1}| .$$

This proves (ii).

Finally, we prove (iii). Let us suppose that (1.17) holds, but $0 < q < q_{n+1}$. Note that since the matrix

$$\begin{pmatrix} p_n & p_{n+1} \\ q_n & q_{n+1} \end{pmatrix}$$

has determinant equal to ± 1, there exists a unique pair of integers μ, ν such that

$$p = \mu p_n + \nu p_{n+1} \quad \text{and} \quad q = \mu q_n + \nu q_{n+1} . \tag{1.19}$$

We claim that μ and ν are both non-zero. For if $\mu = 0$ then $q = \nu q_{n+1}$ and ν is necessarily non-zero, implying $q \geq q_{n+1}$, contrary to assumption. Likewise, if $\nu = 0$, then $p = \mu p_n$, $q = \mu q_n$ with $\mu \neq 0$, and so

$$|q\alpha - p| = |\mu| \cdot |q_n\alpha - p_n| \geq |q_n\alpha - p_n| ,$$

again contrary to assumption. Thus, $\mu \neq 0 \neq \nu$. Next, we claim that μ and ν have opposite signs. Indeed, if they had the same sign, then from the second equality in (1.19) we would have $q = |q| \geq q_{n+1}$, again contrary to assumption. Now, we note that the numbers $q_n\alpha - p_n$ and $q_{n+1}\alpha - p_{n+1}$ have opposite signs (see (1.13)). Therefore the numbers

$$\mu(q_n\alpha - p_n) \quad \text{and} \quad \nu(q_{n+1}\alpha - p_{n+1})$$

have the *same sign*. Therefore we have

$$\begin{aligned}|q\alpha - p| &= |\mu(q_n\alpha - p_n) + \nu(q_{n+1}\alpha - p_{n+1})| \\ &= |\mu|\cdot|q_n\alpha - p_n| + |\nu|\cdot|q_{n+1}\alpha - p_{n+1}| \\ &> |\mu|\cdot|q_n\alpha - p_n| \geq |q_n\alpha - p_n|\,.\end{aligned}$$

This is again a contradiction. This proves that q must be greater than or equal to q_{n+1}, and we are done. □

We close this section with a word on notation. Given any real number x, it is customary to denote by $\|x\|$ the *distance from x to the nearest integer*, that is, $\|x\| = \min\{|x-m| : m \in \mathbb{Z}\}$. It is easy to see, from the above discussion, that

$$\|q_n\alpha\| = |q_n\alpha - p_n| \quad \text{for all } n \geq 0\,.$$

Hence, Theorem 1.2 (ii) tells us that

$$\|q_0\alpha\| > \|q_1\alpha\| > \cdots > \|q_n\alpha\| > \cdots$$

1.3 Weyl's Equidistribution Theorem

The points of a single orbit of an irrational rotation are not just dense in S^1, they are also uniformly distributed in some sense. This fact, although intuitively obvious, requires clarification and proof. This is our purpose in this section.

1.3.1 Equidistribution

Let us start with a definition.

Definition 1.1 A sequence of real numbers $x_0, x_1, \ldots, x_n, \ldots$ is said to be *equidistributed modulo one*[3] if for every interval $\Delta \subseteq [0, 1]$ we have

$$\lim_{N\to\infty} \frac{1}{N}\#\{0 \leq n \leq N-1 \colon \{x_n\} \in \Delta\} = |\Delta| \qquad (1.20)$$

where, as before, $\{x\} = x - \lfloor x \rfloor$ denotes the fractional part of x, and $|\Delta|$ denotes the Euclidean length of the interval Δ.

Alternatively, (1.20) can be written as

[3] Or, equivalently, *uniformly distributed* modulo one.

$$\lim_{N\to\infty} \frac{1}{N} \sum_{n=0}^{N-1} \mathbb{1}_\Delta(\{x_n\}) = \int_0^1 \mathbb{1}_\Delta(x)\, dx \,, \tag{1.21}$$

where $\mathbb{1}_\Delta$ is the characteristic (or indicator) function of the interval Δ.

In a classic paper published in 1914, H. Weyl proved the following criterion for equidistribution.

Theorem 1.3 (Weyl's Criterion) *For a sequence of real numbers $(x_n)_{n\geq 0}$, the following are equivalent.*

(a) The sequence (x_n) is equidistributed modulo one;
(b) For every continuous function $\varphi : \mathbb{R} \to \mathbb{C}$, periodic of period one, we have

$$\lim_{N\to\infty} \frac{1}{N} \sum_{n=0}^{N-1} \varphi(x_n) = \int_0^1 \varphi(x)\, dx \,;$$

(c) For each $m \in \mathbb{Z}^*$, we have

$$\lim_{N\to\infty} \frac{1}{N} \sum_{n=0}^{N-1} e^{2\pi i m x_n} = 0 \,.$$

Proof To see that (a) implies (b), note first of all that it suffices to prove (b) for *real valued* functions (periodic of period one). Given such a φ, approximate it uniformly on the unit interval by means of a sequence of step functions. Since each step function is a linear combination of characteristic functions of intervals, equality (1.21) holds for step functions as well. From the equality in (b) follows, first for the step functions themselves, and then for φ by the uniform approximation. Now, (b) clearly implies (c), for we can simply take $\varphi(x) = \exp\{2\pi i m x\}$. To prove that (c) implies (b), consider the algebra $\mathscr{A}$ of so-called *Laurent polynomials*

$$P(x) = \sum_{m=-k}^{\ell} c_m e^{2\pi i m x}$$

where k, ℓ are non-negative integers and $c_m \in \mathbb{C}$ for all m. It is clear that $\mathscr{A}$ contains the constant functions, separates points of $[0, 1]$ and is invariant under complex conjugation. Therefore, by the Stone–Weierstrass theorem, $\mathscr{A}$ is dense in $C^0_{\mathbb{C}}([0, 1])$. Thus, given φ as in (b) and $\epsilon > 0$, there exists $P_\epsilon \in \mathscr{A}$ such that $\sup_{x\in[0,1]} |\varphi(x) - P_\epsilon(x)| \leq \epsilon$. This implies at once that

$$\left| \int_0^1 \varphi(x)\, dx - \int_0^1 P_\epsilon(x)\, dx \right| \leq \epsilon \,, \tag{1.22}$$

and also, for all $N \geq 1$, that

$$\left| \frac{1}{N} \sum_{n=0}^{N-1} \varphi(x_n) - \frac{1}{N} \sum_{n=0}^{N-1} P_\epsilon(x_n) \right| \leq \epsilon \,. \tag{1.23}$$

If c_0 denotes the constant term of P_ϵ, then applying (c) we deduce that, as $N \to \infty$,

$$\frac{1}{N} \sum_{n=0}^{N-1} P_\epsilon(x_n) \longrightarrow c_0 = \int_0^1 P_\epsilon(x)\, dx \,.$$

Combining this fact with (1.22) and (1.23), we get

$$\lim_{N\to\infty} \left| \frac{1}{N} \sum_{n=0}^{N-1} \varphi(x_n) - \int_0^1 \varphi(x)\, dx \right| \leq 2\epsilon \,.$$

But since ϵ is arbitrary, (b) follows. Finally, to prove that (b) implies (a), let $\Delta \subseteq (0, 1)$ be an interval, and let $\epsilon > 0$. Take two functions φ and ψ, both continuous and periodic of period one, with $\varphi(x) \leq \mathbb{1}_\Delta(x) \leq \psi(x)$ for all $0 \leq x \leq 1$, such that

$$\int_0^1 \psi(x)\, dx - \int_0^1 \varphi(x)\, dx \leq \frac{\epsilon}{3} \,.$$

Note that the integral of $\mathbb{1}_\Delta$ over the unit interval is squeezed between these two. Moreover, for all $N \geq 1$ we have

$$\frac{1}{N} \sum_{n=0}^{N-1} \varphi(x_n) \leq \frac{1}{N} \sum_{n=0}^{N-1} \mathbb{1}_\Delta(\{x_n\}) \leq \frac{1}{N} \sum_{n=0}^{N-1} \psi(x_n) \,.$$

But, by (b), for all sufficiently large N we have

$$\left| \frac{1}{N} \sum_{n=0}^{N-1} \varphi(x_n) - \int_0^1 \varphi(x)\, dx \right| \leq \frac{\epsilon}{3} \,,$$

as well as

$$\left| \frac{1}{N} \sum_{n=0}^{N-1} \psi(x_n) - \int_0^1 \psi(x)\, dx \right| \leq \frac{\epsilon}{3} \,,$$

Combining these facts we deduce (a), and this completes the proof. □

Corollary 1.1 *Every orbit of an irrational rotation* $R_\theta \colon x \mapsto x + \theta \pmod 1$ *is equidistributed modulo one.*

Proof Since $R_\theta^n(x) = \{x + n\theta\}$ for all n, we must prove that the sequence $x_n = x + n\theta$ is equidistributed modulo one. This we do using part (c) of Weyl's criterion; given $m \in \mathbb{Z}^*$, we see that

$$\frac{1}{N}\sum_{n=0}^{N-1} e^{2\pi i m(x+n\theta)} = \frac{e^{2\pi i m x}}{N}\sum_{n=0}^{N-1}\left(e^{2\pi i m\theta}\right)^n = \frac{e^{2\pi i m x}}{N}\,\frac{1-e^{2\pi i m N\theta}}{1-e^{2\pi i m\theta}}$$

Therefore

$$\lim_{N\to\infty}\frac{1}{N}\left|\sum_{n=0}^{N-1} e^{2\pi i m(x+n\theta)}\right| = \lim_{N\to\infty}\frac{1}{N}\left|\frac{1-e^{2\pi i m N\theta}}{1-e^{2\pi i m\theta}}\right| = 0\,,$$

and we indeed deduce from (c) that the orbit is equidistributed modulo one as asserted. □

1.3.2 A Simple Application

Following Arnold and Avez (1968), let us illustrate the usefulness of Weyl's criterion by solving a simple problem in Number Theory. Write down the list of all powers of 2 in base 10, in ascending order:

$$1,\ 2,\ 4,\ 8,\ 16,\ 32,\ 64,\ 128,\ 256,\ 512,\ 1024,\ 2048,\ \ldots$$

Consider the sequence consisting of the *left-most digits* of the above numbers, namely

$$1,\ 2,\ 4,\ 8,\ 1,\ 3,\ 6,\ 1,\ 2,\ 5,\ 1,\ 2,\ \ldots \tag{1.24}$$

Then ask a couple of natural questions:

(i) Does the number 7 appear in the above list of first digits?[4]
(ii) If so, with what frequency?

Let us see how Weyl's criterion—or rather, Corollary 1.1—can aid us in providing answers to these questions. First, we need to express the first digit d_n of 2^n written in base 10 as a function of n. If we take the logarithm in base 10 of each term in our sequence of powers of two, we get the sequence

[4] In the old days when Arnold and Avez discussed the problem, pocket calculators were not available. Today anybody with one at hand and a bit of patience can check by brute force that the answer to the first question is *yes*. Indeed, the first occurrence of 7 as first digit happens in $2^{46} = 70368744177664$.

$$x_n = n \log_{10} 2 .$$

The simple but crucial observation here is the following:

$$d_n = k \in \{1, \ldots, 9\} \iff \{x_n\} = \{n \log_{10} 2\} \in [\log_{10} k, \log_{10}(k+1)) .$$

But $\theta = \log_{10} 2$ is irrational (why?). Hence, by Corollary 1.1, the sequence (x_n) is equidistributed modulo one. In particular, we have

$$\lim_{N\to\infty} \frac{1}{N} \#\{0 \leq n < N : d_n = k\} = \log_{10}\left(1 + \frac{1}{k}\right) > 0 .$$

In other words, all digits from 1 to 9 appear with positive (asymptotic) frequency in the sequence (1.24). The (asymptotic) frequency for the specific case of 7 is

$$\log_{10} \frac{8}{7} = 0.05799\ldots$$

This, of course, answers questions (i) and (ii) posed above: the digit 7 does appear in (1.24), with an asymptotic frequency of about 5.8%.

1.4 Ergodicity of Irrational Rotations

Throughout this book, we denote by m the normalized Lebesgue measure on the unit circle. More precisely, if $A \subset S^1$ is an interval, then $m(A)$ is just the Lebesgue measure of $\pi^{-1}(A) \cap [0, 1)$ in the real line, where $\pi : \mathbb{R} \to S^1$ is the standard covering map. Since it lifts to a translation, any rotation preserves the Lebesgue measure on S^1. We finish Chap. 1 by proving *ergodicity* of *irrational* rotations.

Lemma 1.5 *The Lebesgue measure is ergodic under any irrational rotation.*

Recall that if (X, μ) is a measure space and $\phi : X \to X$ is a measurable map that preserves μ, we say that μ is *ergodic* under ϕ if, for every measurable set $A \subseteq X$ which is invariant under ϕ (meaning $\phi^{-1}(A) = A$), we have either $\mu(A) = 0$ or $\mu(X \setminus A) = 0$.

Note that Lemma 1.5 above is certainly *not* true for rational rotations (why?). As it turns out, Lemma 1.5 follows from what we have done in Sect. 1.3.1. Indeed, if the sequence $\{x_n\}_{n\in\mathbb{N}}$ in Theorem 1.3 is given by an orbit, i.e., $x_n = x_0 + n\alpha \pmod 1$ for some initial condition $x_0 \in \mathbb{R}$, then part (b) of Weyl's criterion is saying that the *Birkhoff averages* of any continuous function $\varphi : \mathbb{R} \to \mathbb{C}$, periodic of period one, along the given orbit $\{x_n\}$ converges to $\int_0^1 \varphi(x)\, dx$. This establishes the ergodicity of the Lebesgue measure under R_α. However, the proof given below (which uses the notion of *Lebesgue density point*) is easier to adapt to more general situations (see for instance Theorem 3.10 in Chap. 3, see also Sect. 8.4).

Proof of Lemma 1.5 Let $\alpha \in [0, 1] \setminus \mathbb{Q}$ and let R_α be the rotation of angle α in $\boldsymbol{S}^1$. Let us assume, by contradiction, that there exist two disjoint R_α-invariant Borel sets A and B in the circle, both having positive Lebesgue measure. Let $x_0 \in \boldsymbol{S}^1$ be a *density point* of A. Recall that this means that

$$\lim_{\varepsilon \to 0} \left\{ \frac{m\big((x_0 - \varepsilon, x_0 + \varepsilon) \cap A\big)}{2\varepsilon} \right\} = 1 .$$

Since A is R_α-invariant and R_α is an isometry, we have that

$$\begin{aligned} m\big(\big(R_\alpha^n(x_0) - \varepsilon, R_\alpha^n(x_0) + \varepsilon\big) \cap A\big) &= m\big(R_\alpha^n(x_0 - \varepsilon, x_0 + \varepsilon) \cap A\big) \\ &= m\big(R_\alpha^n\big((x_0 - \varepsilon, x_0 + \varepsilon) \cap A\big)\big) \\ &= m\big((x_0 - \varepsilon, x_0 + \varepsilon) \cap A\big). \end{aligned}$$

Therefore, $R_\alpha^n(x_0)$ is a density point of A for all $n \in \mathbb{N}$. In the same way, let $y_0 \in \boldsymbol{S}^1$ be a density point of B. For any given $\delta \in (3/4, 1)$, let $\varepsilon > 0$ be sufficiently small so that

$$m\big((x_0 - \varepsilon, x_0 + \varepsilon) \cap A\big) > 2\varepsilon\delta \quad \text{and} \quad m\big((y_0 - \varepsilon, y_0 + \varepsilon) \cap B\big) > 2\varepsilon\delta.$$

By Proposition 1.1, the positive orbit $\big\{R_\alpha^n(x_0)\big\}_{n \in \mathbb{N}}$ is dense in the circle. Moreover, as we just observed, all its points are density points of A. This allows us to assume that $x_0 \in (y_0 - \varepsilon, y_0 + \varepsilon)$. Finally, since A and B are disjoint to each other, we obtain

$$\begin{aligned} 2\varepsilon\delta &< m\big((y_0 - \varepsilon, y_0 + \varepsilon) \cap B\big) \\ &\leq m\big((y_0 - \varepsilon, y_0 + \varepsilon) \cup (x_0 - \varepsilon, x_0 + \varepsilon)\big) - m\big((x_0 - \varepsilon, x_0 + \varepsilon) \cap A\big) \\ &< 3\varepsilon - 2\varepsilon\delta . \end{aligned}$$

This implies that $\delta < 3/4$, a contradiction. □

Remark 1.3 Yet another proof of Lemma 1.5 can be given by means of *Fourier series*. The equivalence between what we will prove below and ergodicity is a criterion proved in almost every book on Ergodic Theory; see for instance Walters (1982, p. 28). Just as before, let $\alpha \in [0, 1] \setminus \mathbb{Q}$ and let R_α be the rotation of angle α in $\boldsymbol{S}^1$. We claim that if $\varphi : \boldsymbol{S}^1 \to \mathbb{C}$ belongs to the Hilbert space $L^2(\boldsymbol{S}^1)$ of *square-integrable* functions (with respect to Lebesgue, i.e., $\int |\varphi|^2 \, dm$ is finite) and satisfies $\varphi \circ R_\alpha(x) = \varphi(x)$ at Lebesgue almost every point, then φ is constant almost everywhere. To prove this, we proceed as follows. It is well known that every such φ has a unique orthogonal expansion in $L^2(\boldsymbol{S}^1)$ of the form

$$\varphi = \sum_{n\in\mathbb{Z}} a_n(\varphi)\, e_n$$

where the functions $e_n(x) = e^{2\pi i n x}$, $n \in \mathbb{Z}$, form an orthonormal basis for $L^2(S^1)$, and the bi-infinite sequence $\{a_n(\varphi)\}_{n\in\mathbb{Z}}$ of complex numbers, the so-called *Fourier coefficients* of φ, is given by

$$a_n(\varphi) = \int_{S^1} \varphi(x)\, e^{-2\pi i n x}\, dx\ , \quad \forall n \in \mathbb{Z}\ .$$

But the function $\varphi \circ R_\alpha$ is also in $L^2(S^1)$, and a simple calculation shows that its Fourier coefficients are given by

$$\begin{aligned} a_n(\varphi \circ R_\alpha) &= \int_{S^1} \varphi(x+\alpha) e^{-2\pi i n x}\, dx \\ &= \int_{S^1} \varphi(x) e^{-2\pi i n(x-\alpha)}\, dx = a_n(\varphi) e^{2\pi i n \alpha}\ , \quad \forall n \in \mathbb{Z}\ . \end{aligned}$$

Now suppose that φ is R_α-invariant, in the sense that $\varphi \circ R_\alpha = \varphi$ at Lebesgue almost every point. By uniqueness of the Fourier coefficients, we must have $a_n(\varphi)\, e^{2\pi i n\alpha} = a_n(\varphi)$ for all $n \in \mathbb{Z}$. Since α is an irrational number, $e^{2\pi i n\alpha} \neq 1$ for all $n \neq 0$. This implies that $a_n(\varphi) = 0$ for all $n \neq 0$. In other words, $\varphi(x) = a_0(\varphi)$ for Lebesgue almost every $x \in S^1$. Thus, we have proved that if $\varphi \in L^2(S^1)$ satisfies $\varphi \circ R_\alpha = \varphi$ almost everywhere, then it is constant almost everywhere. This implies the ergodicity of the Lebesgue measure under R_α.

In Exercise 1.13 below we outline a proof of the fact that the Lebesgue measure is the *unique* invariant measure of an irrational rotation. Dynamical systems preserving *only one* probability measure are called *uniquely ergodic*. As we will see in Sect. 2.3, any circle homeomorphism without periodic points is uniquely ergodic.

Exercises

Exercise 1.1

(i) Let G be an additive sub-group of the real numbers. Show that G is either discrete or dense in $\mathbb{R}$ (*Hint:* Discuss on $\alpha = \inf\{g \in G : g > 0\}$).

(ii) Let $\rho \in [0, 1]$, and note that the set $G_\rho = \{n\rho + m : n, m \in \mathbb{Z}\}$ is a sub-group of $\mathbb{R}$. Show that G_ρ is discrete if, and only if, $\rho \in \mathbb{Q}$.

(iii) Let $R_\rho : S^1 \to S^1$ be the rigid rotation of angle $\rho \in [0, 1]$, and note that

$$\mathcal{O}_{R_\rho}\big(\pi(x)\big) = \pi(x + G_\rho)$$

for any $x \in \mathbb{R}$, where $\pi : \mathbb{R} \to \partial\mathbb{D}$ is the usual covering map $\pi(t) = e^{2\pi i t}$. With this and the previous items, describe the dynamics of any rotation, as in Sect. 1.1.1.

Exercise 1.2 Let $0 \leq \alpha, \beta < 1$ be real numbers, and suppose that there is a continuous monotone map $h : S^1 \to S^1$ such that $h \circ R_\alpha = R_\beta \circ h$. Show that $\alpha = \beta$.

Exercise 1.3 If q_n, $n \geq 0$ are the denominators of the convergents of an irrational number α, prove that this sequence always grows at least as fast as the Fibonacci numbers. Deduce that

$$q_n \geq \frac{1}{\sqrt{5}}\left(\frac{1+\sqrt{5}}{2}\right)^n, \text{ for all } n \geq 0 .$$

Exercise 1.4 For any $n \in \mathbb{N}$, note that $\left(\sqrt{n^2+1}-n\right)\left(\sqrt{n^2+1}+n\right) = 1$ implies

$$\sqrt{n^2+1}-n = \frac{1}{2n + \left(\sqrt{n^2+1}-n\right)},$$

and conclude that the continued fraction expansion of

$$\sqrt{n^2+1}-n = \sqrt{n^2+1} \pmod{1}$$

is given by $[2n, 2n, 2n, \ldots]$.

Exercise 1.5 Prove the identity (1.15).

Exercise 1.6 Given an irrational number α, let a_n, $n \geq 0$, be the partial quotients of its continued-fraction development, and let q_n, $n \geq 0$, be the denominators of the corresponding convergents.

(i) Show that

$$\frac{q_n}{q_{n-1}} = [a_n, a_{n-1}, \ldots, a_1], \text{ for all } n \geq 1 .$$

(ii) Show that, for all $n \geq 2$, we have

$$\|q_{n-1}\alpha\| = a_n\|q_n\alpha\| + \|q_{n+1}\alpha\| ,$$

and deduce that

$$a_n = \left\lfloor \frac{\|q_{n-1}\alpha\|}{\|q_n\alpha\|} \right\rfloor .$$

Exercise 1.7 Let the sequence $(x_n)_{n\geq 0}$ be equidistributed modulo one, and let $(\alpha_n)_{n\geq 0}$ be a sequence that converges to zero in the Cesàro sense, i.e.

$$\lim_{n\to\infty} \frac{1}{n}(\alpha_0 + \alpha_1 + \cdots + \alpha_{n-1}) = 0\,.$$

Prove that the sequence $(x_n + \alpha_n)_{n\geq 0}$ is also equidistributed modulo one.

Exercise 1.8 Let θ be a positive irrational number, and let $(x_n)_{n\geq 0}$ be the sequence given by $x_0 = 0$, $x_1 = \theta$, and $x_n = (n + (\log n)^{-1})\theta$ for all $n \geq 2$. Show that (x_n) is equidistributed modulo one.

Exercise 1.9 Show that the sequence

$$x_n = \left\{\left(\frac{1+\sqrt{5}}{2}\right)^n\right\}$$

is *not* equidistributed modulo one.

Exercise 1.10 Suppose $f : \mathbb{R}^+ \to \mathbb{R}$ is a differentiable function such $f'(x) \to 0$ as $x \to \infty$. Show that

$$\frac{1}{N}\sum_{n=0}^{N-1} e^{2\pi i f(n)} - \frac{1}{N}\int_0^N e^{2\pi i f(t)}\,dt \to 0$$

as $N \to \infty$.

Exercise 1.11 Let f be as in the previous exercise, and suppose in addition that

$$\lim_{x\to\infty} x f'(x) = A \in [-\infty, +\infty]\,.$$

(i) If A is finite, show that the sequence $x_n = f(n)$ is *not* equidistributed modulo one.
(ii) If $A = \pm\infty$ and f' is monotone, show that $x_n = f(n)$ is equidistributed modulo one.

Exercise 1.12 Using the criterion provided by the previous exercise, show that

(i) The sequence $x_n = n^\sigma$ is equidistributed modulo one provided $0 < \sigma < 1$.
(ii) The sequence $x_n = \log n$ is not equidistributed modulo one.

Exercise 1.13 Let $\alpha \in [0, 1] \setminus \mathbb{Q}$ and let R_α be the rotation of angle α in $\boldsymbol{S}^1$. Given a continuous function $\varphi : \boldsymbol{S}^1 \to \mathbb{R}$, consider the sequence $\{\varphi_n : \boldsymbol{S}^1 \to \mathbb{R}\}_{n\in\mathbb{N}}$ of its Birkhoff averages, i.e., $\varphi_n(x) = n^{-1}\sum_{j=0}^{n-1}\varphi\big(R_\alpha^j(x)\big)$ for all $x \in \boldsymbol{S}^1$.

(i) Endowing the space $C^0(\boldsymbol{S}^1)$ with the uniform convergence topology, show that the sequence $\{\varphi_n\}$ is pre-compact (*Hint:* Apply the Arzelà–Ascoli Theorem).

(ii) Show that $\{\varphi_n\}$ converges (uniformly) to the constant $\int \varphi\, dm$ (*Hint:* Combine Lemma 1.5 with Birkhoff's Ergodic Theorem, and then apply the previous item).

(iii) Deduce that R_α is uniquely ergodic (*Hint:* Using again Birkhoff's Ergodic Theorem, show that if μ is an R_α-invariant probability measure, then $\int \varphi\, d\mu = \int \varphi\, dm$ for any $\varphi \in C^0(S^1)$).

Exercise 1.14 A vector $\alpha = (\alpha_1, \ldots, \alpha_n) \in \mathbb{R}^n$ is said to be *rationally independent* if whenever a linear combination $\sum_{j=1}^n k_j \alpha_j$ with integer coefficients $k_1, \ldots, k_n$ belongs to $\mathbb{Z}$, then $k_1 = k_2 = \cdots = k_n = 0$ (in other words, the $n+1$ numbers $\alpha_1, \ldots, \alpha_n, 1$ are rationally independent). Now let $\mathbb{T}^n = \mathbb{R}^n/\mathbb{Z}^n = S^1 \times S^1 \times \cdots \times S^1$ be the *n-dimensional torus*. The purpose of this exercise is to guide the reader to a proof that the Lebesgue measure on $\mathbb{T}^n$ is ergodic under the rotation $R_\alpha : \mathbb{T}^n \to \mathbb{T}^n$ given by

$$R_\alpha(x_1, \ldots, x_n) = (x_1 + \alpha_1, \ldots, x_n + \alpha_n) \pmod 1,$$

provided the vector α is rationally independent. To do this, we will proceed as in Remark 1.3 and prove that if $\varphi : \mathbb{T}^n \to \mathbb{C}$ belongs to $L^2(\mathbb{T}^n)$ and satisfies $\varphi \circ R_\alpha = \varphi$ at Lebesgue almost every point of $\mathbb{T}^n$, then it is constant almost everywhere.

(i) Just as in Remark 1.3, write

$$\varphi(x_1, \ldots, x_n) \sim \sum_{(k_1,\ldots,k_n)\in\mathbb{Z}^n} a_{k_1,\ldots,k_n}\, e^{2\pi i \sum_{j=1}^n k_j x_j}$$

for the Fourier series representation of φ, where the numbers $a_{k_1,\ldots,k_n}$ are the Fourier coefficients[5] of φ.

(ii) Show that

$$a_{k_1,\ldots,k_n}\left(1 - e^{2\pi i \sum_{j=1}^n k_j \alpha_j}\right) = 0$$

for all $(k_1, \ldots, k_n) \in \mathbb{Z}^n$.

(iii) Show that $\varphi(x_1, \ldots, x_n) = a_{0,\ldots,0}$ for Lebesgue almost every $(x_1, \ldots, x_n)$.

(iv) Conclude that the Lebesgue measure on $\mathbb{T}^n$ is ergodic under R_α.

Exercise 1.15 Arguing as in Exercise 1.13, prove that $R_\alpha : \mathbb{T}^n \to \mathbb{T}^n$ is uniquely ergodic, provided α is rationally independent.

Exercise 1.16 Let $\alpha \in [0, 1] \setminus \mathbb{Q}$, and let R_α be the rotation of angle α in S^1. As we have seen in Exercise 1.13, the Lebesgue measure m is the unique probability measure invariant under R_α. By the Riesz Representation Theorem, this amounts to saying that any continuous linear functional T defined in $C^0(S^1)$, which is invariant[6]

---

[5] Given by $a_{k_1,\ldots,k_n} = \int_{\mathbb{T}^n} \varphi(x_1, \ldots, x_n) e^{-2\pi i (k_1 x_1 + \cdots + k_n x_n)}\, dx_1 \cdots dx_n$.

[6] In the sense that $\langle T, u \circ R_\alpha \rangle = \langle T, u \rangle$ for any $u \in C^0(S^1)$.

under R_α, is a scalar multiple of Lebesgue measure. In this exercise we will extend this result, by showing that any continuous linear functional defined in $C^\infty(S^1)$, invariant under R_α, also has to be a scalar multiple of Lebesgue measure. With this goal, let $\varphi \in C^\infty(S^1)$ be such that $\int \varphi\, dm = 0$.

(i) Just as in Remark 1.3, write

$$\varphi = \sum_{n\in\mathbb{Z}} a_n(\varphi)\, e_n\,,$$

where $e_n(x) = e^{2\pi i n x}$ for each $n \in \mathbb{Z}$, and the a_n's are the Fourier coefficients of φ. Note that $a_0(\varphi) = \int \varphi\, dm = 0$.

(ii) Show that there exists a sequence $\{u_n\}_{n\geq 1} \subset C^\infty(S^1)$ such that the sequence $\{u_n \circ R_\alpha - u_n\}$ converges to φ in the C^∞ topology (*Hint:* For each $n \geq 1$, let $A_n = [-n, n] \cap \mathbb{Z} \setminus \{0\} = \{-n, \ldots, -1, 1, \ldots, n\}$. Consider the sequence of trigonometric polynomials given by

$$u_n = \sum_{j\in A_n} \frac{a_j(\varphi)}{e^{2\pi i j\alpha} - 1}\, e_j\,,$$

and note that $u_n \circ R_\alpha - u_n = \sum_{j\in A_n} a_j(\varphi)\, e_j$ for all $n \geq 1$).

(iii) Now let T be a *distribution* on S^1, i.e., T is a continuous linear functional defined in $C^\infty(S^1)$. Assume that T is invariant under R_α, i.e., $\langle\, T, u\circ R_\alpha \,\rangle = \langle\, T, u \,\rangle$ for any $u \in C^\infty(S^1)$. Show that if $\varphi \in C^\infty(S^1)$ is such that $\int \varphi\, dm = 0$, then $\langle\, T, \varphi \,\rangle = 0$.

(iv) Conclude that any R_α-invariant distribution T is a scalar multiple of Lebesgue measure. [*Hint:* By (iii), $\left\langle\, T, \varphi - \int \varphi\, dm \right\rangle = 0$ for any given $\varphi \in C^\infty(S^1)$.]

Invariant distributions for general circle diffeomorphisms will be discussed in Sect. 3.4.3, see also Sect. 8.4.

Chapter 2
Homeomorphisms of the Circle

We will study the orbit structure of orientation-preserving homeomorphisms of the unit circle. As is customary, we will identify the boundary of the unit disk $\partial\mathbb{D} = \{z \in \mathbb{C} : |z| = 1\}$ with the one-dimensional torus $\boldsymbol{S}^1 = \mathbb{R}/\mathbb{Z}$.

Every orientation-preserving homeomorphism $f : \boldsymbol{S}^1 \to \boldsymbol{S}^1$ lifts to an increasing homeomorphism $F : \mathbb{R} \to \mathbb{R}$ such that $F(x+1) = F(x) + 1$ for all $x \in \mathbb{R}$, i.e., such that

$$\begin{array}{ccc} \mathbb{R} & \xrightarrow{\;F\;} & \mathbb{R} \\ {\scriptstyle\pi}\big\downarrow & & \big\downarrow{\scriptstyle\pi} \\ \boldsymbol{S}^1 & \xrightarrow[\;f\;]{} & \boldsymbol{S}^1 \end{array}$$

is a commutative diagram, where $\pi : t \mapsto e^{2\pi i t}$ is the exponential covering map. The lift F is not unique, but any two choices differ by an integer; if we require that, say, $F(0) \in [0, 1)$, then F is uniquely determined. Note that we can write $F(x) = x + \varphi(x)$ for all x, where $\varphi : \mathbb{R} \to \mathbb{R}$ is a continuous periodic function with period one.

The natural order structure of $\mathbb{R}$ (or $\boldsymbol{S}^1$) makes it "easy" to understand and classify f's as above up to topological equivalence, as we shall see.

2.1 Translation and Rotation Numbers

Rotation numbers were first introduced by Poincaré. We will give here three equivalent definitions of rotation numbers and, via lifts, of translation numbers as well.

E. de Faria, P. Guarino, *Dynamics of Circle Mappings*, IMPA Monographs,
https://doi.org/10.1007/978-3-031-67495-2_2

2.1.1 The Classical Definition

Following Herman (1979), let us denote by $\text{Diff}_+^0(S^1)$ the class of orientation-preserving homeomorphisms of the circle, and by $D^0(S^1)$ the class of all lifts of such homeomorphisms to the real line.

We define the *translation number* of $F \in D^0(S^1)$ to be the limit

$$\tau(F) = \lim_{n\to\infty} \frac{F^n(x) - x}{n} . \tag{2.1}$$

Thus, $\tau(F)$ measures the limiting average amount by which F translates points on the real line. For this definition to make sense, we need of course the following basic result.

Proposition 2.1 *For every $F \in D^0(S^1)$, the limit in* (2.1) *exists and is independent of x. Moreover, the convergence is uniform in x.*

The proof will come in a moment. For now, note that if we are given two lifts F_1 and F_2 of the same $f \in \text{Diff}_+^0(S^1)$, then $\tau(F_1) - \tau(F_2)$ is an integer. This motivates Poincaré's definition, namely the following.

Definition 2.1 The *rotation number* of $f \in \text{Diff}_+^0(S^1)$, denoted $\rho(f)$, is the residue class modulo one of $\tau(F)$, where F is any lift of f.

In other words, $\rho(f)$ measures the limiting average amount by which f rotates points on the circle.

Let us now prove Poincaré's fundamental result concerning rotation (and translation) numbers, namely Proposition 2.1 above. Consider the periodic function $\varphi_F(x) = F(x) - x$. The basic fact to observe is that

$$\sup_{x\in\mathbb{R}} \varphi_F(x) < \inf_{x\in\mathbb{R}} \varphi_F(x) + 1 , \tag{2.2}$$

and that this is true for every $F \in D^0(S^1)$. In particular, (2.2) holds if we replace F by any iterate F^n. Therefore it is clear that in order to prove that the limit (2.1) exists for all x, it suffices to prove that

$$\lim_{n\to\infty} \frac{1}{n} \sup \varphi_{F^n}(x) \tag{2.3}$$

exists, and this will be the common limit for all x. There is no loss of generality in assuming from the beginning that $F(0) \geq 0$, so that $\alpha_n = \sup_{x\in\mathbb{R}} \varphi_{F^n}(x) \geq 0$ for all $n \geq 0$. Hence the existence of the limit in (2.3) is reduced to the following simple but extremely useful lemma.

Lemma 2.1 *Let $(\alpha_n)_{n\geq 0}$ be a sequence of non-negative numbers. If (α_n) is sub-additive, i.e. if $\alpha_{m+n} \leq \alpha_m + \alpha_n$ for all $m, n \geq 0$, then the sequence α_n/n converges to a limit.*

Proof Given $n > 0$ fixed and $m > n$ arbitrary, we can write $m = kn + r$, where $0 \le r < n$. Hence, by sub-additivity, $\alpha_m \le k\alpha_n + \alpha_r$. This gives us

$$\frac{\alpha_m}{m} \le \frac{\alpha_n}{n + \frac{r}{k}} + \frac{\alpha_r}{m} \le \frac{\alpha_n}{n} + \frac{\alpha_r}{m} ,$$

and since α_r ranges over a finite set of values, we get $\limsup_{m\to\infty}(\alpha_m/m) \le \alpha_n/n$ for all $n > 0$, or yet

$$\limsup_{m\to\infty} \frac{\alpha_m}{m} \le \liminf_{n\to\infty} \frac{\alpha_n}{n} ,$$

and so the limit exists. □

This completes the proof of Proposition 2.1.

2.1.2 *The Order Definition*

An alternative way to define the rotation number of $f \in \mathrm{Diff}^0_+(S^1)$ is to use the relative order of points of orbits of f along the circle. As before, let $F \in D^0(S^1)$ be a lift of f. Consider the sets

$$\mathbb{Q}^+(F) = \left\{ \frac{p}{q} \in \mathbb{Q} : \quad F^q(x) < p + x, \quad \text{for all} \quad x \in \mathbb{R} \right\}$$

$$\mathbb{Q}^-(F) = \left\{ \frac{p}{q} \in \mathbb{Q} : \quad F^q(x) > p + x, \quad \text{for all} \quad x \in \mathbb{R} \right\}$$

These sets determine a *Dedekind cut* of the rational numbers; this is left as an exercise to the reader. Hence we can define the translation number of F as the real number α determined by this Dedekind cut. To see that α agrees with the number given by the first definition, consider two sequences of rational numbers $p_n/q_n \in \mathbb{Q}^-(F)$ and $P_n/Q_n \in \mathbb{Q}^+(F)$, both converging to α; in particular,

$$\frac{p_n}{q_n} < \alpha < \frac{P_n}{Q_n} .$$

By induction, we have for all $x \in \mathbb{R}$

$$x + Q_n p_n < F^{q_n Q_n}(x) < x + q_n P_n .$$

Dividing these inequalities by $q_n Q_n$ yields

$$\frac{p_n}{q_n} < \frac{F^{q_n Q_n}(x) - x}{q_n Q_n} < \frac{P_n}{Q_n}.$$

Letting $n \to \infty$ and applying Proposition 2.1, we see that indeed $\alpha = \tau(F)$.

2.1.3 The Measure-Theoretic Definition

Yet another way to define the rotation number of $f \in \mathrm{Diff}^0_+(\mathbf{S}^1)$ is to consider, for any given lift $F \in D^0(\mathbf{S}^1)$ of f, the real function $\psi_F : \mathbf{S}^1 \to \mathbb{R}$ whose lift under π is $F - \mathrm{Id}$, that is: $F(x) = x + \psi_F(e^{2\pi i x})$ for all $x \in \mathbb{R}$. We claim that for any given f-invariant Borel probability measure μ we have

$$\rho(f) = \int_{\mathbf{S}^1} \psi_F \, d\mu \pmod 1. \tag{2.4}$$

Indeed, the point here is that the real function $\widetilde{\psi_F}$ given by

$$\widetilde{\psi_F} = \lim_{n\to+\infty} \frac{1}{n} \sum_{j=0}^{n-1} \psi_F \circ f^j$$

is well defined on the whole circle and it is *constant*, equal to $\tau(F)$ (note that this implies at once the identity (2.4), since $\int_{\mathbf{S}^1} \psi_F \, d\mu = \int_{\mathbf{S}^1} \widetilde{\psi_F} \, d\mu$ by the f-invariance of μ). To see that $\widetilde{\psi_F}$ is well defined and constant, let $x \in \mathbf{S}^1$ and let $x_0 \in \mathbb{R}$ be such that $\pi(x_0) = x$. Then

$$\begin{aligned}
\sum_{j=0}^{n-1} \psi_F\big(f^j(x)\big) = \sum_{j=0}^{n-1} \psi_F\big(\pi(F^j(x_0))\big) &= \sum_{j=0}^{n-1} \big(F - \mathrm{Id}\big)\big(F^j(x_0)\big) \\
= \sum_{j=0}^{n-1} \big(F^{j+1}(x_0) - F^j(x_0)\big) &= F^n(x_0) - x_0 .
\end{aligned}$$

By Proposition 2.1, we have $\widetilde{\psi_F}(x) = \tau(F)$, as desired. The equivalent definition of the rotation number of a circle homeomorphism given by (2.4) will not be further mentioned in this book, but it is important in its own right.

2.1.4 Properties of the Rotation Number

Let us now take stock of some useful properties of the rotation number. The first one establishes the fact that the rotation number is a *topological invariant*: two topologically conjugate (or even semi-conjugate) circle homeomorphisms have the same rotation number. In other words, the equivalence classes of $\mathrm{Diff}_+^0(S^1)$ under topological conjugacies are contained in the level sets of ρ.

Lemma 2.2 *Let $f, g \in \mathrm{Diff}_+^0(S^1)$ and let $h : S^1 \to S^1$ be a continuous, degree one monotone map such that $h \circ f = g \circ h$. Then $\rho(f) = \rho(g)$.*

Proof Let $F, G \in D^0(S^1)$ and $H : \mathbb{R} \to \mathbb{R}$ be lifts of f, g and h respectively, so that $H \circ F = G \circ H$ in the real line. Given $y \in \mathbb{R}$ let $x \in \mathbb{R}$ be such that $H(x) = y$. By induction, one easily obtain that $G^n(y) = H\big(F^n(x)\big)$ for all $n \in \mathbb{N}$. Then we write

$$\frac{G^n(y)}{n} = \frac{\big(H - \mathrm{Id}\big)\big(F^n(x)\big)}{n} + \frac{F^n(x)}{n} \quad \text{for all } n \in \mathbb{N}\,.$$

Since H is a lift of h, the difference $H - \mathrm{Id}$ is bounded in the whole real line, and then Proposition 2.1 implies Lemma 2.2. □

The next result states that the rotation number behaves additively over any family of commuting homeomorphisms.

Proposition 2.2 *If $f, g : S^1 \to S^1$ are homeomorphisms such that $f \circ g = g \circ f$, then*

$$\rho(f \circ g) = \rho(f) + \rho(g) \pmod 1\,.$$

Proof Let $F, G : \mathbb{R} \to \mathbb{R}$ be lifts of f and g, respectively, with the property that $F \circ G = G \circ F$ (the existence of lifts with this property is left as an exercise). Let us fix $x \in \mathbb{R}$. From the definition of translation number, we see that

$$\begin{aligned} \tau(F \circ G) &= \lim_{n\to\infty} \frac{1}{n}\big[(F \circ G)^n(x) - x\big] \\ &= \lim_{n\to\infty} \left\{\frac{1}{n}\big[F^n(G^n x) - G^n x\big] - \frac{1}{n}\big[G^n x - x\big]\right\}\,. \end{aligned} \tag{2.5}$$

Note that we have used that $(F \circ G)^n = F^n \circ G^n$, which is true because F and G commute.

Now recall that the periodic function $\phi_n = F^n - I$ satisfies $\sup \phi_n - \inf \phi_n < 1$ for each $n \geq 0$. In particular, we have $-1 < \phi_n(G^n(x)) - \phi_n(x) < 1$, and therefore

$$-\frac{1}{n} < \frac{1}{n}\big[F^n(G^n x) - G^n x\big] - \frac{1}{n}\big[F^n x - x\big] < \frac{1}{n}\,,$$

for all $n \geq 1$. Taking this information back to (2.5), we get $\tau(F \circ G) = \tau(F) + \tau(G)$. Reducing modulo 1, we deduce that $\rho(f \circ g) = \rho(f) + \rho(g)$ as desired. □

We remark that Proposition 2.2 implies the useful formula

$$\rho(f^n) = n\,\rho(f) \pmod 1,$$

for any circle homeomorphism f.

The third property we wish to establish tells us that the topological invariant $\rho(f)$ *varies continuously with* f, in a sense to be made precise below.

Let us introduce a topology on the set of lifts $D^0(S^1)$. Given $F, G \in D^0(S^1)$, let

$$d'(F,G) \;=\; \sup_{x\in\mathbb{R}} |F(x) - G(x)| \;.$$

This defines a metric in $D^0(S^1)$, as the reader can easily check (exercise). The topology in $D^0(S^1)$, thus, is the topology induced by this metric. We may also consider the space $\mathscr{P}$ of continuous functions $\psi : \mathbb{R} \to \mathbb{R}$ which are *periodic of period 1*, endowed with the metric given by the same expression as above. The topology given by this metric has the following property: For each $x \in \mathbb{R}$, the *evaluation map* $\widehat{x} : \mathscr{P} \to \mathbb{R}$ given by $\widehat{x}(\psi) = \psi(x)$ is continuous.

Likewise, we may consider in $\mathrm{Diff}^0(S^1)$ the metric given by

$$d''(f,g) \;=\; \sup_{x\in S^1} \|f(x) - g(x)\| \;.$$

The reader is invited to check that this is indeed a metric, and that the natural map $(D^0(S^1), d') \to (\mathrm{Diff}^0(S^1), d'')$, associating to each lift the corresponding circle homeomorphism, is continuous.

Proposition 2.3 *The translation number function $\tau : D^0(S^1) \to \mathbb{R}$ is continuous.*

Proof For each fixed $n > 0$, the map $\beta_n : D^0(S^1) \to \mathscr{P}$ given by

$$F \;\mapsto\; \frac{1}{n}\left[F^n - I\right]$$

is continuous. Fix $x \in \mathbb{R}$. As the reader can check (see Exercise 2.4), we have

$$\left|\frac{1}{n}\left[F^n(x) - x\right] - \tau(F)\right| \;\le\; \frac{1}{n}\,. \tag{2.6}$$

Since

$$\widehat{x}\circ\beta_n(F) \;=\; \frac{1}{n}\left[F^n(x) - x\right]\,,$$

it follows that the sequence of continuous functions $\{\widehat{x}\circ\beta_n\}$ converges uniformly to τ. Therefore τ is continuous.

□

Corollary 2.1 *The rotation number map $\rho : \mathrm{Diff}^0_+(S^1) \to S^1$ is continuous.*

Proof This follows at once from Proposition 2.3 and reduction modulo 1. □

2.2 Topological Dynamics of Homeomorphisms

As already mentioned, the study of the dynamics of circle mappings began at the end of the nineteenth century, with the pioneering work of Henri Poincaré. In this section we discuss the differences in dynamical behavior between homeomorphisms with rational and irrational rotation number, as noted by Poincaré himself.

Let us briefly recall some simple notions from topological dynamics. If $T : X \to X$ is a homeomorphism of a compact metric space, the *ω-limit set* of a point $x \in X$ under T, denoted $\omega_T(x)$, is the set of all accumulation points of the *forward* orbit of x. In other words, $y \in \omega_T(x)$ if and only if there is a sequence $n_i \to \infty$ such that $T^{n_i}(x) \to y$. Similarly, we define the *α-limit set* of x under T, denoted $\alpha_T(x)$, to be the set of all accumulation points of the *backward* orbit of x. Thus, $y \in \alpha_T(x)$ if and only if there is a sequence $n_i \to -\infty$ such that $T^{n_i}(x) \to y$. Both $\alpha_T(x)$ and $\omega_T(x)$ are closed, non-empty, and totally invariant (in the sense that $T(\alpha_T(x)) = \alpha_T(x)$, and similarly for $\omega_T(x)$). A compact invariant set Λ is said to be *minimal* under T if $\omega_T(x) = \Lambda$ for any given $x \in \Lambda$. The *non-wandering set* of T, denoted $\Omega(T)$, is the set of all $x \in X$ such that for all neighborhoods $U \ni x$ we have $T^n(U) \cap U \neq \emptyset$ for arbitrarily large $n \in \mathbb{N}$. The non-wandering set is also totally invariant. As the following sections show, and in marked contrast with other one-dimensional dynamical systems, these sets have a very simple description in the case of homeomorphisms of the circle.

2.2.1 *Rational Rotation Number*

We now take up the task of showing that, for a circle homeomorphism, having rational rotation number is equivalent to possessing periodic orbits.

Proposition 2.4 *For any $f \in \mathrm{Diff}_+^0(S^1)$ we have that $\rho(f) \in \mathbb{Q}$ if, and only if, f admits at least one periodic orbit. In this case, if $\rho(f) = p/q$, all periodic orbits of f have period q.*

Proof Let $F \in D^0(S^1)$ be a lift of f. If f has a periodic point $\pi(x)$ (say of period q), then there exists $p \in \mathbb{Z}$ such that $F^q(x) = x + p$. This implies that $F^{nq}(x) = x + np$, and then

$$\lim_{n\to\infty} \frac{F^{nq}(x)}{nq} = \lim_{n\to\infty} \frac{x+np}{nq} = \frac{p}{q}\,.$$

Therefore, $\rho(f) = p/q \pmod 1$. On the other hand, by Proposition 2.2, we know that $\rho(f^m) = m\,\rho(f) \pmod 1$ for any $m \in \mathbb{N}$. In particular, if $\rho(f) = p/q$, we have $\rho(f^q) = 0$. This shows that it is enough to prove that if $\rho(f) = 0$, then f has at least one fixed point. To see this, let $F \in D^0(S^1)$ be the lift of f given by $\tau(F) = 0$. If F has no fixed points (and since $F - \mathrm{Id}$ is periodic), there exists

δ such that $|F(x) - x| \geq \delta$ for all $x \in \mathbb{R}$. Say that $F(x) > x$ for all $x \in \mathbb{R}$ (the case $F(x) < x$ can be treated in exactly the same way). Then $F(0) > \delta$, $F^2(0) > F(0) + \delta > 2\delta, \ldots, F^n(0) > n\delta, \ldots$ and so forth. But then $\delta < \frac{F^n(0)}{n}$, which goes to zero. This contradiction shows that F (and then f) has at least one fixed point.

Finally, suppose that $\rho(f) = p/q$, and let us prove that all periodic orbits of f have period q. Let F be the lift of f such that $\tau(F) = p/q$, and let $\pi(x)$ be a periodic point for f. Then there exist integers r, s such that $F^r(x) = x + s$. Now,

$$\rho(f) = \frac{p}{q} = \lim_{k \to \infty} \frac{F^{kr}(x)}{kr} = \frac{s}{r},$$

and then $s = mp$ and $r = mq$ for some m. If $F^q(x) - p > x$, then

$$F^{2q}(x) - 2p = F^q(F^q(x) - p) - p \geq F^q(x) - p > x.$$

Proceeding inductively in this fashion, we deduce that $F^{jq}(x) - jp > x$ for all $j \geq 1$. In particular, this gives us $x < F^{mq}(x) - mp = F^r(x) - s$, which is impossible. In the same way, assuming $F^q(x) - p < x$ leads to a contradiction. Therefore $\pi(x)$ is periodic for f if, and only if, $F^q(x) = x + p$. In particular, all periodic orbits of f have period q. □

Let us present another proof of the fact that if $\rho(f) = p/q$, then all periodic orbits of f have period q. Let $\pi(x)$ be a periodic point of period q. Then the set $\mathbf{S}^1 \setminus \mathcal{O}_f(\pi(x))$ is made up by q pairwise disjoint intervals $I_1, \ldots, I_q$ which are permuted by f. Moreover, $f^j(I_i) = I_i$ if, and only if, $j = q$. In particular, $f^q|_{I_1}$ is an orientation-preserving self-homeomorphism of the interval I_1, and then its dynamics are quite simple: any point is asymptotic, both forwards and backwards, to a fixed point. Now if $\pi(y)$ is a periodic point for f different from $\pi(x)$, then $\mathcal{O}_f(\pi(y)) \cap I_1 \neq \emptyset$. Say that $\pi(y) \in I_1$. Then $f^q(\pi(y)) = \pi(y)$, that is, $\pi(y)$ is periodic with period q. Note that this argument also implies the following fact.

Lemma 2.3 *Let $f \in \mathrm{Diff}^0_+(\mathbf{S}^1)$ with rational rotation number. For any given $x \in \mathbf{S}^1$, the set $\alpha_f(x)$ is a periodic orbit of f, and the same for $\omega_f(x)$.*

2.2.2 *Irrational Rotation Number*

Proposition 2.5 *Let $f : \mathbf{S}^1 \to \mathbf{S}^1$ be a homeomorphism without periodic points. Then $\Omega(f)$ is a non-empty, compact perfect set, and in fact $\alpha_f(x) = \omega_f(x) = \Omega(f)$ for all $x \in \mathbf{S}^1$. Moreover, if $\Omega(f)$ is not the whole circle, then it is also totally disconnected, i.e., a Cantor set.*

Proof Let x and y be any two points of $\mathbf{S}^1$. If $y \in \omega_f(x)$, then by total invariance we have $\omega_f(y) \subseteq \omega_f(x)$. If $y \notin \omega_f(x)$, then let J be the connected component of $S^1 \setminus \omega_f(x)$ that contains y. Then J is a wandering interval, i.e. its images are pairwise disjoint, because f has no periodic points. In particular, $\sum_{n\in\mathbb{Z}} |f^n(J)| \leq 1 =$ length of S^1, which implies $|f^n(J)| \to 0$ as $|n| \to \infty$. Therefore, if a denotes any of the endpoints of J, we have $\text{dist}(f^n(a), f^n(y)) \to 0$ as $|n| \to \infty$. Thus, if $f^{n_i}(y)$ converges to some point $z \in S^1$, so does $f^{n_i}(a)$. Since $a \in \omega_f(x)$, this shows that $\omega_f(y) \subseteq \omega_f(x)$ in this case also. Interchanging x and y we see that in fact $\omega_f(x) = \omega_f(y)$, so the ω-limit set of any point of S^1 under f is the same. Now we claim that $\Omega(f)$ agrees with $\omega_f(x)$, for any $x \in S^1$. Indeed, if $y \in \Omega(f)$, then $y \in \omega_f(y) = \omega_f(x)$, so $\Omega(f) \subseteq \omega_f(x)$. Conversely, if a point y belongs to $\omega_f(x)$, then it belongs to its own ω-limit set, and therefore it belongs to $\Omega(f)$. One shows similarly that $\alpha_f(x) = \Omega(f)$ for all x. Also, an isolated point of $\Omega(f)$ would necessarily have to be periodic, contrary to assumption, so $\Omega(f)$ is indeed a perfect set. To prove the last assertion, if $\Omega(f) \neq S^1$, then its boundary $\partial\Omega(f)$ is non-empty, closed and totally invariant. Hence, if $z \in \partial\Omega(f)$, we have $\partial\Omega(f) \supseteq \omega_f(z) = \Omega(f)$, so $\Omega(f) = \partial\Omega(f)$, and this is only possible if $\Omega(f)$ is totally disconnected. □

A natural question at this point is: Which Cantor sets can be realized as $\Omega(f)$ for some circle homeomorphism f without periodic orbits? The answer is given by the following result.

Theorem 2.1 *Let $K \subset S^1$ be a Cantor set, and let α be an irrational number. Then there exists a homeomorphism $f : S^1 \to S^1$ such that $\rho(f) = \alpha$ and $\Omega(f) = K$.*

Proof We give a sketch of the proof and leave the details as an exercise. Take any point $x_0 \in S^1$ and let $x_n = R_\alpha^n(x_0)$, $n \in \mathbb{Z}$, be its orbit under the rotation R_α. Let $\mathscr{G}$ denote the countable set consisting of all connected components of $S^1 \setminus K$ (the gaps of the Cantor set K). Then, using a simple inductive procedure, one shows that there exists a bijection $\sigma : \{x_n : n \in \mathbb{Z}\} \to \mathscr{G}$ which is order-preserving in the sense that (in the counter-clockwise orientation of S^1) whenever x_k lies between x_m and x_n, the gap $\sigma(x_k)$ lies between the gaps $\sigma(x_m)$ and $\sigma(x_n)$. Accordingly, write $I_n = \sigma(x_n)$, so that $\mathscr{G} = \{I_n : n \in \mathbb{Z}\}$. Next, define $f : S^1 \setminus K \to S^1 \setminus K$ by letting $f|_{I_n}$ be an affine, orientation-preserving bijection onto I_{n+1}, for all $n \in \mathbb{Z}$. At the same time, let $h : S^1 \setminus K \to \{x_n : n \in \mathbb{Z}\}$ be given by $h(x) = x_n$ whenever $x \in I_n$. Then h is order-preserving, and by construction we have $h \circ f = R_\alpha \circ h$ in $S^1 \setminus K$. Hence f is also order-preserving. Since its image is dense on the circle, f extends to a continuous monotone map of the entire circle, which we still denote by f. Then f must be injective. Otherwise f would be constant on some interval J, but this is not possible because J must intersect some I_n. This shows that f is a homeomorphism. The map h also extends to a continuous monotone map $h : S^1 \to S^1$, and by continuity the equation $h \circ f = R_\alpha \circ h$ now holds everywhere, i.e., h is a semi-conjugacy between f and the rotation R_α. Hence, by Lemma 2.2, we have $\rho(f) = \alpha$. Finally, since by construction we also have $f(K) = K$, it follows that $\Omega(f) = K$, as desired. □

Theorem 2.2 *Let $f : S^1 \to S^1$ be a homeomorphism of the circle with irrational rotation number α. Then f is* semi-conjugate *to the rotation by α, i.e., there exists a continuous, degree one monotone map $h : S^1 \to S^1$ such that $h \circ f = R_\alpha \circ h$.*

Proof Let F be a lift of f to the real line. Define $H : \mathbb{R} \to \mathbb{R}$ by

$$H(x) = \sup \left\{ m\alpha + n : \ F^m(0) + n < x \right\} .$$

Since $\{m\alpha + n : m, n \in \mathbb{Z}\}$ is dense in $\mathbb{R}$, H is continuous and monotone. It clearly satisfies $H(x + 1) = H(x) + 1$ for all x, so it is the lift of a continuous, degree one monotone map $h : S^1 \to S^1$. Now, we have

$$\begin{aligned} H \circ F(x) &= \sup \left\{ m\alpha + n : \ F^m(0) + n < F(x) \right\} \\ &= \sup \left\{ m\alpha + n : \ F^{m-1}(0) + n < x \right\} \\ &= \alpha + H(x) . \end{aligned}$$

Hence we have the commutative diagram

$$\begin{array}{ccc} \mathbb{R} & \xrightarrow{F} & \mathbb{R} \\ {\scriptstyle H}\downarrow & & \downarrow{\scriptstyle H} \\ \mathbb{R} & \xrightarrow[T_\alpha]{} & \mathbb{R} \end{array}$$

where T_α is the translation by α. Descending to the quotient space S^1 yields

$$\begin{array}{ccc} S^1 & \xrightarrow{f} & S^1 \\ {\scriptstyle h}\downarrow & & \downarrow{\scriptstyle h} \\ S^1 & \xrightarrow[R_\alpha]{} & S^1 \end{array}$$

Therefore f is semi-conjugate to the rotation by α, as was to be proved. □

Proposition 2.6 *Let $f : S^1 \to S^1$ be a homeomorphism without periodic points. If there is a sequence $n_i \to \infty$ such that $\{f^{n_i}\}$ is equicontinuous on the circle, then f is topologically conjugate to a rotation.*

Proof Let J be a connected component of the complement of the non-wandering set $\Omega(f)$. Then the intervals $f^n(J)$, $n \in \mathbb{Z}$, are pairwise disjoint. In particular, $|f^{-n_i}(J)| \to 0$ as $i \to \infty$. Since $\{f^{n_i}\}$ is equicontinuous, for each $\epsilon > 0$ there exists $\delta > 0$ such that each interval of length $\leq \delta$ is mapped by each f^{n_i} onto an interval of length $\leq \epsilon$. In particular, since f^{n_i} maps $f^{-n_i}(J)$ onto J for all i, it follows that $|J| \leq \epsilon$. But ϵ is arbitrary, so $|J| = 0$. This shows that $\Omega(f) = S^1$, and therefore, by Theorem 2.2, f is topologically conjugate to an irrational rotation. □

2.3 Invariant Measures and Semi-Conjugacies

From the measure-theoretic point of view, there is a dichotomy between homeomorphisms of the circle with rational and irrational rotation numbers. This dichotomy is presented in Table 2.1. In this section we supply the missing proofs of the statements that are implicit in that table.

It is useful at this point to introduce the following auxiliary notion. If $h : S^1 \to S^1$ is a monotone map, we say that $J \subseteq S^1$ is a *plateau* for h if J is an *open* interval, $h|_J$ is constant and J is maximal with respect to these properties. We have the following two easy lemmas.

Lemma 2.4 *Let $f \in \mathrm{Diff}_+^0(S^1)$ have no periodic points and let h be a continuous monotone map of S^1 such that $h \circ f = R_\alpha \circ h$ (where α is irrational). Then, if J is a plateau of h, we have*

(a) The intervals $f^n(J)$, $n \in \mathbb{Z}$, are pairwise disjoint;
(b) $\mu(J) = 0$ for every f-invariant probability measure μ.

Proof Let h map J to a point $p \in S^1$. If m and n are integers such that $f^m(J) \cap f^n(J) \neq \emptyset$ then, since f is a homeomorphism, we have $J \cap f^{m-n}(J) \neq \emptyset$. If $x = f^{m-n}(y)$ is a point in this last intersection, then on the one hand we have $h(x) = p$ (because $x \in J$), and on the other hand $h \circ f^{m-n}(y) = R_\alpha^{m-n}(h(y)) = R_\alpha^{m-n}(p)$ (because $y \in J$). Therefore $R_\alpha^{m-n}(p) = p$, and this can only happen if $m = n$, for R_α is an irrational rotation. This proves (a). But now we know that $\sum_{n\in\mathbb{Z}} \mu(f^n(J)) \leq 1$, and since the measure is invariant, we also know that $\mu(f^n(J)) = \mu(J)$ for all n. This forces $\mu(J) = 0$, and part (b) is proved. □

Lemma 2.5 *If f and h are as in Lemma 2.4, and if μ_1 and μ_2 are invariant probability measures for f, then $h_*\mu_1 = h_*\mu_2$ if and only if $\mu_1 = \mu_2$.*

Proof Let $P \subseteq S^1$ be the union of all closed intervals of the form $\overline{J}$, where J is a plateau of h. If $A \subseteq S^1$ is any Borel set, we have

$$A \subseteq h^{-1}(h(A)) \subseteq A \cup P$$

Table 2.1 Rational *vs* irrational rotation numbers

$\rho(f) = \dfrac{p}{q} \in \mathbb{Q}$	$\rho(f) = \alpha \in \mathbb{R} \setminus \mathbb{Q} \pmod 1$
There is a periodic orbit of period q	There are no periodic orbits
Many invariant measures in general	Unique invariant Borel probability measure μ μ is ergodic
μ ergodic, invariant probability measure $\Longrightarrow$ supp(μ) = periodic orbit	$\Lambda = \mathrm{supp}(\mu)$ is compact, perfect, invariant and $f\|_\Lambda$ is minimal

By Lemma 2.4 (b) we have $\mu_i(P) = 0$ for $i = 1, 2$. Therefore, writing $A' = h(A)$, we see that

$$h_*\mu_i(A') = \mu_i(h^{-1}(h(A))) = \mu_i(A) ,$$

for $i = 1, 2$ If $h_*\mu_1 = h_*\mu_2$, this forces $\mu_1(A) = \mu_2(A)$, so $\mu_1 = \mu_2$. The converse is obvious. □

With the help of this last lemma we can now prove the following important result.

Theorem 2.3 *Every homeomorphism of the circle without periodic points is uniquely ergodic.*

Proof Let $f \in \mathrm{Diff}_+^0(\mathbf{S}^1)$ be without periodic points and let $\alpha = \rho(f)$. Let $h : \mathbf{S}^1 \to \mathbf{S}^1$ be the semi-conjugacy of f to R_α given by Theorem 2.2. If ν is a probability measure on $\mathbf{S}^1$ such that $f_*\nu = \nu$, then $\mu = h_*\nu$ is an invariant measure for R_α. Moreover, if ν is ergodic for f then μ is ergodic for R_α. Hence, by Lemma 2.5, it suffices to show that R_α is uniquely ergodic (which has been proved by the reader in Exercise 1.13. However, we provide a proof here just for the sake of completeness).

Now, we know that R_α leaves Lebesgue measure m invariant, and that m is ergodic for R_α (recall Lemma 1.5). Suppose then that μ is another invariant measure for R_α, and that μ is ergodic for R_α. Let $A \subseteq \mathbf{S}^1$ be any Borel set, and let $\mathbb{1}_A$ denote its characteristic function. Then by Birkhoff's ergodic theorem we have

$$\frac{1}{n}\sum_{j=0}^{n-1} \mathbb{1}_A \circ R_\alpha^j(x) \xrightarrow{n\to\infty} \int_{\mathbf{S}^1} \mathbb{1}_A \, d\mu = \mu(A)$$

for μ-almost every $x \in \mathbf{S}^1$, whereas by Weyl's equidistribution theorem we also have

$$\frac{1}{n}\sum_{j=0}^{n-1} \mathbb{1}_A \circ R_\alpha^j(x) = \frac{1}{n}\,\mathrm{card}\left\{0 \le j < n : R_\alpha^j(x) \in A\right\} \xrightarrow{n\to\infty} m(A)$$

for *every* $x \in \mathbf{S}^1$. Therefore $\mu(A) = m(A)$ for all Borel sets A, and so $m \equiv \mu$. □

A proof of Theorem 2.3 without using Birkhoff's ergodic theorem and Weyl's equidistribution theorem can be found in the recent survey de Faria and Guarino (2022a, Prop. 2.7). Here is a simple consequence of Theorem 2.3. Let us agree to say that a circle homeomorphism f is *non-singular with respect to Lebesgue measure* m, or simply *measurably non-singular*, if the push-forward measure f_*m is equivalent to m, in the sense that they have the same sets of zero measure. In other words, f is non-singular if $m(f^{-1}(B)) = 0 \iff m(B) = 0$ whenever B is a Borel set. For example, every *diffeomorphism* of the circle is non-singular.

Corollary 2.2 *If $f : S^1 \to S^1$ is a measurably non-singular homeomorphism of the circle without periodic points, then its unique invariant probability measure is either absolutely continuous or purely singular with respect to Lebesgue measure.*

Proof Let μ be the unique Borel probability measure invariant under f. By the Lebesgue–Radon–Nikodým theorem, we can write $\mu = \mu_1 + \mu_2$, where μ_1 and μ_2 are positive Borel measures with $\mu_1 \ll m$ and $\mu_2 \perp m$. Hence $f_*\mu = f_*\mu_1 + f_*\mu_2$, and since $f_*\mu = \mu$, we have

$$\mu(B) = \mu(f^{-1}(B)) = \mu_1(f^{-1}(B)) + \mu_2(f^{-1}(B)) \tag{2.7}$$

for every Borel measurable set B. There are now two cases to consider:

(i) If $m(B) = 0$, then $\mu_1(B) = 0$ (because $\mu_1 \ll m$). Since f is measurably non-singular, we also have $m(f^{-1}(B)) = 0$, and therefore $\mu_1(f^{-1}(B)) = 0$ (again because $\mu_1 \ll m$). Thus, $\mu_1(B) = \mu_1(f^{-1}(B))$ in this case.
(ii) If $m(B) > 0$, then $\mu_2(B) = 0$ (because $\mu_2 \perp m$); in particular $\mu(B) = \mu_1(B)$. Since we also have $m(f^{-1}(B)) = 0$ (again because f is measurably non-singular), it follows that $\mu_2(f^{-1}(B)) = 0$. From this and (2.7) we deduce that $\mu_1(B) = \mu_1(f^{-1}(B))$ in this case as well.

This shows that $f_*\mu_1 = \mu_1$. Therefore either $\mu_1 \equiv 0$ or the normalized measure $\nu = \mu_1/\mu_1(S^1)$ is an invariant probability under f. In the first case, we have $\mu = \mu_2$, and so $\mu \perp m$; in the second case, by unique ergodicity we must have $\nu = \mu$, and so $\mu \ll \mu_1 \ll m$. □

Exercises

Exercise 2.1 Let $f : S^1 \to S^1$ be an orientation *reversing* homeomorphism.

(i) Show that f has exactly two fixed points.
(ii) Show that, for any given $x \in S^1$, the ω-limit set $\omega_f(x)$ is either a fixed point or a periodic orbit of period 2.

Exercise 2.2 Given $f \in \mathrm{Diff}^0_+(S^1)$, the *centralizer* of f in $\mathrm{Diff}^0_+(S^1)$ is the group (under composition) given by $Z_0(f) = \{h \in \mathrm{Diff}^0_+(S^1) : h \circ f = f \circ h\}$.

(i) Show that if f is a rigid rotation, then $Z_0(f)$ is just the group of rotations (and then $Z_0(f)$ is topologically a circle).
(ii) Show that if f is a minimal circle homeomorphism, then $Z_0(f)$ is also homeomorphic to a circle.

Exercise 2.3 Let $f \in \mathrm{Diff}^0_+(S^1)$ with irrational rotation number ρ. Show that any semi-conjugacy between f and R_ρ is unique up to post-composition with rotations.

Exercise 2.4 Prove the inequality (2.6).

Exercise 2.5 Fill in the details of the proof of Theorem 2.1.

Exercise 2.6 Let $f : S^1 \to S^1$ be a circle homeomorphism with irrational rotation number, and let μ be the unique invariant probability measure for f. Let $x \in S^1$, and let (m_i) and (n_i) be two sequences of integers such that $m_i - n_i \to +\infty$ as $i \to \infty$. Writing

$$\mu_i = \frac{1}{m_i - n_i} \sum_{j=n_i+1}^{m_i} \delta_{f^j(x)} ,$$

where δ_a denotes the Dirac point mass (probability measure) at $a \in S^1$, prove that the sequence of measures μ_i converges to μ in the weak*-topology. (See Katznelson 1977, p. 5.)

Exercise 2.7 Let $1 < p < q$ be two integers and denote by Δ_q the interval $\left[\frac{1}{q}, 1\right]$ with the endpoints $\frac{1}{q}$ and 1 identified. Consider the piecewise affine map $T_{p,q} : \Delta_q \to \Delta_q$ given by

$$T_{p,q}(x) = \begin{cases} px \ , \text{ if } \dfrac{1}{q} \leq x < \dfrac{1}{p} \ , \\ \\ \dfrac{p}{q}x \ , \text{ if } \dfrac{1}{p} \leq x \leq 1 \ . \end{cases}$$

The map $T_{p,q}$ is built out of two linear maps of the real line: one expanding (multiplication by $p > 1$), the other contracting (multiplication by $p/q < 1$); see Fig. 2.1. Through the identification of the endpoints of Δ_q, the map $T_{p,q}$ is a piecewise affine homeomorphism of the circle.

(i) Show that $T_{p,q}$ is topologically conjugate to the rotation $R_\alpha : S^1 \to S^1$, with $\alpha = \log_q p$.
(ii) Show that the conjugating map is differentiable except at one point.
(iii) Show that $T_{p,q}$ leaves invariant the absolutely continuous measure μ given by

$$\mu(A) = \int_A \frac{dx}{x \log q} .$$

(iv) Show that if $\gcd(p, q) = 1$ then $T_{p,q}$ is minimal, and therefore uniquely ergodic.
(v) Let $1 < p_1, p_2 < q$ be integers; show that $T_{p_1,q}$ and $T_{p_2,q}$ commute. [Hint: use (i) and (iii)]

[References: de Faria and Tresser 2014; Liousse 2004]

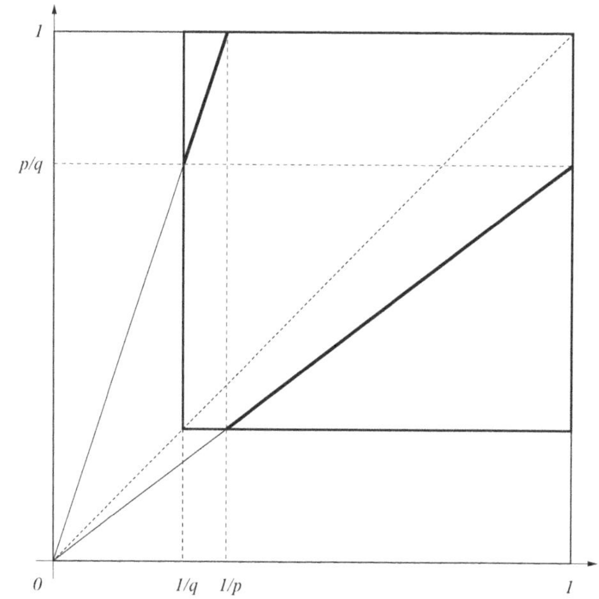

Fig. 2.1 A piecewise linear circle map built out of two linear maps, one expanding with slope p, the other contracting with slope $\frac{p}{q}$

Exercise 2.8 Construct a piecewise affine homeomorphism $f : S^1 \to S^1$ with the following properties: (a) f has irrational rotation number; (b) f leaves invariant the set of rational angles, i.e., $f(\mathbb{Q}/\mathbb{Z}) = \mathbb{Q}/\mathbb{Z}$. The first example of this type was constructed by Boshernitzan (1993). [See also de Faria and Tresser 2014]

Exercise 2.9 Consider $F : \mathbb{R} \to \mathbb{R}$ given by

$$F(x) = x + \frac{1}{100} \sin^2(\pi x) ,$$

and note that F is the lift of a real-analytic diffeomorphism $f : S^1 \to S^1$. Show that $\rho(f) = 0$ and that f is uniquely ergodic.

Exercise 2.10 Let $f \in \mathrm{Diff}^0_+(S^1)$ with irrational rotation number ρ. As we saw in Sect. 2.3, f is uniquely ergodic and its unique invariant measure μ is given by $\mu(A) = m\big(h(A)\big)$ for any Borel set $A \subset S^1$, where h is any semi-conjugacy between f and the rigid rotation R_ρ, and m denotes the normalized Lebesgue measure in S^1 (recall that, by Exercise 2.3, the semi-conjugacy h is unique up to post-composition with rotations, so the measure μ is well-defined). Conversely, given the unique f-invariant measure μ, fix some $x \in S^1$ and consider $h : S^1 \to S^1$ defined by

$$h(y) = \exp\left(2\pi i \int_x^y d\mu\right),$$

where, by convention, we measure the arc (x, y) starting from x in the counter-clockwise sense. Show that h is the unique semi-conjugacy between f and R_ρ which identifies the point x with the point 1.

Exercise 2.11 A self-homeomorphism $f : X \to X$ of a compact metric space (X, d) is said to be *expansive* if there exists a constant $\delta > 0$ such that for every pair of points x and y in X we have that

$$d\big(f^n(x), f^n(y)\big) < \delta \quad \forall\, n \in \mathbb{Z} \;\Rightarrow\; x = y\,.$$

Show that there are no expansive homeomorphisms on the circle.

Part II
Diffeomorphisms

Chapter 3
Diffeomorphisms: Denjoy Theory

The topological theory of *diffeomorphisms* of the circle was started by A. Denjoy, almost fifty years after H. Poincaré introduced the concept of rotation number. In a seminal paper published in 1932, Denjoy (1932) proved that every sufficiently smooth circle diffeomorphism f without periodic points is topologically equivalent to an irrational rotation. Here, the expression "sufficiently smooth" means that f is C^1 and $\log Df$ is a function of bounded variation. In that same work, he also showed that such result is essentially optimal by constructing C^1 diffeomorphisms of the circle without periodic points that leave invariant a Cantor set in S^1—which therefore cannot be conjugate to a rotation—and have the additional property that $\log Df$ is Hölder continuous. An example of this type, without this extra property, had been constructed 16 years earlier by Bohl (1916).

In this chapter we shall prove Denjoy's theorem and construct his examples, the latter with some improvements due to later authors, such as Katznelson (1977) and Herman (1979). In the chapter's last section, we shall examine circle diffeomorphisms from the ergodic-theoretic viewpoint.

3.1 The Naive Distortion Lemma

We will prove first a weaker version of Denjoy's theorem, in which we assume that the circle diffeomorphism is C^2. This we will do primarily for pedagogical reasons. The weaker version provides us with a good opportunity to introduce a very simple, yet extremely useful tool in one-dimensional dynamics: the *nonlinearity* of a map.

Definition 3.1 The nonlinearity of a C^2 diffeomorphism f is

$$\mathscr{N}f \;=\; D\log Df \;=\; \frac{D^2 f}{Df}\,.$$

E. de Faria, P. Guarino, *Dynamics of Circle Mappings*, IMPA Monographs,
https://doi.org/10.1007/978-3-031-67495-2_3

Note that the nonlinearity of f vanishes identically if and only if f is linear (or rather, *affine*). Thus, as the name suggests, the nonlinearity measures how far a map is from being linear. The nonlinearity satisfies a chain rule: if f and g are C^2 diffeomorphisms and $f \circ g$ is well-defined, then

$$\mathcal{N}(f \circ g) = \mathcal{N}f(g)\,Dg + \mathcal{N}g\,.$$

One can see from this chain rule that, under C^2 changes of coordinates, the nonlinearity transforms like a 1-form.

In many situations, the nonlinearity can be used to control the geometric distortion of a long composition of maps. Suppose we have a sequence of intervals $I_0, I_1, \ldots, I_n, \ldots$ on the real line or on the circle, and diffeomorphisms $f_n : I_{n-1} \to I_n$, $n = 1, 2, \ldots$ Let us also assume that

(i) Each f_n is an increasing C^2 diffeomorphism;
(ii) There exists a constant $B > 0$ such that $\sup_{x \in I_n} |\mathcal{N}f_i(x)| \leq B$ for all $n = 1, 2, \ldots$

Let us write $F_n = f_n \circ f_{n-1} \circ \cdots \circ f_1$, for all n. Then we have the following result, known as the *naive distortion lemma.*

Theorem 3.1 *Under the hypotheses (i) and (ii) above, if $x, y \in I_0$ are any two points then for all $n \geq 1$ we have*

$$\exp\left\{-B\sum_{i=0}^{n-1}|I_i|\right\} \leq \frac{F_n'(x)}{F_n'(y)} \leq \exp\left\{B\sum_{i=0}^{n-1}|I_i|\right\}. \tag{3.1}$$

Proof The proof uses the chain rule for the nonlinearity together with the change of variables formula. For all $t \in I_0$ we have

$$\mathcal{N}F_n(t) = \sum_{i=1}^{n}\mathcal{N}f_i(f_{i-1} \circ \cdots \circ f_1(t))\,D\{f_{i-1} \circ \cdots \circ f_1\}(t)\,,$$

by the chain rule. Integrating over the interval J_0 with endpoints x and y contained in I_0, and writing $J_i = f_i \circ f_{i-1} \circ \cdots \circ f_1(J_0)$, we see that

$$\int_{J_0}\mathcal{N}F_n(t)\,dt = \sum_{i=1}^{n}\int_{J_0}\mathcal{N}f_i(F_{i-1}(t))\,DF_{i-1}(t)\,dt = \sum_{i=1}^{n}\int_{J_{i-1}}\mathcal{N}f_i(t)\,dt$$

using over each summand the change of variables $t \mapsto F_{i-1}(t)$. Since $|\mathcal{N}f_i(t)| \leq B$ and $J_{i-1} \subseteq I_{i-1}$, we get

$$\left|\int_{J_0}\mathcal{N}F_n(t)\,dt\right| \leq B\sum_{i=0}^{n-1}|J_i| \leq B\sum_{i=0}^{n-1}|I_i| \tag{3.2}$$

But the integral in the left-hand side of (3.2) is equal to $\pm \log(F_n'(x)/F_n'(y))$, and exponentiating the resulting inequality yields (3.1), as was to be proved.

□

A typical application of the distortion lemma occurs in cases where the f_i's are restrictions of a single map f and the intervals I_i in its domain are pairwise disjoint (or quasidisjoint). In such cases, the estimate offered by the distortion lemma is *uniform* in n, and can be combined with the mean-value theorem to force contradictions, *e.g.* in ruling out the existence of wandering intervals for f. This is precisely what happens in the proof of the C^2 version of Denjoy's theorem, presented below.

3.2 Denjoy's Theorem

The theorem of Denjoy is such an important result that, in this book, we prove it in three different ways. Two of these are given in this section. The third proof will be given in Chap. 6 (see Remark 6.2).

3.2.1 *The C^2 Version*

As promised earlier, we will prove first a weak version of Denjoy's theorem. In this version we assume that the given diffeomorphism is C^2 smooth.

Theorem 3.2 (Weak Denjoy) *If $f : S^1 \to S^1$ is a C^2 diffeomorphism whose rotation number α is irrational, then f is topologically conjugate to the rotation R_α.*

Proof Let $(q_n)_{n\geq 0}$ be the sequence of closest return times of f (denominators of the convergents of α). We already know that there exists a *semi-conjugacy* between f and R_α, that is to say a continuous monotone map $h : S^1 \to S^1$ such that $h \circ f = R_\alpha \circ h$. The pre-image $h^{-1}(p)$ of a point $p \in S^1$ is either a point or the closure of a *plateau* of h. If we can rule out plateaux, h will be one-to-one, and therefore a homeomorphism – in other words, a *conjugacy*.

We argue by contradiction. Suppose $h^{-1}(p) = \overline{J}$, where J is a plateau of h. Then by Lemma 2.4 (a), J is a wandering interval for f, i.e. the intervals $J_m = f^m(J)$, $m \in \mathbb{Z}$, are pairwise disjoint. Fix a small neighborhood of p, and take n so large that $R^{-q_n}(p)$ belongs to that neighborhood. Let I' be the closed interval with endpoints p and $R^{-q_n}(p)$ contained in such neighborhood. Define $I = h^{-1}(I')$, and note that I is an interval because h is monotone. Note also that I contains both J_0 and J_{-q_n}. From our study of the combinatorics of rotations, we already know that the intervals $R_\alpha^i(I')$, $0 \leq i \leq q_n - 1$ are pairwise disjoint. Therefore the intervals $I_i = f^i(I)$, $0 \leq i \leq q_n - 1$ are also pairwise disjoint. In particular, we have $\sum_{i=0}^{q_n-1} |I_i| \leq 1$.

By the mean-value theorem, there exist $x \in J_0$ and $y \in J_{-q_n}$ such that

$$Df^{q_n}(x) = \frac{|J_{q_n}|}{|J_0|} \quad \text{and} \quad Df^{q_n}(y) = \frac{|J_0|}{|J_{-q_n}|} . \tag{3.3}$$

We are now in a position to apply the distortion lemma (Theorem 3.1) to the diffeomorphisms $f_i = f|_{I_{i-1}}$, $1 \leq i \leq q_n$, and the points x and y. From that theorem and (3.3), we deduce that

$$e^{-B} \leq \frac{|J_{q_n}| \cdot |J_{-q_n}|}{|J_0|^2} \leq e^B , \tag{3.4}$$

where $B = \sup|\mathcal{N}f| < \infty$. These inequalities are valid for every sufficiently large n, and the lower and upper bounds are independent of n. But since we have $\sum_{m\in\mathbb{Z}} |J_m| \leq 1$, it follows that $\lim_{m\to\infty} |J_m| = \lim_{m\to\infty} |J_{-m}| = 0$, which contradicts (3.4). Therefore h has no plateaux, and the proof is complete. □

3.2.2 The Bounded Variation Version

As it turns out, in Denjoy's theorem it is not necessary to assume that the diffeomorphism f is C^2. It suffices to assume that $\log Df$ is a function of *bounded variation* on the circle. A function $\varphi : \mathbf{S}^1 \to \mathbb{R}$ is said to be of bounded variation if its *total variation* is finite, that is to say

$$\mathrm{Var}(\varphi) = \sup \sum_{i=1}^{n} |\varphi(a_i) - \varphi(b_i)| < \infty ,$$

where the supremum is taken over all finite collections of pairwise disjoint intervals $(a_i, b_i) \subset \mathbf{S}^1$, $1 \leq i \leq n$. The space of all such functions is denoted by $\mathrm{BV}(\mathbf{S}^1)$.[1]

The fundamental tool for the bounded variation version of Denjoy's theorem is the following.

Theorem 3.3 (Denjoy–Koksma Inequality) *Let $f : \mathbf{S}^1 \to \mathbf{S}^1$ be a homeomorphism with irrational rotation number α, and let μ be the unique Borel probability measure invariant under f. If the rational number p/q is such that*

$$\left|\alpha - \frac{p}{q}\right| < \frac{1}{q^2} ,$$

then for every function $\varphi : \mathbf{S}^1 \to \mathbb{R}$ of bounded variation we have

[1] It is easy to see that $\mathrm{BV}(\mathbf{S}^1)$ is a vector space. It is in fact a Banach space under the norm $\|\varphi\|_{\mathrm{BV}} = |\varphi(0)| + \mathrm{Var}(\varphi)$.

$$\sup_{x\in S^1}\left|\sum_{j=0}^{q-1}\varphi\circ f^j(x)-q\int_{S^1}\varphi\,d\mu\right|\;\leq\;\mathrm{Var}(\varphi)\,.$$

Proof First we remark that the points $\xi_j=\left\{j\frac{p}{q}\right\}\in S^1$ for $j=0,1,\ldots,q-1$ are precisely the q-roots of unity, and therefore they partition S^1 into q arcs of equal length $1/q$. Since for each $j=1,2,\ldots,q$ we have

$$\left|j\alpha-j\frac{p}{q}\right|\;\leq\;\frac{j}{q^2}\;\leq\;\frac{1}{q}\,,$$

it follows that each such arc contains *exactly one* of the points $R_\alpha^j(0)=\{j\alpha\}$, $j=1,\ldots,q$. We label these arcs $\gamma_1,\ldots,\gamma_q$ so that $R_\alpha^j(0)\in\gamma_j$.

Now let $x\in S^1$ and let $h:S^1\to S^1$ be the primitive of μ given by

$$h(t)\;=\;\int_{f^{-1}(x)}^{t}d\mu(s)\,.$$

This monotone map, which is normalized so that $h(f^{-1}(x))=0$, is a semi-conjugacy between f and R_α. Let $\Delta_j=h^{-1}(\gamma_j)$. Then the intervals Δ_j, $j=1,\ldots,q$ form a partition of S^1, and $f^{j-1}(x)\in\Delta_j$ for all j. Note also that $\mu(\Delta_j)=1/q$ for all j.

With these facts at hand, we are ready to prove the Denjoy–Koksma inequality. Let $\varphi\in BV(S^1)$ be of bounded variation. Then we have

$$\begin{aligned}\left|\sum_{j=0}^{q-1}\varphi\circ f^j(x)-q\int_{S^1}\varphi\,d\mu\right| &=\left|\sum_{j=0}^{q-1}\left(\varphi\circ f^j(x)-q\int_{\Delta_{j+1}}\varphi\,d\mu\right)\right|\\ &=q\left|\sum_{j=0}^{q-1}\int_{\Delta_{j+1}}\Big(\varphi(f^j(x))-\varphi(t)\Big)\,d\mu(t)\right|\\ &\leq q\sum_{j=0}^{q-1}\mathrm{Var}(\varphi;\Delta_{j+1})\cdot\mu(\Delta_{j+1})\\ &=\sum_{j=0}^{q-1}\mathrm{Var}\,(\varphi;\Delta_{j+1})\;\leq\;\mathrm{Var}(\varphi)\,.\end{aligned}$$

This concludes the proof, because $x\in S^1$ is arbitrary. □

Theorem 3.4 (Denjoy's Theorem) *Let $f : S^1 \to S^1$ be a diffeomorphism whose rotation number α is irrational. If $\log Df$ is a function of bounded variation, then f is topologically conjugate to the rotation R_α.*

Proof We apply the Denjoy–Koksma inequality to the function $\varphi = \log Df$. Let (q_n) be the sequence of closest returns of f. The chain rule tells us that

$$\log Df^{q_n}(x) = \sum_{j=0}^{q_n-1} \log Df \circ f^j(x) .$$

Hence, by Theorem 3.3, we have

$$\left|\log Df^{q_n}(x) - q_n \int_{S^1} \log Df \, d\mu\right| \leq \mathrm{Var}(\log Df) . \tag{3.5}$$

Since f^{q_n} maps S^1 diffeomorphically onto itself, there is a point x^* such that $Df^{q_n}(x^*) = 1$. Using this point in the above inequality, we see that

$$\left|\int_{S^1} \log Df \, d\mu\right| \leq \frac{\mathrm{Var}(\log Df)}{q_n} .$$

But as $n \to \infty$, the right-hand side of this last inequality goes to zero, and therefore

$$\int_{S^1} \log Df \, d\mu = 0 .$$

Taking this back to the inequality (3.5), we deduce that $|\log Df^{q_n}(x)| \leq V$, where $V = \mathrm{Var}(\log Df)$. Exponentiating this inequality gives us

$$e^{-V} \leq \left|Df^{q_n}(x)\right| \leq e^V .$$

This implies that the sequence of iterates f^{q_n} is equicontinuous. By Proposition 2.6, f must therefore be topologically conjugate to the rotation R_α. □

Remark 3.1 Yet another version of Denjoy's theorem was proved by Hu and Sullivan (1997), for C^1 maps whose first derivative satisfies a *Zygmund* condition. We say that a function $\varphi : S^1 \to \mathbb{R}$ is Zygmund if for all x and h we have

$$\varphi(x+h) + \varphi(x-h) - 2\varphi(x) = O(|h|) .$$

The (linear) space of all Zygmund functions is denoted by Z. Although the Zygmund and bounded variation classes have non-empty intersection, neither class is contained in the other. Hu and Sullivan showed that if $f \in \mathrm{Diff}^{1+Z}(S^1)$, in other words, if f is a C^1 diffeomorphism and $\log Df \in Z$, and $\mathrm{Per}(f) = \emptyset$, then f has no wandering intervals—and therefore it is conjugate to the corresponding rotation. For more on the uses of the Zygmund class in one dimensional dynamics, see de Melo and van Strien (1993, Ch. IV).

3.3 Denjoy's Examples

We have seen in Theorem 2.1 that any perfect, totally disconnected subset of S^1 is the exceptional minimal set of some *homeomorphism* of the circle, and we can even take the rotation number of such homeomorphism to be an arbitrary irrational. The situation is considerably more rigid for *diffeomorphisms*, although still sufficiently flexible to allow for a plethora of examples. We expand on this point a bit.

To start with, Denjoy's theorem rules out exceptional minimal sets for diffeomorphisms whose degree of smoothness is $C^{1+\mathrm{BV}}$ or higher.[2] Thus, if we are looking for diffeomorphisms with exceptional minimal sets, we have to content ourselves with lower smoothness. Let us agree once and for all on the following definition.

Definition 3.2 A *Denjoy example* is a diffeomorphism of the circle having an exceptional minimal set.

In this section, following Herman (1979) and Katznelson (1977), we will construct for each given irrational rotation number, a Denjoy example with that rotation number that is of class $C^{1+\epsilon}$ for *every* $0 < \epsilon < 1$. In fact, for fixed rotation number there are even a countable infinity of topological conjugacy classes of such Denjoy examples. The question of *which* Cantor sets on the circle are minimal sets of C^1 Denjoy examples is a difficult one, and is still open. We will have more to say about that later.

3.3.1 The Basic Construction

The intuitive idea behind the construction of Denjoy examples is to cut the unit circle along one or more orbits of an irrational rotation and introduce a small interval, or *gap*, at each cut. This surgery procedure yields a new, larger circle. In this enlarged circle, the complement of the union of all gaps is a Cantor set (made up of points of the old circle plus the endpoints of the gaps). We define a self-map—a homeomorphism—of the enlarged circle by letting it agree with the irrational rotation on the Cantor set, and by defining it on gaps so that each gap is taken homeomorphically onto another gap, following their exact order on the

[2] We denote by $C^{1+\mathrm{BV}}$ the class of all C^1 maps f such that $\log Df \in \mathrm{BV}(S^1)$.

circle coming from the irrational rotation. The Cantor set becomes the exceptional minimal set of this homeomorphism.

Since we want a *diffeomorphism*, not merely a homeomorphism, the sizes of the gaps have to be carefully chosen. Moreover, if we follow the above procedure to the script, we find that it always produces exceptional minimal sets having positive Lebesgue measure – roughly speaking, equal to the size of the old circle divided by the size of the enlarged circle. We would like to construct examples having zero Lebesgue measure also. Hence, the above procedure will have to be slightly modified.

Let us move to the detailed construction. To start it, we fix an irrational number α with $0 < \alpha < 1$. We want to produce a C^1 diffeomorphism $f : \mathbf{S}^1 \to \mathbf{S}^1$ with $\rho(f) = \alpha$ having an exceptional minimal set $K \subseteq \mathbf{S}^1$. We proceed by steps, as follows.

Step 1: Construction of K We will construct a perfect nowhere dense set $\widehat{K} \subseteq \mathbb{R}$ such that $K = \pi(\widehat{K}) \subseteq \mathbf{S}^1$ is the desired minimal Cantor set. Let us consider a bi-infinite sequence $(\lambda_n)_{n\in\mathbb{Z}}$ of positive real numbers such that

(D1) $\ell = \sum_{n\in\mathbb{Z}} \lambda_n \leq 1.$

(D2) $\lim_{|n|\to\infty} \dfrac{\lambda_{n+1}}{\lambda_n} = 1.$

(D3) $\sup_{n\in\mathbb{Z}} \left| \dfrac{\lambda_{n+1}}{\lambda_n} - 1 \right| = \eta < \dfrac{1}{2}$

For each $n \in \mathbb{Z}$, let $\alpha_n = \{n\alpha\} \in \mathbf{S}^1$; this sequence is dense in $\mathbf{S}^1$, because α is irrational. Define also the sequences

$$\begin{cases} a_n = (1-\ell)\alpha_n + \displaystyle\sum_{i:0\leq\alpha_i<\alpha_n} \lambda_i \,, \\[2ex] b_n = a_n + \lambda_n = (1-\ell)\alpha_n + \displaystyle\sum_{i:0\leq\alpha_i\leq\alpha_n} \lambda_i \,. \end{cases} \tag{3.6}$$

Now let $I_n = (a_n, b_n) \subseteq \mathbb{R}$; note that $I_n \subseteq [0, 1]$, by (D1) above.

Lemma 3.1 *The intervals I_n, $n \in \mathbb{Z}$, are pairwise disjoint, the set*

$$K_0 = (0, 1] \setminus \bigcup_{n\in\mathbb{Z}} I_n$$

is perfect and nowhere dense, and its Lebesgue measure equals $1 - \ell$.

Proof Suppose $m, n \in \mathbb{Z}$ are such that $\alpha_m < \alpha_n$. Then we have

$$a_n = (1-\ell)\alpha_n + \sum_{i:0\leq\alpha_i\leq\alpha_m} \lambda_i + \sum_{i:\alpha_m<\alpha_i<\alpha_n} \lambda_i$$

$$= b_m + (1-\ell)(\alpha_n - \alpha_m) + \sum_{i:\alpha_m<\alpha_i<\alpha_n} \lambda_i > b_m . \tag{3.7}$$

This last inequality holds even if $\ell = 1$, because (α_i) is dense in $[0, 1]$. Thus, we have $a_m < b_m < a_n < b_n$, and therefore $\overline{I}_m \cap \overline{I}_n = \emptyset$. This shows at once that the intervals I_n are pairwise disjoint and that the set K_0 has no isolated points. Moreover, since K_0 is closed, it is Lebesgue measurable, and

$$m(K_0) = 1 - m(\bigcup_{n\in\mathbb{Z}} I_n) = 1 - \ell .$$

It remains to prove that K_0 has empty interior. This is obvious if $\ell = 1$, so we assume that $\ell < 1$. Note that (3.7) tells us that

$$\delta = \operatorname{dist}(I_n, I_m) = (1-\ell)(\alpha_n - \alpha_m) + \sum_{i:\alpha_m<\alpha_i<\alpha_n} \lambda_i$$

and also that

$$\nu = m(K_0 \cap [b_m, a_n]) - (1-\ell)(\alpha_n - \alpha_m)$$

We claim that there exists a gap I_k contained in $[b_m, a_n]$ that intersects the middle third of $[b_m, a_n]$. This is clear if $\nu < \delta/3$, so we assume that $\nu \geq \delta/3$. Let $\alpha_k \in (\alpha_m, \alpha_n)$ be such that

$$\left|\alpha_k - \frac{\alpha_m + \alpha_n}{2}\right| < \frac{1}{18}(\alpha_n - \alpha_m) .$$

Then we have

$$\begin{aligned} a_k - b_m &= (1-\ell)(\alpha_k - \alpha_m) + \sum_{i:\alpha_m<\alpha_i<\alpha_k} \lambda_i \\ &> \frac{4}{9}(1-\ell)(\alpha_n - \alpha_m) \geq \frac{\delta}{3} . \end{aligned}$$

Similarly, $a_n - b_k > \delta/3$. These inequalities show that the gap I_k is contained in the middle third of $[b_m, a_n]$, proving the claim. From the claim it follows that the union of all gaps is dense in $[0, 1]$, and therefore K_0 has empty interior. □

We now define $\widehat{K}$ as the union of all integral translates of K_0, in other words,

$$\widehat{K} = \bigcup_{n\in\mathbb{Z}} (n + K_0) .$$

This set enjoys the same topological properties as we stated for K_0: it is closed, perfect, and nowhere dense.

Step 2: Construction of the Lift of f Let us now construct an increasing diffeomorphism $F : \mathbb{R} \to \mathbb{R}$ with $F(\widehat{K}) = \widehat{K}$, which will be the lift to the real line of f, the Denjoy example we seek to construct. We write $F = \mathrm{Id} + \phi$, where ϕ is periodic of period one. We will construct the derivative $\varphi = D\phi$ first, and then integrate.

We want to have $f(J_n) = J_{n+1}$ for all n, where $J_n = \pi(I_n) \subseteq S^1$. Hence F must map each gap I_n onto (an integral translate of) of I_{n+1}, and therefore

$$\int_{I_n} F'(t)\, dt \;=\; \lambda_{n+1} \;=\; |I_{n+1}|\,.$$

This is the same as requiring that

$$\int_{a_n}^{b_n} [1 + \varphi(t)]\, dt \;=\; \lambda_{n+1}\,,$$

or yet

$$\int_{a_n}^{b_n} \varphi(t)\, dt \;=\; \lambda_{n+1} - \lambda_n\,. \tag{3.8}$$

There are many ways to define φ inside I_n, vanishing at the endpoints, so that the above equality holds. One way is to write

$$\varphi(t) = \frac{1}{2}\, c_n\, (\lambda_n - |2t - a_n - b_n|)$$

for all $t \in I_n$. Here the constant c_n is chosen so that (3.8) holds. A simple computation yields

$$c_n = \frac{4}{\lambda_n}\left(\frac{\lambda_{n+1}}{\lambda_n} - 1\right)\,.$$

In other words, define φ so that for each $t \in I_n$ we have

$$\varphi(t) \;=\; 2\left(\frac{\lambda_{n+1}}{\lambda_n} - 1\right)\left(1 - \frac{1}{\lambda_n}\,|2t - a_n - b_n|\right)\,. \tag{3.9}$$

Furthermore, let $\varphi(t) = 0$ for all $t \in K_0$. So far we have φ defined on the unit interval only. It is clearly continuous in the union of all gaps I_n. Since by (3.9) and (D2) we have

$$\lim_{|n|\to\infty}\, \sup_{t\in I_n} |\varphi(t)| = \lim_{|n|\to\infty} 2\left|\frac{\lambda_{n+1}}{\lambda_n} - 1\right| = 0\,,$$

it follows that φ is continuous at all points of K_0 as well. Now extend φ outside the unit interval by making it periodic of period one, so that $\varphi(t+k) = \varphi(t)$, for all $k \in \mathbb{Z}$ and all $0 \leq t \leq 1$. The extended function $\varphi : \mathbb{R} \to \mathbb{R}$ is continuous everywhere, and it vanishes at all points of $\widehat{K}$.

Now let $F : \mathbb{R} \to \mathbb{R}$ be given by

$$F(t) = a_1 + t + \int_0^t \varphi(s)\, ds \,. \tag{3.10}$$

We summarize the essential facts about F in our next lemma.

Lemma 3.2 *The map F is an increasing C^1 diffeomorphism of the real line, and it has the following properties.*

(a) For all $t \in \mathbb{R}$, $F(t+1) = F(t) + 1$;
(b) For every $n \in \mathbb{Z}$,

$$F(I_n) = \begin{cases} I_{n+1}\,, & \text{if } \alpha_n < 1 - \alpha \\ 1 + I_{n+1}\,, & \text{if } \alpha_n > 1 - \alpha\,; \end{cases}$$

(c) We have $F(\widehat{K}) = \widehat{K}$;
(d) The translation number of F is equal to α.

Proof Note that $F'(t) = 1 + \varphi(t)$, which is continuous, so F is C^1. Moreover, by (3.9) and (D3), we have $\varphi(t) \geq -2\eta > -1$, and this implies $F'(t) \geq 1 - 2\eta > 0$ for all t. Therefore F is an increasing diffeomorphism. Property (a) is immediate from

$$\int_0^1 \varphi(t)\, dt = 0\,,$$

which in turn follows from (3.8) and the fact that φ vanishes on K_0. To prove property (b), it suffices to show that

$$F(a_n) = \begin{cases} a_{n+1}\,, & \text{if } \alpha_n < 1 - \alpha \\ 1 + a_{n+1}\,, & \text{if } \alpha_n > 1 - \alpha\,. \end{cases}$$

From (3.8) and (3.10), we have

$$\begin{aligned} F(a_n) &= a_1 + a_n + \sum_{0 \leq \alpha_i < \alpha_n} \int_{a_i}^{b_i} \varphi(t)\, dt \\ &= a_1 + a_n + \sum_{0 \leq \alpha_i < \alpha_n} (\lambda_{i+1} - \lambda_i)\,. \end{aligned}$$

Using the expressions of a_1 and a_n given in (3.6), this last equality becomes

$$F(a_n) = (1-\ell)(\alpha + \alpha_n) + \sum_{0\le\alpha_j<\alpha} \lambda_j + \sum_{0\le\alpha_i<\alpha_n} \lambda_{i+1}\,. \tag{3.11}$$

Assume first that $\alpha_n < 1-\alpha$. In this case, for all $\alpha_i \in [0, \alpha_n]$ we have $\alpha_i+\alpha = \alpha_{i+1}$. Using this fact in (3.11), we deduce that

$$F(a_n) = (1-\ell)\alpha_{n+1} + \sum_{0\le\alpha_j<\alpha} \lambda_j + \sum_{\alpha\le\alpha_{i+1}<\alpha_{n+1}} \lambda_{i+1} = a_{n+1}\,.$$

Now assume instead that $\alpha_n > 1-\alpha$. In this case $\alpha_{n+1} = \alpha_n + \alpha - 1$, and therefore we can write

$$\begin{aligned}\sum_{0\le\alpha_i<\alpha_n} \lambda_{i+1} &= \sum_{0\le\alpha_i<1-\alpha} \lambda_{i+1} + \sum_{1-\alpha\le\alpha_i<\alpha_n} \lambda_{i+1}\\ &= \sum_{\alpha\le\alpha_j<1} \lambda_j + \sum_{0\le\alpha_j<\alpha_{n+1}} \lambda_j\,.\end{aligned} \tag{3.12}$$

Substituting (3.12) into (3.11), we get

$$F(a_n) = (1-\ell)(1+\alpha_{n+1}) + \ell + \sum_{0\le\alpha_j<\alpha_{n+1}} \lambda_j = 1 + a_{n+1}$$

This proves property (b). Thus F leaves invariant the union of all gaps I_n and its translates, and since it is a homeomorphism, the complement $\widehat{K}$ is kept invariant also, which proves (c).

Finally, to prove (d), let $H : \mathbb{R} \to \mathbb{R}$ be defined as follows. For each $t \in [0, 1]$, put $H(t) = \sup_{I_n\subseteq[0,t]} \alpha_n$, and extend H to the whole real line writing $H(t+k) = H(t) + k$ for all $k \in \mathbb{Z}$. This function is clearly non-decreasing. Hence its only possible discontinuities are jump discontinuities. To rule out jumps, it suffices to show that the image of H is dense in $\mathbb{R}$. To do this, we first calculate $H(t)$ for $t \in I_n$, for each $n \in \mathbb{Z}$. Suppose $\alpha_m < \alpha_n$; then it follows easily that $a_m < a_n$, and therefore $b_m < a_n$ as well. Therefore $I_m \subset [0, a_n] \subset [0, t]$, and from this it follows that $H(t) = \alpha_n$ (see Fig. 3.1). But $\{\alpha_n : n \in \mathbb{Z}\}$ is dense in $[0, 1]$, so $H([0, 1])$ is dense in $[0, 1]$. Since $H(t+k) = H(t) + k$ for all $k \in \mathbb{Z}$, we deduce that $H(\mathbb{R})$ is dense in $\mathbb{R}$. Therefore H is indeed continuous.

It remains to check that the semi-conjugacy equation $H \circ F(t) = T_\alpha \circ H(t)$ holds for all $t \in \mathbb{R}$. Once again, we only need to check this for $t \in [0, 1]$. Let $t \in I_n$, and suppose first that $\alpha_n < 1-\alpha$. Since $H(t) = \alpha_n$, we have in this case $T_\alpha \circ H(t) = \alpha_n + \alpha = \alpha_{n+1}$. We also have $F(t) \in I_{n+1}$, and therefore $H \circ F(t) = \alpha_{n+1}$. This shows that the semi-conjugacy equation holds in this case. The case $\alpha_n > 1-\alpha$ is proved in the same way, *mutatis mutandis*. Summarizing, we have proved that for all $t \in U = \bigcup_{n\in\mathbb{Z}} I_n$ we have $H \circ F(t) = T_\alpha \circ H(t)$. But since

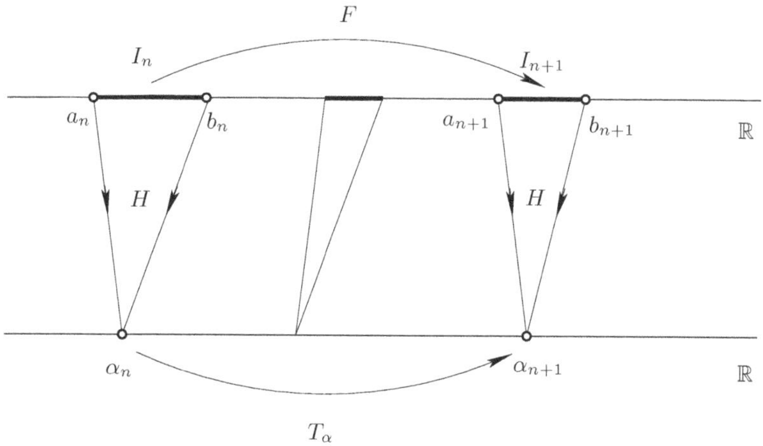

Fig. 3.1 The semi-conjugacy H and its plateaux

U is dense in $[0, 1]$ and H, F, T_α are continuous, it follows that $H \circ F(t) = T_\alpha \circ H(t)$ for all $t \in [0, 1]$, which proves what we wanted. In particular, the translation number of F is α, as asserted in (d). □

We can now quotient everything down to the circle S^1. Thus, F is the lift of a C^1 diffeomorphism $f : S^1 \to S^1$, while H is the lift of a continuous monotone map $h : S^1 \to S^1$, and we deduce that $h \circ f = R_\alpha \circ h$. Moreover, we have $\Omega(f) = K$, a Cantor set—in other words, f is a Denjoy example.

Remark 3.2 A bi-infinite sequence $\Lambda = (\lambda_n)_{n\in\mathbb{Z}}$ of positive numbers satisfying properties (D1)-(D3) stated above is called a *Denjoy sequence*. Given a Denjoy sequence Λ and an irrational $\alpha \in (0, 1)$, the Denjoy example constructed with this data is denoted by $f_{\alpha,\Lambda}$. In what follows, we will denote by $\boldsymbol{W}^0 \subset \mathrm{Diff}^0_+(\boldsymbol{S}^1)$ the class of all homeomorphisms of the circle without periodic points that possess a wandering interval (in other words, $f \in \boldsymbol{W}^0$ if and only if $\mathrm{Per}(f) = \emptyset$ and $\Omega(f) \neq \boldsymbol{S}^1$). We will also write $\boldsymbol{W}^1 = \boldsymbol{W}^0 \cap \mathrm{Diff}^1_+(\boldsymbol{S}^1)$, and will denote by $\boldsymbol{D} \subset \boldsymbol{W}^1$ the class of all Denjoy examples constructed by the procedure described above (in other words, $f \in \boldsymbol{D}$ if and only if $f = f_{\alpha,\Lambda}$ for some irrational $\alpha \in (0, 1)$ and some Denjoy sequence Λ).

3.3.2 *Moduli of Continuity*

Now we go a bit further and show that our Denjoy examples can be made *almost Lipschitz*; more precisely, they can be made $C^{1+\epsilon}$ for every $0 < \epsilon < 1$, with no restriction on the (irrational) rotation number.

Theorem 3.5 *For each* $0 < \ell \leq 1$, *each* $\beta > 0$ *and each* $0 < \alpha < 1$ *irrational, there exists a* C^1 *diffeomorphism* $f : \mathbf{S}^1 \to \mathbf{S}^1$ *such that*

(i) The rotation number of f *is equal to* α.
(ii) The non-wandering set $\Omega(f) \subset \mathbf{S}^1$ *is a Cantor set with Lebesgue measure equal to* $1 - \ell$.
(iii) The function $\log Df$ *has modulus of continuity* $w_\beta(t) = t\left(\log \frac{1}{t}\right)^{1+\beta}$.

This last property implies that $\log Df$ *is* δ*-Hölder continuous for each* $0 < \delta < 1$.

Proof Given the construction performed above and Lemma 3.2, the proof boils down to taking $f = f_{\alpha,\Lambda}$ for a smart choice of the Denjoy sequence $\Lambda = (\lambda_n)_{n\in\mathbb{Z}}$. We take

$$\lambda_n = \frac{b}{(|n|+1)\left[\log(|n|+2)\right]^{1+\beta}}\,, \quad \forall\, n \in \mathbb{Z}\,, \tag{3.13}$$

where $0 < b < 1$ is chosen so that $\sum_{n\in\mathbb{Z}} \lambda_n = \ell$. This choice of the sequence (λ_n) guarantees that

$$\left|\frac{\lambda_{n+1}}{\lambda_n} - 1\right| = O\left(\frac{1}{|n|+1}\right) \tag{3.14}$$

Let $F : \mathbb{R} \to \mathbb{R}$ be given by Lemma 3.2 for this choice of (λ_n) and the given irrational α. Then we already know that F is the lift of a C^1 diffeomorphism $f : \mathbf{S}^1 \to \mathbf{S}^1$ which is topologically semi-conjugate to the rotation R_α, and we also know that $\Omega(f)$ is Cantor set with $m(\Omega(f)) = 1 - \ell$. Thus, properties (i) and (ii) are satisfied, and we only need to check property (iii).

Since $DF = 1 + \varphi$ stays bounded away from 0 and ∞, to verify the validity of (iii) it suffices to show that φ itself has modulus of continuity w_β. And since φ is $\mathbb{Z}$-periodic, all we need to do is to check that

$$|\varphi(x) - \varphi(y)| < C|x-y|\left(\log\frac{1}{|x-y|}\right)^{1+\beta} \tag{3.15}$$

for $x, y \in [0, 1]$. Recall the notation introduced earlier: $K_0 = \widehat{K} \cap [0, 1]$, where $\widehat{K} = \pi^{-1}(\Omega(f))$ is the lift of the minimal set of f to the real line. There are three cases to consider: (a) $x, y \in K_0$; (b) $x \in K_0$ and $y \in I_n$ for some $n \in \mathbb{Z}$; and (c) $x \in I_m$ and $y \in I_n$ for some $m, n \in \mathbb{Z}$. Case (a) is trivial because, by construction, $\varphi|_{K_0} \equiv 0$. We prove the inequality (3.15) for case (b) and leave case (c) to the reader.

Without loss of generality, we may assume that $x < y$, so that $x \leq a_n < y$ (where as before a_n is the left endpoint of I_n). We may also assume that $|x - y| < t_\beta$, where $0 < t_\beta < 1$ is the point where the concave function $w_\beta|_{[0,1]}$ assumes its maximum. Since $x \in K_0$, we have $\varphi(x) = 0$ and therefore

$$|\varphi(x) - \varphi(y)| = |\varphi(y)| \leq L_n|y - a_n|\,, \tag{3.16}$$

where L_n is the Lipschitz constant of $\varphi|_{I_n}$. But we know from (3.9) that

$$L_n = \frac{4}{\lambda_n}\left|\frac{\lambda_{n+1}}{\lambda_n} - 1\right| . \tag{3.17}$$

From (3.17), combined with (3.13) and (3.14), we deduce after some tedious computations that

$$\frac{1}{\lambda_n}\left|\frac{\lambda_{n+1}}{\lambda_n} - 1\right| \leq C_0\left[\log\left(|n|+2\right)\right]^{1+\beta} , \tag{3.18}$$

where $C_0 > 0$ is a constant. Moreover, since

$$\frac{b}{|n|+1} > \frac{b}{(|n|+1)\left[\log\left(|n|+2\right)\right]^{1+\beta}} = |I_n| > |y - a_n| ,$$

we have

$$\log\left(|n|+2\right) < C_1 \log\left(\frac{1}{|y-a_n|}\right) , \tag{3.19}$$

where $C_1 > 0$ is a constant. Combining (3.19), (3.18) and (3.17) and putting the resulting inequality for L_n back into (3.16), we arrive at

$$\begin{aligned} |\varphi(x) - \varphi(y)| &< C_2|y - a_n|\left[\log\left(\frac{1}{|y-a_n|}\right)\right]^{1+\beta} \\ &< C_2|x - y|\left[\log\left(\frac{1}{|x-y|}\right)\right]^{1+\beta} , \end{aligned}$$

where in the last inequality we have used the fact that $w_\beta(t) = t\left(\log\frac{1}{t}\right)^{1+\beta}$ is increasing for $0 < t < t_\beta$. This shows that (3.15) holds in case (b). The proof of that inequality in case (c) is similar, and is left as an exercise. □

3.3.3 *Further Results*

The class of general Denjoy examples, i.e., non-minimal C^1 circle diffeomorphisms without periodic points, has been thoroughly investigated. There are various questions one can ask about these maps, some of which are still unsolved. In this section we address some of these questions, limiting ourselves to simply stating results that are currently known. No proofs will be given; instead, we will refer the reader to the original sources.

3.3.3.1 Classification

The topological classification of Denjoy examples is not particularly difficult, and has been accomplished by Markeley (1970). Before stating the result, we make some simple observations.

Given $f \in \boldsymbol{W}^0$, let $h : \boldsymbol{S}^1 \to \boldsymbol{S}^1$ be a monotone map that semi-conjugates f to the rotation by $\alpha = \rho(f)$, i.e., $h \circ f = R_\alpha \circ h$. Then $E_{f,h} = h(\boldsymbol{S}^1 \setminus \Omega(f))$ is a countable dense subset of the circle, and it is invariant under R_α. Thus, $E_{f,h}$ is the union of a collection of full orbits of the rotation R_α, and such collection can be either finite or countably infinite. If $\phi : \boldsymbol{S}^1 \to \boldsymbol{S}^1$ is another semi-conjugacy between f and R_α, it is easy to see that $E_{f,h}$ and $E_{f,\phi}$ differ by a rotation of the circle (exercise). In particular, the cardinalities of the sets of orbits of R_α contained in $E_{f,h}$ and $E_{f,\phi}$ are the same. Hence this common cardinality is a topological invariant of f. The theorem proved by Markeley can be stated as follows.

Theorem 3.6 *Let $f_1, f_2 \in \boldsymbol{W}^0$ with $\rho(f_1) = \rho(f_2) = \alpha$, and let $h_1, h_2 : \boldsymbol{S}^1 \to \boldsymbol{S}^1$ be monotone maps such that $h_i \circ f_i = R_\alpha \circ h_i$, $i = 1, 2$. Then f_1 is topologically conjugate to f_2 if an only if there exists a circle rotation R such that $E_{f_1,h_1} = R(E_{f_2,h_2})$.*

Note in particular that if in the theorem above $f_1, f_2 \in \boldsymbol{D}$, then E_{f_1,h_1} and E_{f_2,h_2} both consist of a single orbit of R_α. Since any two orbits of R_α differ by a rotation, it follows that f_1 and f_2 are topologically conjugate. But one can say a bit more.

Theorem 3.7 *If $f_1, f_2 \in \boldsymbol{D}$ have the same rotation number, then they are topologically conjugate. Moreover, if both $\Omega(f_1)$ and $\Omega(f_2)$ have Lebesgue measure zero and $h : \boldsymbol{S}^1 \to \boldsymbol{S}^1$ is a homeomorphism conjugating f_1 and f_2, then both h and h^{-1} are absolutely continuous.*

For a discussion of this theorem (with an indication of proof), look up the *magnum opus* of Herman (1979, p. 146).

By contrast, the classification of Denjoy examples up to C^1 conjugacy still has not been completely worked out. The problem can be reformulated as follows.

Problem 3.1 If $f, g : \boldsymbol{S}^1 \to \boldsymbol{S}^1$ are C^1 circle diffeomorphisms which are topologically conjugate to each other, find necessary and sufficient conditions for f to be C^1 conjugate to g.

3.3.3.2 Which Cantor Sets Are Denjoy?

As we saw in Chap. 2, any Cantor set on the circle is the non-wandering set of some *homeomorphism* of $\boldsymbol{S}^1$, and the (irrational) rotation number can be arbitrarily prescribed. Once we ask for more smoothness, however, things become much more complicated. It is still unknown which Cantor sets appear as minimal sets of Denjoy examples. The partial results that are known are mostly negative results, stating that certain families of Cantor sets do not appear as $\Omega(f)$ for any C^1 diffeomorphism f.

For instance, regular Cantor sets such as the standard middle-thirds Cantor set are *not* C^1 minimal. This was first proved by McDuff (1981). In order to state her result, we need the following definition.

Definition 3.3 Let $K \subset S^1$ be a Cantor set, and let $\mathscr{G}(K)$ be the set of all *gaps* of K—the elements of $\mathscr{G}(K)$ are the connected components of $S^1 \setminus K$. The *spectrum* of K, denoted $\sigma(K)$, is the set of all lengths of gaps in $\mathscr{G}(K)$ ordered as a decreasing sequence. Thus, $\sigma(K) = \{\ell_n : n \geq 1\}$, where (i) $\ell_{n+1} < \ell_n$ for all $n \geq 1$; (ii) for each $n \geq 1$, there exists $I \in \mathscr{G}(K)$ (possibly non-unique) such that $\ell_n = |I|$.

Note that $\sum_{n=1}^{\infty} \ell_n \leq 1$.

Theorem 3.8 *If* $f \in W^1$ *and* $\sigma(\Omega(f)) = \{\ell_n : n \in \mathbb{N}\}$ *is the spectrum of its minimal set, then the sequence* $\dfrac{\ell_n}{\ell_{n+1}}$ *is bounded and has* 1 *as a limit point.*

An immediate consequence of this result is the fact stated before that the standard middle-thirds Cantor set is not C^1 minimal. Indeed, in this case the ratios ℓ_n/ℓ_{n+1} are constant and equal to 3.

Besides the original paper by McDuff, the reader interested in the full proof of Theorem 3.8 should look up the nice exposition by Athanassopoulos (2015). It is worth remarking that the fact that the sequence of ratios $\dfrac{\ell_n}{\ell_{n+1}}$ is bounded is easy to prove. Given $n \in \mathbb{N}$, let $I \in \mathscr{G}(\Omega(f))$ be such that $\ell_n = |I|$. Then there exists $m \geq 1$ such that $\ell_n \leq |f^m(I)|$ and $|f^k(I)| \leq \ell_{n+1}$ for all $k > m$. Thus, if we set $J = f^m(I)$, we have

$$\frac{\ell_n}{\ell_{n+1}} \leq \frac{|J|}{|f(J)|} .$$

But by the mean value theorem, this last ratio is bounded above by $1/b$, where $b = \min_{x \in S^1} |Df(x)|$. Hence the real issue in proving Theorem 3.8 is to show that 1 is a limit point of the sequence of ratios.

Given the above theorem, a natural question posed by McDuff (1981) is the following.

Problem 3.2 If $f \in W^1$ and $\sigma(\Omega(f)) = \{\ell_n : n \in \mathbb{N}\}$ is the spectrum of its minimal set, is it always true that

$$\lim_{n\to\infty} \frac{\ell_n}{\ell_{n+1}} = 1\ ?$$

Even after four decades since its formulation, this problem remains open.

McDuff's Theorem 3.8 provides us with a necessary condition for a Cantor subset of the circle to be C^1 minimal. This condition is used to rule out several types of Cantor sets. But many other Cantor sets do satisfy the condition, so they cannot be immediately ruled out. For generalizations of McDuff's theorem and further work, see Norton (2002), Portela (2007), and Portela (2009).

3.3.3.3 Hausdorff Dimension

It turns out that the Hausdorff dimension of the minimal set of a Denjoy example with zero Lebesgue measure depends on the Diophantine nature of its rotation number. An irrational real number α is said to be of *Diophantine class* $\tau > 0$ if the inequality

$$\left|\alpha - \frac{p}{q}\right| < \frac{1}{q^{1+\eta}}$$

has infinitely many rational solutions p/q for $0 \leq \eta < \tau$ and only finitely many for $\eta > \tau$ (the Diophantine condition will appear several times in the present book, see Chap. 4 and Appendix A).

The following theorem was proved by Kra and Schmeling (2002).

Theorem 3.9 *Let* $0 < \delta < 1$ *and let* $\alpha \in (0, 1)$ *be an irrational number of Diophantine class* $\tau > 0$*. If* $f : \boldsymbol{S}^1 \to \boldsymbol{S}^1$ *is a* $C^{1+\delta}$ *diffeomorphism with rotation number* α *and* f *has an exceptional minimal set* $\Omega(f)$*, then*

$$\dim_H(\Omega(f)) \geq \frac{\delta}{\tau} .$$

The lower bound given in this theorem is sharp. Indeed, Kra and Schmeling showed that the lower bound is achieved by a classical Denjoy example of the form $f = f_{\alpha,\Lambda} \in \boldsymbol{D}$, the bi-infinite Denjoy sequence $\Lambda = (\lambda_n)_{n\in\mathbb{Z}}$ being given by

$$\lambda_n = \frac{b}{(|n| + 1)^{1/\delta}} ,$$

where b is chosen so that $\sum_{n\in\mathbb{Z}} \lambda_n = 1$. In the same article, Kra and Schmeling also provide a lower bound for the *box dimension* of Denjoy minimal Cantor sets, extending previous work by Norton (1999).

3.4 Ergodic Properties

We have seen already, in Chap. 2 (Theorem 2.3), that every circle homeomorphism without periodic orbits is uniquely ergodic. In this section, we examine a few additional ergodic properties of circle diffeomorphisms.

3.4.1 Ergodicity with Respect to Lebesgue Measure

It is possible to talk about ergodicity of a map with respect to a measure in phase space even when the map is not measure preserving. If (X, μ) is a measure space and $\phi : X \to X$ is a (measurable) map, we say that ϕ is *ergodic with respect to* μ if, for every measurable set $A \subseteq X$ which is invariant under ϕ (meaning $\phi^{-1}(A) = A$), we have either $\mu(A) = 0$ or $\mu(X \setminus A) = 0$. Note that, for this to make sense, it is not necessary to assume that μ is a finite measure.

Our goal here is to prove that $C^{1+\mathrm{BV}}$ circle diffeomorphisms without periodic points are always ergodic with respect to Lebesgue measure. Let us state more formally this result, and then prove it.

Theorem 3.10 *If $f : \mathbf{S}^1 \to \mathbf{S}^1$ is a C^1 diffeomorphism without periodic points and $\log Df \in BV(\mathbf{S}^1)$, then f is ergodic with respect to Lebesgue measure.*

The proof will use the following lemmas.

Lemma 3.3 *Given f as in Theorem 3.10, let $M \subset T \subset \mathbf{S}^1$ be two intervals, and let $n \geq 1$. Suppose the intervals $T, f(T), f^2(T), \ldots, f^n(T)$ have multiplicity of intersection $k \geq 1$, i.e., every point in $\mathbf{S}^1$ belongs to at most k of these intervals. Then we have*

$$\frac{|f^n(M)|}{|f^n(T)|} \leq e^{kV}\,\frac{|M|}{|T|}\,, \tag{3.20}$$

where $V = \mathrm{Var}(\log Df)$.

Proof By the mean value theorem, there exist $x_0 \in M$ and $y_0 \in T$ such that $Df^n(x_0) = |f^n(M)|/|M|$ and $Df^n(y_0) = |f^n(T)|/|T|$. From this and the chain rule, we get

$$\frac{|f^n(M)|}{|f^n(T)|} = \frac{|M|}{|T|}\prod_{i=0}^{n-1}\frac{Df(x_i)}{Df(y_i)}\,,$$

where $x_i = f^i(x_0)$ and $y_i = f^i(y_0)$, for each $0 \leq i \leq n-1$. Now write

$$\prod_{i=0}^{n-1}\frac{Df(x_i)}{Df(y_i)} = \exp\left\{\sum_{i=0}^{n-1}(\log Df(x_i) - \log Df(y_i))\right\}\,. \tag{3.21}$$

Since the n intervals with endpoints x_i and y_i are k-quasidisjoint, we have (see Exercise 3.7):

$$\sum_{i=0}^{n-1}(\log Df(x_i) - \log Df(y_i)) \leq k\,\mathrm{Var}(\log Df)\,.$$

Putting this back into (3.21), we deduce (3.20) as desired. □

Lemma 3.4 *Let f be as in Theorem 3.10. Then for each $\varepsilon > 0$ and each $x \in \mathbf{S}^1$ there exist an interval $\Delta \subset \mathbf{S}^1$ containing x and a positive integer N such that $|\Delta| < \epsilon$ and the intervals $\Delta, f(\Delta), \ldots, f^N(\Delta)$ cover the circle and are 3-quasidisjoint.*

Proof By Denjoy's theorem, f is topologically conjugate to an irrational rotation R. Let $h : \mathbf{S}^1 \to \mathbf{S}^1$ be a homeomorphism such that $h \circ f = R \circ h$. Since h^{-1} is uniformly continuous, given $\epsilon > 0$ there exists $\delta > 0$ such that, for each interval $I \subset \mathbf{S}^1$ with $|I| < \delta$ we have $|h^{-1}(I)| < \epsilon$. Given $x \in \mathbf{S}^1$, let $z = h^{-1}(x)$, and consider for each $n \geq 0$ the interval $J_n \subset \mathbf{S}^1$ with endpoints $R^{-q_n}(z)$ and $R^{q_{n+1}}(z)$ that contains z—where $\{q_n\}_{n\geq 0}$ is the sequence of closest returns for R (or f). Since $R^{q_k}(z) \to z$ as $k \to \infty$, we can choose n large enough that $|J_n| < \delta$. It is not difficult to see that the intervals $J_n, R(J_n), \ldots, R^{q_{n+1}-1}(J_n)$ cover the circle and are 3-quasidisjoint. The reader can either prove this as an exercise or else look up the proof (of a slightly stronger fact) in Sect. 6.4. Hence we can take $\Delta = h^{-1}(J_n)$ and $N = q_{n+1} - 1$. □

Proof of Theorem 3.10 The proof uses a Lebesgue density argument (akin to the one used in Lemma 1.5). Let $A \subset \mathbf{S}^1$ be a measurable set invariant under f, and suppose that $m(A) > 0$. Write $B = \mathbf{S}^1 \setminus A$, and note that B is also invariant under f. Let $x \in A$ be a Lebesgue density point for A. Given $\delta > 0$, let $\epsilon > 0$ be so small that for all $0 < \eta < \frac{1}{2}\epsilon$ we have

$$\frac{m(A \cap (x-\eta, x+\eta))}{2\eta} > (1-\delta)\,,$$

or, equivalently, $m(B \cap (x-\eta, x+\eta)) < 2\eta\delta$. Let $\Delta \subset (x - \frac{1}{2}\epsilon, x + \frac{1}{2}\epsilon)$ be as in Lemma 3.4, and write $\Delta = (x - \eta_1, x + \eta_2)$. Then we have

$$m(B \cap \Delta) = m(B \cap (x-\eta_1, x)) + m(B \cap (x, x+\eta_2)) < 2\delta(\eta_1 + \eta_2) = 2\delta|\Delta|\,.$$

Now, using efficient covers of $B \cap \Delta$ by intervals contained in Δ and applying Lemma 3.3, it follows that

$$m(f^i(B \cap \Delta)) < 2C\delta|f^i(\Delta)|\,,\quad i = 0, 1, \ldots, N\,,$$

where $C = \exp\{3\,\mathrm{Var}(Df)\}$. But since B is f-invariant, we have $f^i(B \cap \Delta) = B \cap f^i(\Delta)$ for each i, and since the intervals $f^i(\Delta)$ cover the circle, we deduce that

$$m(B) \leq \sum_{i=0}^{N} m(B \cap f^i(\Delta)) < 2C\delta \sum_{i=0}^{N} |f^i(\Delta)| < 6C\delta\,.$$

But δ is arbitrary, so $m(B) = 0$. This concludes the proof. □

3.4.2 Zero Lyapunov Exponents

When studying ergodic properties of a differentiable dynamical system, an important concept is that of *Lyapunov exponent* (sometimes also called *characteristic exponent*). Rather than defining this concept in broad generality, we focus our attention to one-dimensional maps. For a one-dimensional map f, the Lyapunov exponent at a point x is a number that essentially measures the exponential growth rate of the sequence $|Df^n(x)|$. More precisely, we have the following formal definition.

Definition 3.4 The Lyapunov exponent at x, denoted $\chi_f(x)$, is given by

$$\chi_f(x) = \lim_{n\to\infty} \frac{1}{n} \log|Df^n(x)|\,,$$

provided the limit exists. If the forward orbit of x hits a critical point of f, we set $\chi_f(x) = -\infty$.

When the Lyapunov number $\chi_f(x)$ is non-zero, this means that there is *asymptotic hyperbolicity* along the orbit of x: asymptotic *contraction* when $\chi_f(x)$ is *negative*, and asymptotic *expansion* when $\chi_f(x)$ is *positive*. Examples of such situations occur when f has an attracting or expanding periodic orbit, respectively (see Exercise 3.6).

It is perhaps intuitively obvious that, in the absence of periodic points, a circle diffeomorphism must have zero Lyapunov exponents everywhere, because there should be no asymptotic contraction or expansion. This is indeed the case, as the following theorem shows.

Theorem 3.11 *If f is an orientation-preserving C^1 circle diffeomorphism with irrational rotation number, then $\chi_f(x) = 0$ for all $x \in S^1$.*

Proof The function $\psi : S^1 \to \mathbb{R}$ defined by $\psi = \log Df$ is a continuous function and therefore, by the unique ergodicity of f, the sequence of continuous functions

$$\frac{1}{n}\sum_{j=0}^{n-1} \psi \circ f^j$$

converges uniformly to a constant, and this constant must be $\int_{S^1} \log Df \, d\mu$. By the chain rule, $\sum_{j=0}^{n-1} \psi \circ f^j = \log Df^n$ and, therefore, the sequence of continuous functions $\frac{1}{n}\log Df^n$ converges to the constant $\ell = \int_{S^1} \log Df \, d\mu$ *uniformly* in S^1. If $\ell > 0$, then there exists $n_0 \geq 1$ such that, for all $n \geq n_0$, we have $Df^n(x) > 1$ for all x. But this is impossible, since f^n is a diffeomorphism of S^1 onto itself. The same argument rules out $\ell < 0$. Therefore we must have $\ell = 0$ □

3.4.3 Further Ergodicity Results

In recent years, the study of circle maps from the measurable or ergodic viewpoint has been considerably expanded, even in the case of diffeomorphisms. At least two new objects have emerged from this study: *automorphic measures* and *invariant distributions*. We end this chapter with a brief discussion of both of them.

3.4.3.1 Automorphic Measures

As we saw in Sect. 3.4.1, every sufficiently smooth circle diffeomorphism $f : S^1 \to S^1$ without periodic points is ergodic with respect to Lebesgue measure. It turns out that there are plenty of other (non-invariant) Borel probability measures which are dynamically relevant and with respect to which f is also ergodic. An important class of such measures is the class of so-called *automorphic measures*.

Definition 3.5 Given a C^1 homeomorphism $f : S^1 \to S^1$ and a real number s, we say that a Borel probability measure ν on S^1 is an *automorphic measure of exponent s for f*—or an *s-automorphic measure for f*—if for every continuous function $\varphi : S^1 \to \mathbb{R}$ we have

$$\int_{S^1} \varphi\, d\nu = \int_{S^1} \varphi \circ f\, (Df)^s\, d\nu\, .$$

Alternatively,[3] a Borel probability measure $\nu \in \mathscr{P}(S^1)$ is s-automorphic for f iff its pullback $f^*\nu$ under f is equivalent to ν, and the Radon–Nikodým derivative of $f^*\nu$ with respect to ν is given by $(Df)^s$. It will certainly be clear to the reader that a 0-automorphic measure for f is simply an invariant (probability) measure for f—and therefore, when f has irrational rotation number, it coincides with the unique invariant probability measure for f.

In the context of circle diffeomorphisms, the concept of automorphic measure was introduced by Douady and Yoccoz (1999). However, this concept makes perfect sense for C^1 self-maps of smooth compact manifolds of any dimension, provided the one-dimensional derivative $Df(x)$ is replaced by the Jacobian of the map f at x, i.e., the absolute value of the determinant of the matrix $Df(x)$. For complex one-dimensional systems this is exactly the same as the notion of *conformal measure* introduced by Sullivan (1983)[4]

In their paper, Douady and Yoccoz (1999) proved the following result.

Theorem 3.12 (Douady–Yoccoz) *Let $f : S^1 \to S^1$ be a C^1 circle diffeomorphism with irrational rotation number. Then for every $s \in \mathbb{R}$ there exists an automorphic measure of exponent s for f, denoted μ_s. Such measure is unique and ergodic if f is $C^{1+\mathrm{BV}}$.*

[3] We denote by $\mathscr{P}(S^1)$ the space of Borel probability measures on the circle. Also, we say that two such measures are *equivalent* if they are mutually absolutely continuous.

[4] Which in turn is inspired by a similar notion introduced by Patterson (1976) and Sullivan (1979) himself in the context of Fuchsian and Kleinian groups.

In the exercises at the end of this chapter, the reader will be guided to a proof of this theorem.

Remark 3.3 What are automorphic measures good for? In their paper, Douady and Yoccoz were particularly concerned with building the tangent space to the space of C^2 circle diffeomorphisms with a given irrational rotation number at a given diffeomorphism f. They showed that this tangent space arises precisely as the *kernel* of the automorphic measure μ_{-1} for f (viewed, by Riesz duality, as a linear functional on the space $C^0(S^1)$). Another use of automorphic measures was made by de Melo and Pugh (1994) in their study of the so-called C^1 *Brunovsky hypothesis*.

3.4.3.2 Invariant Distributions

In this subsection, we assume that the reader is familiar with some basic facts about distributions.

In order to motivate the discussion, let us consider first the general case of a continuous self-map f of a compact Hausdorff space M. An invariant measure for f can be seen, via the Riesz representation theorem, as a continuous linear functional on $C^0(M)$—i.e., an element of the dual space $C^0(M)^*$—which is also invariant under the so-called Koopman operator $\varphi \mapsto \varphi \circ f$. If M happens to be a compact smooth manifold, it seems natural to consider also invariant linear functionals on the space $C^k(M)$ of C^k functions, for each $1 \leq k \leq \infty$. However, in order to define distributions of finite order below, we need an additional structure: we assume that M is endowed with a Riemannian metric.

Recall that the space $\mathscr{D}'(M)$ of (Schwartz) distributions on M is the dual of the Fréchet space $C^\infty(M)$. We consider in each space $C^k(M)$ with finite k the C^k norm $\|u\|_k = \max_{0\leq j\leq k} \|D^j u\|_{C^0}$ induced by the Riemannian metric on M.

Definition 3.6 (Finite-Order Distribution) We say $T \in \mathscr{D}'(M)$ has *order at most* $k \geq 0$ if there exists $C > 0$ such that $|\langle T, u\rangle| \leq C\|u\|_k$ for all $u \in C^\infty(M)$. The smallest such k is the *order* of T.

If $T \in \mathscr{D}'(M)$ has order at most k, then it extends uniquely to an element of $\mathscr{D}'_k(M) = C^k(M)^*$ (because $C^\infty(M)$ is dense in every $C^k(M)$). Every element of $\mathscr{D}'_k(M)$ arises in this way. Hence we can think that each $\mathscr{D}'_k(M)$ is (linearly) embedded in $\mathscr{D}'(M)$. By a slight abuse of language we think of such embeddings as inclusions, writing

$$\mathscr{D}'_0(M) \subset \mathscr{D}'_1(M) \subset \cdots \subset \mathscr{D}'_k(M) \subset \cdots \subset \mathscr{D}'(M)$$

It is a well-known fact that on a compact manifold every distribution has finite order. Hence we have $\mathscr{D}'(M) = \bigcup_{k\geq 0} \mathscr{D}'_k(M)$.

We are now ready to define the notion of invariant distribution. Let $f : M \to M$ be a C^r map, and let $0 \leq k \leq r$.

Definition 3.7 (Invariant Distribution) We say $T \in \mathscr{D}_k'(M)$ is *f-invariant* if $\langle T, u\rangle = \langle T, u \circ f\rangle$ for all $u \in C^k(M)$; i.e., if $u \circ f - u \in \ker T$ for all $u \in C^k(M)$.

For each $k \geq 0$, let $\mathscr{D}_k'(f) = \{T \in \mathscr{D}_k'(M) : T \text{ is } f\text{-invariant}\}$. Then we have, of course,

$$\mathscr{D}_0'(f) \subseteq \mathscr{D}_1'(f) \subseteq \cdots \subseteq \mathscr{D}_k'(f) \subseteq \cdots .$$

Also, let

$$B\left(f, C^k(M)\right) = \left\{\varphi \in C^k(M) : \exists u \in C^k(M) \text{ s.t. } u \circ f - u = \varphi\right\} .$$

This is a linear subspace of $C^k(M)$, and its elements are called C^k *coboundaries* for f. The Hahn–Banach separation theorem implies that

$$\mathrm{cl}_k\ B\left(f, C^k(M)\right) = \bigcap_{T \in \mathscr{D}_k'(f)} \ker T \tag{3.22}$$

where cl_k denotes closure in the C^k topology. This fact yields the following result.

Proposition 3.1 *Let $f : M \to M$ be a C^r endomorphism of a compact smooth Riemannian manifold M, where $0 \leq r \leq \infty$, and let μ be an f-invariant Borel probability measure on M. Let $0 \leq k \leq r$ be an integer. Then*

$$\mathscr{D}_k'(f) = \mathbb{R}\mu$$

if, and only if, the following holds. For any $\varphi \in C^k(M)$ with $\int_M \varphi\, d\mu = 0$, there is a sequence $\{\varphi_n = u_n \circ f - u_n\}_{n \geq 1} \subset B\left(f, C^k(M)\right)$ of C^k-coboundaries for f converging to φ in the C^k topology.

We should remark at this point that an equation of the form $u \circ f - u = \varphi$ in which φ is given and u is the unknown is called a *cohomological equation*. Equations of this type are rather ubiquitous in the study of dynamical systems. In the present book, cohomological equations appear in a natural way in the solution to the problem of linearization of circle diffeomorphisms. See the proof of Arnold's conjugacy theorem (Theorem 4.4) in Chap. 4, and also the proof of the Khanin–Teplinsky theorem (Theorem 4.11) presented in Appendix B.

Note that we could form, for each $0 \leq k < \infty$, the quotient of $C^k(M)$ by the closed subspace in (3.22). The resulting quotient (Banach) space is called the *first reduced cohomology group of f in C^k*, sometimes denoted by $\tilde{H}^1(f, C^k(M))$.[5] We

[5] In contradistinction to the 'ordinary' cohomology group $H^1(f, C^k(M)) = C^k(M)/B\left(f, C^k(M)\right)$. The latter is not as nice: it is usually non-Hausdorff, because $B\left(f, C^k(M)\right)$ is usually not closed in $C^k(M)$.

will have no use for such cohomology groups in the present book, but the reader should keep in mind that they are, in some sense, a measure of the obstruction to the solvability of cohomological equations in each $C^k(M)$.

The problem of finding and describing all invariant distributions in $\mathscr{D}'_k(f)$ for a given map f can be rather daunting. Indeed, even $\mathscr{D}'_0(f)$ can be a large space, because a given map f may have many distinct invariant measures (think of a map with infinitely many periodic orbits, for example). The smaller the dimension of $\mathscr{D}'_k(f)$ is, the more manageable the problem becomes. Thus, the best case scenario is when $\mathscr{D}'_k(f)$ is one-dimensional. This leads naturally to the following definition, which seems to have been first proposed by Katok (see for instance Katok and Robinson 2001).

Definition 3.8 Let $f : M \to M$ be a C^r endomorphism. We say that f is *distributionally uniquely ergodic* if for each $0 \leq k \leq r$ the linear space $\mathscr{D}'_k(f)$ is one-dimensional (hence spanned by the unique f-invariant probability measure).

For example, every rigid rotation of the circle with irrational rotation number is distributionally uniquely ergodic: this fact was already established in Exercise 1.16. Moreover, the property of distributional unique ergodicity is invariant under smooth conjugacies; see Exercise 3.18.

In general, distributional unique ergodicity is strictly stronger than unique ergodicity. For an example of a circle diffeomorphism which is uniquely ergodic but not distributionally uniquely ergodic, see Exercise 3.20. However, the two notions agree for minimal circle diffeomorphisms with very high smoothness, as shown by the following theorem due to Avila and Kocsard (2011).

Theorem 3.13 (Avila–Kocsard) *Every C^∞ diffeomorphism of the circle with irrational rotation number is distributionally uniquely ergodic.*

The analogous result for circle diffeomorphisms with low smoothness was proved shortly afterwards by Navas and Triestino (2013).

Theorem 3.14 (Navas–Triestino) *Every $C^{1+\mathrm{BV}}$ diffeomorphism of the circle with irrational rotation number is distributionally uniquely ergodic.*

The criterion given by Proposition 3.1 is used in the proofs of both these theorems.

Remark 3.4 We have introduced above two new objects: automorphic measures and invariant distributions. Is there a relationship between these objects, at least in the one-dimensional setting? The answer is *yes*: every 1-automorphic measure for a map f gives rise to an f-invariant distribution of order at most 1 (see Exercise 3.19).

Exercises

Exercise 3.1 Suppose $\varphi : \boldsymbol{S}^1 \to \mathbb{R}$ is a function that has modulus of continuity $w(t) = t|\log t|^\eta$ for some $\eta > 0$. Show that φ is δ-Hölder continuous for every $0 < \delta < 1$.

Exercise 3.2 Let $f : \boldsymbol{S}^1 \to \boldsymbol{S}^1$ be a homeomorphism with irrational rotation number α, and let $\varphi : \boldsymbol{S}^1 \to \mathbb{R}$ be a function of bounded variation. Suppose $\frac{p}{q}$ is a good rational approximation to α, that is, $\left|\alpha - \frac{p}{q}\right| < \frac{1}{q^2}$. As we saw in Sect. 3.2.2, the Denjoy–Koksma inequality states that, for every $x \in \boldsymbol{S}^1$, the Birkhoff sum $\sigma_q(x, f) = \sum_{j=0}^{q-1} \varphi \circ f^j(x)$ stays at a bounded distance away from $q \int_{\boldsymbol{S}^1} \varphi \, d\mu$, where μ is the unique invariant probability measure under f. What can be said about the Birkhoff sums $\sigma_n(x, f)$ for arbitrary values of n? Not much in general, but something *can* be said if we impose certain restrictions on the rotation number α. Let us assume that the partial quotients a_k of the continued fraction development of α satisfy the condition $a_k < k^{1+\delta}$ for some fixed $\delta > 0$, for every sufficiently large k. We note *en passant* that the set of all numbers α satisfying this condition has full Lebesgue measure in $[0, 1]$ (this follows easily from Lemma A.2 in Appendix A).

(i) Given $n \geq 1$, let k_n be the unique non-negative integer such that $q_{k_n} \leq n < q_{k_n+1}$ (where, as usual, $(q_n)_{n \geq 0}$ is the sequence of denominators of the best rational approximations to α). Show that we can write

$$n = \sum_{i=0}^{k_n} b_i q_i \ ,$$

where $0 \leq b_i \leq a_{i+1}$ for all $0 \leq i \leq k_n$, and $b_{k_n} \geq 1$.

(ii) Using (i) and the Denjoy–Koksma inequality, show that for all $x \in \boldsymbol{S}^1$ we have

$$\left| \sum_{j=0}^{n-1} \varphi \circ f^j(x) - n \int_{\boldsymbol{S}^1} \varphi \, d\mu \right| < \mathrm{Var}(\varphi) \sum_{i=0}^{k_n} a_{i+1} \ .$$

(iii) Deduce from (ii) that there exists a constant $C > 0$ such that

$$\sup_{x \boldsymbol{S}^1} \left| \sum_{j=0}^{n-1} \varphi \circ f^j(x) - n \int_{\boldsymbol{S}^1} \varphi \, d\mu \right| \leq C \, \mathrm{Var}(\varphi) \, (\log n)^{1+\delta} \ .$$

[*Reference*: Guillotin-Plantard and Schott 2006, pp. 236–237]

Exercise 3.3 Prove that if $f : \boldsymbol{S}^1 \to \boldsymbol{S}^1$ is an orientation preserving C^1 diffeomorphism with irrational rotation number and μ denotes its unique invariant Borel probability measure, then

$$\lim_{n\to\infty}\left(\int_{S^1} Df^n\,d\mu\right)^{1/n} = 1\,.$$

Exercise 3.4 Let $f : S^1 \to S^1$ be an orientation preserving C^1 diffeomorphism. Show that there exists a point $x_0 \in S^1$ such that the bi-infinite sequence $(Df^n(x_0))_{n\in\mathbb{Z}}$ remains bounded.

Exercise 3.5 Show that Lyapunov exponents are C^1 conjugacy invariants. That is, let $f, g : S^1 \to S^1$ be C^1 maps, and suppose there exists a C^1 diffeomorphism h such that $h \circ f = g \circ h$. Prove that if $x \in S^1$ is such that $\chi_f(x)$ exists, then so does $\chi_g(h(x))$, and these numbers are equal, i.e., $\chi_f(x) = \chi_g(h(x))$.

Exercise 3.6 Let $\ell \in \mathbb{R}$ be an arbitrary number. For each rational number $r \in (0, 1)$, find a smooth diffeomorphism $f : S^1 \to S^1$ whose rotation number is equal to r and a point $x_0 \in S^1$ such that $\chi_f(x_0) = \ell$.

Exercise 3.7 Let $\varphi \in \mathrm{BV}(S^1)$ be a function of bounded variation. Given an interval $\Delta \subset S^1$ with endpoints a and b, write $v_\varphi(\Delta) = |\varphi(a) - \varphi(b)|$.

(i) If $\Delta, \Delta_1, \ldots, \Delta_m \subset S^1$ are intervals with $\Delta = \bigcup_{i=1}^m \Delta_i$ and the Δ_i's have pairwise disjoint interiors, show that

$$v_\varphi(\Delta) \le \sum_{i=1}^m v_\varphi(\Delta_i)\,.$$

(ii) Deduce from (i) that, if $I_1, I_2, \ldots, I_N \subset S^1$ is a collection of k-quasidisjoint intervals (for some $k \ge 1$), then

$$\sum_{j=1}^N v_\varphi(I_j) \le k\,\mathrm{Var}(\varphi)\,.$$

This fact was implicitly used in the proof of Lemma 3.3.

Exercise 3.8 Let $f : S^1 \to S^1$ be a C^1 homeomorphism, let $s \in \mathbb{R}$, and let $\nu \in \mathscr{P}(S^1)$ be an s-automorphic measure for f. Prove that for all $\varphi \in L^1(\nu)$ and $n \ge 1$, we have $(\varphi \circ f^n)(Df^n)^s \in L^1(\nu)$ and

$$\int_{S^1} \varphi\,d\nu = \int_{S^1} (\varphi \circ f^n)(Df^n)^s\,d\nu\,.$$

[*Hint:* First prove the result for continuous functions using induction and the chain rule. Then use approximation by continuous functions and Lebesgue's dominated convergence theorem.]

Exercise 3.9 Let $f : \mathbf{S}^1 \to \mathbf{S}^1$ be a C^1 homeomorphism, let $s \in \mathbb{R}$, and let $\mathscr{A}_s(f) \subset \mathscr{P}(\mathbf{S}^1)$ be the set of all s-automorphic measures for f. Show that $\mathscr{A}_s(f)$ is convex.

Exercise 3.10 *Existence of automorphic measures.* Let $f : \mathbf{S}^1 \to \mathbf{S}^1$ be a C^1 diffeomorphism with irrational rotation number, and let $s \in \mathbb{R}$. Let $V_s : \mathscr{P}(\mathbf{S}^1) \to \mathscr{P}(\mathbf{S}^1)$ be the operator implicitly defined—via the Riesz representation theorem—by the formula

$$\int_{\mathbf{S}^1} \varphi \, d(V_s \nu) = \frac{1}{\int_{\mathbf{S}^1} (Df)^s \, d\nu} \int_{\mathbf{S}^1} (\varphi \circ f)(Df)^s \, d\nu \ , \forall \varphi \in C^0(\mathbf{S}^1) \, .$$

(i) Show that V_s is well-defined, and that it is continuous if we endow $\mathscr{P}(\mathbf{S}^1)$ with the weak* topology.
(ii) Deduce from (i), as well as the convexity and compactness of $\mathscr{P}(\mathbf{S}^1)$, that V_s has a fixed point. [*Hint:* Use the Schauder–Tychonoff fixed point theorem.]
(iii) Let $\mu \in \mathscr{P}(\mathbf{S}^1)$ be a fixed point of V_s. Show that for each $n \geq 1$ we have

$$\int_{\mathbf{S}^1} (Df)^s \, d\mu = \left(\int_{\mathbf{S}^1} (Df^n)^s \, d\mu \right)^{1/n} .$$

(iv) Combine (iii) with Theorem 3.11 to show that, in fact, $\int_{\mathbf{S}^1} (Df)^s \, d\mu = 1$.
(v) Deduce from (iv) that μ is s-automorphic for f.

This establishes the *existence* part of Theorem 3.12.

Exercise 3.11 Show that Lebesgue (Haar) measure on $\mathbf{S}^1$ is automorphic of exponent $s = 1$ for *every* circle diffeomorphism.

Exercise 3.12 Let $f : \mathbf{S}^1 \to \mathbf{S}^1$ be a C^1 Denjoy example, and let I be one of its gaps (i.e., a connected component of $\mathbf{S}^1 \setminus \Omega(f)$). Let $S : \mathbf{S}^1 \to \mathbb{R}^+ \cup \{\infty\}$ be given by

$$S(x) = \sum_{n \in \mathbb{Z}} Df^n(x) \, ,$$

and let $E = \{x \in I : S(x) < \infty\}$.

(i) Show that E is a (measurable) set of full Lebesgue measure in I.
(ii) Fix a point $x_0 \in E$ and define $\nu \in \mathscr{P}(\mathbf{S}^1)$ by

$$\nu = \frac{1}{S(x_0)} \sum_{n \in \mathbb{Z}} Df^n(x_0) \delta_{f^n(x_0)} \, ,$$

where, as usual, δ_x denotes the Dirac measure concentrated at x. Prove that ν is 1-automorphic for f.

(iii) Deduce from (ii) and Exercise 3.11 that the uniqueness part of Theorem 3.12 breaks down if we do not assume that $\log Df \in \mathrm{BV}(S^1)$.

Exercise 3.13 *Automorphic measures are ergodic.* Let $f : S^1 \to S^1$ be a $C^{1+\mathrm{BV}}$ diffeomorphism with irrational rotation number α, and let $\mu \in \mathscr{P}(S^1)$ be an s-automorphic measure for f. The purpose of this exercise is to show that μ is ergodic for f. Given $x \in S^1$, for each $n \geq 1$ let $J_n(x) \subset S^1$ be the interval with endpoints $f^{q_n}(x)$ and $f^{-q_n}(x)$ that contains x, where q_n is the denominator of the n-th convergent to α.

(i) Show that for each $y \in J_n(x)$ the interval $\Delta_{x,y} \subset J_n(x)$ with endpoints x and y is such that its images up to time $q_{n+1} - 1$, namely

$$\Delta_{x,y}\,,\; f(\Delta_{x,y})\,,\; f^2(\Delta_{x,y})\,,\; \ldots\,,\; f^{q_{n+1}-1}(\Delta_{x,y})\,,$$

have pairwise disjoint interiors.

(ii) Deduce from (i) that for each $k = 0, 1, \ldots, q_{n+1} - 1$ we have

$$\frac{1}{K} \leq \frac{Df^k(x)}{Df^k(y)} \leq K\,,$$

where $K = \exp \mathrm{Var}(\log Df) < \infty$.

(iii) Now let $B \subset S^1$ be a measurable (Borel) set. Using (ii), show that for each $k = 0, 1, \ldots, q_{n+1} - 1$ we have

$$\mu\left(B \cap f^k(J_n(x))\right) \leq K^s (Df^k(x))^s \mu\left(B \cap J_n(x)\right)\,,$$

as well as

$$\mu\left(f^k(J_n(x))\right) \geq K^{-s} (Df^k(x))^s \mu\left(J_n(x)\right)\,.$$

(iv) Deduce from (iii) that

$$\mu(B) \leq \sum_{k=0}^{q_{n+1}-1} \mu\left(B \cap f^k(J_n(x))\right) \leq 2K^{2s} \frac{\mu\left(B \cap J_n(x)\right)}{\mu\left(J_n(x)\right)}\,.$$

(v) Now suppose $\mu(B) < 1$. Show that there exists a point $x \in S^1 \setminus B$ such that

$$\lim_{n\to\infty} \frac{\mu\left(B \cap J_n(x)\right)}{\mu\left(J_n(x)\right)} = 0\,.$$

(vi) Combining (iv) and (v), conclude that μ is indeed ergodic for f.

Exercise 3.14 *Uniqueness of automorphic measures.* Once again, let $f : S^1 \to S^1$ be a $C^{1+\mathrm{BV}}$ diffeomorphism with irrational rotation number, and let $s \in \mathbb{R}$ be given.

Suppose $\mu, \nu \in \mathscr{P}(\mathbf{S}^1)$ are both s-automorphic measures for f. Show that $\mu = \nu$ by arguing as follows.

(i) In the special case when μ is absolutely continuous with respect to ν and $\psi = d\mu/d\nu$ is the corresponding Radon–Nikodým derivative, show that ψ is invariant in the sense that $\psi \circ f = \psi$ at ν-a.e. point.
(ii) Deduce from (i) and Exercise 3.13 that ψ must be constant ν-almost everywhere. The constant must be equal to 1, because both μ and ν are probability measures. Hence $\mu = \nu$ in the special case.
(iii) If neither of the two measures is absolutely continuous with respect to the other, let $\sigma = \frac{1}{2}(\mu+\nu)$. This measure is s-automorphic for f, by Exercise 3.9. Check that $\mu \ll \sigma$ and $\nu \ll \sigma$ both hold, and conclude using (ii) that $\mu = \sigma = \nu$.

This establishes the *uniqueness* part of Theorem 3.12.

Exercise 3.15 Given a $C^{1+\mathrm{BV}}$ diffeomorphism $f : \mathbf{S}^1 \to \mathbf{S}^1$ with irrational rotation number, denote by $\mu_{s,f}$ the unique s-automorphic measure for f, for each $s \in \mathbb{R}$. Prove the following continuity statement: If $s_n \to s$ and $f_n \to f$ in the C^1 topology, then $\mu_{s_n,f_n} \to \mu_{s,f}$ in the weak* topology.

Exercise 3.16 Let $f, g : \mathbf{S}^1 \to \mathbf{S}^1$ be $C^{1+\mathrm{BV}}$ diffeomorphisms. Show that if they commute, i.e., if $f \circ g = g \circ f$, then they share the same automorphic measures. In other words, in the notation of the previous exercise, we have $\mu_{s,f} = \mu_{s,g}$ for each $s \in \mathbb{R}$.

Exercise 3.17 Using the Hahn–Banach theorem, give a detailed proof of (3.22), and explain how that equality implies Proposition 3.1.

Exercise 3.18 Let $f, g : M \to M$ be C^∞ self-maps of a compact smooth Riemannian manifold M, and suppose there exists a C^∞ diffeomorphism $h : M \to M$ such that $h \circ f = g \circ g$. Show that f is distributionally uniquely ergodic if and only if the same happens to g.

Exercise 3.19 Let $f : \mathbf{S}^1 \to \mathbf{S}^1$ be a C^1 map, and suppose that ν is an automorphic measure of exponent 1 for f. Show that $T_\nu : C^1(\mathbf{S}^1) \to \mathbb{R}$ given by

$$\langle T_\nu, u \rangle = \int_{\mathbf{S}^1} u' \, d\nu$$

is an f-invariant distribution of order at most 1. What happens if ν is Lebesgue measure?

Exercise 3.20 Let $f : \mathbf{S}^1 \to \mathbf{S}^1$ be a C^1 Denjoy example.

(i) Combining Exercise 3.12 with Exercise 3.19, construct an f-invariant distribution of order 1.
(ii) Go a bit further than (i) and show that, for each $k \geq 1$, the space $\mathscr{D}'_k(f)$ is infinite-dimensional.

Chapter 4
Smooth Conjugacies to Rotations

We have seen in Chap. 3 that every sufficiently smooth diffeomorphism of the circle without periodic points is topologically conjugate to a rigid rotation. In other words, the *topological* orbit structure of such a diffeomorphism is indistinguishable from that of a rigid rotation. The relative order of points of a given orbit on the circle is the same no matter which orbit we take; everything is determined by a single invariant, the rotation number.

What can be said about the *geometric* orbit structure of such a diffeomorphism? Is it the same, asymptotically at least, as that of the corresponding rotation? As we shall see in this chapter, this is a subtle question, one whose answer depends on the arithmetic nature of the rotation number.

We will not attempt at a formal definition of geometric orbit structure. Intuitively, the geometric structure of an orbit of a circle map can be defined as the set of ratios of distances between the various points of that orbit. When we only care about ratios of distances between points that are close to each other, at smaller and smaller scales, we speak of the orbit's *asymptotic* geometric structure.

When a C^1 diffeomorphism of the circle f is conjugate to a rotation, and the conjugacy h is a C^1 diffeomorphism, then, because h is essentially affine at small scales, the geometric structure of the orbits of f is asymptotically the same as the geometric structure of the orbits of the rotation. Thus, we can rephrase the question posed above as follows: If a C^1 circle diffeomorphism is topologically conjugate to an irrational rotation, when is the conjugacy a C^1 diffeomorphism?

More generally, when f is a C^r diffeomorphism one may consider its *C^r-smooth structure* at small scales. Here, we can have r finite greater than or equal to 1, $r = \infty$, or even $r = \omega$ (i.e., f can be a real-analytic diffeomorphism). Again, we refrain from giving a formal definition of smooth structure, but instead formulate the general problem as follows: Find necessary and sufficient conditions for a C^r circle diffeomorphism f which is topologically conjugate to an irrational rotation to be C^s-conjugate to that rotation, where $s \leq r$ is as large as possible. This problem

E. de Faria, P. Guarino, *Dynamics of Circle Mappings*, IMPA Monographs,
https://doi.org/10.1007/978-3-031-67495-2_4

has been thoroughly investigated by Arnold (in the analytic case), Herman, Yoccoz, among others, and our aim in this chapter is to describe some of their results.

4.1 Herman's Invariants

In this section we will present a fundamental criterion for smoothness of conjugacies that was introduced by Herman (1979, Ch. IV) in his *thèse d'État*. It is very simply stated in terms of what we now call *Herman's conjugacy invariants*.

Definition 4.1 If $f \in \mathrm{Diff}^r(S^1)$, where r is a positive integer, set $\mathcal{H}_r(f) = \sup_{n\in\mathbb{Z}} \|Df^n\|_{C^{r-1}}$.

Here, given a C^k function $\varphi : S^1 \to \mathbb{R}$, where $k \geq 1$, we write $\|\varphi\|_{C^k} = \sum_{j=0}^{k} \|D^j\varphi\|$, where $\|\cdot\|$ denotes the usual sup-norm.

In this chapter, we will only make explicit use of Herman's first invariant $\mathcal{H}_1(f)$. We leave it as an exercise to the reader to prove that $\mathcal{H}_1(f)$ is indeed a C^1 conjugacy invariant, in the sense that $\mathcal{H}_1(h \circ f \circ h^{-1}) < \infty$ if and only if $\mathcal{H}_1(f) < \infty$, whenever $f, h \in \mathrm{Diff}^1(S^1)$ (Exercise 4.1). The proof that $\mathcal{H}_r(f)$ is a C^r conjugacy invariant when $r > 1$ is also not difficult, but depends on the so-called *Faa-di Bruno formula* for the higher derivatives of a composition of C^r maps. We once again refer the reader to Herman (1979, Ch. IV) for details.

With such an invariant at hand, Herman's criterion reads as follows.

Theorem 4.1 (Herman's Criterion) *If $f : S^1 \to S^1$ is a C^r diffeomorphism and $\mathcal{H}_r(f)$ is* finite, *then f is C^r conjugate to a rigid rotation, and conversely.*

We will prove Herman's criterion only for $r = 1$, deriving it as a consequence of the following general result in topological dynamics due to Gottschalk and Hedlund (1955).

Theorem 4.2 (Gottschalk–Hedlund) *Let X be a compact metric space, $f : X \to X$ be a homeomorphism all of whose orbits are dense, and $\varphi : X \to \mathbb{R}$ be a continuous function. Then the following assertions are equivalent.*

(a) There exists $\psi : X \to \mathbb{R}$ continuous such that $\varphi = \psi \circ f - \psi$;
(b) There exists $x_0 \in X$ such that $\sup_{n\geq 1} \left|\sum_{j=0}^{n-1} \varphi \circ f^j(x_0)\right|$ is finite.

Proof That (a) implies (b) is clear, because if $\varphi = \psi \circ f - \psi$ then

$$\sum_{j=0}^{n-1} \varphi \circ f^j(x) = \psi \circ f^n(x) - \psi(x) ,$$

and so taking any $x \in X$ as x_0 will do.

To prove that (b) implies (a), consider the map $H : X \times \mathbb{R} \to X \times \mathbb{R}$ given by $H(x,t) = (f(x), t + \varphi(x))$, an example of what is usually referred to as a skew product. Then H is continuous and invertible; in fact,

$$H^{-1}(y,s) = \left(f^{-1}(y)\,,\, s - \varphi(f^{-1}(y))\right) ,$$

so the inverse is also continuous. Hence H is a homeomorphism. Note that for all $n \geq 0$ we have

$$H^n(x,t) = \left(f^n(x), t + \sum_{j=0}^{n-1} \varphi \circ f^j(x)\right) . \tag{4.1}$$

Now (b) implies that the positive orbit $\mathcal{O}_H^+(x_0, 0) = \{H^n(x_0, 0) : n \geq 0\}$ is *bounded*. Therefore the ω-limit set Ω of $\mathcal{O}_H^+(x_0, 0)$ is a compact subset of $X \times \mathbb{R}$, and obviously H-invariant.

Claim The set Ω is the graph of a continuous function $\psi : X \to \mathbb{R}$.

To prove this claim, we must show that each vertical line $\{x\} \times \mathbb{R}$ cuts Ω at exactly one point. First note that this happens for the vertical line $\{x_0\} \times \mathbb{R}$, its intersection with Ω being the point $(x_0, 0)$. Indeed, if $(x_0, t) \in \Omega$ then there is a sequence $n_i \to \infty$ such that $H^{n_i}(x_0, 0) \to (x_0, t)$, and using (4.1) we see that this implies that $H^{n_i}(x_0, t) \to (x_0, 2t)$. By induction we deduce that $(x_0, nt) \in \Omega$ for all n, but since Ω is bounded this can only happen if $t = 0$. Now, if some vertical line $\{x\} \times \mathbb{R}$ cuts Ω at two points, say (x, t_1) and (x, t_2), then every other vertical line must do so as well: for any $y \in X$, since the orbit of x under f is dense in X, we find a sequence $m_i \to \infty$ such that $H^{m_i}(x, t_1) \to (y, t)$ for some $t \in \mathbb{R}$, and therefore $H^{m_i}(x, t_2) \to (y, t + (t_2 - t_1))$. But this contradicts the fact that $\{x_0\} \times \mathbb{R}$ intersects Ω at $(x_0, 0)$ only. This proves that Ω is the graph of a function $\psi : X \to \mathbb{R}$, necessarily continuous because Ω is closed in $X \times \mathbb{R}$.

Finally, since now we know that every point in Ω is of the form $(x, \psi(x))$ for some $x \in X$, we see that

$$H(x, \psi(x)) = (f(x)\,,\, \psi(x) + \varphi(x)) = (f(x)\,,\, \psi \circ f(x)) ,$$

by the H-invariance of Ω, and therefore $\psi(x) + \varphi(x) = \psi \circ f(x)$, thereby establishing the desired cocycle identity. □

We are ready for the promised special case of Theorem 4.1.

Theorem 4.3 (Herman's Criterion) *A C^1 diffeomorphism $f : S^1 \to S^1$ is C^1-conjugate to a rotation if and only if $\mathcal{H}_1(f) < \infty$.*

Proof First suppose that $\mathcal{H}_1(f) < \infty$, i.e. $\sup_n \|\log Df^n\| < \infty$. Since by the chain rule,

$$\log Df^n = \sum_{j=0}^{n-1} \log Df \circ f^j ,$$

condition (b) of the Gottschalk–Hedlund theorem holds for $\varphi = \log Df$ and x_0 any point in $X = S^1$. We deduce from that theorem that $\log Df = \psi - \psi \circ f$ for some continuous function $\psi : S^1 \to \mathbb{R}$. Adding a suitable constant to ψ if necessary, we may assume that

$$\int_{S^1} \exp\{\psi(t)\}\, dt = 1 . \tag{4.2}$$

Now we define $h : S^1 \to S^1$ by

$$h(x) = \int_{x_0}^{x} \exp\{\psi(t)\}\, dt .$$

Here and throughout, it is implicit that all calculations are performed modulo 1. The normalization (4.2) makes h a well-defined, degree one map. It is a C^1 diffeomorphism because ψ is continuous. Moreover, writing

$$\alpha = \int_{x_0}^{f(x_0)} \exp\{\psi(t)\}\, dt ,$$

we see that

$$h \circ f(x) = \alpha + \int_{f(x_0)}^{f(x)} \exp\{\psi(t)\}\, dt .$$

Applying the change of variables $t = f(s)$, this becomes

$$h \circ f(x) = \alpha + \int_{x_0}^{x} \exp\{\psi \circ f(s)\} Df(s)\, ds . \tag{4.3}$$

Using the cocycle relation $\psi = \psi \circ f + \log Df$ in (4.3), we get

$$\begin{aligned} h \circ f(x) &= \alpha + \int_{x_0}^{x} \exp\{\psi \circ f(s) + \log Df(s)\}\, ds \\ &= \alpha + \int_{x_0}^{x} \exp\{\psi(s)\}\, ds \\ &= \alpha + h(x) \end{aligned}$$

Therefore $h \circ f(x) = R_\alpha \circ h(x)$. The converse is left as an easy exercise to the reader. □

For an interesting use of Herman's criterion in the context of one-parameter families of circle diffeomorphisms, see Sect. 4.3.2.

4.2 Small Denominators: Arnold's Theorem

In this section we present a fundamental theorem due to Arnold (1961) stating that every analytic circle diffeomorphism with "good" rotation number α and which is sufficiently close to the rotation R_α is *analytically conjugate* to R_α.

Arnold's analytic conjugacy theorem can be regarded as a toy model for what is known as *KAM theory*.[1] This theory was developed as an attempt (largely successful) at making rigorous certain perturbation arguments used by physicists in their studies of nearly integrable Hamiltonian systems arising in Celestial Mechanics. The major difficulty in dealing with the perturbative series expansions of the solutions of the differential equations coming from these problems is that the coefficients of these series often involve rational expressions with *small denominators*, rendering the task of proving convergence extremely difficult.

It is fair to say that the taming of small denominators started with Siegel (1942). His paper deals with the problem of linearization of analytic functions near an irrationally indifferent fixed (or periodic) point, say with multiplier $\lambda = e^{2\pi i\alpha}$ for some irrational α. Siegel wrote down the conjugacy equation (in which the unknown is an analytic change of coordinates transforming the given map into the linear map $z \mapsto \lambda z$), expanded everything in power series, and compared coefficients. This resulted in complicated recursive relations for the coefficients of the desired conjugacy; in these relations, factors of the form $\lambda^n - 1$, $n \neq 0$, appeared in the denominators. In order to control such factors (so as to prove convergence) Siegel had to assume that α is a *Diophantine number*. The required estimates are quite difficult to carry out, and Siegel's paper, despite being short, is a real *tour-de-force*. But in some sense it also shows that the method of direct comparison of coefficients (followed by brute force estimates) for perturbative series is not viable in the general KAM setting.

A different approach was proposed by Kolmogorov (1954) in his ICM address. He laid down a strategy to deal with such problems that, roughly speaking, consists of two steps:

(1) linearize the equations of motion and solve the linear problem, obtaining an approximate solution to the original non-linear problem.
(2) Improve the approximate solution obtained in (1) by an iterative procedure akin to Newton's method.

It is in the first step that the small denominators mark their presence. The second step is usually the more difficult one; here the hope is that the successive approximate

[1] The acronym stands for Kolmogorov, Arnold and Moser.

solutions are such that the distance to the exact solution at the $(n+1)$-st step is of the order of the *square* of the corresponding distance at the n-th step. It is this quadratic decay that is meant by the expression “akin to Newton’ method”.

This strategy was first carried out by Arnold (1961) for analytic systems, and later by Moser (1966) for C^k-smooth systems. The first case analysed by Arnold was the one we mentioned in the beginning of this section, namely the problem of analytically conjugating an analytic circle diffeomorphism sufficiently close to a “good" rotation to the rotation itself. Such diffeomorphisms arise as (global) cross-sections for flows on the two-dimensional torus. This problem is the exact analogue for maps of the circle of the linearization problem for local analytic diffeomorphisms studied by Siegel.

Before giving a precise statement of Arnold’s theorem, let us introduce some notation and formulate a definition. For each $r > 1$, let A_r denote the annular region $A_r = \{z \in \mathbb{C} : r^{-1} < |z| < r\}$ in the complex plane. Let us also consider the horizontal strips $S_\sigma = \{z \in \mathbb{C} : |\mathrm{Im}\, z| < \sigma\}$, for each given $\sigma > 0$. We denote by $\exp : \mathbb{C} \to \mathbb{C}/\mathbb{Z} \equiv \mathbb{C}^*$ the exponential covering map $\exp(z) = e^{2\pi i z}$. Note that, in the notation just introduced, $\exp(S_\sigma) = A_{e^{2\pi\sigma}}$.

Definition 4.2 We say that an irrational number α is *Diophantine of type* (K, ν), where $K > 0$ and $\nu > 2$ are given constants, if

$$\left|\alpha - \frac{p}{q}\right| \geq \frac{K}{q^\nu}, \quad \text{for all } \frac{p}{q} \in \mathbb{Q}.$$

Remark 4.1 We note the obvious but useful fact that, if α is Diophantine of type (K, ν) and $K' < K$, then α is Diophantine of type (K', ν). Thus, we can always assume that K is as small as necessary.

We are ready for the statement of Arnold’s theorem.

Theorem 4.4 (Arnold) *Let $r > 1$, $K > 0$ and $\nu > 2$ be given, and let $\alpha \in (0, 1)$ be a Diophantine number of type (K, ν). There exists $\varepsilon = \varepsilon(r, \nu, K) > 0$ with the following property. Suppose $f : \mathbf{S}^1 \to \mathbf{S}^1$ is a diffeomorphism with $\rho(f) = \alpha$ such that*

(i) f has a univalent extension to the annulus A_r (which we still denote by f);
(ii) $\sup_{z\in A_r} |f(z) - R_\alpha(z)| < \varepsilon$.

Then there exists a univalent map $h : A_{\sqrt{r}} \to \mathbb{C}$ with $h(\mathbf{S}^1) = \mathbf{S}^1$ such that h conjugates f to the rotation R_α, i.e., satisfies the conjugacy equation $f \circ h = h \circ R_\alpha$, in an annular region around $\mathbf{S}^1$.

In light of Kolmogorov’s strategy, the proof goes as follows.

(1) First linearize the conjugacy equation $f \circ h = h \circ R_\alpha$ and, expanding everything in Laurent series, get an approximate solution $h_1 : A_{r_1} \to \mathbb{C}$ which is holomorphic univalent in a smaller annulus $(1 < r_1 < r)$ and preserves the unit circle. To prove convergence of the series for h_1 in the smaller annulus, it

is necessary to use that α is Diophantine. Then define $f_1 = h_1^{-1} \circ f \circ h_1$. This new map is holomorphic univalent in a smaller domain than the original f; we refer to this as a *loss of analyticity*. Note that $\rho(f_1) = \rho(f)$.

(2) Repeat step (1) with f_1 replacing f, getting $h_2 : A_{r_2} \to \mathbb{C}$ holomorphic univalent in a yet smaller annulus ($1 < r_2 < r_1$) and preserving the unit circle. Then define $f_2 = h_2^{-1} \circ f_1 \circ h_2$, and so on, inductively. As a result, we obtain two sequences of univalent maps, namely $h_n : A_{r_n} \to \mathbb{C}$ and $f_n : A_{r_n} \to \mathbb{C}$ (where $1 < r_{n+1} < r_n < r$ for all n) such that $f_{n+1} = h_n^{-1} \circ f_n \circ h_n$ for all n (all maps preserving the unit circle). Again, note that $\rho(f_n) = \rho(f)$, and in fact

$$f_n = (h_1 \circ h_2 \circ \cdots \circ h_n)^{-1} \circ f \circ (h_1 \circ h_2 \circ \cdots \circ h_n) \ , \quad \text{for all } n \ .$$

Denoting by $\Delta_n = \|f_n - R_\alpha\|_{C^0(A_{r_n})}$ the C^0-distance between f_n and R_α in the appropriate annular domain, the estimates will show that $\Delta_{n+1} = O(\Delta_n^{1+\epsilon})$ for all $n \geq 1$ and some $\epsilon > 0$. They will also show that $r_n > \sqrt{r} > 1$, and from this it will follow that $h = \lim_{n\to\infty} h_1 \circ h_2 \circ \cdots \circ h_n$ exists as a holomorphic univalent map with domain $A_{\sqrt{r}}$ and is the desired analytic conjugacy.

What will make this inductive procedure work is that the faster-than-linear decay in step (2) beats the loss of analyticity in step (1) at each stage.

Having presented the general idea, we now move to the rather painful details.

4.2.1 The Linearized Equation

It will be much more convenient to deal with the *lifts* of f, R_α through the exponential covering map. The lift of R_α is, of course, the translation $T_\alpha : z \mapsto z+\alpha$. The lift of f is a holomorphic univalent map $F : S_\sigma \to \mathbb{C}$ defined on the strip $S_\sigma = \{z : |\text{Im}(z)| < \sigma\}$ with $e^{2\pi\sigma} = r$, satisfying $F(z+1) = F(z) + 1$ for all $z \in S_\sigma$, and such that the diagram

$$\begin{array}{ccc} S_\sigma & \xrightarrow{\ F\ } & F(S_\sigma) \subset \mathbb{C} \\ {\scriptstyle \exp}\downarrow & & \downarrow{\scriptstyle \exp} \\ A_r & \xrightarrow[\ f\ ]{} & f(A_r) \subset \mathbb{C}^* \end{array}$$

commutes; the restriction $F|_\mathbb{R}$ is the lift of our circle map $f|_{S^1}$. Of course, F is determined only up to addition by an integer, but we choose it so that $0 < F(0) < 1$: this ensures that $\|F - T_\alpha\|_{C^0(S_\sigma)}$ is of the same size as $\|f - R_\alpha\|_{C^0(A_r)}$. Thus, if f is a small perturbation of the rotation R_α, then F is a small perturbation of the translation T_α.

Our ultimate goal is to find a holomorphic univalent map $H : S_{\sigma/2} \to \mathbb{C}$, with $H(z+1) = H(z) + 1$ for all $z \in S_{\sigma/2}$ and $H(\mathbb{R}) = \mathbb{R}$, satisfying the conjugacy equation

$$F \circ H(z) = H \circ T_\alpha \,, \quad \text{for all } z \in S_{\sigma/2} \,. \tag{4.4}$$

In particular, it should be the case that $H(S_{\sigma/2}) \subseteq S_\sigma$.

Let us write $F(z) = z + \alpha + \varphi(z)$ and $H(z) = z + \psi(z)$, where φ, ψ are holomorphic and periodic of period one. Here φ is given, and ψ is the unknown. If a solution to (4.4) exists, then we must have

$$\psi(z+\alpha) - \psi(z) = \varphi(z + \psi(z)) \,.$$

This rather non-linear equation in the unknown ψ is, not surprisingly, too difficult to be solved directly. We try to do the next best thing, which is to find an approximate solution by considering the linearized equation

$$\psi(z+\alpha) - \psi(z) = \varphi(z) \,. \tag{4.5}$$

However, a necessary condition for (4.5) to be solvable is that $\int_0^1 \varphi(x)\, dx = 0$, which is not reasonable to expect. Hence we replace (4.5) by

$$\psi(z+\alpha) - \psi(z) = \varphi(z) - \widehat{\varphi}(0) \,, \tag{4.6}$$

where $\widehat{\varphi}(0) = \int_0^1 \varphi(x)\, dx$. If we solve (4.6), then $H = \mathrm{Id} + \psi$ will not be an exact solution to (4.4), but rather an approximate solution (we will deal with the problem of determining the correct domain strip on which H (or ψ) is defined in due time).

Since we are dealing with periodic functions, it is natural to use Fourier series. Let us write

$$\varphi(z) = \sum_{n\in\mathbb{Z}} \widehat{\varphi}(n) e^{2\pi i n z} \,, \tag{4.7}$$

where the Fourier coefficients $\widehat{\varphi}(n)$ are given by

$$\widehat{\varphi}(n) = \int_0^1 \varphi(x) e^{-2\pi i n x}\, dx \,. \tag{4.8}$$

Note that, since $\varphi(x)$ is real when x is real, we have $\widehat{\varphi}(-n) = \overline{\widehat{\varphi}(n)}$ for all $n \in \mathbb{Z}$. The series in (4.7) is absolutely convergent in the strip S_σ: see Exercise 4.2. Let us also consider the formal expansion of the unknown ψ in Fourier series, namely

$$\psi(z) = \sum_{n\in\mathbb{Z}} \widehat{\psi}(n) e^{2\pi i n z} \,, \tag{4.9}$$

Putting (4.7) and (4.9) back in (4.6) and solving for the coefficients of ψ, we get $\widehat{\psi}(0) = 0$ and

$$\widehat{\psi}(n) = \frac{\widehat{\varphi}(n)}{e^{2\pi i n\alpha} - 1}\,, \quad \text{for all } n \in \mathbb{Z}^*\,. \tag{4.10}$$

Here we see the small denominators making their presence felt. In order to estimate the coefficients of ψ, we need the following two lemmas.

Lemma 4.1 *If α is Diophantine of type (K, ν) then*

$$\left|e^{2\pi i n\alpha} - 1\right| \geq \frac{4K}{|n|^{\nu-1}}\,,$$

for all $n \in \mathbb{Z}^*$.

Proof See Exercise 4.3. □

Remark 4.2 About notation: from now on we shall write $\|\cdot\|_\sigma$ instead of $\|\cdot\|_{C^0(S_\sigma)}$ for the C^0 norm of functions defined on S_σ.

Lemma 4.2 *Let $\xi : S_\sigma \to \mathbb{C}$ be holomorphic and periodic of period one, and let*

$$\widehat{\xi}(n) = \int_0^1 \xi(x)e^{-2\pi i n x}\,dx$$

be its n-th Fourier coefficient. Then

$$\left|\widehat{\xi}(n)\right| \leq e^{-2\pi\sigma|n|}\|\xi\|_\sigma\,, \quad \textit{for all } n \in \mathbb{Z}\,.$$

Proof See Exercise 4.4. □

With these two facts at hand, we now prove the following.

Lemma 4.3 *For each $0 < \delta < \sigma$ the series*

$$\psi(z) = \sum_{n\in\mathbb{Z}^*} \frac{\widehat{\varphi}(n)}{e^{2\pi i n\alpha} - 1}\,e^{2\pi i n z}$$

converges absolutely and uniformly for $|\mathrm{Im}(z)| < \sigma - \delta$, and $\psi(\mathbb{R}) \subseteq \mathbb{R}$. Moreover, there exists $C_0 = C_0(\nu, K) > 0$ such that (i) $\|\psi\|_{\sigma-\delta} \leq C_0\delta^{-\nu}\|\varphi\|_\sigma$, and (ii) $\|\psi'\|_{\sigma-2\delta} \leq C_0\delta^{-\nu-1}\|\varphi\|_\sigma$.

Proof Let $z \in S_{\sigma-\delta}$. Using Lemma 4.1 and applying Lemma 4.2 with $\xi = \varphi$, we have

$$\begin{aligned} |\psi(z)| &\leq \sum_{n\in\mathbb{Z}^*} \frac{|\widehat{\varphi}(n)|}{|e^{2\pi i n\alpha}-1|}\,|e^{2\pi i n z}| \\ &\leq \frac{\|\varphi\|_\sigma}{4K} \sum_{n\in\mathbb{Z}^*} |n|^{\nu-1} e^{-2\pi\sigma|n|}\, e^{-2\pi n(\mathrm{Im}(z))}\,. \end{aligned}$$

But, as the reader can easily check, $e^{-2\pi n(\mathrm{Im}(z))} \leq e^{2\pi|n|(\sigma-\delta)}$ whenever $|\mathrm{Im}(z)| < \sigma-\delta$. Therefore

$$|\psi(z)| \leq \frac{\|\varphi\|_\sigma}{4K} \sum_{n\in\mathbb{Z}^*} |n|^{\nu-1} e^{-2\pi|n|\delta}\,.$$

This last series is convergent, as we see by the integral test:

$$\begin{aligned} \sum_{n\in\mathbb{Z}^*} |n|^{\nu-1} e^{-2\pi|n|\delta} &< 2\int_0^\infty x^{\nu-1} e^{-2\pi\delta x}\,dx \\ &= \frac{2}{(2\pi\delta)^\nu}\int_0^\infty t^{\nu-1}e^{-t}\,dt = \frac{2\Gamma(\nu)}{(2\pi\delta)^\nu}\,, \end{aligned}$$

where Γ denotes, as usual, the standard *gamma function*. This shows at once that the series for $\psi(z)$ converges absolutely and that $|\psi(z)| \leq C_0\delta^{-\nu}\|\varphi\|_\sigma$, where $C_0 = \frac{\Gamma(\nu)}{2K(2\pi)^\nu}$. Hence the convergence is also uniform on $S_{\sigma-\delta}$, and $\|\psi\|_{\sigma-\delta} \leq C_0\delta^{-\nu}\|\varphi\|_\sigma$ as stated in (i).

To prove (ii), note that if $z \in S_{\sigma-2\delta}$ then the closed disk D of center z and radius δ is contained in $S_{\sigma-\delta}$. By Cauchy's integral formula, we have

$$\psi'(z) = \frac{1}{2\pi i}\int_{\partial D} \frac{\psi(\zeta)\,d\zeta}{(\zeta-z)^2}\,.$$

Hence $|\psi'(z)| \leq \delta^{-1}\sup_{\zeta\in\partial D}|\psi(\zeta)|$, and therefore, using (i), we get $\|\psi'\|_{\sigma-2\delta} \leq \delta^{-1}\|\psi\|_{\sigma-\delta} < C_0\delta^{-\nu-1}\|\varphi\|_\sigma$.

Finally, note from (4.10) that the Fourier coefficients of ψ satisfy the relation $\widehat{\psi}(-n) = \overline{\widehat{\psi}(n)}$ for all $n \in \mathbb{Z}$. This shows that $\psi(\overline{z}) = \overline{\psi(z)}$ for all z, and therefore ψ preserves the real axis. This finishes the proof. □

Remark 4.3 Note that, by taking K to be sufficiently small, we can (and will) always assume that $C_0 > 1$ (*cf.* Remark 4.1).

4.2.2 Non-linear Estimates

Now that we have bounds on the solution ψ to the linear equation (4.6), we proceed to the analysis of the holomorphic map $H = \mathrm{Id} + \psi$, which we will show to be univalent on a neighborhood of the real axis. We will derive good estimates on how close H and H^{-1} are to the identity map.

Lemma 4.4 *If* $0 < \delta < \sigma/4$ *and* $\|\varphi\|_\sigma < C_0^{-1}\delta^{\nu+1}$, *then: (i)* H *is univalent in* $S_{\sigma-2\delta}$*; (ii)* $H(S_{\sigma-2\delta}) \subseteq S_{\sigma-\delta}$*; (iii)* $H(S_{\sigma-2\delta}) \supseteq S_{\sigma-3\delta}$.

Proof We already know that $H = \mathrm{Id}+\psi$ is holomorphic in the strip $S_{\sigma-\delta} \supset S_{\sigma-2\delta}$, so we only need to show it is injective in the latter strip. Note that second estimate in Lemma 4.3 and the hypothesis on φ imply that $\|\psi'\|_{\sigma-2\delta} < 1$. Let z_1, z_2 be two distinct points in $S_{\sigma-2\delta}$. Then

$$|H(z_1) - H(z_2)| \geq \Big|\, |z_1 - z_2| - |\psi(z_1) - \psi(z_2)| \,\Big| .$$

But by the mean-value inequality,

$$|\psi(z_1) - \psi(z_2)| \leq \|\psi'\|_{\sigma-2\delta}|z_1 - z_2| < |z_1 - z_2| .$$

Therefore $H(z_1) \neq H(z_2)$, and so H is injective. This proves (i). Next, for each $z \in S_{\sigma-2\delta}$, the first estimate in Lemma 4.3 and the hypothesis on φ imply that

$$|\mathrm{Im}\, H(z)| \leq |\mathrm{Im}\, z| + |\mathrm{Im}\, \psi(z)| < (\sigma - 2\delta) + \delta = \sigma - \delta ,$$

so $H(z) \in S_{\sigma-\delta}$. This proves (ii). Finally, the proof of (iii) is more of the same, since

$$|\mathrm{Im}\, H(z)| \geq \Big|\, |\mathrm{Im}\, z| - |\mathrm{Im}\, \psi(z)| \,\Big| \geq (\sigma - 2\delta) - \delta = \sigma - 3\delta .$$

□

This lemma implies, in particular, that H has an inverse $H^{-1} : H(S_{\sigma-2\delta}) \to S_{\sigma-2\delta}$ which, of course, is also univalent. Let $\vartheta : H(S_{\sigma-2\delta}) \to \mathbb{C}$ be the holomorphic function given by

$$H^{-1}(z) = z - \psi(z) + \vartheta(z) .$$

Lemma 4.5 *We have*

$$\|\vartheta\|_{\sigma-4\delta} < C_0^2\delta^{-2\nu-1}\|\varphi\|_\sigma^2 .$$

Proof From the identity $z = H^{-1} \circ H(z)$, valid for all $z \in S_{\sigma-2\delta}$, we get the equation

$$\vartheta(z + \psi(z)) = \psi(z + \psi(z)) - \psi(z) . \tag{4.11}$$

We would like to bound the right-hand side of (4.11) using the mean-value inequality, but to do that we need $H(z) = z + \psi(z)$ to be a point inside $S_{\sigma-2\delta}$. Hence we assume that $z \in S_{\sigma-3\delta}$, and we get

$$
\begin{aligned}
|\vartheta(z+\psi(z))| &\leq |\psi'\|_{\sigma-2\delta}|\psi(z)| \\
&< \left(C_0\delta^{-\nu}\|\varphi\|_\sigma\right)\left(C_0\delta^{-\nu-1}\|\varphi\|_\sigma\right) = C_0^2\delta^{-2\nu-1}\|\varphi\|_\sigma^2 \,. \qquad (4.12)
\end{aligned}
$$

But $H(S_{\sigma-3\delta}) \supseteq S_{\sigma-4\delta}$ (mimic the proof of assertion (iii) in Lemma 4.4). This means that for each $w \in S_{\sigma-4\delta}$ there exists (a unique) $z \in S_{\sigma-3\delta}$ such that $w = z + \psi(z)$. Using this fact in (4.12) we deduce that

$$
\|\vartheta\|_{\sigma-4\delta} = \sup_{w\in S_{\sigma-4\delta}} |\vartheta(w)| < C_0^2\delta^{-2\nu-1}\|\varphi\|_\sigma^2 \,,
$$

as required. □

Now that we have estimates on H and H^{-1} on strips around the real axis that are narrower than the original strip domain of F, we would like to know how close $G = H^{-1} \circ F \circ H$ is to the translation T_α. Our hope is that G will be much closer to T_α than F. This will indeed be the case, provided we shrink even further the strip domains on which these maps are defined.

Lemma 4.6 *Let* $0 < \delta < \min\{1, \sigma/6\}$ *and, as before, suppose that* $\|\varphi\|_\sigma < C_0^{-1}\delta^{\nu+1}$. *Then the composition* $G = H^{-1} \circ F \circ H$ *is a well-defined univalent map with domain* $S_{\sigma-6\delta}$. *Moreover, if* $\eta : S_{\sigma-6\delta} \to \mathbb{C}$ *is the holomorphic function given by* $\eta(z) = G(z) - z - \alpha$, *then*

$$
\|\eta\|_{\sigma-6\delta} < 8C_0^2\delta^{-2\nu-1}\|\varphi\|_\sigma^2 \,.
$$

Proof Writing down $G(z)$ explicitly in terms of φ, ψ and ϑ, after some straightforward computations we deduce that $\eta(z) = G(z) - z - \alpha$ can be written as a sum of three terms, namely

$$
\eta(z) = A(z) + B(z) + C(z) \,,
$$

where

$$
\begin{cases}
A(z) = \psi(z) - \psi(z+\alpha) + \varphi(z+\psi(z)) \\[2ex]
B(z) = \psi(z+\alpha) - \psi(z+\alpha+\psi(z)+\varphi(z+\psi(z))) \\[2ex]
C(z) = \vartheta(z+\alpha+\psi(z)+\varphi(z+\psi(z)))
\end{cases}
$$

Note that, since ψ is a solution of the linearized equation (4.6), the first term $A(z)$ can be re-written as

$$
A(z) = \varphi(z+\psi(z)) - \varphi(z) + \widehat{\varphi}(0) \,.
$$

We are going to bound these three terms in reverse order.

(1) The term $C(z)$ is easy to estimate from Lemma 4.5. Indeed, if $z \in S_{\sigma-6\delta}$, then $z + \alpha + \psi(z) + \varphi(z + \psi(z)) \in S_{\sigma-4\delta}$, and therefore

$$|C(z)| < C_0^2 \delta^{-2\nu-1} \|\varphi\|_\sigma^2 \, . \tag{4.13}$$

(2) In order to bound the term $B(z)$, we combine the mean-value inequality with Lemma 4.3 and get

$$\begin{aligned} |B(z)| &\le \|\psi'\|_{\sigma-2\delta} \, |\psi(z) + \varphi(z + \psi(z))| \\ &< \left(C_0 \delta^{-\nu-1} \|\varphi\|_\sigma\right) (\|\psi\|_{\sigma-\delta} + \|\varphi\|_\sigma) \\ &< 2C_0^2 \delta^{-2\nu-1} \|\varphi\|_\sigma^2 \, , \end{aligned} \tag{4.14}$$

where in the last line we have used that $C_0 \delta^{-\nu} > 1$.

(3) Finally, let us find an upper-bound for $|A(z)|$. We have, of course,

$$|A(z)| \le |\varphi(z + \psi(z)) - \varphi(z)| + |\widehat{\varphi}(0)| \, .$$

The first absolute value on the right-hand side is estimated using the mean-value inequality. We have

$$|\varphi(z + \psi(z)) - \varphi(z)| \le \|\varphi'\|_{\sigma-2\delta} \cdot \|\psi\|_{\sigma-\delta} \, .$$

But $\|\varphi'\|_{\sigma-2\delta} \le \delta^{-1} \|\varphi\|_\sigma$ (this follows from Cauchy's integral formula for φ' just as in the proof of Lemma 4.3). Also, $\|\psi\|_{\sigma-\delta} < C_0 \delta^{-\nu} \|\varphi\|_\sigma$ (again by Lemma 4.3). Therefore

$$|\varphi(z + \psi(z)) - \varphi(z)| \le C_0 \delta^{-\nu-1} \|\varphi\|_\sigma^2 \, . \tag{4.15}$$

It remains to bound $|\widehat{\varphi}(0)|$. To do this, we use the fact that, since G has translation number α, there exists some $x_0 \in \mathbb{R}$ such that $\eta(x_0) = G(x_0) - x_0 - \alpha = 0$ (see Exercise 4.5). This means that $A(x_0) + B(x_0) + C(x_0) = 0$, that is

$$\widehat{\varphi}(0) = -(\varphi(x_0 + \psi(x_0)) - \varphi(x_0)) - B(x_0) - C(x_0) \, .$$

Using (4.13), (4.14) and (4.15), we get

$$|\widehat{\varphi}(0)| \le \left(3C_0^2 \delta^{-2\nu-1} + C_0 \delta^{-\nu-1}\right) \|\varphi\|_\sigma^2 < 4C_0^2 \delta^{-2\nu-1} \|\varphi\|_\sigma^2 \, .$$

From this and (4.15) we deduce that

$$|A(z)| < 5C_0^2 \delta^{-2\nu-1} \|\varphi\|_\sigma^2 \, . \tag{4.16}$$

Putting together (4.13), (4.14) and (4.16), we finally get the inequality

$$|\eta(z)| < 8C_0^2\delta^{-2\nu-1}\|\varphi\|_\sigma^2 \ .$$

Since this holds for every $z \in S_{\sigma-6\delta}$, the lemma is proved. □

4.2.3 Proof of Arnold's Theorem

We are now in a position to describe the inductive procedure leading to the proof of Arnold's local conjugacy theorem. The key to the induction is Lemma 4.6.

We are given $\sigma > 0$ and want to consider univalent maps $F : S_\sigma \to \mathbb{C}$ of the form $F(z) = z + \alpha + \varphi(z)$, where $\alpha \in (0, 1)$ is a fixed Diophantine number of type (K, ν), which are very close to the translation T_α, preserve the real axis and have translation number equal to α.

We start by defining three sequences of positive numbers $(\delta_n)_{n\geq 0}$, $(\sigma_n)_{n\geq 0}$ and $(\varepsilon_n)_{n\geq 0}$ as follows. First we set $\sigma_0 = \sigma$ and take $\delta_0 = \frac{1}{2}\min\{1, \sigma/6\}$. We also let

$$\varepsilon_0 = \left(\min\left\{\frac{\delta_0^{2\nu+1}}{8C_0^2}, \frac{1}{2^{2\nu+1}}\right\}\right)^4 .$$

The reason for such a strange choice will become apparent later. Then we define recursively, for all $n \geq 0$,

$$\delta_{n+1} = \frac{1}{2}\delta_n \ ; \quad \sigma_{n+1} = \sigma_n - 6\delta_n \ ; \quad \varepsilon_{n+1} = \varepsilon_n^{3/2} \ .$$

Note that $\sigma_n > \frac{1}{2}\sigma_0$ for all n. It is also easy to check that $\varepsilon_n < C_0^{-1}\delta_n^{\nu+1}$ for all n.

Given these preliminaries, we proceed through the following steps.

(1) *Basis of induction*. Suppose that $F_0 : S_{\sigma_0} \to \mathbb{C}$ is given by $F_0(z) = z+\alpha+\varphi_0(z)$ with φ_0 holomorphic, periodic of period one, and such that $\|\varphi_0\|_{\sigma_0} < \varepsilon_0$. Let $\psi_0 : S_{\sigma_0-\delta_0} \to \mathbb{C}$ be the holomorphic solution to the equation

$$\psi_0(z + \alpha) - \psi_0(z) = \varphi_0(z) - \widehat{\varphi}_0(0) \ ,$$

whose existence and uniqueness are guaranteed by Lemma 4.3. By that same lemma, we have $\|\psi_0\|_{\sigma_0-\delta_0} < C_0\delta_0^{-\nu}\|\varphi_0\|_{\sigma_0}$, as well as $\|\psi_0'\|_{\sigma_0-2\delta_0} < C_0\delta_0^{-\nu-1}\|\varphi_0\|_{\sigma_0}$ Let $H_0 = \mathrm{Id} + \psi_0$. By Lemma 4.4, this map is univalent in $S_{\sigma_0-2\delta_0}$, and by Lemma 4.5 the function $\vartheta_0(z) = H_0^{-1}(z) - z + \psi_0(z)$ is holomorphic in $S_{\sigma_0-4\delta_0}$ and satisfies $\|\vartheta_0\|_{\sigma_0-4\delta_0} < C_0^2\delta_0^{-2\nu-1}\|\varphi_0\|_{\sigma_0}^2$.

(2) *Induction step.* Now suppose we have already defined a univalent map $F_n : S_{\sigma_n} \to \mathbb{C}$ and a holomorphic map $H_n : S_{\sigma_n - \delta_n} \to \mathbb{C}$ having the following properties:

(i) The map F_n preserves the real axis and $F_n|_{\mathbb{R}}$ has translation number α.
(ii) If $\varphi_n = F_n - T_\alpha$, then φ_n is periodic of period one and $\|\varphi_n\|_{\sigma_n} < \varepsilon_n$.
(iii) The map H_n is univalent on $S_{\sigma_n - 2\delta_n}$, and we have $H_n(S_{\sigma_n-2\delta_n}) \subseteq S_{\sigma_n-\delta_n}$ and $H_n(S_{\sigma_n-2\delta_n}) \supseteq S_{\sigma_n-3\delta_n}$.
(iv) If $\psi_n = H_n - \mathrm{Id}$, then ψ_n is periodic of period one, and we have $\|\psi_n\|_{\sigma_n-\delta_n} < C_0\delta_n^{-\nu}\|\varphi_n\|_{\sigma_n}$ and $\|\psi_n'\|_{\sigma_n-2\delta_n} < C_0\delta_n^{-\nu-1}\|\varphi_n\|_{\sigma_n}$.

Applying Lemmas 4.5 and 4.6 to $H = H_n$ and $F = F_n$, we define $F_{n+1} = H_n^{-1} \circ F_n \circ H_n$ on the strip $S_{\sigma_n - 6\delta_n} = S_{\sigma_{n+1}}$. Then F_{n+1} is univalent, preserves the real axis, and has translation number equal to α. Moreover, writing $\varphi_{n+1} = F_{n+1} - T_\alpha$, it follows from Lemma 4.6 and (ii) that

$$\|\varphi_{n+1}\|_{\sigma_{n+1}} < 8C_0^2\delta_n^{-2\nu-1}\|\varphi_n\|_{\sigma_n}^2 < \left(8C_0^2\delta_n^{-2\nu-1}\varepsilon_n^{1/2}\right)\varepsilon_n^{3/2} . \tag{4.17}$$

But $\varepsilon_n = \varepsilon_0^{(3/2)^n}$, and our choice of ε_0 implies after some calculation that

$$8C_0^2\delta_n^{-2\nu-1}\varepsilon_0^{\frac{1}{2}\left(\frac{3}{2}\right)^n} < 1 .$$

Putting this back into (4.17) we deduce that $\|\varphi_{n+1}\|_{\sigma_{n+1}} < \varepsilon_n^{3/2} = \varepsilon_{n+1}$. This shows that F_{n+1} satisfies the analogues of properties (i) and (ii) above. Finally, let ψ_{n+1} be the solution to the equation

$$\psi_{n+1}(z+\alpha) - \psi_{n+1}(z) = \varphi_{n+1}(z) - \widehat{\varphi}_{n+1}(0) ,$$

whose existence and uniqueness, once again, are guaranteed by Lemma 4.3 (with $\varphi = \varphi_{n+1}$). That lemma also implies that the analogue of (iv) above holds for ψ_{n+1}, and from this and Lemma 4.4 it follows that $H_{n+1} = \mathrm{Id} + \psi_{n+1}$ satisfies the analogue of (iii) as well. This completes the induction.

(3) *The conjugacy.* Now that we have constructed the sequences $(F_n)_{n\geq 0}$ and $(H_n)_{n\geq 0}$ with the above properties, we know in particular that $\|F_n - T_\alpha\|_{\sigma_n} < \varepsilon_n$ and $\|H_n - \mathrm{Id}\|_{\sigma_n-\delta_n} < C_0\delta_n^{-\nu}\varepsilon_n$, for all n. We also know, applying Lemma 4.5 with $H = H_n$, that $\|H_n^{-1} - \mathrm{Id}\|_{\sigma_n-4\delta_n} < 2C_0^2\delta_n^{-2\nu-1}\varepsilon_n$, for all n. Moreover, the strip $S_{\sigma_0/2}$ is contained in the domain of all of these maps. It follows that, on this strip, we have $F_n \to T_\alpha$, $H_n \to \mathrm{Id}$ and $H_n^{-1} \to \mathrm{Id}$, and the convergence is *uniform* in each case. In addition, for each $z \in S_{\sigma_0/2}$, we have

$$F_n(z) = (H_0 \circ H_1 \circ \cdots \circ H_{n-1})^{-1} \circ F_0 \circ (H_0 \circ H_1 \circ \cdots \circ H_{n-1})(z) \tag{4.18}$$

We claim that the sequence of univalent maps $\Psi_n = H_0 \circ H_1 \circ \cdots \circ H_{n-1} : S_{\sigma_n} \to \mathbb{C}$ when restricted to the strip $S_{\sigma_0/2}$ converges uniformly to a holomorphic map

$H : S_{\sigma_0/2} \to \mathbb{C}$ which is necessarily univalent. To prove this claim, we first estimate $\|\Psi_{n+1} - \Psi_n\|_{\sigma_0/2}$ for all n. Using the mean-value inequality, we have

$$\begin{aligned}\|\Psi_{n+1} - \Psi_n\|_{\sigma_0/2} \leq \|\Psi_{n+1} - \Psi_n\|_{\sigma_n-\delta_n} &= \|\Psi_n \circ H_n - \Psi_n\|_{\sigma_n-\delta_n} \\ &\leq \|\Psi_n'\|_{\sigma_n} \|H_n - \mathrm{Id}\|_{\sigma_n-\delta_n} < C_0\delta_n^{-\nu}\varepsilon_n \|\Psi_n'\|_{\sigma_n} \,. \end{aligned} \tag{4.19}$$

Now, by then chain rule,

$$\Psi_n' = \prod_{j=0}^{n-1} H_j' \circ H_{j+1} \circ \cdots \circ H_{n-1}$$

Since $H_j' = 1 + \psi_j'$ for each j, it follows that

$$\begin{aligned}\|\Psi_n'\|_{\sigma_n} &\leq \prod_{j=0}^{n-1} \left(1 + \|\psi_j'\|_{\sigma_j - 2\delta_j}\right) \\ &< \prod_{j=0}^{n-1} \left(1 + C_0\delta_j^{-\nu-1}\varepsilon_j\right) < \prod_{j=0}^{\infty} \left(1 + C_0\delta_j^{-\nu-1}\varepsilon_j\right) \,.\end{aligned}$$

But the latter product converges, because the series $\sum_{j=0}^{\infty} \delta_j^{-\nu-1}\varepsilon_j$ converges. This shows that there exists a constant $B_0 > 0$ such that $\|\Psi_n'\|_{\sigma_n} < B_0$ for all n. Taking this back to (4.19), we see that $\|\Psi_{n+1} - \Psi_n\|_{\sigma_0/2} \leq B_1\delta_n^{-\nu}\varepsilon_n$, where $B_1 = B_0C_0$. Therefore, for all $m > n \geq 0$, we have

$$\|\Psi_m - \Psi_n\|_{\sigma_0/2} \leq B_1 \sum_{j=n}^{m-1} \delta_j^{-\nu}\varepsilon_j$$

Since the series $\sum_{j=0}^{\infty} \delta_j^{-\nu}\varepsilon_j$ is also convergent, we deduce that $(\Psi_n)_{n\geq 0}$ is a uniform Cauchy sequence in $S_{\sigma_0/2}$. Therefore $H = \lim_{n\to\infty} \Psi_n$ exists and is holomorphic (and univalent) in $S_{\sigma_0/2}$. Going back to (4.18) and letting $n \to \infty$, we finally get the conjugacy equation $T_\alpha = H^{-1} \circ F_0 \circ H$ in the strip $S_{\sigma_0/2}$, which is what we wanted.

(4) Now that all the hard work has been done, to complete the proof of Theorem 4.4, all one needs to do is to quotient everything down to $\mathbb{C}/\mathbb{Z}$ using the exponential covering map. This task is left to the reader as an exercise.

This concludes the proof of Arnold's theorem.

4.3 Counterexamples to Linearizability

In the same paper where he proved his analytic conjugacy theorem, Arnold (1961) also gave examples of analytic circle diffeomorphisms without periodic points which are not C^1 conjugate—in fact, not even absolutely continuously conjugate—to an irrational rotation. Of course, the rotation number of such a diffeomorphism must be an irrational that can be well approximated by rationals.

4.3.1 One-Parameter Families

The examples we seek will be found in suitable one-parameter families of analytic diffeomorphisms. We will in fact show that they are, in a suitable topological sense, abundant.

Let $f : S^1 \to S^1$ be an analytic diffeomorphism, and consider the one-parameter family $f_\alpha = R_\alpha \circ f$, where $\alpha \in \mathbb{R}/\mathbb{Z}$. We will refer to this family as the *standard family generated by* f, or simply the *standard family of* f. We know from Chap. 2 that the rotation number varies continuously and monotonically with the parameter α, i.e., the function $\Theta_f : \mathbb{R}/\mathbb{Z} \to \mathbb{R}/\mathbb{Z}$ given by $\Theta_f(\alpha) = \rho(f_\alpha)$ is continuous and monotone non-decreasing.[2] We call this function the *rotation number function* associated with f.

We say that a surjective, monotone function $\psi : \mathbb{R}/\mathbb{Z} \to \mathbb{R}/\mathbb{Z}$ is a *devil staircase* if ψ is continuous, and there exists a countable dense set $C \subset [0, 1]$ such that (i) $\psi^{-1}(c)$ is a closed interval with non-empty interior, for each $c \in C$; and (ii) $K = (\mathbb{R}/\mathbb{Z}) \setminus \bigcup_{c\in C} \mathrm{int}(\psi^{-1}(c))$ is a Cantor set.

Lemma 4.7 *Let* $f : S^1 \to S^1$ *be an analytic diffeomorphism, and let* $f_\alpha = R_\alpha \circ f$ *be its standard family. Suppose we have* $f_\alpha^n \neq \mathrm{Id}$ *for all* $\alpha \in [0, 1]$ *and all* $n \geq 1$. *Then the rotation number function* Θ_f *is a devil staircase.*

Proof We already know that Θ_f is continuous and monotone, and it is also surjective. Let $C = \mathbb{Q}/\mathbb{Z}$. For each rational $r \in C$, the pre-image $\Delta_r = \Theta_f^{-1}(r)$ is a non-empty closed interval. We claim that this interval has non-empty interior. To see this, write $r = p/q$ in irreducible form, and let $\alpha \in \Delta_{p/q}$. Let $F_\alpha : \mathbb{R} \to \mathbb{R}$ be the lift of f_α with $0 \leq F_\alpha(0) < 1$. Consider the (periodic) function $\varphi_\alpha(x) = F^q(x) - x - p$. Its graph $\mathrm{Gr}(\varphi_\alpha)$ intersects the real axis, because f_α has a periodic orbit of period q. Note that φ_α cannot vanish identically, for if it did, we would have $f_\alpha^q = \mathrm{Id}$, contrary to our hypothesis. Hence there are three cases to consider:

(1) The graph $\mathrm{Gr}(\varphi_\alpha)$ crosses the real axis. In this case, by continuity of the map $\beta \mapsto \varphi_\beta$, we see that there exists $\delta > 0$ small such that , for each $\beta \in (\alpha -$

[2] Here and throughout we think of $\mathbb{R}/\mathbb{Z}$ as the interval $[0, 1]$ with the endpoints identified, and the induced order relation via this identification.

$\delta, \alpha + \delta)$, the graph $\mathrm{Gr}(\varphi_\beta)$ also crosses the real axis, so that $\rho(f_\beta) = p/q$. In other words, we have $\Delta_{p/q} \supset (\alpha - \delta, \alpha + \delta)$.

(2) The graph $\mathrm{Gr}(\varphi_\alpha)$ touches the real axis, but $\varphi_\alpha(x) \geq 0$ for all x. Here, since $\varphi_\alpha(x_0) > 0$ for some x_0, it follows from the continuity of $\beta \mapsto \varphi_\beta$ that there exist $\delta > 0$ small such that, for each $\beta \in (\alpha - \delta, \alpha]$, the graph $\mathrm{Gr}(\varphi_\beta)$ intersects the real axis, so that $\rho(f_\beta) = p/q$. Hence $\Delta_{p/q} \supset (\alpha - \delta, \alpha]$ in this case.

(3) The graph $\mathrm{Gr}(\varphi_\alpha)$ touches the real axis, but this time $\varphi_\alpha(x) \leq 0$ for all x. This case is analogous to case (2). Proceeding as before, we deduce in this case that there exists $\delta > 0$ small such that $\Delta_{p/q} \supset [\alpha, \alpha + \delta)$.

Whichever case occurs, we see that $\Delta_{p/q}$ has non-empty interior, as claimed. Finally, if $y \in (\mathbb{R} \setminus \mathbb{Q})/\mathbb{Z}$ is an irrational point, then $\Theta_f^{-1}(y)$ reduces to a single point. This follows from the fact that Θ_f is strictly increasing at each point α for which $\rho(f_\alpha)$ is irrational. This fact is left as an instructive exercise to the reader (see Exercise 4.6). Putting all these facts together, we deduce that

$$K_f = (\mathbb{R}/\mathbb{Z}) \setminus \bigcup_{r \in \mathbb{Q}/\mathbb{Z}} \mathrm{int}(\Delta_r) \tag{4.20}$$

is compact, totally disconnected and without isolated points, i.e., a Cantor set. Therefore Θ_f is indeed a devil staircase. □

Remark 4.4 The fact that the intervals $\Delta_{p/q}$ have non-empty interior is known as *phase-locking* or *mode-locking* phenomenon. Accordingly, these intervals are called *phase-locking* or *mode-locking intervals*.

The reader may wonder how easy it is to produce examples of (standard) one-parameter families satisfying the hypothesis of Lemma 4.7. It turns out that if f has a lift to the real line which is the restriction of a holomorphic map of the entire complex plane, then the hypothesis in question is always satisfied—see Exercise 4.7. This is what happens with one-parameter families extracted from the so-called *Arnold family*, which depends on two parameters. The maps in the Arnold family have as lifts the restrictions to the real line of the entire maps given by

$$F_{\alpha,\beta}(z) = z + \alpha + \beta \sin 2\pi z\ .$$

Here, we have $0 \leq \alpha < 1$ and $0 < \beta < 1/2\pi$. These entire maps project down to holomorphic self-maps of the cylinder $\mathbb{C}/\mathbb{Z} \simeq \mathbb{C}^*$—call them $f_{\alpha,\beta}$—whose restrictions to the unit circle are analytic diffeomorphisms. Note that $f_{\alpha,\beta} = R_\alpha \circ f_{0,\beta}$. Thus, by fixing β and varying α, we get a one-parameter family satisfying the hypothesis of Lemma 4.7. In Table 4.1 the reader can see the plots for two values of β smaller than $1/2\pi$. When $\beta = 1/2\pi$ we still have a family of circle *homeomorphisms*, but these are not *diffeomorphisms*: $x = 1/2$ is now a critical point. Such maps are called critical circle maps, and will be the main object of study in parts III and IV of this book. For $\beta > 1/2\pi$ the corresponding maps in the Arnold family are no longer invertible; these maps will not be studied in this book.

Table 4.1 Devil staircases in the Arnold family for $\beta = 0.125$ (left) and for $\beta = 0.158$ (right). In each case, $f = f_{0,\beta}$

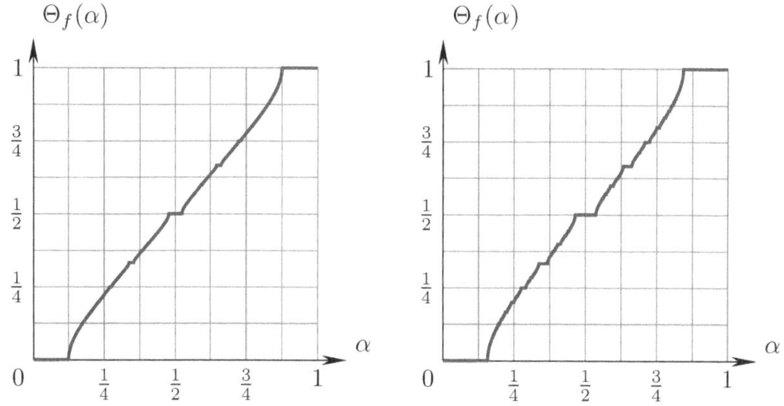

Remark 4.5 For each fixed value of β in the range $0 < \beta \leq 1/2\pi$, we may consider the Cantor set $K_\beta \subset [0, 1]$ obtained as the closure of the complement of the union of all phase-locking intervals in the one-parameter family $\alpha \mapsto f_{\alpha,\beta}$. As shown by Herman (1979), K_β has positive Lebesgue measure when $\beta < 1/2\pi$. By contrast, when $\beta = 1/2\pi$ the corresponding Cantor set has zero Lebesgue measure; this was first proved by Świątek (1988).

Remark 4.6 An interesting picture emerges if one looks at the Arnold family in parameter space. For each rational $p/q \in [0, 1]$, the set of all pairs of parameters (α, β) inside the rectangle $[0, 1] \times [0, \frac{1}{2\pi}]$ for which the map $f_{\alpha,\beta}$ has rotation number p/q is a connected set known as an *Arnold tongue*. See Fig. 4.1 for a computer-generated picture of some of these tongues (for selected values of the rotation number p/q).

4.3.2 *Residual Sets of Non-linearizable Parameters*

We now combine what we learned in Sect. 4.1 about Herman's C^1 conjugacy invariant with a simple Baire category argument to show the existence of minimal analytic circle diffeomorphisms which are *not* C^1 conjugate to a rotation (of course, they are always *topologically* conjugate to a rotation, by Denjoy's theorem).

Recall that a subset E of a complete metric space X is *residual* if it contains a countable intersection of sets which are open and dense in X. Baire's theorem says that residual subsets of a complete metric space X are always dense in X. It is easy to see that the intersection of any finite collection of residual sets is residual.

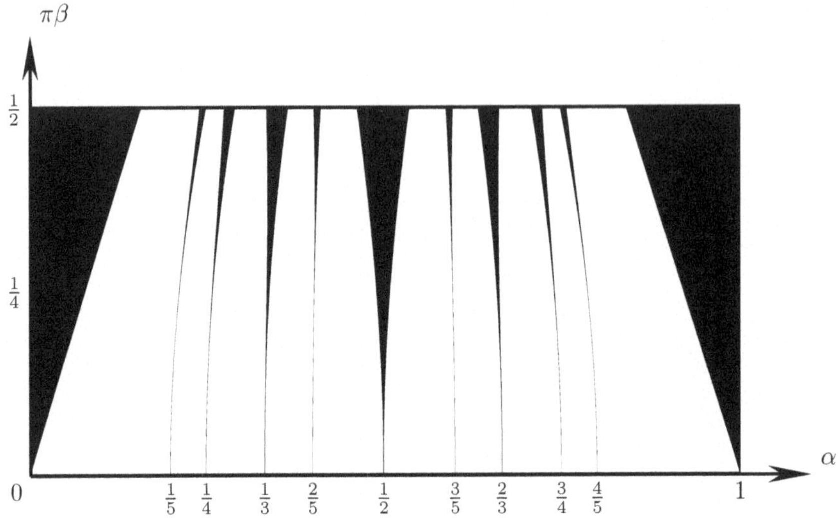

Fig. 4.1 Arnold tongues in the family $x \mapsto x + \alpha + \beta \sin(2\pi x)$ for $0 \leq \alpha \leq 1$ and $0 \leq \beta \leq \frac{1}{2\pi}$

Let $f : S^1 \to S^1$ be an analytic circle diffeomorphism satisfying the hypothesis of Lemma 4.7, let $f_\alpha = R_\alpha \circ f$ be the standard family it generates, and let K_f be the Cantor set in (4.20).

Theorem 4.5 *There exists a residual subset $\Lambda \subset K_f$ such that, for every $\alpha \in \Lambda$, the analytic diffeomorphism f_α is not C^1 conjugate to $R_{\rho(f_\alpha)}$.*

Proof Let $D_f \subset K_f$ denote the set of all endpoints of phase-locking intervals in the complement of K_f in $\mathbb{R}/\mathbb{Z}$. Then D_f is a countable dense subset of the Cantor set K_f. For each $\alpha \in D_f$ the corresponding diffeomorphism f_α has rational rotation number, say $\rho(f_\alpha) = p/q$, but it cannot be C^1 conjugate to the rotation $R_{p/q}$. If it were, then *a fortiori* we would have $f_\alpha^q = \text{Id}$, contrary to hypothesis. Hence, by Theorem 4.3, we must have $\mathscr{H}_1(f_\alpha) = \infty$ for all $\alpha \in D_f$. Here, as in Sect. 4.1, $\mathscr{H}_1(f) = \sup_{n\geq 1} \|Df^n\|$ is Herman's (first) invariant.

Now, for each positive integer k, let $V_k = \{\alpha \in K_f : \mathscr{H}_1(f_\alpha) > k\}$. Then V_k is open, and we clearly have $D_f \subset V_k$, for all $k \geq 1$. In other words, each V_k is an open and dense subset of K_f. Since K_f is a compact subset of the complete metric space $\mathbb{R}/\mathbb{Z}$, it is itself complete, and therefore, by Baire's theorem, $V_\infty = \bigcap_{k\geq 1} V_k$ is residual in K_f. But every $\alpha \in V_\infty$ obviously satisfies $\mathscr{H}_1(f_\alpha) = \infty$, so by Theorem 4.3 the corresponding diffeomorphism f_α is not C^1-linearizable. Hence we can take $\Lambda = V_\infty$. □

4.3.3 Singular Measures and Conjugacies

We now wish to go beyond Theorem 4.5 and show that there are plenty of analytic diffeomorphisms that are minimal but *not absolutely continuously conjugate* to a rotation. The examples can be constructed so as to be as close to a rigid rotation as desired. Rather than a Baire category argument, we will employ an approximation argument.

To achieve our goal, the following criterion will be crucial. In what follows, we denote by m the Lebesgue measure on the circle.

Lemma 4.8 *Let $f : \mathbf{S}^1 \to \mathbf{S}^1$ be a homeomorphism, and let μ be an f-invariant Borel probability measure. Suppose f has the following property: for each $n \geq 1$, there exist a Borel set $A_n \subset \mathbf{S}^1$ and a positive integer k_n such that (i) $f^{k_n}(\mathbf{S}^1 \setminus A_n) \subseteq A_n$; and (ii) $m(A_n) < 2^{-n}$. Then μ is not absolutely continuous with respect to m.*

Proof Since μ is invariant under f, we have

$$\mu(A_n) \geq \mu\left(f^{k_n}(\mathbf{S}^1 \setminus A_n)\right) = \mu(\mathbf{S}^1 \setminus A_n) = 1 - \mu(A_n) ,$$

so $\mu(A_n) \geq \frac{1}{2}$ for all n. Consider the set

$$A_\infty = \limsup A_n = \bigcap_{k=1}^{\infty} \bigcup_{n \geq k} A_n .$$

Since $\mu\left(\bigcup_{n\geq k} A_n\right) \geq \frac{1}{2}$ for every k, we have $\mu(A_\infty) \geq \frac{1}{2}$ as well. But at the same time, for all $k \geq 1$ we have

$$m(A_\infty) \leq m\left(\bigcup_{n\geq k} A_n\right) \leq \sum_{n=k}^{\infty} \frac{1}{2^n} = \frac{1}{2^{k-1}} ,$$

and therefore $m(A_\infty) = 0$. This shows that μ cannot be absolutely continuous with respect to Lebesgue measure. □

Adapting the terminology of Cornfeld et al. (1982, p. 88), we introduce the following definition.

Definition 4.3 Given a rational number p/q in irreducible form, we say that a circle homeomorphism f is *(p, q)-stable* if f has a lift F to the real line such that $F^q(x) \geq x + p$ for all $x \in \mathbb{R}$ and the equality $F^q(x_0) = x_0 + p$ holds for some x_0.

Note that if f is (p, q)-stable[3] then in particular f has a periodic orbit of period q, and in fact $\rho(f) = p/q$. The following lemma states that, in any standard one-parameter family of diffeomorphisms, there are (p, q)-stable diffeomorphisms for all rationals $p/q \in [0, 1]$.

Lemma 4.9 *Let $f : \mathbf{S}^1 \to \mathbf{S}^1$ be a diffeomorphism, and suppose the standard family $f_\alpha = R_\alpha \circ f$ it generates is such that its rotation number function $\Theta_f(\alpha) = \rho(f_\alpha)$ is a devil staircase. If $\Delta_{p/q} = [\alpha_{p/q}, \beta_{p/q}]$ is the phase locking interval corresponding to the rational p/q, then $f_{\beta_{p/q}}$ is (p, q)-stable.*

Proof This is left as an exercise to the reader (Exercise 4.8). □

Our next lemma states in essence that, arbitrarily near any analytic (p, q)-stable diffeomorphism we can find another analytic (p, q)-stable diffeomorphism having *exactly one periodic orbit of period q*. We formulate the result in terms of lifts. We assume these lifts are defined on the horizontal strip $S_1 = \{z : |\mathrm{Im}(z)| < 1\}$. Given two holomorphic maps F, G defined on this strip, we let $d(F, G) = \sup_{z\in S_1} |F(z) - G(z)|$ denote the C^0 distance between them.

Lemma 4.10 *Let $F : S_1 \to \mathbb{C}$ be a univalent map with $F(\mathbb{R}) = \mathbb{R}$ such that $F|_\mathbb{R}$ is the lift of a (p, q)-stable diffeomorphism $f : \mathbf{S}^1 \to \mathbf{S}^1$. For each $\delta > 0$ there exists a univalent map $G : S_1 \to \mathbb{C}$ with the following properties:*

(i) $d(F, G) < \delta$;
(ii) $G(\mathbb{R}) = \mathbb{R}$;
(iii) $G|_\mathbb{R}$ *is the lift of a (p, q)-stable diffeomorphism g;*
(iv) g *has a unique periodic orbit of period q.*

Proof Let $x_0 \in \mathbb{R}$ be such that $F^q(x_0) = x_0 + p$, and write $x_j = F^j(x_0)$ for each $j \in \mathbb{Z}$. Also, let $z_j = e^{2\pi i x_j} \in \mathbf{S}^1$ (so that $z_{j+q} = z_j$ for all j), and note that $\{z_0, z_1, \ldots, z_{q-1}\}$ is a periodic orbit for f. Let $\phi : \mathbb{C} \to \mathbb{C}$ be the holomorphic function given by

$$\phi(z) = \prod_{j=0}^{q-1} \sin^2\left(\pi(z - x_j)\right).$$

This function is periodic of period one. Note that $\phi(x) \geq 0$ for all $x \in \mathbb{R}$, and equality holds only for $x \in \{x_0, x_1, \ldots, x_{q-1}\} + \mathbb{Z}$. Now, consider the holomorphic map $G(z) = F(z) + \epsilon\phi(z)$. Taking $\epsilon > 0$ sufficiently small, G becomes univalent in S_1, and $d(F, G) = \epsilon \sup_{z\in S_1} |\phi(z)| < \delta$. Moreover, its restriction to the real line is the lift of an analytic circle diffeomorphism g. Note also that $G^j(x_0) = F^j(x_0)$

[3] Note that the (p, q)-stability property can always be destroyed by a small perturbation. This suggests that it would be more appropriate to use the moniker (p, q)-*unstable* when referring to such maps. We will nevertheless conform to the terminology used in (Cornfeld et al. 1982, p. 88).

for all $j \in \mathbb{Z}$. In particular, we have $G^q(x_0) = x_0 + p$, and our choice of ϕ implies that $G^q(x) > x + p$ for all $x \notin \{x_0, x_1, \ldots, x_{q-1}\} + \mathbb{Z}$. This shows that g is (p, q)-stable, and also that $\{z_0, z_1, \ldots, z_{q-1}\}$ is its only periodic orbit (of period q). □

Lemma 4.11 *Let $f : \mathbf{S}^1 \to \mathbf{S}^1$ be a (p, q)-stable homeomorphism having a unique periodic orbit of period q. Then for each $\varepsilon > 0$ there exist an open set A_ε containing that periodic orbit and a positive integer k_ε such that $m(A_\varepsilon) < \varepsilon$ and $f^{k_\varepsilon}(\mathbf{S}^1 \setminus A_\varepsilon) \subseteq A_\varepsilon$.*

Proof Let $\mathscr{O} = \{z_0, z_1, \ldots, z_{q-1}\}$ be the periodic orbit in question. For each $j = 0, 1, \ldots, q-1$, write $z_j = e^{2\pi i x_j}$ as before, and let $J(z_j, \varepsilon) \subset \mathbf{S}^1$ denote the arc centered at z_j with endpoints $e^{2\pi i(x_j - \frac{\varepsilon}{3q})}$ and $e^{2\pi i(x_j + \frac{\varepsilon}{3q})}$. Consider the open set

$$A_\varepsilon = \bigcup_{j=0}^{q-1} J(z_j, \varepsilon) \subseteq \mathbf{S}^1 ,$$

whose Lebesgue measure is equal to $\frac{2}{3}\varepsilon$. As we saw in Lemma 2.3, the omega-limit set of every point on the circle is equal to the periodic orbit $\mathscr{O}$. This means that the orbit of every $z \in \mathbf{S}^1 \setminus A_\varepsilon$ enters A_ε after some time $k(z)$ and never leaves it. Since $\mathbf{S}^1 \setminus A_\varepsilon$ is compact and f is continuous, it follows that $k_\varepsilon = \sup_{z \in \mathbf{S}^1 \setminus A_\varepsilon} k(z) < \infty$, and the lemma is proved. □

The next result yields the crucial inductive procedure for the construction of the examples we promised above. For convenience of notation, let us denote by $\mathscr{U}_1$ the class of all univalent maps $F : S_1 \to \mathbb{C}$ defined over the strip $S_1 = \{z \in \mathbb{C} : |\operatorname{Im} z| < 1\}$ such that $F(z) - z$ is periodic of period one and $F(\mathbb{R}) = \mathbb{R}$, so that $F|_\mathbb{R}$ is the lift of an analytic diffeomorphism of the circle.

Proposition 4.1 *Given $p_0/q_0 \in (0, 1)$ and $\delta_0 > 0$, there exist a sequence of univalent maps $(F_n)_{n \geq 0}$ with $F_n \in \mathscr{U}_1$ for all $n \geq 0$ and a sequence $(\delta_n)_{n \geq 0}$ of positive numbers with $\delta_{n+1} \leq \delta_n/2$ for all $n \geq 0$, such that the following properties hold.*

(1) The restriction of F_n to the real line is the lift of an analytic diffeomorphism $f_n : \mathbf{S}^1 \to \mathbf{S}^1$ with rational rotation number $\rho(f_n) = p_n/q_n$.
(2) We have $f_0 = R_{p_0/q_0}$, and for all $n \geq 1$, the diffeomorphism f_n is (p_n, q_n)-stable and has a unique periodic orbit (of period q_n).
(3) We have $d(F_n, F_{n+1}) < \frac{1}{2}\delta_{n+1}$.
(4) For each $n \geq 1$, there exist a positive integer k_n and an open set $A_n \subset \mathbf{S}^1$ with $m(A_n) = 2^{-n}$ such that, for all $G \in \mathscr{U}_1$ with $d(F_n, G) \leq \delta_{n+1}$, we have $g(\mathbf{S}^1 \setminus A_n) \subseteq A_n$, where $g : \mathbf{S}^1 \to \mathbf{S}^1$ is the diffeomorphism with lift $G|_\mathbb{R}$.
(5) We have, for all $n \geq 0$,

$$\left| \frac{p_{n+1}}{q_{n+1}} - \frac{p_n}{q_n} \right| < \frac{1}{2n^2 \max_{0 \leq j \leq n} {q_j}^2} .$$

Proof We start by taking $F_0 = T_{p_0/q_0}$, so that F_0 is the lift of $f_0 = R_{p_0/q_0}$. Now, suppose that $\delta_j > 0$ and $F_j \in \mathscr{U}_1$ have already been constructed for all $j \leq n$. In particular, $F_n|_{\mathbb{R}}$ is the lift of a (p_n, q_n)-stable circle diffeomorphism f_n, and f_n has a unique periodic orbit $\mathscr{O}_n$ of period q_n.

First, we define δ_{n+1}. Applying Lemma 4.11 to $f = f_n$, we know that there exists an open ϵ_n-neighborhood V_n of $\mathscr{O}_n$, where $\epsilon_n = 1/(q_n 2^{n+2})$, and a positive integer k_n such that $f_n(S^1 \setminus V_n) \subset V_n$. Let $\eta_n > 0$ be small enough that, if $G \in \mathscr{U}_1$ is such that $d(F_n, G) \leq \eta_n$, then $d(F_n^{k_n}, G^{k_n}) < \epsilon_n$. Then every such G will have the property that $g^{k_n}(S^1 \setminus V_n) \subset V_n^*$, where g is the circle diffeomorphism of which G is the lift and where V_n^* is the open $2\epsilon_n$-neighborhood of $\mathscr{O}_n$. Thus, if we let $A_n = V_n^*$, then $g^{k_n}(S^1 \setminus A_n) \subset A_n$ for all G with $d(F_n, G) < \eta_n$, and moreover $m(A_n) = 4q_n\epsilon_n = 2^{-n}$. Having done this, we define $\delta_{n+1} = \min\{\eta_n, \frac{1}{2}\delta_n\}$.

Next, we define F_{n+1}. To do this, we first look at the standard one-parameter family $f_{n,\alpha} = R_\alpha \circ f_n$. We know from Lemma 4.7 that the rotation number function $\Theta_{f_n} : \alpha \mapsto \rho(f_{n,\alpha})$ is a devil staircase. Choose a rational number p_{n+1}/q_{n+1} such that

$$\rho(f_n) = \frac{p_n}{q_n} < \frac{p_{n+1}}{q_{n+1}} < \rho(f_{n,\delta_{n+1}/4}) ,$$

and choose it so close to p_n/q_n that the inequality in (5) is satisfied. Then the phase-locking interval $\Delta_{p_{n+1}/q_{n+1}} = \Theta_{f_n}^{-1}(p_{n+1}/q_{n+1})$ is contained in the interval $(0, \frac{1}{4}\delta_{n+1})$. Let α_n be the right endpoint of $\Delta_{p_{n+1}/q_{n+1}}$. Then the map f_{n,α_n} is (p_{n+1}, q_{n+1})-stable, and its lift $F_{n,\alpha_n} \in \mathscr{U}_1$ satisfies $d(F_n, F_{n,\alpha_n}) < \frac{1}{4}\delta_{n+1}$. However, there is no guarantee that f_{n,α_n} has only one periodic cycle. To fix this problem, we need to perturb f_{n,α_n} slightly. Here we apply Lemma 4.10 with $F = F_{n,\alpha_n}$ and $\delta = \frac{1}{4}\delta_{n+1}$. We get a new univalent map $F_{n+1} \in \mathscr{U}_1$ whose restriction to the real line is the lift of a circle diffeomorphism f_{n+1} which is (p_{n+1}, q_{n+1})-stable, and has a unique periodic orbit of period q_{n+1}. We now have

$$d(F_n, F_{n+1}) \leq d(F_n, F_{n,\alpha_n}) + d(F_{n,\alpha_n}, F_{n+1}) < \frac{1}{2}\delta_{n+1} .$$

This completes the induction, and finishes the proof. □

We are finally ready for the main result of this section.

Theorem 4.6 *Given a circle rotation R_α and $\varepsilon > 0$, there exists an analytic diffeomorphism $f : S^1 \to S^1$ without periodic points such that $d(f, R_\alpha) < \epsilon$ and whose unique invariant measure is not absolutely continuous with respect to Lebesgue measure. In particular, no conjugacy between f and the corresponding irrational rotation $R_{\rho(f)}$ can be absolutely continuous.*

Proof We can of course assume that α is rational. Applying Proposition 4.1 to $F_0 = R_\alpha$ and $\delta_0 = \varepsilon$, we get a sequence of univalent maps $F_n \in \mathscr{U}_1$ possessing properties (1)-(5) in the statement of that proposition. In particular, from property (3) and the way the sequence $(\delta_n)_{n\geq 0}$ is constructed, we see that $(F_n)_{n\geq 0}$ is a uniform Cauchy sequence in the strip S_1. Let $F \in \mathscr{U}_1$ be its limit, and let $f : \mathbf{S}^1 \to \mathbf{S}^1$ be the analytic diffeomorphism whose lift is F.

First we claim that $\theta = \rho(f)$ is irrational. We know by continuity of the rotation number that

$$\theta = \lim_{n\to\infty} \rho(f_n) = \lim_{n\to\infty} \frac{p_n}{q_n} .$$

Hence we can write

$$\frac{p_n}{q_n} - \theta = \sum_{j=n}^{\infty} \left(\frac{p_j}{q_j} - \frac{p_{j+1}}{q_{j+1}} \right) ,$$

and from the inequality in property (5) we get

$$\left| \theta - \frac{p_n}{q_n} \right| < \sum_{j=n}^{\infty} \frac{1}{2j^2 \max_{0\le k\le j} q_k{}^2} < \frac{1}{2q_n^2} \sum_{j=n}^{\infty} \frac{1}{j^2} < \frac{\pi^2}{12q_n^2} < \frac{1}{q_n^2}$$

Thus, we have infinitely many rational solutions to the inequality $|\theta - (p/q)| < q^{-2}$, and this means that θ is irrational.[4]

Next, note that from property (3) that

$$d(F, F_n) \le \frac{1}{2} \sum_{j=n+1}^{\infty} \delta_j \le \delta_{n+1} .$$

In particular, $d(F, R_\alpha) \le \delta_1 < \varepsilon$. Then, by property (4), we see that $f(\mathbf{S}^1 \setminus A_n) \subset A_n$, where $A_n \subset \mathbf{S}^1$ is an open set with $m(A_n) = 2^{-n}$. Since this holds for all $n \ge 1$, we deduce from the criterion in Lemma 4.8 that the unique Borel probability measure invariant under f is not absolutely continuous with respect to Lebesgue measure. The last assertion in the statement follows immediately from this. The proof is complete. □

4.4 Further Local Theory: The Brjuno Condition

As before, an irrational number ρ in (0, 1) is said to be *Diophantine of order* $\delta \ge 0$ if there exists a constant $C > 0$ such that

[4] It is easy to see that the inequality in (5) implies that $q_{n+1} > 2n^2 q_n$, so the sequence (q_n) is strictly increasing.

$$\left| \rho - \frac{p}{q} \right| \geq \frac{C}{q^{2+\delta}}$$

for any rational number p/q. As it is not difficult to prove (see Lemma A.4 in Appendix A), for any given $\delta > 0$ the set of Diophantine numbers of order δ has full Lebesgue measure.

In Sect. 4.2 we have proved a *local linearization* result, namely Theorem 4.4, which says that any real-analytic circle diffeomorphism with Diophantine rotation number ρ, which is a small perturbation of the rigid rotation R_ρ, is analytically linearizable (i.e., it is conjugate to R_ρ by a real-analytic diffeomorphism). On the other hand, we have constructed in Theorem 4.6 examples of real-analytic diffeomorphisms with irrational rotation number (as close to a rigid rotation as desired) for which any conjugacy with the corresponding rotation is not even absolutely continuous.

Still dealing with analytic diffeomorphisms close to a rotation, we proceed to state two fundamental results due to J.-C. Yoccoz (2002). For any given $b > 1$, we say that $f \in \mathrm{Diff}_b^\omega(S^1)$ if f is a real-analytic circle diffeomorphism, whose holomorphic extension is defined in the annulus

$$A_b = \left\{ z \in \mathbb{C} : b^{-1} < |z| < b \right\}.$$

Definition 4.4 An irrational number $\rho \in (0, 1)$ satisfies the *Brjuno* condition if

$$\sum_{n \in \mathbb{N}} \frac{\log q_{n+1}}{q_n} < \infty ,$$

where p_n/q_n are the convergents of ρ.

As it is not difficult to prove (see Exercise 4.12), any Diophantine number satisfies the Brjuno condition. Therefore, the following result extends Theorem 4.4.

Theorem 4.7 (Yoccoz (2002)) *For any Brjuno number ρ and any $b > 1$ there exists $\varepsilon = \varepsilon(\rho, b) > 0$ with the following property. If $f \in \mathrm{Diff}_b^\omega(S^1)$ has rotation number ρ and satisfies $\| f - R_\rho \|_{C^0(A_b)} < \varepsilon$, then any topological conjugacy between f and R_ρ belongs to $\mathrm{Diff}_{b/2}^\omega(S^1)$.*

Yoccoz also proved that the Brjuno condition in Theorem 4.7 is sharp in the analytic class, as expressed by the following result.

Theorem 4.8 (Yoccoz (2002)) *If $\rho \in (0, 1)$ is an irrational number which is not Brjuno, the following holds. For any given $b > 1$ and $\varepsilon > 0$ there exists $f \in \mathrm{Diff}_b^\omega(S^1)$ with rotation number ρ and satisfying $\| f - R_\rho \|_{C^0(A_b)} < \varepsilon$, which is not analytically linearizable.*

4.5 Global Theory: Herman–Yoccoz Results and Beyond

The linearization results of the previous sections are *local*, in the sense that they hold for real-analytic dynamics whose holomorphic extensions are *small perturbations* of a linear rotation. In this final section we survey, without proofs, some of the most relevant *global* linearization results, starting with the seminal works of Herman (1979) and Yoccoz (1984a).

Theorem 4.9 (Herman–Yoccoz) *If f is a C^r diffeomorphism of $\mathbf{S}^1$, with $r \geq 3$, whose rotation number is Diophantine of order δ then,* provided $r > 2\delta + 1$, *f is $C^{r-1-\delta-\varepsilon}$-conjugate to the corresponding rigid rotation, for every $\varepsilon > 0$.*

Note that no assumption on being close to a rotation is needed here. Herman proved that such a global linearization result holds for Lebesgue almost every rotation number, while Yoccoz proved that it holds in fact for any Diophantine number. A proof of Theorem 4.9 can be found in de Melo and van Strien (1993, Section I.3). Let us mention that Herman's proof was simplified by Khanin and Sinai (1987) and Stark (1988), through the use of renormalization methods.

Theorem 4.9 was subsequently sharpened by Katznelson and Ornstein (1989b), who proved the following result.

Theorem 4.10 (Katznelson–Ornstein) *If $f \in \mathrm{Diff}^r(\mathbf{S}^1)$ and its rotation number ρ is Diophantine of order δ, then any topological conjugacy between f and the rigid rotation of angle ρ is a $C^{r-1-\delta-\varepsilon}$ diffeomorphism for any $\varepsilon > 0$, provided $r > \delta + 2$.*

In this statement $r > 2$ belongs to $\mathbb{R}$, and the condition $f \in \mathrm{Diff}^r(\mathbf{S}^1)$ means that f is a $C^{\lfloor r \rfloor}$ diffeomorphism whose $\lfloor r \rfloor$-th derivative satisfies a Hölder condition with exponent $\{r\}$.

More recently, Khanin and Teplinsky (2009) were able to prove that, in the particular case $2 < r < 3$, rigidity holds without the need of any ε. More precisely, they proved the following result.

Theorem 4.11 (Khanin–Teplinsky) *If $f \in \mathrm{Diff}^{2+\alpha}(\mathbf{S}^1)$ and its rotation number ρ is Diophantine of order δ, then any topological conjugacy between f and the rigid rotation of angle ρ is a $C^{1+\alpha-\delta}$ diffeomorphism, provided $0 \leq \delta < \alpha < 1$.*

A detailed proof of Theorem 4.11, following the original work of Khanin and Teplinsky (2009), will be provided in Appendix B.

The previous statements are given for the lowest possible smoothness and are sharp, as the examples constructed in Katznelson and Ornstein (1989b, App. 3) show. The case of highest possible smoothness has been completely solved by Herman and Yoccoz.

Theorem 4.12 (Herman–Yoccoz) *Every C^∞ circle diffeomorphism with irrational rotation number $\rho \in (0, 1)$ is C^∞-conjugate to a rotation if, and only if, ρ is Diophantine.*

Theorem 4.13 (Herman–Yoccoz) *Any real analytic circle diffeomorphism with Diophantine rotation number is real analytically conjugate to the corresponding rigid rotation.*

Finally, we remark that in Yoccoz (2002, Section 2.5), Yoccoz introduced a set $\mathscr{H} \subset (0, 1)$ of irrational numbers, that contains all Diophantine numbers and is contained in the Brjuno class, which is sufficient and, in some sense, necessary to solve the global linearization problem in the real-analytic case. More precisely, Yoccoz (2002, Th. 1.4) proved the following result.

Theorem 4.14 (Yoccoz) *Any real-analytic diffeomorphism with irrational rotation number in $\mathscr{H}$ is real analytically conjugate to the corresponding rigid rotation. Moreover, given $\rho \notin \mathscr{H}$, there exists a real-analytic diffeomorphism with rotation number ρ which is not analytically linearizable.*

We refer the reader to the survey by Eliasson et al. (2018) for much more on Yoccoz's seminal contributions to the theory of circle diffeomorphisms (see also Yoccoz 1984a, 1995b,a, 2002).

Exercises

Exercise 4.1 If $f : S^1 \to S^1$ and $h : S^1 \to S^1$ are both C^1 diffeomorphisms, prove that $\mathscr{H}_1(h \circ f \circ h^{-1}) < \infty$ if and only if $\mathscr{H}_1(f) < \infty$.

Exercise 4.2 Let $\varphi : S_\sigma \to \mathbb{C}$ be holomorphic and periodic of period one.

(i) Show that there exists a unique holomorphic function $\phi : A_r \to \mathbb{C}$, where $r = e^{2\pi\sigma}$, such that $\varphi(z) = \phi(\exp(z))$ for all z.
(ii) Deduce from (i) and the Laurent series for ϕ that φ has a Fourier series expansion

$$\varphi(z) = \sum_{n \in \mathbb{Z}} \widehat{\varphi}(n) e^{2\pi i n z}$$

which is absolutely convergent in the strip S_σ, and that its Fourier coefficients are precisely the Laurent coefficients of ϕ.

Exercise 4.3 Prove Lemma 4.1. [*Hint:* Note that for all $n, m \in \mathbb{Z}$ we have $|e^{2\pi i(n\alpha - m)} - 1| = 2|\sin(\pi(n\alpha - m))$.]

Exercise 4.4 Prove Lemma 4.2. [*Hint:* Apply Cauchy's theorem to the holomorphic function $g_n(z) = \xi(z)e^{-2\pi i n z}$ in a suitable rectangle.]

Exercise 4.5 Let $F : \mathbb{R} \to \mathbb{R}$ be a homeomorphism given by $F(x) = x + \alpha + \varphi(x)$, and suppose that F has translation number α. Show that there exists $x_0 \in \mathbb{R}$ such that $\varphi(x_0) = 0$.

Exercise 4.6 Let $f : S^1 \to S^1$ be a homeomorphism with irrational rotation number. Show that for all $\alpha > 0$ small we have $\rho(R_\alpha \circ f) > \rho(f)$.

Exercise 4.7 Let $F : \mathbb{C} \to \mathbb{C}$ be an entire holomorphic map. Show that, if there exists $n \geq 1$ such that $F^n = \text{Id}$, then F is complex affine, i.e., it has the form $F(z) = az + b$.

Exercise 4.8 Prove Lemma 4.9.

Exercise 4.9 Recall from Dirichlet's Lemma 1.1 that for any irrational number $\rho \in (0, 1)$ there exist infinitely many rational numbers p/q such that

$$\left| \rho - \frac{p}{q} \right| < \frac{1}{q^2} . \tag{4.21}$$

Conclude from (4.21) that there are no Diophantine numbers of order $\delta < 0$.

Exercise 4.10 Show that an irrational number ρ is Diophantine of order δ if, and only if, there exists a constant $M > 0$ such that $q_{n+1} \leq M\, q_n^{1+\delta}$ for all $n \in \mathbb{N}$ (*Hint:* From (1.16) in Theorem 1.2 we have the estimates

$$\frac{1}{2\, q_{n+1}} < |q_n\, \rho - p_n| < \frac{1}{q_{n+1}} .$$

In particular, $|q_n\, \rho - p_n|\, q_n^{1+\delta} < q_n^{1+\delta}/q_{n+1}$. If ρ is Diophantine of order δ, then $|q_n\, \rho - p_n|\, q_n^{1+\delta} \geq C$ and we are done taking $M = 1/C$. On the other hand, consider $q \in \mathbb{Z}$ with $q_n < q < q_{n+1}$. As also showed in Theorem 1.2 (see (1.17)), $|q\rho - p| > |q_n\, \rho - p_n|$ for all $p \in \mathbb{Z}$. Since $\delta \geq 0$, we have $q^{1+\delta} > q_n^{1+\delta}$ and then $|q\rho - p|\, q^{1+\delta} > |q_n\, \rho - p_n|\, q_n^{1+\delta} > q_n^{1+\delta}/2\, q_{n+1}$. By assumption, this last ratio is bounded from below by the positive constant $1/2M$, and then ρ is Diophantine of order δ).

Exercise 4.11 Conclude from the previous exercise that an irrational number ρ is Diophantine of order 0 if, and only if, ρ is of *bounded type*: $\sup\{a_n(\rho)\}$ is finite (*Hint:* From the identity $q_{n+1} = a_n\, q_n + q_{n-1}$ we know that $a_n = \lfloor q_{n+1}/q_n \rfloor$).

Exercise 4.12 Show that any Diophantine number satisfies the Brjuno condition given in Definition 4.4 (*Hint:* Use Exercise 4.10 and the fact that the sequence $\{q_n\}$ grows at least exponentially fast as n goes to infinity).

Exercise 4.13 Fix some constant $\sigma \in (0, 1)$ and consider an irrational number $\rho = [a_0, a_1, \dots]$ such that

$$e^{a_n^\sigma} \leq a_{n+1} \leq e^{a_n}$$

for all $n \in \mathbb{N}$. Show that ρ is a Liouville number that satisfies the Brjuno condition (in other words, the inclusion given by Exercise 4.12 is a proper inclusion).

Part III
Multicritical Circle Maps

Chapter 5
Cross-Ratios and Distortion Tools

This chapter is to be regarded as an intermezzo. We want to move on to the study of homeomorphisms of the circle having one or more critical points.

The distortion techniques we used in our study of diffeomorphisms (bounded variation, boundedness of nonlinearity, the naive distortion lemma) are not immediately applicable to the study of maps having critical points. For instance, the nonlinearity of a map clearly explodes at a critical point.

A major breakthrough in one-dimensional dynamics achieved in the early eighties was the discovery that one could oftentimes understand the topology and/or the geometry of a one-dimensional map through a careful analysis of the way such map distorts *cross-ratios*. Several tools were introduced to control the distortion of cross-ratios. In the present chapter we will introduce some of these tools, which will then be used extensively in the next chapters.

5.1 Cross-Ratios

There are several types of cross-ratios used in one-dimensional dynamics. We describe here two of the most ubiquitous.

Let us denote by N either the unit circle S^1 or the real line $\mathbb{R}$. Given two intervals $M \subset T \subset N$ with M compactly contained in the interior of T, let us denote by L and R the two connected components of $T \setminus M$. We define the a-cross-ratio and the b cross-ratio of the pair (M, T), respectively, as follows:

$$a(M, T) = \frac{|M||T|}{|L||R|} \ ; \quad b(M, T) = \frac{|L||R|}{|L \cup M||M \cup R|} \ .$$

One easily checks that $b(M, T)^{-1} = 1 + a(M, T)$. Both cross-ratios are preserved by Möbius transformations; the latter is weakly contracted by maps with negative

E. de Faria, P. Guarino, *Dynamics of Circle Mappings*, IMPA Monographs,
https://doi.org/10.1007/978-3-031-67495-2_5

Schwarzian derivative (see below), whereas the former is weakly expanded (see Exercise 5.4)

Unlike, say, de Faria and de Melo (1999), where the a-cross-ratio was used throughout, in the present text it will often be more convenient to use the b-cross-ratio. The latter has the advantage that its logarithm is given by the *Poincaré length* of M inside T. More precisely,

$$\log b(M,T) = -\int_M \rho_T(x)\,dx ,$$

where $\rho_T(x)$ is the *Poincaré density* of $T = [\alpha, \beta]$, given by

$$\rho_T(x) = \frac{\beta - \alpha}{(x-\alpha)(\beta - x)} .$$

From now on, since the b-cross-ratio will be the cross-ratio most used in this book, we will simplify the notation a bit and write $[M, T]$ instead of $b(M, T)$.

We end this section with the following useful observation. Suppose $M = (b, c)$ and $T = (a, d)$ are such that $M \subset T$, and let ϕ be the Möbius transformation determined by $\phi(a) = 0$, $\phi(c) = 1$ and $\phi(d) = \infty$. Then

$$[M,T] = \phi(b) = \left(\frac{d-c}{c-a}\right)\left(\frac{b-a}{d-b}\right) .$$

5.2 The Schwarzian

The Schwarzian derivative is a somewhat mysterious object discovered at the end of the nineteenth century by H. A. Schwarz, in the context of complex-analytic function theory. Its use in one-dimensional dynamics was initiated by D. Singer (1978).

5.2.1 *Definition*

In Chap. 3 (see Definition 3.1) we introduced the concept of *nonlinearity* of a C^2 one-dimensional map f, namely $\mathcal{N} f = D \log Df = D^2 f / Df$. When f is C^3, we define its *Schwarzian derivative* to be

$$Sf = D(\mathcal{N} f) - \frac{1}{2}(\mathcal{N} f)^2 .$$

A simple computation yields the alternative formula

$$Sf = \frac{D^3 f}{Df} - \frac{3}{2}\left(\frac{D^2 f}{Df}\right)^2 .$$

Just as $\log Df$ and the nonlinearity, the Schwarzian derivative is a *dynamical cocycle*, i.e., it satisfies a chain rule: if f, g are C^3 maps for which $f \circ g$ makes sense, then

$$S(f \circ g) = Sg + Sf \circ g\,[Dg]^2 . \tag{5.1}$$

The chain rule (5.1) indicates that Sf behaves as a *quadratic differential* under smooth changes of coordinates; indeed the expression $g^*(Sf) = Sf \circ g\,[Dg]^2$ appearing in the right-hand side of (5.1) is the pull-back of Sf by g as a quadratic differential.

From (5.1) we easily deduce a chain rule for iterates, namely,

$$Sf^n = \sum_{j=0}^{n-1} Sf \circ f^j \left[Df^j\right]^2$$

Now, since the Schwarzian derivative is a differential operator, it is important to identify its kernel.

Proposition 5.1 *The kernel of the Schwarzian derivative is the group of Möbius transformations. In addition, if ϕ is a Möbius transformation and f is a C^3 map, then $S(\phi \circ f) = Sf$.*

Proof The fact that the Schwarzian derivative vanishes at Möbius transformations is a straightforward computation. On the other hand, given an increasing C^3 diffeomorphism ϕ, consider the C^2 map $g = (D\phi)^{-1/2}$. An easy computation shows that $S\phi = -2\,D^2g/g$. Hence ϕ has zero Schwarzian derivative if and only if $D^2 g$ vanishes identically. In other words, g must be affine, say $g(x) = ax + b$. But then $D\phi(x) = (ax+b)^{-2}$, and integrating we get

$$\phi(x) = -\frac{1}{a}\,\frac{1}{ax+b} + c ,$$

where c is a constant. This shows that ϕ is a fractional linear (i.e., Möbius) transformation, and the first assertion is proved. To prove the second assertion, it suffices to apply the chain rule for the Schwarzian, namely

$$S(\phi \circ f) = Sf + S\phi \circ f\,[Df]^2 .$$

If ϕ is Möbius, then $S\phi \equiv 0$, and therefore $S(\phi \circ f) = Sf$ as asserted. □

5.2.2 *Koebe's Nonlinearity Principle*

As we will see shortly, when the Schwarzian derivative of a C^3 one-dimensional map ϕ has a definite *sign*, then ϕ has a monotonic behaviour with respect to its action on cross-ratios, and one can control its distortion in certain places. The first result in this direction is known as *Koebe's nonlinearity principle.* It states that if the Schwarzian derivative of ϕ is non-negative, then the nonlinearity of ϕ on any interval sitting in the domain of ϕ with definite *space* on both sides is bounded by a constant that depends only on said space.

Let us be more precise. First, let us define what we mean by *space*. Given two intervals M, T in the domain of ϕ, with M compactly contained in the interior of T, let $L, R \subset T$ be the connected components of $T \setminus M$. The space of M inside T is defined to be the number

$$\tau = \min\left\{\frac{|L|}{|M|}, \frac{|R|}{|M|}\right\} .$$

Now we can state Koebe's nonlinearity principle as follows:

Proposition 5.2 (Koebe's Nonlinearity Principle) *Let $\phi : T \to \phi(T)$ be a C^3-diffeomorphism. If $S\phi(x) \geq 0$ for all $x \in T$, then $|\mathscr{N}\phi(x)| \leq 2/\tau$, where τ is the space of M inside T.*

Here, we will prove the following generalization of this principle, which first appeared in de Faria and de Melo (1999, Lem. A.3).

Proposition 5.3 *Given constants $B > 0$ and $\tau > 0$, there exists $K_{\tau,B} > 0$ such that the following holds. If ϕ is a C^3-diffeomorphism mapping an interval $I \supseteq [-\tau, 1+\tau]$ into the reals, and if $S\phi(t) \geq -B$ for all $t \in I$, then for all $t \in [0, 1]$ we have*

$$\left|\frac{\phi''(t)}{\phi'(t)}\right| \leq K_{\tau,B} . \tag{5.2}$$

Proof Writing $y = \phi''/\phi'$, so that $S\phi = y' - \frac{1}{2}y^2$, we have the differential inequality

$$y' \geq \frac{1}{2}y^2 - B .$$

Let $0 \leq t_0 \leq 1$ be a point where $|y(t)|$ attains its maximum in $[0, 1]$ and suppose that $y_0 = y(t_0)$ is such that $|y_0| > \sqrt{2B} = \beta$. If $z(t)$ is the solution of the differential equation

$$z' = \frac{1}{2}z^2 - B$$

with initial condition $z(t_0) = y_0$, then by a well-known comparison theorem[1] we must have $y(t) \geq z(t)$ for all $t \geq t_0$ and $y(t) \leq z(t)$ for all $t \leq t_0$. Now, if $y_0 > \beta$ then integration of the ODE yields the explicit formula

$$z(t) = \beta \frac{(y_0 + \beta) + (y_0 - \beta)e^{\beta(t-t_0)}}{(y_0 + \beta) - (y_0 - \beta)e^{\beta(t-t_0)}} .$$

Since this solution explodes at time

$$t_1 = t_0 + \frac{1}{\beta} \log\left(\frac{y_0 + \beta}{y_0 - \beta}\right),$$

so does $y(t)$. Hence $t_1 \notin I$, i.e., $t_1 - t_0 > \tau$, which gives us

$$\frac{\phi''(t_0)}{\phi'(t_0)} = y_0 < \beta \, \frac{e^{\beta\tau} + 1}{e^{\beta\tau} - 1} .$$

If instead $y_0 < -\beta$, then we get

$$z(t) = \beta \frac{(\beta + y_0) - (\beta - y_0)e^{\beta(t-t_0)}}{(\beta + y_0) + (\beta - y_0)e^{\beta(t-t_0)}} ,$$

and arguing as before for $t \leq t_0$ gives us

$$\frac{\phi''(t_0)}{\phi'(t_0)} = y_0 > -\beta \, \frac{e^{\beta\tau} + 1}{e^{\beta\tau} - 1} .$$

Therefore the lemma is proved if we take

$$K_{\tau,B} = \beta \frac{e^{\beta\tau} + 1}{e^{\beta\tau} - 1} .$$

□

Proof of Proposition 5.2 If $S\phi \geq 0$, then of course $S\phi \geq -B$ for every $B > 0$. Applying Proposition 5.3, we get the bound in (5.2) for each $B > 0$. But now it suffices to note that $K_{\tau,B} \to 2/\tau$ as $B \to 0$. This finishes the proof. We have recovered the classical Koebe principle. □

---

[1] Look up *Gronwall's inequality* in any good book on differential equations.

5.2.3 *The Minimum Principle*

Another important consequence of assuming that the Schwarzian of a given map has a definite sign is the following result, known as the Minimum Principle (*cf.* de Melo and van Strien 1993, Section II.6, Lemma 6.1).

Lemma 5.1 (Minimum Principle) *Let $\phi : T \to N$ be a C^3 diffeomorphism onto its image, where $T = [a, b] \subset N$ is a closed interval, and suppose ϕ has negative Schwarzian at all points of T. Then, for any given x in the interior of T, we have*

$$|D\phi(x)| > \min\{|D\phi(a)|, |D\phi(b)|\} . \tag{5.3}$$

In other words, $x \mapsto |D\phi(x)|$ does not have a local minimum inside T.

Proof Note that, since $D\phi$ never vanishes, the function $\varphi(x) = |D\phi(x)|$ is in fact smooth. Suppose this function has a point of local minimum x_0 lying in the interior of T. Then we must have $D^2\phi(x_0) = 0$, and this tells us that

$$0 > S\phi(x_0) = \frac{D^3\phi(x_0)}{Df(x_0)}$$

Hence $D\phi(x_0)$ and $D^3\phi(x_0)$ have opposite signs, and so there are two possibilities:

(i) If $D\phi(x_0) > 0$, then $D^3\phi(x_0) < 0$ and therefore x_0 is a point of local maximum for $x \mapsto D\phi(x)$; but since in this case we have $\varphi(x) = D\phi(x)$ for all x, it follows that x_0 is a point of local *maximum* for ϕ.
(ii) If $D\phi(x_0) < 0$, then $D^3\phi(x_0) > 0$ and therefore x_0 is a point of local minimum for $x \mapsto D\phi(x)$; but since in this case $\varphi(x) = -D\phi(x)$ for all x, it follows that x_0 is a point of local *maximum* for φ.

Therefore $\varphi(x) = |D\phi(x)|$ has no local minimum inside T, and this in particular implies (5.3). □

5.3 Distortion and Cross-Ratio Distortion

5.3.1 *Koebe's Distortion Principle*

Let $f : N \to N$ be a smooth map, and suppose we have an interval $T \subset N$ whose iterates up to a certain time k stay away from the critical points of f. The *Koebe distortion principle* states that the distortion of f^k restricted to a slightly smaller interval $M \subset T$ is bounded independently of k, where the bound depends solely on f, the amount of space that M has inside T, and the total sum of the lengths of the images of T up to time k. This principle is one of the most important tools in one-dimensional dynamics, and it will be used quite a few times in the chapters to come. To state it in a precise way, let us agree to say that an interval T contains a *τ-scaled neighborhood* of M if the space of M inside T is at least τ. Here is the formal statement.

Lemma 5.2 (Koebe Distortion Principle) *For each $\ell, \tau > 0$ and each map $f : N \to N$ there exists a constant $K = K(\ell, \tau, f) > 1$ with the following property. If $T \subset N$ is an interval such that $f^k|_T$ is a diffeomorphism onto its image and if it satisfies the summability condition*

$$\sum_{j=0}^{k-1} |f^j(T)| \leq \ell \,,$$

then for each interval $M \subset T$ for which $f^k(T)$ contains a τ-scaled neighborhood of $f^k(M)$ one has

$$\frac{1}{K} \leq \frac{|Df^k(x)|}{|Df^k(y)|} \leq K$$

for all $x, y \in M$.

The proof of this lemma can be found in de Melo and van Strien (1993, p. 295).

5.3.2 *Distortion and the Schwarzian*

The concept of *cross-ratio distortion* we are about to introduce has become fundamental in one-dimensional dynamics.

Let $f : N \to N$ be a continuous map, and let $U \subseteq N$ be an open set such that $f|_U$ is a homeomorphism onto its image. If $M \subset T \subset U$ are intervals, with M *compactly contained* in T (written $M \Subset T$), the *cross-ratio distortion* of the map f on the pair of intervals (M, T) is defined to be the ratio

$$\mathrm{CrD}(f; M, T) = \frac{[f(M), f(T)]}{[M, T]}.$$

If $f|_T$ is the restriction of a projective (Möbius) transformation, then one can easily see that $\mathrm{CrD}(f; M, T) = 1$.

Let us examine a few important properties of cross-ratio distortion. The first is that it satisfies a *chain rule*.

Lemma 5.3 (Chain Rule) *Let $f : N \to N$ and $U \subset N$ be as before. Given two intervals $M \Subset T \subset U$, and given $n \in \mathbb{N}$, we have*

$$\mathrm{CrD}(f^n; M, T) = \prod_{i=0}^{n-1} \mathrm{CrD}\left(f; f^i(M), f^i(T)\right).$$

Proof The proof is by direct computation using a simple telescoping trick—the details are left as an exercise. □

Also, when $f|_T$ is a diffeomorphism onto its image and $\log Df|_T$ has *bounded variation* in T, then an easy calculation using the mean value theorem shows that $\mathrm{CrD}(f; M, T) \leq e^{2V}$, where $V = \mathrm{Var}(\log Df|_T)$.

Now, if $f|_U$ is a diffeomorphism onto its image, we define $\delta_f : U \times U \to \mathbb{R}$ by

$$\delta_f(x, y) = \begin{cases} \log \dfrac{f(x) - f(y)}{x - y}, & \text{if } x \neq y \\ \\ \log Df(x), & \text{if } x = y \end{cases}$$

If f is C^3 then δ_f is C^2, and the following facts are straightforward (see also Exercise 5.5).

(i) For all $M \subset T \subset U$,

$$\log \mathrm{CrD}(f; M, T) = \iint_{M \times T} \frac{\partial^2 \delta_f}{\partial x \partial y} \, dxdy. \tag{5.4}$$

(ii) For all $x \in U$ we have

$$\lim_{y \to x} \frac{\partial^2 \delta_f}{\partial x \partial y}(x, y) = \frac{1}{6} Sf(x),$$

where Sf is the Schwarzian derivative of f.

Remark 5.1 The mixed partial derivative appearing in (5.4) is, up to a multiplicative constant, what one calls the *bi-Schwarzian* of f. More precisely, the bi-Schwarzian B_f is defined as

$$B_f(x, y) = 6 \frac{\partial^2 \delta_f}{\partial x \partial y}(x, y).$$

Clearly, $B_f(x, y) \to Sf(x)$ as $y \to x$, hence the name. The bi-Schwarzian is a cocycle, in the sense that it satisfies a chain rule: if f, g are C^3 maps for which $f \circ g$ makes sense, then $B_{f \circ g}(x, y) = g'(x)g'(y)B_f(g(x), g(y)) + B_g(x, y)$. This is entirely consistent with the chain rule for the Schwarzian, to wit,

$$S(f \circ g) = Sf \circ g \cdot [Dg]^2 + Sg.$$

Unlike the Schwarzian, which is used extensively, the bi-Schwarzian will not be used in the present book.

The cross-ratio is preserved by maps with zero Schwarzian derivative (since these are Moebius transformations, as we have seen in Proposition 5.1). As it turns out, it is weakly contracted by maps with negative Schwarzian derivative. This is the content of our next lemma.

Lemma 5.4 *If f is a C^3 diffeomorphism with $Sf < 0$, then for any two intervals $M \subset T$ contained in the domain of f we have* $\mathrm{CrD}(f; M, T) < 1$, *that is,* $\big[f(M), f(T)\big] < [M, T]$.

Proof The proof is the one given in de Melo and van Strien (1993, Section IV.1). Let $M = [b, c] \subseteq T = [a, d]$. Let us call L and R the two connected components of $T \setminus M$. Let ϕ be the (unique) Möbius transformation such that $\phi(f(a)) = a$, $\phi(f(b)) = b$ and $\phi(f(d)) = d$. Note that $\phi \circ f$ is a C^3 diffeomorphism with negative Schwarzian derivative, since $S(\phi \circ f) = Sf < 0$ by Proposition 5.1.

We claim that $\phi(f(c)) > c$. Indeed, if this is not true, then by the Mean Value Theorem there exist $z_0 \in [a, b]$, $z_1 \in [b, c]$ and $z_2 \in [c, d]$ such that

$$D(\phi \circ f)(z_0) = \frac{\phi(f(a)) - \phi(f(b))}{a - b} = 1,$$

$$D(\phi \circ f)(z_1) = \frac{\phi(f(c)) - \phi(f(b))}{c - b} \leq 1 \quad \text{and}$$

$$D(\phi \circ f)(z_2) = \frac{\phi(f(d)) - \phi(f(c))}{d - c} \geq 1.$$

If $z_1 \in (z_0, z_2)$, the previous inequalities contradict the Minimum Principle for diffeomorphisms with negative Schwarzian derivative.[2] Therefore, $\phi(f(c)) > c$ as claimed. With this at hand we get:

$$\mathrm{CrD}(\phi \circ f; M, T) = \frac{\big[\phi(f(M)), \phi(f(T))\big]}{[M, T]} = \frac{|M \cup L|\, \big|\phi\big(f(c)\big) - d\big|}{|R|\, \big|a - \phi\big(f(c)\big)\big|} < 1\,.$$

Since ϕ is a Möbius transformation, $\mathrm{CrD}(\phi \circ f; M, T) = \mathrm{CrD}(f; M, T)$ and the lemma is proved. □

[2] In the special case $z_1 = z_0$, we obtain $z_1 = z_0 = b$, and then $D(\phi \circ f)(b) = 1$ and $\phi(f(c)) = c$. This implies that $D(\phi \circ f)(c) < 1$ (otherwise, the Minimum Principle would imply that $D(\phi \circ f)(x) > 1$ for all $x \in (b, c)$, which is impossible since $\phi \circ f$ fixes both b and c). Again, this contradicts the Minimum Principle since $c \in (b, z_2)$. The remaining case $z_1 = z_2$ is analogous.

5.3.3 Behavior Near Critical Points

The circle maps we are interested in from now onwards possess critical points—more specifically, *non-flat* critical points. Here is what we mean by non-flat.

Definition 5.1 We say that a critical point c of a C^r one-dimensional map f is *non-flat* of degree $d > 1$ if there exists a neighborhood W of the critical point such that $f(x) = f(c) + \phi(x)\left|\phi(x)\right|^{d-1}$ for all $x \in W$, where $\phi : W \to \phi(W)$ is a C^r diffeomorphism such that $\phi(c) = 0$. The number d is also called the *criticality*, the *type* or the *order* of c.

Example 5.1 Every critical point of a real-analytic map is non-flat, and its criticality must be a positive integer.

The following proposition clarifies the geometric behavior of a map near a non-flat critical point. It shows, among other things, that the Schwarzian derivative is always negative around such a critical point.

Proposition 5.4 *Given a C^3 map f with a non-flat critical point c of criticality $d > 1$, there exists a neighborhood $U \subseteq W$ of c such that*

(i) f has negative Schwarzian derivative on $U \setminus \{c\}$. More precisely, there exists $K = K(f) > 0$ such that for all $x \in U \setminus \{c\}$ we have

$$Sf(x) < -\frac{K}{(x-c)^2}\,.$$

(ii) There exist constants $0 < \alpha < \beta$ such that for all $x \in U$

$$\alpha|x-c|^{d-1} < Df(x) < \beta|x-c|^{d-1}.$$

Moreover, α and β can be chosen so that $\beta < (3/2)\alpha$.

(iii) Given a non-empty interval $J \subseteq U$ and $x \in J$ we have

$$Df(x) \leq 3d\,\frac{|f(J)|}{|J|}\,.$$

(iv) Given two non-empty intervals $M \subseteq T \subseteq U$ we have

$$\mathrm{CrD}(f; M, T) \leq 9d^2\,.$$

Proof From Definition 5.1, there exists a neighborhood of the critical point c such that $f(x) = g\big(\phi(x)\big) + f(c)$, where g is the map given by

$$g(x) = \begin{cases} x^d & \text{if } x > 0 \\ -(-x)^d & \text{if } x < 0 \,, \end{cases}$$

and ϕ is a C^3 diffeomorphism with $\phi(c) = 0$. A simple computation shows that for all $x \neq 0$ we have

$$Sg(x) = -\frac{(d^2-1)}{2x^2} \,. \tag{5.5}$$

We proceed to the proof of all four assertions in the statement of our proposition.

(i) The chain rule for the Schwarzian derivative gives $Sf = Sg(\phi)(D\phi)^2 + S\phi$. From (5.5), we get:

$$Sg(\phi(x))(D\phi(x))^2 = -\frac{1}{2}(d-1)(d+1)\left(\frac{D\phi(x)}{\phi(x)}\right)^2 \leq -\frac{A}{(\phi(x))^2},$$

where $A = \frac{1}{2}(d^2-1)\min_x \left|D\phi(x)\right| > 0$. In particular:

$$Sf(x) < \frac{-A + S\phi(x)\big(\phi(x)\big)^2}{\big(\phi(x)\big)^2} \,.$$

On the other hand, since ϕ is a diffeomorphism, $|S\phi(x)| < M$ for some $M > 0$. Then we can choose $\delta > 0$ such that for all $x \in (c-\delta, c+\delta)$ we have $|\phi(x)| < \sqrt{\frac{A}{M}}$, and this implies that $Sf < 0$ in $(c-\delta, c+\delta) \setminus \{c\}$. Finally, since ϕ is bi-Lipschitz we have $|\phi(x)| \asymp |x-c|$ and this proves (i).

(ii) This follows at once from Taylor's formula, since:

$$\lim_{x\to c}\left(\frac{Df(x)}{|x-c|^{d-1}}\right) = d(D\phi(c))^d > 0 \,.$$

(iii) With (ii) at hand the proof of (iii) goes as follows. Let $J = (a, b) \subseteq U$. By symmetry it is enough to consider the following two cases:

(a) We have $c \leq a < b$. In this case, given any $x \in (a, b)$, we see that

$$\begin{aligned} \frac{Df(x)|J|}{|f(J)|} &\leq \frac{\beta(x-c)^{d-1}(b-a)}{\alpha\int_a^b (t-c)^{d-1}dt} \\ &\leq \left(\frac{\beta d}{\alpha}\right)\frac{(b-c)^{d-1}(b-c-a+c)}{(b-c)^d-(a-c)^d} \end{aligned}$$

$$= \left(\frac{\beta d}{\alpha}\right)\left(1 + \frac{(a-c)^d - (b-c)^{d-1}(a-c)}{(b-c)^d - (a-c)^d}\right)$$
$$\leq \frac{\beta d}{\alpha} < 3d/2 \,.$$

(b) We have $a < c < b$. Without loss of generality, we may assume that $|a-c| < |c-b|$. If $x \in J$, then

$$\frac{Df(x)|J|}{|f(J)|} \leq \frac{\beta|x-c|^{d-1}|b-a|}{\int_c^b Df(t)\,dt}$$
$$\leq \frac{2\beta|b-c|^d}{\int_c^b \alpha(t-c)^{d-1}dt} = \frac{2\beta d}{\alpha} < 3d \,.$$

(iv) Finally, let us call L, R the two connected components of $T \setminus M$. By the Mean Value Theorem there exist $z_0 \in L$ and $z_1 \in R$ such that

$$\mathrm{CrD}(f; M, T) = \frac{Df(z_0)\,Df(z_1)\,|L \cup M|\,|M \cup R|}{\big|f(L \cup M)\big|\,\big|f(M \cup R)\big|} \,.$$

Since $z_0 \in L \cup M$ and $z_1 \in R \cup M$ we deduce from (iii) that

$$\mathrm{CrD}(f; M, T) \leq (3d)^2.$$

□

Remark 5.2 Using property (ii) above, it is not difficult to see that, when f is injective, there exists a constant $\gamma = \gamma(f) > 0$ such that, for any two points in the domain of f with $|x-c| \leq |y-c|$, we have

$$\frac{|f(x)-f(c)|}{|f(y)-f(c)|} \leq \gamma \left(\frac{|x-c|}{|y-c|}\right)^d \,.$$

This remark will be used in the proof of Proposition 6.1.

5.4 The Cross-Ratio Inequality

One of the main reasons why cross-ratio distortion is a useful tool in one-dimensional dynamics is the *Cross-ratio Inequality*. Various essentially equivalent formulations of this tool were given during the eighties. The reader will find extensive material on this topic in de Melo and van Strien (1993, Ch. IV).

Our purpose in this section is to prove the following version of the Cross-ratio Inequality which, apart from notational differences, is essentially the one in Świątek (1988). First, let us introduce a useful terminology. As before, we denote by N either the unit circle or the real line. Given a family of intervals $\mathscr{F}$ in N and a positive integer m, we say that $\mathscr{F}$ has *multiplicity of intersection at most* m if each $x \in \mathbf{S}^1$ belongs to at most m elements of $\mathscr{F}$.

Theorem 5.1 (Cross-Ratio Inequality) *If $f : \mathbf{S}^1 \to \mathbf{S}^1$ is a C^3 strictly monotone smooth map all of whose critical points are non-flat, there exists a constant $C > 1$, depending only on f, such that the following holds. If $M_i \Subset T_i \subset \mathbf{S}^1$, where i runs through some finite set of indices $\mathscr{I}$, are intervals on the circle such that the family $\{T_i : i \in \mathscr{I}\}$ has multiplicity of intersection at most m, then*

$$\prod_{i \in \mathscr{I}} \mathrm{CrD}(f; M_i, T_i) \leq C^m . \tag{5.6}$$

This theorem was first obtained by Yoccoz in a slightly different form involving a certain degenerate cross-ratio, see Yoccoz (1984b, Section 4). The specific version stated above can be found in Świątek (1988, Section 2). We provide only a sketch of the proof, and the reader is invited to fill in the details as an exercise.

Proof of Theorem 5.1 Let $\mathscr{U} = \bigcup W_i$, where the W_i's are as in Definition 5.1, and let $\mathscr{V}$ be an open set with $\mathscr{U} \cup \mathscr{V} = \mathbf{S}^1$ whose closure does not contain any critical point of f. We assume without loss of generality that the maximum length of the T_i's is smaller than the Lebesgue number of the covering $\{\mathscr{U}, \mathscr{V}\}$. Write the product on the left-hand side of (5.6) as $P_1 \cdot P_2$, where

$$P_1 = \prod_{T_i \subseteq \mathscr{V}} \mathrm{CrD}(f; M_i, T_i) \quad , \quad P_2 = \prod_{T_i \subseteq \mathscr{U}} \mathrm{CrD}(f; M_i, T_i) .$$

Then on the one hand we claim that $P_1 \leq e^{2mV}$, where $V = \mathrm{var}(\log Df|_{\mathscr{V}})$. Indeed:

$$\begin{aligned} \log P_1 &= \sum_{T_i \subseteq \mathscr{V}} \log \mathrm{CrD}(f; M_i, T_i) \qquad (5.7) \\ &= \sum_{T_i \subseteq \mathscr{V}} \log Df(w_i) - \log Df(x_i) + \log Df(y_i) - \log Df(z_i) \leq 2mV , \end{aligned}$$

where the points w_i, x_i, y_i and z_i belong to T_i and are given by the Mean Value Theorem. On the other hand, the factors making up P_2 are of two types: those such that $f|_{T_i}$ is a diffeomorphism onto its image, and those such that T_i contains some critical point of f. By Proposition 5.4, all factors of the first type have negative Schwarzian and therefore, by Lemma 5.4, satisfy $\mathrm{CrD}(f; M_i, T_i) < 1$. Factors of the second type are easily controlled by the non-flatness condition: $\mathrm{CrD}(f; M_i, T_i) \leq 9d^2$, where $d > 1$ is the order of the critical point that belongs to T_i (again, see Proposition 5.4). Since there are at most mN such factors (where N is the number of critical points of f), the result follows. For more details, see Świątek (1988, Section 2). □

When used in combination with the chain rule (Lemma 5.3), Theorem 5.1 is a great tool for estimating the cross-ratio distortion of large iterates of multicritical circle maps (to be defined in the next chapter).

5.5 A Cancellation Lemma

In this final section of Chap. 5 we state and prove a technical result called *the cancellation lemma* (see Lemma 5.7 below), which is due to Świątek (1992). We will not provide specific applications of Lemma 5.7 in this book, rather we refer to the original paper by Świątek (but see some remarks after the proof of Lemma 5.7). Its proof is a nice illustration of the power of some of the tools we have presented in this chapter, such as the Schwarzian derivative and cross-ratio distortion.

Let $\mathscr{X} = \mathrm{Diff}^3_+\big([0,1]\big)$ be the group (under composition) of orientation-preserving C^3 diffeomorphisms of $[0,1]$ fixing the boundary, and let $\mathscr{Y} = \mathrm{Diff}^3_+\big(\mathbb{R}\big)$ be the group (under composition) of orientation-preserving C^3 diffeomorphisms of the real line. There is a natural group isomorphism between $\mathscr{X}$ and $\mathscr{Y}$. Indeed, consider first the real-analytic diffeomorphism $\psi : \mathbb{R} \to (0,1)$ given by $\psi(x) = 1/(1+e^{-x})$, whose inverse $\psi^{-1} : (0,1) \to \mathbb{R}$ is given by $\psi^{-1}(y) = \log\big(y/(1-y)\big)$, and then consider the isomorphism $\Psi : \mathscr{X} \to \mathscr{Y}$ given by $\Psi(f) = \psi^{-1} \circ f \circ \psi$.

$$\begin{array}{ccc} \mathbb{R} & \xrightarrow{\Psi(f)} & \mathbb{R} \\ {\scriptstyle\psi}\big\downarrow & & \big\downarrow{\scriptstyle\psi} \\ (0,1) & \xrightarrow[f]{} & (0,1) \end{array}$$

The isomorphism Ψ is natural from the *hyperbolic geometry* viewpoint: given $x < y$ in $(0,1)$ write $(0,1) = L \cup M \cup R$, where $M = (x,y)$ and L and R are the connected components of $(0,1) \setminus M$, and define a distance d_{hyp} between x and y by

$$d_{hyp}(x,y) = -\log\,[M,T] = -\log\left(\frac{|L|}{|L \cup M|}\,\frac{|R|}{|M \cup R|}\right).$$

In other words,

$$d_{hyp}(x,y) = -\log\left(\frac{x}{1-x}\,\frac{1-y}{y}\right).$$

It is easy to see that ψ is an *isometry*[3] between the Euclidean distance in the real line and the hyperbolic distance in $(0, 1)$. By Lemma 5.4, elements in $\mathscr{X}$ with non-negative Schwarzian derivative weakly expand the cross-ratio $[M, T]$, and then they weakly contract the hyperbolic distance. This gives us the following fact.

Lemma 5.5 *Let $f \in \mathscr{X}$ with $Sf \geq 0$. Then $0 < D\big(\Psi(f)\big)(x) \leq 1$ for all $x \in \mathbb{R}$.*

The isomorphism Ψ identifies the family $\{T_\lambda\}_{\lambda \in \mathbb{R}} \subset \mathscr{Y}$ of translations of the real line, $T_\lambda(x) = x + \lambda$ for any $x \in \mathbb{R}$, with the family $\{M_\lambda\}_{\lambda \in \mathbb{R}} \subset \mathscr{X}$ of Möbius transformations

$$M_\lambda(x) = \frac{x}{(1 - e^{-\lambda})\, x + e^{-\lambda}} \quad \text{for } x \in (0, 1).$$

Indeed, note that $\big(M_\lambda \circ \psi\big)(x) = \psi(x + \lambda)$ for any $x \in \mathbb{R}$ and any $\lambda \in \mathbb{R}$. In particular, $\{M_\lambda\}_{\lambda \in \mathbb{R}}$ is *abelian* under composition, and $M_{\lambda_1} \circ M_{\lambda_2} = M_{\lambda_1 + \lambda_2}$ for all $\lambda_1, \lambda_2 \in \mathbb{R}$, as well as $M_\lambda^{-1} = M_{-\lambda}$ for all $\lambda \in \mathbb{R}$.

Let us point out another nice property of Ψ.

Lemma 5.6 *For any $f, g \in \mathscr{X}$ we have $\|f - g\|_{C^0([0,1])} \leq \|\Psi(f) - \Psi(g)\|_{C^0(\mathbb{R})}$.*

Note that Lemma 5.6 follows at once from the fact that

$$D\psi(x) = \frac{1}{(1 + e^x)(1 + e^{-x})} \in (0, 1/4] \quad \text{for all } x \in \mathbb{R}\,.$$

The following *cancellation lemma* is due to Świątek (1992), and is the main result of Sect. 5.5.

Lemma 5.7 *Let $\{\lambda_n\} \subset \mathbb{R}$ and let $\{\phi_n\} \subset \mathscr{X}$ be such that $S\phi_n \geq 0$ for all $n \in \mathbb{N}$. Then we have*

$$\big\| M_{\lambda_n} \circ \phi_n \circ \cdots \circ M_{\lambda_1} \circ \phi_1 \;-\; \phi_n \circ \cdots \circ \phi_1 \big\|_{C^0([0,1])} \leq 2 \max_{1 \leq j \leq n} \left| \sum_{i=1}^{j} \lambda_i \right|$$

for all $n \in \mathbb{N}$.

Our proof of Lemma 5.7 follows the original paper by Świątek (1992, Section 3, pages 91–93).

Proof of Lemma 5.7 By Lemma 5.6, it is enough to prove that

[3] Actually, the diffeomorphism ψ extends to a *biholomorphism* between the $\mathbb{R}$-symmetric strip $\big\{z \in \mathbb{C} : -\pi/2 < \operatorname{Im} z < \pi/2\big\}$ and the $\mathbb{R}$-symmetric open disc with diameter $(0, 1)$. In particular, the distance d_{hyp} defined above coincides with the standard Poincaré distance on $(0, 1)$.

$$\left\|\Psi(M_{\lambda_n}\circ\phi_n\circ\cdots\circ M_{\lambda_1}\circ\phi_1)-\Psi(\phi_n\circ\cdots\circ\phi_1)\right\|_{C^0(\mathbb{R})}\leq 2\max_{1\leq j\leq n}\left|\sum_{i=1}^{j}\lambda_i\right| .$$

For each $t\in[0,1]$ consider $\eta_n^t\in\mathscr{Y}$ defined by

$$\eta_n^t=T_{t\,\lambda_n}\circ\Psi(\phi_n)\circ T_{t\,\lambda_{n-1}}\circ\Psi(\phi_{n-1})\circ\cdots\circ T_{t\,\lambda_2}\circ\Psi(\phi_2)\circ T_{t\,\lambda_1}\circ\Psi(\phi_1) .$$

In other words, for any $x\in\mathbb{R}$ we have

$$\eta_1^t(x)=\Psi(\phi_1)(x)+t\,\lambda_1\quad\text{and}\quad\eta_{n+1}^t(x)=\Psi(\phi_{n+1})\big(\eta_n^t(x)\big)+t\,\lambda_{n+1} .$$

Note that

$$\eta_n^0=\Psi(\phi_n\circ\phi_{n-1}\circ\cdots\circ\phi_2\circ\phi_1)$$

and also that

$$\eta_n^1=\Psi(M_{\lambda_n}\circ\phi_n\circ M_{\lambda_{n-1}}\circ\phi_{n-1}\circ\cdots\circ M_{\lambda_2}\circ\phi_2\circ M_{\lambda_1}\circ\phi_1) .$$

In particular, for all $x\in\mathbb{R}$ and $n\in\mathbb{N}$ we have

$$\begin{aligned}&\left|\big(\Psi(M_{\lambda_n}\circ\phi_n\circ\cdots\circ M_{\lambda_1}\circ\phi_1)-\Psi(\phi_n\circ\cdots\circ\phi_1)\big)(x)\right|\\&\qquad=\left|\left(\eta_n^1-\eta_n^0\right)(x)\right|\leq\max_{t\in[0,1]}\left|\frac{\partial\,\eta_n^t}{\partial\,t}(x)\right| .\end{aligned}$$

To bound these derivatives, note that

$$\frac{\partial\,\eta_1^t}{\partial\,t}(x)=\lambda_1\quad\text{and}\quad\frac{\partial\,\eta_{n+1}^t}{\partial\,t}(x)=D\Psi(\phi_{n+1})\big(\eta_n^t(x)\big)\,\frac{\partial\,\eta_n^t}{\partial\,t}(x)+\lambda_{n+1} ,$$

from which it follows that

$$\frac{\partial\,\eta_{n+1}^t}{\partial\,t}(x)=\lambda_{n+1}+\sum_{j=1}^{n}\lambda_j\prod_{i=j}^{n}D\Psi(\phi_{i+1})\left(\eta_i^t(x)\right)$$

for all $x\in\mathbb{R}$, $t\in[0,1]$ and $n\geq 1$. Now, for each $x\in\mathbb{R}$, $t\in[0,1]$ and $n\in\mathbb{N}$, define $\beta_1,\beta_2,\ldots,\beta_{n+1}\in(0,1]$ by setting $\beta_{n+1}=1$ and

$$\beta_j=\prod_{i=j}^{n}D\Psi(\phi_{i+1})\left(\eta_i^t(x)\right) ,\quad\forall\,1\leq j\leq n .$$

With this notation we have

$$\frac{\partial \eta_{n+1}^{t}}{\partial t}(x) = \sum_{j=1}^{n+1} \lambda_j \beta_j .$$

Therefore, we need to prove that $\left|\sum_{j=1}^{n} \lambda_j \beta_j\right|$ is bounded by $2M$, where

$$M = \max\left\{\left|\sum_{i=1}^{j} \lambda_i\right| : 1 \leq j \leq n\right\} .$$

To do that, let us write

$$\sum_{j=1}^{n} \lambda_j \beta_j = \beta_n \sum_{i=1}^{n} \lambda_i - \sum_{j=1}^{n-1} \left[(\beta_{j+1} - \beta_j) \sum_{i=1}^{j} \lambda_i \right] .$$

Since $\beta_n \in (0, 1]$, we have $|\beta_n \sum_{i=1}^{n} \lambda_i| = \beta_n \left|\sum_{i=1}^{n} \lambda_i\right| \leq \left|\sum_{i=1}^{n} \lambda_i\right| \leq M$. Moreover, since $S\phi_n \geq 0$ for all $n \in \mathbb{N}$, we know from Lemma 5.5 that the sequence $\{\beta_1, \beta_2, \ldots, \beta_n, \beta_{n+1}\} \subset (0, 1]$ is non-decreasing:

$$\beta_j(x, t, n) = D\Psi(\phi_{j+1})\big(\eta_j^t(x)\big)\, \beta_{j+1}(x, t, n) \leq \beta_{j+1}(x, t, n) .$$

Therefore,

$$\begin{aligned}
\left|\sum_{j=1}^{n-1} \left[(\beta_{j+1} - \beta_j) \sum_{i=1}^{j} \lambda_i \right]\right| &\leq \sum_{j=1}^{n-1} \left| (\beta_{j+1} - \beta_j) \sum_{i=1}^{j} \lambda_i \right| \\
&= \sum_{j=1}^{n-1} (\beta_{j+1} - \beta_j) \left| \sum_{i=1}^{j} \lambda_i \right| \\
&\leq M \sum_{j=1}^{n-1} \big(\beta_{j+1} - \beta_j\big) = M\,(\beta_n - \beta_1) \leq M.
\end{aligned}$$

□

Remark 5.3 The *monotonicity* of $\{\beta_n\}$ is crucial in the proof. Indeed, consider $\lambda_n = (-1)^n/\sqrt{n}$ and $\beta_n = 1+\lambda_n$. Then $\{\beta_n\} \to 1$ and $\sum \lambda_n$ is finite, but $\sum \lambda_n \beta_n$ is unbounded. This is the reason why the non-negative Schwarzian condition is needed in Lemma 5.7.

Let $A = C^0\big([0, 1]\big)$ be the space of continuous functions from $[0, 1]$ to the real line, and recall that A is a Banach space when endowed with the sup norm. We can consider a homeomorphism from $\mathscr{X}$ onto A, called the *nonlinearity* function (see also Sect. 12.4), defined by

$$\mathcal{N} f = \frac{D^2 f}{Df} = D \log Df \,.$$

We then define the *weight* $\omega(f)$ of any given $f \in \mathscr{X}$ as

$$\omega(f) = \int_0^1 \mathcal{N} f = \log \frac{Df(1)}{Df(0)}\,,$$

which is a homomorphism from $\mathscr{X}$ onto (the additive group) $\mathbb{R}$, i.e.,

$$\omega(f_1 \circ f_2 \circ \cdots \circ f_n) = \sum_{i=1}^{n} \omega(f_i)$$

for $\{f_1, f_2, \ldots, f_n\} \subset \mathscr{X}$. The weight of an element of $\mathscr{X}$ carries its *signed* total distortion. The main point of the cancellation lemma (Lemma 5.7) is that it provides a bound in terms of the sum of the weights of the Möbius transformations involved (note that $\omega(M_\lambda) = -2\,\lambda$ for any $\lambda \in \mathbb{R}$), thus *allowing cancellations*. For instance, similar to what we did in the proof of the *Cross-ratio Inequality* above, we may consider a long composition of *backwards* iterates of a certain map f (under the same hypothesis of Theorem 5.1). Iterates around the critical point (those related to the product P_2) will have non-negative Schwarzian derivative, while iterates disjoint from the critical neighborhoods (related to the product P_1) might be close to Möbius transformations (identifying the same intervals, and having the same weight). If the weights of these iterates almost cancel (even if the sum of their absolute values is not small), the cancellation lemma says that we still get an efficient approximation of the whole composition if we replace the Möbius transformations involved just by affine maps (identifying the same intervals). This is a rather technical but useful result, and we refer the reader to the original paper by Świątek (1992) for the implementation of these ideas.

Exercises

Exercise 5.1 Let f be a C^3 map into the reals defined in a neighborhood of 0, which we assume is a regular point for f.

(i) Prove that there exists a unique fractional linear transformation ψ such that

$$\lim_{x \to 0} \frac{\psi \circ f(x) - x}{x^3}$$

exists and is finite.

(ii) Show that the limit in (i) is in fact equal to $\frac{1}{6} Sf(0)$.

(iii) For each $h > 0$, write $M_h = [h, 2h]$ and $T_h = [0, 3h]$, and let $A_f(h)$ be given by

$$A_f(h) = \frac{a(f(M_h), f(T_h))}{a(M_h, T_h)} .$$

Show that

$$Sf(0) = -\frac{3}{2} \lim_{h \to 0} \frac{A_f(h) - h}{h^2} .$$

(iv) Find a similar formula to the one in (iii) in terms of distortion of the b-cross-ratio rather than that of the a-cross-ratio.

Exercise 5.2 (Constant Negative Schwarzian) Given a positive constant α, consider $f_\alpha : \mathbb{R} \to \mathbb{R}$ defined as

$$f_\alpha(x) = \frac{\tanh\left(\sqrt{\alpha/2}\, x\right)}{\tanh\left(\sqrt{\alpha/2}\right)} = \frac{1}{\tanh\left(\sqrt{\alpha/2}\right)} \frac{e^{\sqrt{2\alpha}\, x} - 1}{e^{\sqrt{2\alpha}\, x} + 1} .$$

Show that f_α is a real analytic diffeomorphism onto its image, fixing -1, 0 and 1, and such that $Sf_\alpha = -\alpha$ on the whole real line.

Exercise 5.3 Let $f : S^1 \to S^1$ be a C^3 circle diffeomorphism with irrational rotation number, and let μ_2 be its (unique) 2-automorphic measure (recall Sect. 3.4.3). Show that the sequence

$$\left\{ \frac{1}{n} \int_{S^1} Sf^n \, d\mu_2 \right\}_{n \in \mathbb{N}}$$

is constant, equal to $\int_{S^1} Sf \, d\mu_2$.

Exercise 5.4 Prove that cross-ratios are preserved by fractional linear (i.e., Möbius) transformations.

Exercise 5.5 Let f be a C^3 diffeomorphism between intervals, and recall from Sect. 5.3.2 that the bi-Schwarzian B_f has been defined as

$$B_f(x, y) = 6 \frac{\partial^2 \delta_f}{\partial x \partial y}(x, y).$$

Show that $B_f(x, y) \to Sf(x)$ as $y \to x$ (*Hint:* A straightforward computation gives

$$\frac{\partial^2 \delta_f}{\partial x \partial y}(x, y) = \frac{f'(x)\, f'(y)}{\left(f(x) - f(y)\right)^2} - \frac{1}{(x - y)^2} .$$

Write $f(y)$ and $f'(y)$ as Taylor expansions around x, and take limit).

Exercise 5.6 Prove the chain rule for the bi-Schwarzian.

Exercise 5.7 Let $n \geq 1$ and let f be a polynomial of degree $n + 1$ with real coefficients. Suppose that all zeros of Df are real, so that $Df(x) = c \prod_{i=1}^{n}(x - \alpha_i)$, where $c, \alpha_1, \ldots, \alpha_n \in \mathbb{R}$.

(i) Show that

$$Sf(x) = 2 \sum_{1 \leq i < j \leq n} \frac{1}{(x - \alpha_i)(x - \alpha_j)} - \frac{3}{2} \left[\sum_{i=1}^{n} \frac{1}{x - \alpha_i} \right]^2 .$$

(ii) Deduce from (i) that $Sf < 0$.

Exercise 5.8 Consider the sequence $\{f_n\}_{n \geq 1} \subset \mathscr{X} = \mathrm{Diff}_+^3([0, 1])$ of Möbius transformations given by

$$f_n(x) = \frac{x}{(1 - e^{1/n})\, x + e^{1/n}} .$$

(i) Note that $\{f_n\}$ converges to the identity in $\mathscr{X}$.
(ii) Show that the sequence $\left\{ \bigcirc_{n=1}^{N} f_n \right\}_{N \geq 1}$ has no limit in $\mathscr{X}$.

Chapter 6
Topological Classification and the Real Bounds

In this chapter we go beyond the theory of circle diffeomorphisms and begin the study of topological and geometric properties of smooth circle homeomorphisms having *critical points*. These dynamical systems are called *multicritical circle maps* (see Definition 6.1 below), and will be the main object of study in the remainder of this book.

After introducing some classical examples, we will prove that multicritical circle maps with irrational rotation number are topologically conjugate to a rotation (Theorem 6.2). This theorem is due to J.-C. Yoccoz (1984b), and is an extension of Denjoy's Theorem from Chap. 3. The proof of this result, to be given in Sect. 6.2, relies on the distortion tools presented in Chap. 5.

In Sect. 6.3 we state and prove one of the most fundamental results in this book: *the real a-priori bounds* (Theorem 6.3), first proved in the eighties by Herman (1988) and Świątek (1988). We would like to remark that the Cross-ratio Inequality, namely Theorem 5.1, will play a major role in our proof of the real bounds. Theorem 6.3 (see also Theorem 6.4) is a cornerstone in the geometrical study of multicritical circle maps, and it will be invoked throughout the book.

We will close Chap. 6 with some of the first consequences of the real bounds, such as the C^1-bounds and the negative Schwarzian property (see Sects. 6.4 and 6.5).

6.1 Definition and Examples of Multicritical Circle Maps

Let us start by defining the maps which will be the main object of study in the present chapter and beyond. The reader should make sure to recall the notion of *non-flat critical point* introduced in Chap. 5 (Definition 5.1).

E. de Faria, P. Guarino, *Dynamics of Circle Mappings*, IMPA Monographs,
https://doi.org/10.1007/978-3-031-67495-2_6

Definition 6.1 A *multicritical circle map* is an orientation preserving C^3 circle homeomorphism having $N \geq 1$ critical points, all of which are non-flat.

Being a homeomorphism, a multicritical circle map f has a well defined rotation number $\rho \in (0, 1)$. We will assume that ρ is irrational, in which case it follows from Theorem 2.3 that there exists a *unique* f-invariant Borel probability measure μ.

Definition 6.2 We define the *signature* of f to be the $(2N + 2)$-tuple

$$(\rho\,;N;\ d_0, d_1, \ldots, d_{N-1};\ \delta_0, \delta_1, \ldots, \delta_{N-1}),$$

where d_i is the criticality of the critical point c_i for $0 \leq i \leq N - 1$, and $\delta_i = \mu[c_i, c_{i+1})$ (with the convention that $c_N = c_0$).

In this section we provide some interesting families of real-analytic critical circle maps.

6.1.1 Blaschke Products

Conforming with standard notation, we denote by $\widehat{\mathbb{C}} = \mathbb{C} \cup \{\infty\}$ the Riemann sphere. Consider the two-parameter family $f_{a,\omega} : \widehat{\mathbb{C}} \to \widehat{\mathbb{C}}$ of *Blaschke products* in the Riemann sphere $\widehat{\mathbb{C}}$ given by:

$$f_{a,\omega}(z) = e^{2\pi i \omega} z^2 \left(\frac{z - a}{1 - az}\right) \quad \text{for } a \geq 3 \text{ and } \omega \in [0, 1). \tag{6.1}$$

As it happens with any Blaschke product, every map in this family commutes with the geometric involution around the unit circle $\Phi(z) = 1/\bar{z}$ (note that Φ is the identity in the unit circle), and therefore it leaves invariant the unit circle (in fact, every rational map leaving invariant the unit circle *is* a Blaschke product). Moreover, its restriction to $\boldsymbol{S}^1$ is a real-analytic homeomorphism (the fact that $f_{a,\omega}$ has topological degree one, when restricted to the unit circle, follows from the Argument Principle since it has two zeros and one pole in the unit disk). When $a > 3$, each $f_{a,\omega}$ has four critical points in the Riemann sphere, which are all different and non-degenerate (quadratic), given by $0, \infty$,

$$w_a = \frac{a^2+3}{4a} + \frac{\sqrt{(a+3)(a+1)(a-1)(a-3)}}{4a} > 1 \quad \text{and} \tag{6.2}$$

$$1/w_a = \frac{a^2+3}{4a} - \frac{\sqrt{(a+3)(a+1)(a-1)(a-3)}}{4a} \in (0, 1)\,. \tag{6.3}$$

In particular, the restriction of $f_{a,\omega}$ to the unit circle is a real-analytic diffeomorphism for any $a > 3$. When $a \to 3$, both critical points $w_a > 1$ and $1/w_a \in (0, 1)$

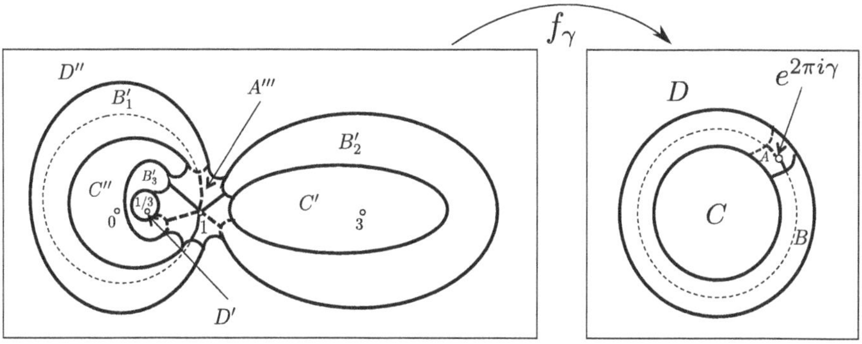

Fig. 6.1 Topological behaviour of the Blaschke product $f_{3,\omega}$ (6.1) around the unit circle, for ω approximately equal to $1/8$. At the left of Fig. 6.1 we see the preimage under $f_{3,\omega}$ of the annulus around the unit circle drawn at the right (in both planes, the unit circle is dashed). The complement of the annulus $A \cup B$ in the complex plane has two connected components, C and D. The preimage of C is the union $C' \cup C''$, where the notation C' means that $f_{3,\omega} : C' \to C$ has topological degree 1 (equivalently $f_{3,\omega} : C'' \to C$ has topological degree 2). In the same way, the preimage of D is the union $D' \cup D''$, the preimage of B is $B_1' \cup B_2' \cup B_3'$ and the preimage of A is A'''

collapse to the point $w = 1$, as we can see from (6.2) and (6.3). In other words, when $a \to 3$, the family $f_{a,\omega}$ converges to the *boundary* of the space of circle diffeomorphisms: for any $\omega \in [0, 1)$, the restriction of $f_{3,\omega}$ to $\mathbf{S}^1$ is a real-analytic multicritical circle map with a single critical point at 1, which is of cubic type, and with critical value $e^{2\pi i\omega}$ (Fig. 6.1).

Now let $p, q \in \mathbb{C}$ with $|p| > 1$, $|q| > 1$, let $\omega \in [0, 1)$ and consider $g_{p,q,\omega} : \widehat{\mathbb{C}} \to \widehat{\mathbb{C}}$ given by

$$g_{p,q,\omega}(z) = e^{2\pi i\omega} z^3 \left(\frac{z-p}{1-\overline{p}z}\right)\left(\frac{z-q}{1-\overline{q}z}\right). \tag{6.4}$$

Just as before, every map in this family leaves invariant the unit circle. The following fact was proved by Zakeri (1999, Section 7).

Theorem 6.1 *For any given $\rho \in (0, 1)\setminus\mathbb{Q}$ and $\delta \in (0, 1)$ there exists a unique $g_{p,q,\omega}$ of the form* (6.4) *such that $g_{p,q,\omega}|_{S^1}$ is a bi-critical circle map with signature* $(\rho\,; 2; 3, 3;\ \delta, 1-\delta)$.

Remark 6.1 It would be interesting to extend Zakeri's construction in order to obtain representative families of Blaschke products that restrict to multicritical circle maps with $N \geq 3$ critical points. Such construction should be useful to understand rigidity and renormalization problems for multicritical circle maps with any given number of critical points (to be discussed in the fourth and last part of this book).

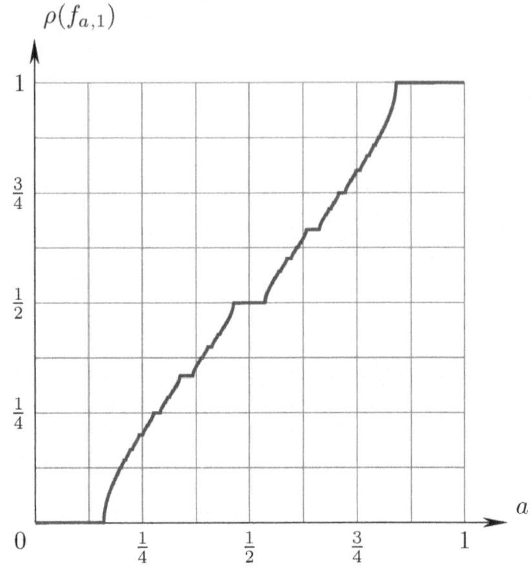

Fig. 6.2 The rotation number, as it varies in the one parameter family $f_{a,1} : x \mapsto x + a - \frac{1}{2\pi}\sin 2\pi x$, produces a devil staircase

6.1.2 *The Arnold Family*

Consider the two-parameter family $F_{a,b} : \mathbb{C} \to \mathbb{C}$ of entire maps in the complex plane given by

$$F_{a,b}(z) = z + a - \frac{b}{2\pi}\sin(2\pi z) \quad \text{for } a \in [0, 1) \text{ and } b \geq 0.$$

Since each $F_{a,b}$ commutes with unitary horizontal translation, it is the lift of a holomorphic map of the punctured plane $f_{a,b} : \mathbb{C}\setminus\{0\} \to \mathbb{C}\setminus\{0\}$ under the universal cover $z \mapsto e^{2\pi i z}$. Since $F_{a,b}$ preserves the real axis, $f_{a,b}$ preserves the unit circle. This classical two-parameter family of real-analytic circle maps was introduced by Arnold (1961), and it is known as the *Arnold family*.

For $b = 0$, the family $f_{a,b} : S^1 \to S^1$ is just the family of rigid rotations $z \mapsto e^{2\pi i a} z$. As it is easy to check, for $b \in (0, 1)$ the Arnold family is still contained in the space of real-analytic circle diffeomorphisms. For $b = 1$, however, the Arnold family belongs to the *boundary* of the space of circle diffeomorphisms: each $F_{a,1}$ projects to an orientation preserving real-analytic circle homeomorphism $f_{a,1}$, which has a critical point (of cubic type) at the point $z = 1$. The rotation number of $f_{a,1}$ varies with the parameter a in a continuous, monotone, non-decreasing way, and as we saw in Chap. 4 the resulting graph is a *devil staircase*; see Fig. 6.2. Each interval $\{a \in [0, 1) : \rho(f_{a,1}) = \theta\}$ degenerates to a point whenever θ is irrational and moreover, the set $\{a \in [0, 1) : \rho(f_{a,1}) \in \mathbb{R} \setminus \mathbb{Q}\}$ has zero Lebesgue measure (Świątek 1988). For integers $0 \leq p < q$, the set $\{a \in [0, 1) : \rho(f_{a,1}) = p/q\}$

is a non-degenerate closed interval (a phase-locking interval, in the language of Chap. 4). Its interior is made up of parameters whose corresponding critical circle maps have two periodic orbits (both of period q), one attracting and one repelling, which collapse to a single parabolic orbit when the parameter reaches the boundary of this interval, see Epstein et al. (1995).

Finally, we remark that for $b > 1$ the maps $f_{a,b} : S^1 \to S^1$ are no longer invertible (they possess two quadratic critical points). The dynamics of these maps is much richer than the case of homeomorphisms: the rotation number becomes a rotation *interval*, and typical dynamics here have positive topological entropy, infinitely many periodic orbits (coexisting with dense orbits) and, under certain conditions on the combinatorics, they preserve an absolutely continuous probability measure (see Boyland 1986; Chenciner et al. 1984; Crovisier et al. 2019; Misiurewicz 1986 and references therein).

The examples presented in both Sects. 6.1.1 and 6.1.2 show how multicritical circle maps arise as bifurcations from circle diffeomorphisms to endomorphisms, and in particular, from zero to positive topological entropy (compare with infinitely renormalizable unimodal maps, de Melo and van Strien 1993, Ch. VI). This is one of the main reasons why multicritical circle maps attracted the attention of physicists and mathematicians interested in the *boundary of chaos*, see Dixon et al. (1996), Feigenbaum et al. (1982), Kadanoff and Shenker (1982), Lanford (1987, 1988), MacKay (1983, 1993), Ostlund et al. (1983), Rand (1987, 1988, 1992), and Shenker (1982).

6.2 Topological Classification

Being a homeomorphism, a multicritical circle map f has a well defined rotation number. Just as before, we will focus on the case when f has no periodic orbits. In the early eighties, Yoccoz (1984b) proved that f has no wandering intervals. More precisely, we have the following fundamental result.

Theorem 6.2 (Yoccoz) *Let f be a multicritical circle map with irrational rotation number ρ. Then f is topologically conjugate to the rigid rotation R_ρ, i.e., there exists a homeomorphism $h : S^1 \to S^1$ such that $h \circ f = R_\rho \circ h$.*

It is not possible to remove the non-flatness condition on the critical points (recall Definitions 5.1 and 6.1). Indeed, Hall (1981) was able to construct C^∞ homeomorphisms of the circle with no periodic points and no dense orbits.

As we have already observed in Chap. 5, in the presence of critical points, the standard distortion tools used for diffeomorphisms no longer apply, at least not directly, since $\log Df$ is unbounded (see Fig. 6.3). We will need instead the tools introduced in Chap. 5, especially the Cross-ratio Inequality (Theorem 5.1).

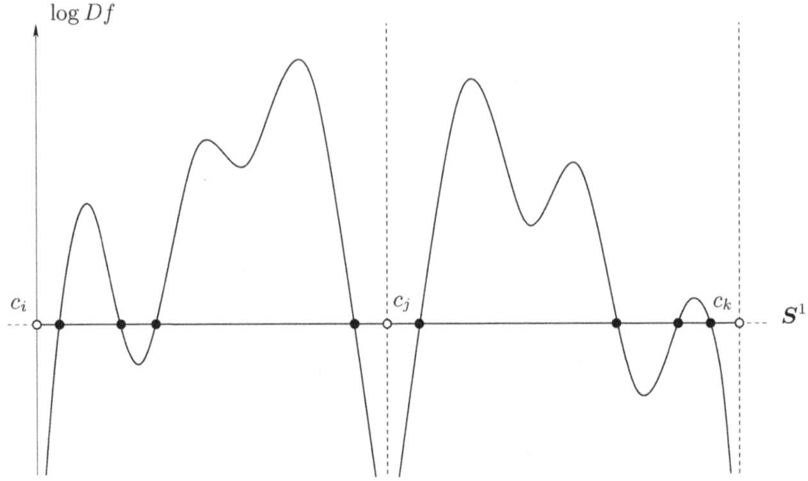

Fig. 6.3 The cocycle $\log Df$ is unbounded for a multicritical circle map f

6.2.1 Dynamically Symmetric Intervals

The proof of Theorem 6.2 that we wish to present differs considerably from Yoccoz's original proof in Yoccoz (1984b) (which uses a certain degenerate cross-ratio instead of the one we use here).

The key to our proof is a comparability result for general dynamically symmetric intervals, that is, any pair of intervals with an endpoint in common $x \in S^1$, the other endpoints being $f^{q_n}(x)$ and $f^{-q_n}(x)$, for some $n > 0$. This comparability result—Lemma 6.3 below—is also a crucial step in the proof of the *real bounds* to be presented in Sect. 6.3.

In order to accomplish our goal, we need the following two lemmas. The first lemma is proved by what is called the *seven-point argument* in Estevez and de Faria (2018). The reader may find the name a bit puzzling, since only five points appear in the statement, but in fact seven points are used in the proof.

Lemma 6.1 *There exists a constant $C_1 > 1$ depending only on f satisfying the following. For each $n \geq 0$ there exist z_1, z_2, z_3, z_4 and z_5 points in S^1 with $z_{j+1} = f^{q_n}(z_j)$ such that*

$$C_1^{-1} \leq \frac{|z_{i-1} - z_i|}{|z_{i+1} - z_i|} \leq C_1, \quad \textit{for } i = 2, 3, 4. \tag{6.5}$$

Proof Let $z \in S^1$ be a point such that, for all $x \in S^1$,

$$|f^{q_n}(z) - z| \leq |f^{q_n}(x) - x|.$$

Then consider the seven points

$$z_0 = f^{-4q_n}(z)\,,\ z_1 = f^{-3q_n}(z)\,,\ z_2 = f^{-2q_n}(z)\,,\ z_3 = f^{-q_n}(z)\,,$$

$$z_4 = z\,,\ z_5 = f^{q_n}(z)\,,\ z_6 = f^{2q_n}(z)\,.$$

Note that, by our choice of z,

$$|z_4 - z_5| \leq |z_i - z_{i+1}|\,, \quad \text{for all } 0 \leq i \leq 5\,. \tag{6.6}$$

These seven points are cyclically ordered as given (either in clockwise or counter-clockwise order in the circle), provided n is sufficiently large. Let $J \subset S^1$ be the closed interval with endpoints z_0 and z_6 that contains $z = z_4$. For each $0 \leq i \leq 3$, let $T_i = [z_i, z_{i+3}] \subset J$ and $M_i = [z_{i+1}, z_{i+2}] \subset T_i$. Then the homeomorphism f^{q_n} maps T_i onto T_{i+1} and M_i onto M_{i+1}, for $0 \leq i \leq 2$. Moreover, the collection of intervals $\{T_i\,,\ f(T_i)\,,\ \ldots\,,\ f^{q_n}(T_i)\}$ has intersection multiplicity equal to 3.

(i) Let us first prove (6.5) for $i = 4$. Applying the Cross-ratio Inequality to f^{q_n} and the pair (M_2, T_2), we have

$$\mathrm{CrD}(f^{q_n}; M_2, T_2) = \frac{[M_3, T_3]}{[M_2, T_2]} = \frac{|z_3 - z_4||z_5 - z_6||z_2 - z_4|}{|z_4 - z_6||z_2 - z_3||z_4 - z_5|} \leq B\,,$$

where $B > 1$ is a constant that depends only on f. But then, using (6.6), we see that

$$\frac{|z_3 - z_4|}{|z_4 - z_5|} \leq B\frac{|z_4 - z_6|}{|z_5 - z_6|} = B\left(\frac{|z_4 - z_5|}{|z_5 - z_6|} + 1\right) \leq 2B.$$

Therefore, defining $B_1 = 2B$ and again using (6.6), we get

$$B_1^{-1} \leq \frac{|z_3 - z_4|}{|z_4 - z_5|} \leq B_1\,. \tag{6.7}$$

(ii) Let us now prove (6.5) for $i = 3$. Applying the Cross-ratio Inequality to f^{q_n} and the pair (M_1, T_1), we have

$$\mathrm{CrD}(f^{q_n}; M_1, T_1) = \frac{[M_2, T_2]}{[M_1, T_1]} = \frac{|z_2 - z_3||z_4 - z_5||z_1 - z_3|}{|z_3 - z_5||z_1 - z_2||z_3 - z_4|} \leq B\,,$$

or equivalently, using (6.6) and the upper bound in (6.7),

$$\frac{|z_2 - z_3|}{|z_3 - z_4|} \leq B\frac{|z_3 - z_5|}{|z_4 - z_5|} \leq B\left(\frac{|z_3 - z_4|}{|z_4 - z_5|} + 1\right) \leq B(B_1 + 1).$$

On the other hand, using (6.6) once again,

$$\frac{|z_3 - z_4|}{|z_2 - z_3|} \leq \frac{|z_3 - z_4|}{|z_4 - z_5|} \leq B_1 .$$

Taking $B_2 = B(B_1 + 1)$ and putting the last two inequalities together, we get

$$B_2^{-1} \leq \frac{|z_2 - z_3|}{|z_3 - z_4|} \leq B_2 . \tag{6.8}$$

(iii) Finally, let us prove (6.5) for $i = 2$. As before, applying the Cross-ratio Inequality to f^{q_n} and the pair (M_0, T_0), we have

$$\mathrm{CrD}(f^{q_n}; M_0, T_0) = \frac{[M_1, T_1]}{[M_0, T_0]} = \frac{|z_1 - z_2||z_3 - z_4||z_0 - z_2|}{|z_2 - z_3||z_0 - z_1||z_2 - z_3|} \leq B ,$$

From this, using (6.6) and (6.8), we get on the one hand

$$\frac{|z_1 - z_2|}{|z_2 - z_3|} \leq B\frac{|z_2 - z_4|}{|z_3 - z_4|} \leq B\left(\frac{|z_2 - z_3|}{|z_3 - z_4|} + 1\right) \leq B(B_2 + 1). \tag{6.9}$$

On the other hand, the inequalities (6.7) and (6.8) tell us that

$$\frac{|z_2 - z_3|}{|z_1 - z_2|} \leq B_2\frac{|z_3 - z_4|}{|z_1 - z_2|} \leq B_2 B_1\frac{|z_4 - z_5|}{|z_1 - z_2|} \leq B_2 B_1 . \tag{6.10}$$

Defining $B_3 = \max\{B(B_2 + 1), B_2 B_1\} = B_1 B_2$, and using inequalities (6.9) and (6.10), we obtain

$$B_3^{-1} \leq \frac{|z_1 - z_2|}{|z_2 - z_3|} \leq B_3 .$$

Summarizing, we have proved (6.5) with $C_1 = \max\{B_1, B_2, B_3\} = B_3 > 1$, a constant that indeed depends only on f. □

Lemma 6.2 *There exists a constant $C_2 > 1$ depending only on f satisfying the following. Let z_1, z_2, z_3, z_4 and z_5 be the points given by Lemma 6.1. If w_0, w_1, w_2, w_3 and w_4 are points on the circle such that $w_{j+1} = f^{q_n}(w_j)$ and such that w_1 lies in the interval with endpoints z_1 and z_2 that does not contain z_3, then*

$$\frac{|w_1 - w_2|}{|w_0 - w_1|} \leq C_2 \quad \text{and} \quad C_2^{-1} \leq \frac{|w_{i-1} - w_i|}{|w_i - w_{i+1}|} \leq C_2 , \quad \text{for } i = 2, 3 . \tag{6.11}$$

Proof To prove the first inequality, we consider the interval T with endpoints w_0 and w_3 containing z_1, w_1, z_2, w_2, z_3, and the subinterval $M = [w_1, w_2] \subset T$. Note that $\{T, f(T), \ldots, f^{q_n}(T)\}$ has intersection multiplicity equal to 3. Hence, applying the Cross-ratio Inequality to f^{q_n} and the pair (M, T), we get

$$[f^{q_n}(M), f^{q_n}(T)] \leq B[M, T]$$

or equivalently

$$\frac{|w_1 - w_2||w_3 - w_4|}{|w_1 - w_3||w_2 - w_4|} \leq B \frac{|w_0 - w_1||w_2 - w_3|}{|w_0 - w_2||w_1 - w_3|} . \tag{6.12}$$

Since the points $w_0, z_1, w_1, \ldots, z_4, w_4, z_5$ are cyclically ordered as given, we have the inequalities $|z_1 - z_2| \leq |w_0 - w_2|$, $|w_2 - w_3| \leq |z_2 - z_4|$, and $|w_2 - w_4| \leq |z_2 - z_5|$. Moreover, we have $|z_4 - z_5| \leq |w_3 - w_4|$, by our choice of $z = z_4$ in Lemma 6.1. These facts, when put back into (6.12), yield

$$\frac{|w_1 - w_2|}{|w_0 - w_1|} \leq B \frac{|z_2 - z_4||z_2 - z_5|}{|z_1 - z_2||z_4 - z_5|} \leq B(C_1 + C_1^2)(1 + C_1 + C_1^2) ,$$

where we have used the inequalities of Lemma 6.1.

To prove the upper bound in the last two inequalities in (6.11), we simply note that $|w_i - w_{i+1}| \geq |z_4 - z_5|$ and that $|w_{i-1} - w_i| \leq |z_{i-1} - z_{i+1}|$. Using the inequalities (6.5), we deduce that

$$\frac{|w_{i-1} - w_i|}{|w_i - w_{i+1}|} \leq \frac{|z_{i-1} - z_i|}{|z_4 - z_5|} + \frac{|z_i - z_{i+1}|}{|z_4 - z_5|} \leq 2C_1^3$$

The lower bound for the same inequalities in (6.11) is proven in exactly the same way (the value obtained is $(2C_1^3)^{-1}$). Thus, (6.11) is established, provided we take $C_2 = \max\{2C_1^3 , B(C_1 + C_1^2)(1 + C_1 + C_1^2)\}$. □

We are now in a position to show that dynamically symmetric intervals are always comparable. In the lemma below, we make use of the following simple remark. Given $\xi \in \boldsymbol{S}^1$, let $J_n(\xi) \subset \boldsymbol{S}^1$ be the interval with endpoints $f^{-q_n}(\xi)$ and $f^{q_n}(\xi)$ that contains ξ. Then $\bigcup_{i=0}^{q_{n+1}} f^{-i}(J_n(\xi)) = \boldsymbol{S}^1$.

Lemma 6.3 *There exists a constant* $C_3 > 1$ *depending only on* f *such that, for all* $n \geq 0$ *and all* $x \in \boldsymbol{S}^1$, *we have*

$$C_3^{-1}|x - f^{-q_n}(x)| \leq |f^{q_n}(x) - x| \leq C_3|x - f^{-q_n}(x)|. \tag{6.13}$$

Proof Note that it suffices to prove the second of the two inequalities in (6.13) for all x (to get the first inequality from the second, just replace x by $f^{-q_n}(x)$).

Thus, let $x \in \boldsymbol{S}^1$ and let $0 \leq i \leq q_{n+1}$ such that $f^i(x)$ lies on the interval J with endpoints z_1 and z_3 that contains z_2, where $z_1, z_2, \ldots, z_5$ are the points given by Lemma 6.1. Such an i exists because of the simple remark preceding the present lemma, applied to $\xi = z_2$ (so that $J_n(z_2) = J$). Then either $f^i(x) \in [z_1, z_2] \subset J$, or $f^i(x) \in (z_2, z_3] \subset J$. We prove the lemma assuming the former case (the proof in the latter case being similar).

Let us consider the points $w_0 = f^{i-q_n}(x)$, $w_1 = f^i(x)$, $w_2 = f^{i+q_n}(x)$ and $w_3 = f^{i+2q_n}(x)$. Then we are in the situation of Lemma 6.2. Consider the interval T with endpoints $f^{-q_n}(x)$ and $f^{2q_n}(x)$ that contains x, and let $M = [x, f^{q_n}(x)] \subset T$. Note that

$$[M, T] = \frac{|x - f^{-q_n}(x)||f^{q_n}(x) - f^{2q_n}(x)|}{|f^{q_n}(x) - f^{-q_n}(x)||x - f^{2q_n}(x)|} \leq \frac{|x - f^{-q_n}(x)|}{|f^{q_n}(x) - x|} . \tag{6.14}$$

From the inequalities (6.11) in Lemma 6.2, we also have

$$[f^i(M), f^i(T)] = \frac{|w_0 - w_1||w_2 - w_3|}{|w_0 - w_2||w_1 - w_3|} \geq \frac{1}{(1 + C_2)^2} . \tag{6.15}$$

Since $\{T, f(T), \ldots, f^i(T)\}$ has intersection multiplicity at most equal to 3, the Cross-ratio Inequality tells us that $[f^i(M), f^i(T)] \leq B[M, T]$, where the constant B is the same as in the previous lemmas. Combining this fact with (6.14) and (6.15), we deduce that

$$|f^{q_n}(x) - x| \leq B(1 + C_2)^2 |x - f^{-q_n}(x)|.$$

This proves (6.13), provided we take $C_3 = B(1 + C_2)^2$. □

6.2.2 *Proof of Yoccoz's Theorem*

Yoccoz's Theorem 6.2 is now a straightforward consequence of Lemma 6.3.

Proof of Theorem 6.2 Suppose, by contradiction, that there exists a wandering interval $J = (a, b)$, which we can assume to be maximal. For each $n \in \mathbb{N}$, let $\Delta_n \subset S^1$ be the open interval with endpoints $f^{-q_n}(a)$, $f^{q_n}(a)$ that contains a. Since J is a wandering interval, its iterates are pairwise disjoint, so $f^{\pm q_n}(a) \notin J$, and from this it follows that Δ_n must contain J for all $n \in \mathbb{N}$. Hence the sequence $\{|\Delta_n|\}_{n\in\mathbb{N}}$ is bounded away from zero. However, since J is maximal, the point a is recurrent, and therefore there exists a subsequence $n_i \to \infty$ such that $f^{q_{n_i}}(a) \to a$ as $i \to \infty$. But then, by Lemma 6.3, we have also $f^{-q_{n_i}}(a) \to a$ as $i \to \infty$, and this tells us that $|\Delta_{n_i}| \to 0$ as $i \to \infty$. This contradiction shows that no such wandering interval J exists, and the proof is complete. □

Remark 6.2 The argument above gives a new proof of Denjoy's Theorem 3.4, since the Cross-ratio Inequality (Theorem 5.1) certainly holds whenever f is a C^1 diffeomorphism and $\log Df$ has bounded variation (note that the Schwarzian derivative is not needed in this case: estimate (5.7) holds on the whole circle).

6.3 Real a Priori Bounds

Now that we understand the topology of a multicritical circle map f, we move to the more delicate task of understanding its geometry.

Our ultimate goal is to understand the geometry of f at fine scales, in other words the *asymptotic scaling structure* of f. A general, informal principle in one-dimensional dynamics is that, in order to understand the geometry of a map at fine scales, it suffices to understand the asymptotic geometry of the orbits of the *critical points* of the map. The first step towards this goal is to get some bounds on finite pieces of the orbit of a given critical point $c \in \mathbf{S}^1$, say up to a closest return time: $c, f(c), \ldots, f^{q_n-1}(c)$. The bounds we look for are bounds on the ratios of distances between (some of) these points, with constants that are independent of n. In fact, we will see that these constants are even asymptotically independent of f itself.

The above description is admittedly rather vague, since we have not explained what we mean by expressions such as "geometry at fine scales" or "asymptotic scaling structure", but precise statements (and proofs) will be given below.

6.3.1 Dynamical Partitions

Let f be a homeomorphism without periodic points, i.e., with irrational rotation number $\rho \in (0, 1)$, and let $\{q_n\}_{n\geq 0}$ be the corresponding sequence of return times (the denominators of the best rational approximations to ρ; see Chap. 1).

Let us fix some base point $x \in \mathbf{S}^1$. For each non-negative integer n, let $I_n(x)$ be the closed interval with endpoints x and $f^{q_n}(x)$ that contains $f^{q_{n+2}}(x)$. Consider the following collection of closed intervals:

$$\mathscr{P}_n(x) = \left\{f^i(I_n(x)) : 0 \leq i \leq q_{n+1} - 1\right\} \cup \left\{f^j(I_{n+1}(x)) : 0 \leq j \leq q_n - 1\right\}$$

The following fact is fundamental.

Lemma 6.4 *For each $n \geq 0$, the collection $\mathscr{P}_n(x)$ is a partition of the circle* modulo endpoints.

Proof Since the families $\mathscr{P}_n(x)$ are dynamically defined, we may assume by Yoccoz's Theorem 6.2 that f is the rigid rotation of the unit circle of angle ρ. Let $\{p_n/q_n\}$ be the sequence of best rational approximations to ρ. As we saw in Chap. 1, Eq. (1.6), for all $n \in \mathbb{N}$ we have

$$q_n p_{n+1} - q_{n+1} p_n = (-1)^n . \tag{6.16}$$

The arithmetical properties of the continued fraction expansion of ρ described in Chap. 1 imply that, for any point $x \in \mathbf{S}^1$, the iterates $\{f^{q_n}(x)\}_{n\in\mathbb{N}}$ are the *closest returns* of the orbit of x under the rigid rotation f, in the following sense:

$$d\big(x, f^{q_n}(x)\big) < d\big(x, f^j(x)\big) \quad \text{for any} \quad j \in \{1, \ldots, q_n - 1\}$$

where d denote the standard distance in $\boldsymbol{S}^1$. In particular, all members of the family

$$\Big\{I_n(x), f(I_n(x)), \ldots, f^{q_{n+1}-1}(I_n(x))\Big\}$$

are pairwise disjoint, and all members in the family

$$\big\{I_{n+1}(x), f(I_{n+1}(x)), \ldots, f^{q_n-1}(I_{n+1}(x))\big\}$$

are pairwise disjoint too. Moreover, we claim that any two members in the union of these families (which is precisely $\mathscr{P}_n$) are disjoint. Indeed, suppose, by contradiction, that there exist $i < q_{n+1}$ and $j < q_n$ such that $f^i(I_n) \cap f^j(I_{n+1}) \neq \emptyset$. Without loss of generality, we may assume that $i < j = i + l$, for some $l < q_n$, and that the q_n-th iterate of every point $z \in \boldsymbol{S}^1$ is on the right-hand side of z, and consequently the q_{n+1}-th iterate is on the left-hand side of z. We have three possible cases to consider:

- If $f^i(I_n(x)) \subseteq f^j(I_{n+1}(x))$, then $f^j(I_{n+1}(x))$ intersects $f^i(I_{n+1}(x))$ and this is impossible as explained above.
- If $f^j(I_{n+1}(x)) \subseteq f^i(I_n(x))$, then the point $f^j(x) = f^{i+l}(x)$ is closer to $f^i(x)$ than $f^{i+q_n}(x)$, which is impossible since $l < q_n$.
- If both differences between $f^j(I_{n+1}(x))$ and $f^i(I_n(x))$ are non-empty and connected, then we have two sub-cases:

$$\text{either } f^j(x) \in f^i(I_n(x)) \text{ or } f^{j+q_{n+1}}(x) \in f^i(I_n(x))\,.$$

  In the first case, $f^j(x) = f^{i+l}(x)$ is closer to $f^i(x)$ than $f^{i+q_n}(x)$, and since $l < q_n$ this is a contradiction. In the second case, the point $f^{i+q_n}(x) = f^j(f^{q_n+i-j}(x))$ is closer to $f^j(x)$ than $f^{j+q_{n+1}}(x)$, which again is impossible since $q_n + i - j < q_{n+1}$.

Therefore, any two members of $\mathscr{P}_n(x)$ are disjoint, as claimed.

Finally, since we are assuming that f is the rigid rotation of angle ρ in the (normalized) unit circle, the lengths of the intervals $I_n(x)$ and $I_{n+1}(x)$ are $|q_n\rho - p_n| = q_n|\rho - p_n/q_n|$ and $q_{n+1}|p_{n+1}/q_{n+1} - \rho|$ respectively. Therefore, the total length of the union of the members of $\mathscr{P}_n(x)$ is equal to

$$\left|q_n q_{n+1}\left(\frac{p_{n+1}}{q_{n+1}} - \frac{p_n}{q_n}\right)\right| = |q_n p_{n+1} - p_n q_{n+1}|.$$

By (6.16), this absolute value is equal to 1, that is, the union of the members of $\mathscr{P}_n$ is a compact set of full Lebesgue measure, and therefore it covers the whole circle.

□

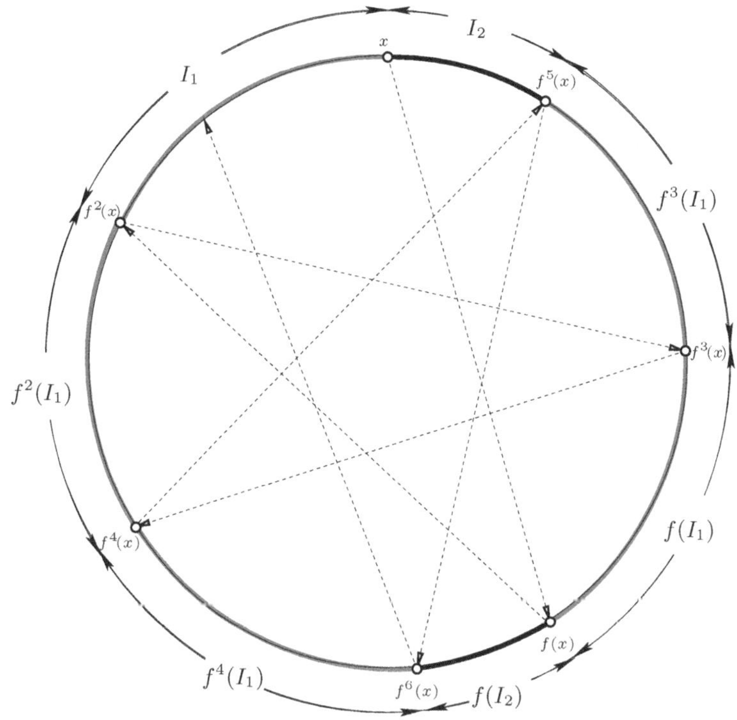

Fig. 6.4 Dynamical partition $\mathscr{P}_1(x)$ of a circle homeomorphism with rotation number $\rho(f) = \sqrt{2} - 1 = [2, 2, 2, \ldots]$

We call $\mathscr{P}_n(x)$ the *n-th dynamical partition* associated with x. The intervals of the form $f^i(I_n(x))$ are called *long*, whereas those of the form $f^j(I_{n+1}(x))$ are called *short*. The initial partition $\mathscr{P}_0(x)$ is given by

$$\mathscr{P}_0(x) = \Big\{\big[f^i(x), f^{i+1}(x)\big] : i \in \{0, \ldots, a_0 - 1\}\Big\} \cup \big\{\big[f^{a_0}(x), x\big]\big\},$$

where a_0 is the integer part of $1/\rho$.

Example 6.1 Figure 6.4 shows the dynamical partition $\mathscr{P}_1(x)$ associated to a circle homeomorphism with rotation number $\rho(f) = \sqrt{2} - 1 = [2, 2, 2, \ldots]$, for which $q_1 = 2$ and $q_2 = 5$. Explicitly, writing $I_1 = I_1(x)$ and $I_2 = I_2(x)$, we have

$$\mathscr{P}_1(x) = \Big\{I_1, f(I_1), f^2(I_1), f^3(I_1), f^4(I_1)\Big\} \cup \{I_2, f(I_2)\} .$$

Remark 6.3 We end this section with the simple but very important observation that the dynamical partitions $\mathscr{P}_n(x)$ of a given point x are *nested*. Indeed, it follows directly from the definition that every *short* atom of $\mathscr{P}_n(x)$ becomes a *long* atom of $\mathscr{P}_{n+1}(x)$, whereas each *long* atom of $\mathscr{P}_n(x)$ is partitioned into a disjoint union of *short* atoms of $\mathscr{P}_{n+1}(x)$.

6.3.2 *The Real Bounds*

We are now in a position to state and prove the following absolutely fundamental result in the theory of critical circle maps, known in the literature as the *real a priori bounds theorem*, or simply the *real bounds theorem*. This theorem, in slightly different formulation, was first proved in the eighties by Herman (1988) and Świątek (1988). Our exposition here follows very closely the one in Estevez and de Faria (2018, § 3). See also Petersen (2000) for a different treatment.

Theorem 6.3 (Real A-Priori Bounds) *Let f be a multicritical circle map. There exists a constant $C > 1$ depending only of f such that the following holds for every critical point c of f. For all $n \geq 0$ and for each pair of adjacent atoms $I, J \in \mathscr{P}_n(c)$ we have*

$$C^{-1}|J| \leq |I| \leq C|J|.$$

Intuitively, this theorem is saying that, in every dynamical partition $\mathscr{P}_n(c)$, any two consecutive atoms are *comparable*. See Sect. 6.3.3 below, where the notion of comparability will be made more precise.

Note that for a rigid rotation (and any point $x \in \mathbf{S}^1$) we have $|I_n(x)| = a_{n+1}\,|I_{n+1}(x)| + |I_{n+2}(x)|$. If a_{n+1} is very large, then $|I_n(x)|$ is much larger than $|I_{n+1}(x)|$. Thus, even for rigid rotations, real bounds do not hold in general.

The main tools to be used in the proof of Theorem 6.3 are the Cross-ratio Inequality (Theorem 5.1) and Lemma 6.3. All constants appearing in the proof, including constant C_3 of Lemma 6.3, can be traced back to the constant appearing in the Cross-ratio Inequality. We will denote these constants $C_4, C_5, \ldots$ in succession, keeping track of how each constant being introduced depends on the previous ones.

6.3.2.1 Comparability of Closest Returns and Beyond

The major step in the proof of Theorem 6.3 states that the atoms of the partition $\mathscr{P}_n(c)$ that are closest to the critical point c, including the closest return intervals $I_n(c)$ and $I_{n+1}(c)$, are pairwise comparable. This is the contents of Proposition 6.1 below. In order to simplify the notation a bit, from now until the end of this section we write $I_n = I_n(c)$ and $I_{n+1} = I_{n+1}(c)$, as well as $I_n^i = f^i(I_n)$ for all i and $I_{n+1}^j = f^j(I_{n+1})$ for all j.

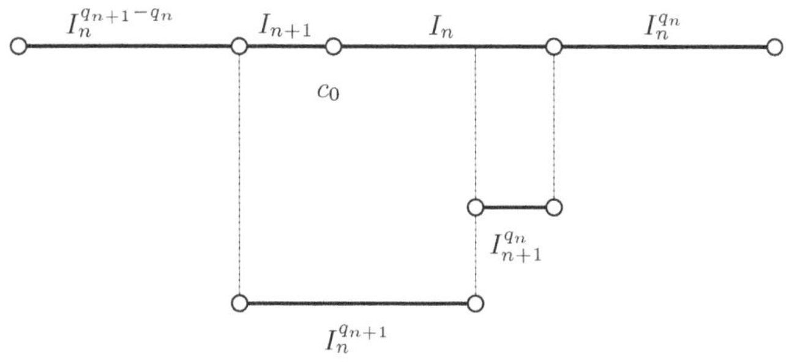

Fig. 6.5 The six intervals of Proposition 6.1

Proposition 6.1 *The six intervals in Fig. 6.5 are pairwise comparable. More precisely, there exists a constant $C_4 > 1$ depending only on f such that, for all $n \geq 1$ and for all $I, J \in \{I_n, I_{n+1}, I_n^{q_n}, I_n^{q_{n+1}}, I_{n+1}^{q_n}, I_n^{q_{n+1}-q_n}\}$, we have*

$$C_4^{-1} \leq \frac{|I|}{|J|} \leq C_4 \,. \tag{6.17}$$

Proof We break up the proof into several steps, as follows.

(i) *The intervals I_n and $I_n^{q_n}$ are comparable.* Indeed, these two intervals are dynamically symmetric with respect to their common endpoint $f^{q_n}(c)$. Hence, by Lemma 6.3 we have

$$C_3^{-1}|I_n| \;\leq\; |I_n^{q_n}| \;\leq\; C_3|I_n| \tag{6.18}$$

(ii) *The intervals $I_n^{q_{n+1}}$ and $I_n^{q_{n+1}-q_n}$ are comparable.* Indeed, these two intervals are dynamically symmetric with respect to their common endpoint $f^{q_{n+1}}(c)$. Hence, again by Lemma 6.3 we have

$$C_3^{-1}|I_n^{q_{n+1}}| \leq |I_n^{q_{n+1}-q_n}| \leq C_3|I_n^{q_{n+1}}|. \tag{6.19}$$

(iii) *The intervals $I_n^{q_{n+1}-q_n}$ and I_n are comparable.* Consider the interval $I_n^{-q_n}$, with endpoints c and $f^{-q_n}(c)$. Since such interval is dynamically symmetric to the interval I_n, we have by the Lemma 6.3

$$C_3^{-1}|I_n^{-q_n}| \leq |I_n| \leq C_3|I_n^{-q_n}|. \tag{6.20}$$

From the right-hand side of (6.20), the inclusion $I_n^{-q_n} \subseteq I_n^{q_{n+1}-q_n} \cup I_n^{q_{n+1}}$ and the left-hand side of (6.19), we deduce that

$$|I_n| \leq C_3(C_3+1)|I_n^{q_{n+1}-q_n}|. \tag{6.21}$$

Now, we have $I_n^{q_{n+1}} \subseteq I_{n+1} \cup I_n$, and also $|I_{n+1}| \leq C_3|I_{n+1}^{-q_{n+1}}|$, because the intervals I_{n+1} and and $I_{n+1}^{-q_{n+1}}$ are dynamically symmetric. Moreover, we have the inclusion $I_{n+1}^{-q_{n+1}} \subseteq I_n$. Combining these facts with the right-hand side of (6.19), we get

$$|I_n^{q_{n+1}-q_n}| \leq C_3(C_3+1)|I_n| \ .$$

From this and (6.21), we arrive at

$$C_3^{-1}(C_3+1)^{-1}|I_n| \leq |I_n^{q_{n+1}-q_n}| \leq C_3(C_3+1)|I_n|. \tag{6.22}$$

(iv) *The intervals I_n and I_{n+1} are comparable.* It is here that we use the power-law at the critical point c in an essential way. First note that $I_{n+1}^{-q_{n+1}} \subseteq I_n$ and that the intervals $I_{n+1}^{-q_{n+1}}$ and I_{n+1} are dynamically symmetric with respect to their common endpoint c. Hence, using Lemma 6.3 we get

$$|I_{n+1}| \leq C_3|I_n| \ .$$

The real issue here, thus, is to prove an inequality in the opposite direction. Let us consider the interval $T = I_{n+1} \cup I_n \cup I_n^{q_n}$ and its image $f(T)$ under f, which contains the critical value $f(c)$; note that the family $\{T, f(T), \ldots, f^{q_{n+1}-1}(T)\}$ has intersection multiplicity equal to 3. We look at the cross-ratio distortion of $f^{q_{n+1}-1}$ on the pair $(I_n^1, f(T))$. By the Cross-ratio Inequality, we have

$$\mathrm{CrD}(f^{q_{n+1}-1}; I_n^1, f(T)) = \frac{[I_n^{q_{n+1}}, f^{q_{n+1}}(T)]}{[I_n^1, f(T)]} \leq B \ . \tag{6.23}$$

But

$$[I_n^{q_{n+1}}, f^{q_{n+1}}(T)] = \frac{|I_{n+1}^{q_{n+1}}|}{|I_{n+1}^{q_{n+1}}| + |I_n^{q_{n+1}}|} \cdot \frac{|I_n^{q_{n+1}+q_n}|}{|I_n^{q_{n+1}}| + |I_n^{q_{n+1}+q_n}|} \ . \tag{6.24}$$

Since the intervals $I_n^{q_{n+1}+q_n}$ and $I_n^{q_{n+1}}$ are dynamically symmetric with respect to their common endpoint, we see from Lemma 6.3 that the second fraction on the right-hand side of (6.24) is bounded from below by $C_3^{-1}/(1+C_3)$. The intervals $I_{n+1}^{q_{n+1}}$ and I_{n+1} are also dynamically symmetric with respect to their common endpoint, so again by Lemma 6.3 we have $C_3^{-1}|I_{n+1}| \leq |I_{n+1}^{q_{n+1}}| \leq C_3|I_{n+1}|$; in addition, $I_{n+1}^{q_{n+1}} \subset I_{n+1} \cup I_n$, so that $|I_{n+1}^{q_{n+1}}| \leq |I_{n+1}| + |I_n|$. Putting all these facts back into (6.24), we deduce that

$$[I_n^{q_{n+1}}, f^{q_{n+1}}(T)] \geq \theta_1 \frac{|I_{n+1}|}{|I_n|} \ , \tag{6.25}$$

where $\theta_1 = C_3^{-2}(1+C_3)^{-1}(1+C_3+C_3^2+C_3^3)^{-1}$. This bounds the numerator of (6.23) from below, so we proceed to bound the denominator from above. We have

$$[I_n^1, f(T)] = \frac{|I_{n+1}^1|}{|I_{n+1}^1| + |I_n^1|} \cdot \frac{|I_n^{1+q_n}|}{|I_n^1| + |I_n^{1+q_n}|} . \tag{6.26}$$

Since the intervals I_n^1 and $I_n^{1+q_n}$ are also dynamically symmetric with respect to their common endpoint, applying Lemma 6.3 yet again yields

$$[I_n^1, f(T)] \leq \frac{C_3}{1+C_3} \frac{|I_{n+1}^1|}{|I_n^1|} . \tag{6.27}$$

Here, using the power-law at the critical point (at last!) we see that

$$\frac{|I_{n+1}^1|}{|I_n^1|} \leq \gamma_0 \left(\frac{|I_{n+1}|}{|I_n|} \right)^{s_0} ,$$

where $\gamma_0 = \gamma_0(f) > 0$ is a constant as in Remark 5.2, and $s_0 > 1$ is the criticality of the critical point c. Carrying this information back to (6.27) gives us

$$[I_n^1, f(T)] \leq \theta_2 \left(\frac{|I_{n+1}|}{|I_n|} \right)^{s_0} , \tag{6.28}$$

where $\theta_2 = \gamma_0 C_3/(1+C_3)$. Combining (6.25) and (6.28) we get the inequality

$$\frac{|I_{n+1}|}{|I_n|} \geq \left(\frac{\theta_1}{B\theta_2} \right)^{\frac{1}{s_0-1}} = \theta_3 .$$

Summarizing, we have proved that

$$\theta_3 |I_n| \leq |I_{n+1}| \leq C_3 |I_n| . \tag{6.29}$$

(v) *The intervals I_n and $I_{n+1}^{q_n}$ are comparable.* Note that $I_{n+1}^{q_n} \subset I_n$, so $|I_{n+1}^{q_n}| \leq |I_n|$. We must prove an inequality in the opposite direction. For this purpose, let us consider the interval $T^* = I_{n+1}^{q_n} \cup I_n^{q_n} \cup I_n^{2q_n}$. We shall look at the cross-ratio distortion of the pair $(I_n^{q_n}, T^*)$ under the map $f^{q_{n+1}-q_n}$. Clearly, the family $\{T^*, f(T^*), \ldots, f^{q_{n+1}-q_n}(T^*)\}$ has intersection multiplicity equal to at most 3. By the Cross-ratio Inequality, we have

$$\mathrm{CrD}(f^{q_{n+1}-q_n}; I_n^{q_n}, T^*) = \frac{[I_n^{q_{n+1}}, f^{q_{n+1}-q_n}(T^*)]}{[I_n^{q_n}, T^*]} \leq B \tag{6.30}$$

Now, the intervals $I_{n+1}^{q_{n+1}}$ and I_{n+1} are dynamically symmetric with respect to their common endpoint $f^{q_{n+1}}(c)$. Also, the intervals $f^{q_{n+1}-q_n}(I_n^{2q_n}) = I_n^{q_{n+1}+q_n}$ and $I_n^{q_{n+1}}$ are dynamically symmetric with respect to their common endpoint $f^{q_{n+1}+q_n}(c)$. Moreover, we have $I_n^{q_{n+1}} \subset I_n \cup I_{n+1}$. Combining these facts with (6.29) and Lemma 6.3, we deduce after some computations that

$$
\begin{aligned}
[I_n^{q_{n+1}}, f^{q_{n+1}-q_n}(T^*)] &= \frac{|I_{n+1}^{q_{n+1}}|}{|I_{n+1}^{q_{n+1}}| + |I_n^{q_{n+1}}|} \frac{|I_n^{q_{n+1}+q_n}|}{|I_n^{q_{n+1}}| + |I_n^{q_{n+1}+q_n}|} \\
&\geq \frac{C_3^{-2}\theta_3}{(1+C_3)(1+C_3+C_3^2)} .
\end{aligned} \tag{6.31}
$$

We proceed to bound the denominator in (6.30) from above in similar fashion. Since the intervals $I_n^{q_n}$ and $I_n^{2q_n}$ are dynamically symmetric with respect to their common endpoint $f^{q_n}(c)$, applying Lemma 6.3 one final time yields

$$
\begin{aligned}
[I_n^{q_n}, T^*] &= \frac{|I_{n+1}^{q_n}|}{|I_{n+1}^{q_n}| + |I_n^{q_n}|} \frac{|I_n^{2q_n}|}{|I_n^{q_n}| + |I_n^{2q_n}|} \leq \frac{|I_{n+1}^{q_n}|}{|I_n^{q_n}|} \frac{C_3}{1+C_3^{-1}} \\
&\leq \frac{C_3^2}{1+C_3^{-1}} \frac{|I_{n+1}^{q_n}|}{|I_n|} .
\end{aligned} \tag{6.32}
$$

Putting (6.31) and (6.32) back into (6.30), we deduce at last that

$$
\theta_4 |I_n| \leq |I_{n+1}^{q_n}| \leq |I_n| , \tag{6.33}
$$

where

$$
\theta_4 = \frac{(1+C_3^{-1})C_3^{-4}\theta_3}{B(1+C_3)(1+C_3+C_3^2)}
$$

The above estimates—more precisely the inequalities (6.18), (6.19), (6.22), (6.29) and (6.33)—provide bounds for 5 of the 15 comparability ratios involved in (6.17). Each of the remaining 10 comparability ratios is obtained by suitable telescoping products of at most 4 of these 5 ratios. Thus, define K to be the largest of all constants greater than 1 appearing as bounds in the above estimates, namely $K = \max\{C_3(C_3+1), \theta_3^{-1}, \theta_4^{-1}\}$. With this choice, all 15 inequalities involved in (6.17) are established provided we take $C_4 = K^4$. □

6.3.2.2 Proof of Theorem 6.3

Finally, to obtain Theorem 6.3, we use the Cross-ratio Inequality to propagate the information in Proposition 6.1 to any pair of adjacent intervals in the dynamical partition $\mathscr{P}_n(c)$. Fix an atom $M \in \mathscr{P}_n(c)$, and let $L, R \in \mathscr{P}_n(c)$ be its two immediate neighbors; write $T = L \cup M \cup R$. It suffices to show that the cross-ratio $[M, T]$ is bounded from below by a constant depending only on the constant C_4 of Proposition 6.1. There are two cases to consider, depending on whether M is a *short* or a *long* atom of the dynamical partition $\mathscr{P}_n(c)$. If M is a short atom, say $M = I_{n+1}^j$ with $j < q_n$, then L and R are both long atoms. In fact, the combinatorics tells us that one of them, say R, is the interval I_n^j, whereas the other, L, is the interval $I_n^{j+q_{n+1}-q_n}$. But then the homeomorphism f^{q_n-j} maps M onto $M^* = I_{n+1}^{q_n}$ and T onto $T^* = I_n^{q_{n+1}} \cup I_{n+1}^{q_n} \cup I_n^{q_n}$. By Proposition 6.1, the cross-ratio $[M^*, T^*]$ is bounded from below (by a constant depending only on C_4). Since the intervals $T, f(T), \ldots, f^{q_n-j}(T) = T^*$ have multiplicity of intersection at most 3, it follows from the Cross-ratio Inequality that

$$\mathrm{CrD}(f^{q_n-j}; M, T) = \frac{[M^*, T^*]}{[M, T]} \leq B .$$

Therefore $[M, T]$ is also bounded from below (by a constant depending only on C_4). The same argument applies, *mutatis mutandis*, when M is a long atom. This finishes the proof. □

6.3.3 *On the Notion of Comparability*

The proof of the real bounds was given in such a way as to allow us to keep track of the constants involved in all the estimates—in other words, so that one could actually write down the constant C in Theorem 6.3 explicitly, if necessary. For most of what we do from now on, however, it will not be necessary to keep track of such constants. Instead, we will adopt the same notion and notation of comparability introduced in de Faria and de Melo (1999). To wit, given two positive real numbers α and β, we will say that α is *comparable* to β *modulo* f (or simply that α and β are *comparable*) if there exists a constant $K > 1$ depending only on the real bounds constant $C = C(f)$ such that $K^{-1}\beta \leq \alpha \leq K\beta$. This relation will be denoted $\alpha \asymp \beta$. As observed in de Faria and de Melo (1999, p. 350), comparability modulo f is reflexive and symmetric but not transitive: if we are given a comparability chain $\alpha_1 \asymp \alpha_2 \asymp \cdots \asymp \alpha_k$, we can only say that $\alpha_1 \asymp \alpha_k$ if the length k of the chain is bounded by a constant that depends only on f. In everything we do in this chapter, the lengths of all comparability chains are in fact universally bounded.

6.4 First Consequences

The real bounds given in Theorem 6.3 have many important consequences. In this section we present two of the most basic such consequences.

6.4.1 C^1 Bounds

The first corollary to Theorem 6.3 is the fact that the first returns of a multicritical circle map (to any one of its critical points) are uniformly bounded in the C^1 topology. This is a consequence of the following lemma.

Lemma 6.5 *Given a multicritical circle map f there exists $K = K(f) > 1$ such that, for each $c \in \mathrm{Crit}(f)$, $n \in \mathbb{N}$, $x \in I_n(c)$ and $j \in \{0, 1, \ldots, q_{n+1}\}$, we have*

$$Df^j(x) \leq K \frac{|f^j(I_n(c))|}{|I_n(c)|} . \tag{6.34}$$

The detailed proof will be given below. Let us first show how this lemma implies the C^1 bounds we mentioned above.

Corollary 6.1 *The sequence $\left\{f^{q_{n+1}}|_{I_n(c)}\right\}$ is bounded in the C^1 metric.*

This statement is perhaps a bit too informal. To be really precise, what we mean to say is that, if $\Lambda_n : \mathbb{R} \to \boldsymbol{S}^1 = \mathbb{R}/\mathbb{Z}$ is the unique affine map with $\Lambda_n(0) = c$ and $\Lambda_n([0, 1]) = I_n(c)$, then the sequence of normalized maps $\Lambda_n^{-1} \circ f^{q_{n+1}} \circ \Lambda_n|_{[0,1]}$ is bounded in the C^1 topology.

Proof of Corollary 6.1 By combinatorics, $I_{n+1}(c) \subset f^{q_{n+1}}(I_n(c)) \subset I_n(c) \cup I_{n+1}(c)$. Then:

$$\frac{|I_{n+1}(c)|}{|I_n(c)|} \leq \frac{\left|f^{q_{n+1}}(I_n(c))\right|}{|I_n(c)|} \leq 1 + \frac{|I_{n+1}(c)|}{|I_n(c)|} .$$

By the real bounds (Theorem 6.3) we have $|I_{n+1}(c)| \asymp |I_n(c)|$, and from this it follows that $\left|f^{q_{n+1}}(I_n(c))\right| \asymp |I_n(c)|$. Therefore Corollary 6.1 follows from Lemma 6.5. □

The remainder of this section is devoted to proving Lemma 6.5. For ease of notation, in the proof we adopt the same convention we used in the proof of Theorem 6.3: we drop the dependency on c and write $I_n = I_n(c)$, etc.

Proof of Lemma 6.5 For each $n \in \mathbb{N}$ consider $L_n = I_{n+1}$, $R_n = f^{q_n}(I_n)$ and $T_n = I_n^* = L_n \cup I_n \cup R_n$. We have three preliminary facts:

Fact 6.1 *The family $\{T_n, f(T_n), \ldots, f^{q_{n+1}-1}(T_n)\}$ has intersection multiplicity bounded by 3.*

Fact 6.1 follows from the following general fact: given $z \in S^1$ and $n \in \mathbb{N}$ let $I = \big[z, R_\rho^{3q_n}(z)\big]$, where R_ρ is the rigid rotation of angle $2\pi\rho$ in the unit circle. Then the multiplicity of intersection of the family $\big\{I, R_\rho(I), \dots, R_\rho^{q_{n+1}-1}(I)\big\}$ is 3 for any $n \in \mathbb{N}$.

Fact 6.2 *There exists a constant $\tau > 0$ (depending only on the real bounds of f) such that*

$$|L_n^j| > \tau |I_n^j| \quad \textit{and} \quad |R_n^j| > \tau |I_n^j|$$

for each $j \in \{0, \dots, q_{n+1}\}$ and for all $n \in \mathbb{N}$.

Proof of Fact 6.2 For $j = 0$, observe that the intervals L_n, I_n and R_n are adjacent and belong to the dynamical partition $\mathscr{P}_n$, then by the real bounds they are comparable by a constant that only depends on f. Let us prove now that for $j = q_{n+1}$ the three intervals L_n^j, I_n^j and R_n^j are comparable too.

On the one hand, the intervals I_{n+1} and $I_{n+1}^{q_{n+1}}$ are adjacent and belong to $\mathscr{P}_{n+1}$, hence they are comparable (again by the real bounds). Moreover $I_{n+1} \subset I_n^{q_{n+1}} \subset I_{n+1} \cup I_n$. By the real bounds $|I_n| \asymp |I_{n+1}|$ and then $|I_n^{q_{n+1}}| \asymp |I_{n+1}^{q_{n+1}}|$, that is:

$$|L_n^{q_{n+1}}| \asymp |I_n^{q_{n+1}}|\,. \tag{6.35}$$

On the other hand, the intervals I_n and $I_n^{q_n}$ are adjacent and belong to $\mathscr{P}_n$, hence they are comparable. Moreover:

$$I_{n+1}^{q_n} \subset I_n^{q_n+q_{n+1}} \subset I_n \cup I_n^{q_n}\,.$$

From Fig. 6.5 we know that $|I_{n+1}^{q_n}| \asymp |I_n|$ and then $|I_n^{q_n+q_{n+1}}| \asymp |I_n|$. But $I_{n+1} \subset I_n^{q_{n+1}} \subset I_n \cup I_{n+1}$ and then by the real bounds:

$$|R_n^{q_{n+1}}| = |I_n^{q_n+q_{n+1}}| \asymp |I_n| \asymp |I_n^{q_{n+1}}|\,. \tag{6.36}$$

Therefore, for $j = q_{n+1}$, the three intervals L_n^j, I_n^j and R_n^j are comparable. Now, let $1 \leq j \leq q_{n+1} - 1$. Consider the intervals $|L_n^j|$, $|I_n^j|$, $|R_n^j|$ and their images by the map $f^{q_{n+1}-j}$. By the Cross-ratio Inequality (combined with Fact 6.1) we have that there exists a constant $K_0 = K_0(f) > 1$ such that

$$\frac{|L_n^{q_{n+1}}||R_n^{q_{n+1}}||L_n^j \cup I_n^j||I_n^j \cup R_n^j|}{|L_n^j||R_n^j||L_n^{q_{n+1}} \cup I_n^{q_{n+1}}||I_n^{q_{n+1}} \cup R_n^{q_{n+1}}|} \leq K_0\,.$$

Using (6.35) and (6.36) in the last inequality, we get

$$\left(1+\frac{|I_n^j|}{|L_n^j|}\right)\left(1+\frac{|I_n^j|}{|R_n^j|}\right) \le K\,,$$

and we are done. □

Remark 6.4 We can always assume, whenever necessary, that $n_0 = n_0(f)$ given by Lemma 6.5 is such that for all $n \ge n_0$ and $j \in \{0, \dots, q_{n+1}\}$ we have $\mathrm{Card}(f^j(T_n) \cap \mathrm{Crit}(f)) \le 1$, where Card denotes the cardinality of a finite set, and $\mathrm{Crit}(f)$ is the set of critical points of f (this is because, by minimality, $\left|f^j(T_n)\right|$ goes to zero as n goes to infinity).

Definition 6.3 (Critical Times) We say that $j \in \{1, \dots, q_{n+1}\}$ is a *critical time* if $f^j(T_n) \cap \mathrm{Crit}(f) \ne \emptyset$.

Remark 6.5 Note that $\mathrm{Card}(\{\text{critical times}\}) \le 3N$.

Fact 6.3 *Let $1 \le j_1 < j_2 \le q_{n+1}$ be two consecutive critical times. Then for all $x \in f^{j_1+1}(I_n)$ we have:*

$$Df^{j_2-j_1-1}(x) \asymp \frac{|f^{j_2}(I_n)|}{|f^{j_1+1}(I_n)|}\,,$$

with universal constants (depending only on the real bounds).

Proof of Fact 6.3 Note that $f^{j_2-j_1-1} : f^{j_1+1}(T_n) \to f^{j_2}(T_n)$ is a diffeomorphism. Fact 6.1 implies that $\sum_{i=0}^{j_2-j_1-1} |f^i(f^{j_1+1}(T_n))| < 3$, and by Fact 6.2 the interval $f^{j_2-j_1-1}(f^{j_1+1}(T_n))$ contains a τ−scaled neighborhood of the interval $f^{j_2-j_1-1}(f^{j_1+1}(I_n))$. By Koebe Distortion Principle (Lemma 5.2) there exists a constant $K_0 = K_0(f) > 1$ such that for all $x, y \in f^{j_1+1}(I_n)$ we have that

$$\frac{1}{K_0} \le \frac{Df^{j_2-j_1-1}(x)}{Df^{j_2-j_1-1}(y)} \le K_0\,.$$

Let $y \in I_n^{j_1+1}$ be given by the Mean Value Theorem such that

$$Df^{j_2-j_1-1}(y) = \frac{|f^{j_2}(I_n)|}{|f^{j_1+1}(I_n)|}\,.$$

Then for all $x \in f^{j_1+1}(I_n)$,

$$\frac{1}{K_0}\frac{|f^{j_2}(I_n)|}{|f^{j_1+1}(I_n)|} \le Df^{j_2-j_1-1}(x) \le K_0\frac{|f^{j_2}(I_n)|}{|f^{j_1+1}(I_n)|}\,.$$

□

We finish the proof of Lemma 6.5 by combining Fact 6.3 and item (iii) in Proposition 5.4 with the help of the chain rule:

$$Df^j(x) \leq (3d)^{3N} K_0^{3N} \frac{|f^j(I_n)|}{|I_n|} \quad \text{for any } x \in I_n \text{ and } j \in \{1, \ldots, q_{n+1}\},$$

where $N = \text{Card}\big(\text{Crit}(f)\big)$ is the number of critical points of f, d is the maximum of its criticalities and $K_0 = K_0(f)$ is given by Fact 6.3.

6.4.2 Sums of Polar Ratios

Here is a purely geometric property of dynamical partitions that also follows from the real bounds. It is very useful in situations that require bounding the Schwarzian derivative of first returns—for example in the study of Lyapunov exponents of multicritical circle maps, see Chap. 8.

Let f be a multicritical circle map, and let c be one of its critical points. Let $I \in \mathscr{P}_n(c)$ be an atom of the n-th dynamical partition of f associated to c. If I does not contain c, i.e., if $I \neq I_n(c), I_{n+1}(c)$, we define the *polar ratio* of I with respect to c to be the ratio $|I|/\text{dist}(c, I)$, where $\text{dist}(\cdot, \cdot)$ denotes the usual distance in the circle.

The result we have in mind states that the sum of all polar ratios for atoms at level n grows at most linearly with n. It holds under the general assumptions of Theorem 6.3, for maps with an arbitrary number of critical points. For each $n \geq 1$ let:

$$S_n(c) = \sum_{I \in \mathscr{P}_n(c) \setminus \{I_n(c), I_{n+1}(c)\}} \frac{|I|}{d(c, I)},$$

where $d(c, I)$ denotes the Euclidean distance between an interval $I \subset S^1$ and the critical point c.

Lemma 6.6 *For each critical point c of f, the sequence* $\left\{\dfrac{S_n(c)}{n}\right\}_{n \geq 1}$ *is bounded.*

Proof As before, for simplicity of notation we write $\mathscr{P}_k$, I_k instead of $\mathscr{P}_k(c)$, $I_k(c)$ respectively, for each $k \in \mathbb{N}$. Note that the transition from $\mathscr{P}_n$ to $\mathscr{P}_{n+1}$ can be described in the following way: the interval $I_n = [c, f^{q_n}(c)]$ is subdivided by the points $f^{jq_{n+1}+q_n}(c)$ with $1 \leq j \leq a_{n+1}$ into $a_{n+1} + 1$ subintervals. This sub-partition is spread by the iterates of f to yield sub-partitions of each long atom $f^j(I_n) = f^j([c, f^{q_n}(c)])$ with $0 \leq j < q_{n+1}$. The other elements of the partition $\mathscr{P}_n$, namely the intervals $f^j(I_{n+1})$ with $0 \leq j < q_n$, remain unchanged. Now, on one hand, for any $I \in \mathscr{P}_n \setminus \{I_n, I_{n+1}\}$ we have:

$$\sum_{I\supset J\in\mathscr{P}_{n+1}} \frac{|J|}{d(c,J)} \leq \frac{1}{d(c,I)} \sum_{I\supset J\in\mathscr{P}_{n+1}} |J| = \frac{|I|}{d(c,I)} .$$

On the other hand:

$$\sum_{\mathscr{P}_{n+1}\ni J\subset I_n\setminus I_{n+2}} \frac{|J|}{d(c,J)} \leq \frac{1}{|I_{n+2}|} \sum_{\mathscr{P}_{n+1}\ni J\subset I_n\setminus I_{n+2}} |J| = \frac{|I_n \setminus I_{n+2}|}{|I_{n+2}|} .$$

This gives us:

$$0 \leq S_{n+1} - S_n \leq \frac{|I_n \setminus I_{n+2}|}{|I_{n+2}|} \quad \text{for all } n \geq 1.$$

But, by the real bounds, we have

$$\frac{|I_n \setminus I_{n+2}|}{|I_{n+2}|} \leq \frac{|I_n|}{I_{n+2}|} \leq C^2 ,$$

for all $n \geq 1$, where $C = C(f)$ is the constant in Theorem 6.3. Telescoping, we deduce that $S_n \leq S_0 + C^2 n$, as desired. □

Remark 6.6 More generally, we may consider sums of *powers* of polar ratios; see Exercise 6.4. Such sums appear in several places in the study of renormalization of one-dimensional maps, e.g., in de Faria and de Melo (1999). Similar sums (with weights) are used in the study of unimodal maps: see de Faria et al. (2006), and also Clark et al. (2022).

6.5 A Negative Schwarzian Property

The study of the fine geometry of a smooth one-dimensional map is usually facilitated if the Schwarzian derivative of said map happens to be negative (see Chap. 5). Such negative Schwarzian property is therefore certainly desirable.

A general (C^3-smooth) multicritical circle map does not have, in general, negative Schwarzian, but in some sense this property emerges as we iterate the map. This is expressed in more precise terms through the following result.

Proposition 6.2 *Given a multicritical circle map f there exists a constant $n_0 = n_0(f) \in \mathbb{N}$ such that, for all $n > n_0$ and each $c \in \mathrm{Crit}(f)$ the following facts hold.*

(i) For all $j \in \{1, \ldots, q_{n+1}\}$ and each $x \in I_n(c)$ regular point of f^j, we have $Sf^j(x) < 0$.
(ii) For all $j \in \{1, \ldots, q_n\}$ and each $x \in I_{n+1}(c)$ regular point of f^j, we have $Sf^j(x) < 0$.

Remark 6.7 Later in this book (see Chap. 10) we will introduce the notion of *renormalization* of a multicritical circle map (around one of its critical points). Roughly speaking, given a map f and a point x in its domain, a *renormalization of f around x* is simply a first return map to a neighborhood of x (linearly rescaled to unit size, say). In this language, Proposition 6.2 is saying in particular that every sufficiently deep renormalization of a multicritical circle map has the negative Schwarzian property.

Remark 6.8 The fact that $Sf^{q_n+1}(x) < 0$ is most likely true for any regular point x of $f^{q_{n+1}}$, not necessarily contained in $I_n(c)$ (and the same with the second assertion in Proposition 6.2). For bounded combinatorics, a proof of this fact can be found in Sect. 8.2.3.

Proof of Proposition 6.2 In the proof we adapt the exposition in de Faria and de Melo (1999, pp. 380–381). We give the proof only for the case $x \in I_n$ regular point of f^j for some $j \in \{1, \ldots, q_{n+1}\}$ (the other case being entirely analogous).

From item (i) in Proposition 5.4, we know that for each critical point c_i there exist a neighborhood $U_i \subseteq \mathbf{S}^1$ of c_i and a positive constant K_i such that for all $x \in U_i \setminus \{c_i\}$ we have

$$Sf(x) < -\frac{K_i}{(x - c_i)^2} < 0\,. \tag{6.37}$$

Let us call $\mathscr{U} = \bigcup_{i=0}^{i=N-1} U_i$, and let $\mathscr{V} \subset \mathbf{S}^1$ be an open set that contains none of the critical points of f and such that $\mathscr{U} \cup \mathscr{V} = \mathbf{S}^1$. Since f is C^3, $M = \sup_{y \in \mathscr{V}} |Sf(y)|$ is finite. Let $\delta_n = \max_{0 \le j < q_{n+1}} |I_n^j|$. We know that $\delta_n \to 0$ as $n \to \infty$, because f is topologically conjugate to a rotation. We choose $n_0 = n_0(f)$ so large that δ_n is smaller than the Lebesgue number of the covering $\{\mathscr{U}, \mathscr{V}\}$ of the circle for all $n \geq n_0$. Using the *chain rule* for the Schwarzian derivative, we have for all $n \geq n_0$ and all $x \in I_n(c)$ regular point of f^j

$$Sf^j(x) = \sum_{k=0}^{j-1} Sf(f^k(x)) \left[Df^k(x)\right]^2 . \tag{6.38}$$

We can decompose this sum as $\Sigma_1^{(n)}(x) + \Sigma_2^{(n)}(x)$ where

$$\Sigma_1^{(n)}(x) = \sum_{k: I_n^k \subset \mathscr{U}} Sf(f^k(x)) \left[Df^k(x)\right]^2 , \tag{6.39}$$

and $\Sigma_2^{(n)}(x)$ is the sum over the remaining terms.
Now we proceed through the following steps:

(i) Since $I_n \subset \mathscr{U}$, the sum in the right-hand side of (6.39) includes the term with $k = 0$, namely $Sf(x)$. Since all the other terms in (6.39) are negative as well, and since $|x - c| \leq |I_n|$, we deduce from (6.37) that

$$\Sigma_1^{(n)}(x) \; < \; -\frac{K_1}{|I_n|^2} \,. \tag{6.40}$$

(ii) Observe that,

$$\left|\Sigma_2^{(n)}(x)\right| \; \leq \; \sum_{I_n^k \subset \mathcal{V}} |Sf(f^k(x))| \left[Df^k(x)\right]^2 . \tag{6.41}$$

By choosing n_0 large enough, we know from Eq. (6.34) that there exists $K = K(f) > 1$ such that

$$\begin{aligned}\left|\Sigma_2^{(n)}(x)\right| &\leq \sum_{I_n^k \subset \mathcal{V}} |Sf(f^k(x))| K^2 \frac{|I_n^k|^2}{|I_n|^2} \\ &\leq M \frac{K^2}{|I_n|^2} \sum_{I_n^k \subset \mathcal{V}} |I_n^k|^2 \\ &\leq M \frac{K^2}{|I_n|^2} \max_{0 \leq k \leq j-1} |I_n^k| \sum_{I_n^k \subset \mathcal{V}} |I_n^k| \\ &\leq M \frac{K^2}{|I_n|^2} \delta_n .\end{aligned} \tag{6.42}$$

Choosing n_0 so large that $K^2 M \delta_n < K_1$ for all $n \geq n_0$, we deduce from (6.40) and (6.42) that, indeed, $Sf^j(x) < 0$ for all $j \in \{1, \ldots, q_{n+1}\}$ and for $n \geq n_0$. □

6.6 Beau Bounds

As we have already observed, the comparability constant C we obtained in Theorem 6.3 depends on the map f. In this section we will show that, asymptotically, we can replace $C = C(f)$ by a universal constant. Uniform bounds of this type are called *beau* by Sullivan (1992). The precise result is the following.

Theorem 6.4 (Beau Bounds) *Given $N \geq 1$ in $\mathbb{N}$ and $d > 1$ there exists a universal constant $C = C(N, d) > 1$ with the following property. For any given multicritical circle map f with irrational rotation number, and with at most N critical points whose criticalities are bounded by d, there exists $n_0 = n_0(f) \in \mathbb{N}$ such that for each critical point c of f, for all $n \geq n_0$, and for every pair I, J of adjacent atoms of $\mathcal{P}_n(c)$ we have*

$$C^{-1}|I| \leq |J| \leq C|I|.$$

The proof of this theorem is the same as the proof of Theorem 6.3, but we must replace Theorem 5.1 with the following result (originally given in Estevez et al. 2018, Th. B).

Theorem 6.5 *Given $N \geq 1$ in $\mathbb{N}$ and $d > 1$ there exists a constant $B = B(N, d) > 1$ with the following property. Given a multicritical circle map f, with at most N critical points whose criticalities are bounded by d, there exists $n_0 = n_0(f)$ such that for all $n \geq n_0$, $\Delta \in \mathscr{P}_n(c)$ and $k \in \mathbb{N}$ such that $f^i(\Delta)$ is contained in an element of $\mathscr{P}_n(c)$ for all $1 \leq i \leq k$, we have that*

$$\mathrm{CrD}(f^k; \Delta, \Delta^*) \leq B,$$

where Δ^* denotes the union of Δ with its left and right neighbours in $\mathscr{P}_n(c)$.

The following *decomposition* will be crucial in the proof of Theorem 6.5 given below (recall that, for a given $J \in \mathscr{P}_n$, we denote by J^* the union of J with its left and right neighbours in $\mathscr{P}_n$). For each critical point c_i we consider its neighborhood U_i given by Proposition 5.4. Moreover, let $n_1 \in \mathbb{N}$ be given by Proposition 6.2.

Lemma 6.7 *Given $\varepsilon > 0$ there exists $n_2 \in \mathbb{N}$, $n_2 = n_2(\varepsilon, f) > n_1$, with the following property: given $n \geq n_2$, $\Delta \in \mathscr{P}_n$ and $k \in \mathbb{N}$ such that $f^j(\Delta)$ is contained in an element of $\mathscr{P}_n$ for all $1 \leq j \leq k$, we can write*

$$f^k|\Delta^* = \phi_k \circ \phi_{k-1} \circ \cdots \circ \phi_1,$$

where:

1. *For at most $3N + 1$ values of $i \in \{1, \ldots, k\}$, ϕ_i is a diffeomorphism with distortion bounded by $1 + \varepsilon$.*
2. *For at most $3N$ values of $i \in \{1, \ldots, k\}$, ϕ_i is the restriction of f to some interval contained in U_i.*
3. *For the remainder values of i, ϕ_i is either the identity or a diffeomorphism with negative Schwarzian derivative.*

The above statement and its proof below are borrowed from Estevez et al. (2018, pp. 853–855), which in turn is an adaptation of the argument given in de Faria and de Melo (1999, pp. 352–353).

Proof of Lemma 6.7 Let $C_0 = C_0(f) \geq 1$ be given by the Koebe distortion principle (Lemma 5.2). Let $C > 1$ and $\mu \in (0, 1)$ given by Theorem 6.3. Let $\delta \in (0, 1)$ be such that $(1 + \delta)^2 \exp(C_0 \delta) < 1 + \varepsilon$, and let $n_2 \in \mathbb{N}$ be such that

$$n_2 > n_1 + \frac{4 \log(\delta \mu^{3/2}/C)}{\log \mu}.$$

Note that $0 < (\mu^{1/4})^{n_2 - n_1} < \delta\mu^{3/2}/C$. Given $n \geq n_2$ consider

$$m = m(n) = \left\lfloor \frac{n + n_1}{2} \right\rfloor,$$

the *integer part* of $\frac{1}{2}(n + n_1)$. Let Δ and k as in the statement, and consider $J_m \in \mathscr{P}_m$ such that $\Delta \subseteq J_m$, and consider also $J_{n_1} \in \mathscr{P}_{n_1}$ with $J_m \subseteq J_{n_1}$. Taking n sufficiently large, we may assume that $\Delta^* \subset J_m$.

Let $s \geq 0$ be the smallest natural number such that $f^s(J_{n_1})$ contains a critical point of f.

Claim 6.6.1 *The distortion of f^s on Δ^* is bounded by $1 + \varepsilon$.*

□

Proof of Claim 6.6.1 The proof uses the Koebe Distortion Principle (Lemma 5.2). Replacing n_1 by $n_1 + 1$ if necessary, we may assume that $f^j(J_{n_1}) \in \mathscr{P}_{n_1}$ for all $j \in \{0, \ldots, s-1\}$. By the real bounds, the space τ of Δ^* inside J_m^* is bounded from below by

$$\tau \geq \frac{1}{C}\frac{|J_m|}{|\Delta^*|} \geq \frac{1}{C}\left(\frac{1}{\mu}\right)^{\lfloor (n-m)/2 \rfloor} > \frac{\mu}{C}\left(\frac{1}{\mu}\right)^{(n-m)/2}.$$

Since $m \leq \frac{n+n_1}{2}$, we have $n - m \geq n - \frac{n+n_1}{2} = \frac{n-n_1}{2}$, and then

$$\frac{1}{\tau} \leq \frac{C}{\mu}\,\mu^{(n-m)/2} \leq \frac{C}{\mu}(\mu^{1/4})^{n-n_1} < \sqrt{\mu}\,\delta < \delta\,. \tag{6.43}$$

Now we estimate the sum ℓ of the lengths of the iterates of J_m^* between 1 and $s - 1$. Since $\frac{n+n_1}{2} < m+1$, we have $m - n_1 > \frac{n-n_1}{2} - 1$, and then for all $j \in \{0, \ldots, s-1\}$:

$$\begin{aligned}\left|f^j(J_m^*)\right| &\leq \mu^{\lfloor (m-n_1)/2 \rfloor}\left|f^j(J_{n_1}^*)\right| \leq \\ &\leq (\mu^{1/4})^{n-n_1}\left(\frac{1}{\mu}\right)^{3/2}\left|f^j(J_{n_1}^*)\right| \leq \frac{\delta}{C}\left|f^j(J_{n_1}^*)\right|.\end{aligned}$$

Therefore:

$$\ell = \sum_{j=0}^{s-1} |f^j(J_m^*)| < \frac{3\delta}{C} < \delta\,, \tag{6.44}$$

since $\sum_{j=0}^{s-1} |f^j(J_{n_1}^*)| < 3$ by combinatorics (and assuming $C > 3$). From inequalities (6.43), (6.44) and Koebe distortion principle (Lemma 5.2) we get that the distortion on Δ^* is bounded from above by

$$(1+\delta)^2 \exp(C_0\,\delta) < 1+\varepsilon\,.$$

□

To prove Lemma 6.7 we decompose the orbit of Δ^* under f according to the following algorithm. For each $i \in \{0, 1, \ldots, k-1\}$ we have two cases to consider:

1. If $f^i(J_{n_1})$ does not contain any critical point of f, we define the corresponding ϕ to be f^s, where $s \geq 1$ is the smallest natural such that $f^{i+s}(J_{n_1})$ contains a critical point of f. Arguing as in Claim 6.6.1 above, we see that this case belongs to the first type of components in the statement.
2. If $f^i(J_{n_1})$ contains a critical point c of f we may assume, by taking n_2 large enough, that $f^i(\Delta^*) \subset I_{n_1}(c) \cup I_{n_1+1}(c)$. We have two sub-cases to consider:
   (i) If $f^i(\Delta^*)$ does not contain c (and therefore no other critical point) let $s \geq 1$ be the smallest natural such that $f^{i+s}(\Delta^*)$ contains a critical point of f, and we define the corresponding ϕ to be f^s. By Proposition 6.2 (and the fact that composition of diffeomorphisms with negative Schwarzian derivative is a diffeomorphism with negative Schwarzian derivative too) this case belongs to the third type of components in the statement.
   (ii) If the critical point belongs to $f^i(\Delta^*)$ we define the corresponding ϕ to be just a single iterate of f (and this sub-case belongs to the second type of components in the statement).

Note finally that, by combinatorics, the first case happens at most $3N+1$ times, while the second case occurs at most $3N$ times.

With Lemma 6.7 at hand, we are ready to prove our main results.

Proof of Theorem 6.5 Theorem 6.5 follows at once from the decomposition obtained in Lemma 6.7, by combining Lemma 5.4 and item (iv) of Proposition 5.4. The constant B depends only on the number and order of the critical points of f, but not on f itself. It is in fact enough to consider $B = (1+1/2)^{2(3N+1)}(9d^2)^{3N}$. □

Proof of Theorem 6.4 As we have already explained, the proof of Theorem 6.4 is the same as the proof of the real bounds (Theorem 6.3), the only difference being that the Cross-ratio Inequality is replaced by Theorem 6.5. □

Exercises

Exercise 6.1 Give another proof of Lemma 6.4 as follows.

(i) Show that

$$\mathscr{P}_0(x) = \Big\{\big[f^i(x), f^{i+1}(x)\big] : i \in \{0, \ldots, a_0-1\}\Big\} \cup \big\{\big[f^{a_0}(x), x\big]\big\}$$

is a partition of the circle (modulo endpoints), where a_0 is the integer part of $1/\rho$.

(ii) Using Remark 6.3, show that if $\mathscr{P}_n(x)$ is a partition of the circle, then $\mathscr{P}_{n+1}(x)$ is a partition as well.

Exercise 6.2 Let $f : \mathbf{S}^1 \to \mathbf{S}^1$ be a circle homeomorphism with irrational rotation number ρ, and with unique invariant measure μ. Show that for any $x \in \mathbf{S}^1$ and any $n \in \mathbb{N}$ we have

$$\mu(I_n) = \prod_{j=0}^{n} G^j(\rho) = \rho\, G(\rho)\, G^2(\rho) \cdots G^n(\rho), \tag{6.45}$$

where I_n is the interval with endpoints x and $f^{q_n}(x)$ containing $f^{q_{n+2}}(x)$, and G denotes the Gauss map from Chap. 1 (see also Appendix A). (*Hint:* see de Faria and Guarino 2022b, Lem. 2.3).

Exercise 6.3 The following consequence of the real bounds is extracted from (de Faria and de Melo 2000, Lem. 2.2). Let f be a unicritical circle map with arbitrary rotation number and critical point c, let $n \geq 1$, and let $J_{-i} = f^{q_{n+1}-i}(I_n(c))$ for $0 \leq i < q_{n+1}$. Given $m < n$, let $i_1 < i_2 < \cdots < i_\ell$ be the moments in the backward orbit $\{J_{-i}\}$ before the first return to $I_{m+1}(c)$ such that $J_{-i_k} \subseteq I_m(c)$.

(i) Show that $\ell = a_{m+1}$, and that

$$J_{-i_k} \subseteq f^{q_m+(a_{m+1}-k+1)q_{m+1}}(I_{m+1}(c))\ .$$

(ii) Show that, given an integer $M \geq 1$, there exists $C_M > 0$ such that for all sufficiently large n we have $C_M^{-1}|I_n| \leq |J_{-i_k}| \leq C_M|I_n|$, provided $1 \leq k \leq M$ or $a_{m+1} - M + 1 \leq k \leq a_{m+1}$.

[*Hint:* The largest $j < q_{n+1}$ such that $f^j(I_n(c)) \subseteq I_{m+1}(c)$ is easily computed as $j = q_{n+1} - q_{m+2}$. Since $q_{m+2} = q_m + a_{m+1}q_{m+1}$, there are exactly a_{m+1} subsequent moments $j < i < q_{n+1}$ such that $f^i(I_n(c)) \subseteq I_m(c)$. The rest follows from the real bounds (Theorem 6.3) and the Koebe distortion principle (Lemma 5.2).]

Exercise 6.4 Let f be a multicritical circle map, let $c \in \mathrm{Crit}(f)$, and fix $p > 1$. For each $n \geq 1$, let

$$S_n^{(p)}(c) = \sum_{I \in \mathscr{P}_n(c) \setminus \{I_n(c), I_{n+1}(c)\}} \left(\frac{|I|}{d(c, I)}\right)^p .$$

Show that the sequence $\left\{S_n^{(p)}(c)\right\}_{n \geq 1}$ is bounded. [*Hint:* Imitate the proof of Lemma 6.6.]

Exercise 6.5 Let f be a C^3 critical circle map with irrational rotation number and a *unique* critical point $c \in S^1$. Show that there exists a constant $K_1 > 1$ such that the following facts hold for each $n \geq n_0$:

(i) For all $x, y \in f(I_{n+1}(c))$, we have

$$\frac{1}{K_1} \leq \frac{\left|Df^{q_n-1}(x)\right|}{\left|Df^{q_n-1}(y)\right|} \leq K_1 .$$

(ii) For all $x, y \in f(I_n(c))$, we have

$$\frac{1}{K_1} \leq \frac{\left|Df^{q_{n+1}-1}(x)\right|}{\left|Df^{q_{n+1}-1}(y)\right|} \leq K_1 .$$

Do these statements remain true if f has two or more critical points? Explain.

Exercise 6.6 Let f be as in Exercise 6.5. Prove that there exists $C > 1$ such that

$$\frac{1}{C} \leq Df^{q_{n+1}}(x) \leq C .$$

for all $x \in I_n(c) \setminus I_{n+2}(c)$ and all $n \in \mathbb{N}$.

Exercise 6.7 In Ergodic Theory, a famous lemma due to Kakutani and Rokhlin[1] states that, if $(X, \mathscr{B}, \mu)$ is a non-atomic probability measure space and $T : X \to X$ is an ergodic measure-preserving invertible transformation, then for each $n \in \mathbb{N}$ and each $\varepsilon > 0$ there exists $B \in \mathscr{B}$ such that $B, TB, \ldots, T^{n-1}B$ are pairwise disjoint and $\mu(B \cup TB \cup \cdots \cup T^{n-1}B) > 1 - \varepsilon$. Using the dynamical partitions $\mathscr{P}_n(x)$ of Sect. 6.3.1 and Yoccoz's Theorem 6.2, prove the Kakutani–Rokhlin lemma in the special case when X is the unit circle, $\mathscr{B}$ is its Borel σ-algebra, T is a multicritical circle map f with $\mathrm{Per}(f) = \emptyset$, and μ is the unique f-invariant Borel probability measure.

[1] Later generalized by Halmos (1956, p. 71).

Chapter 7
Quasisymmetric Rigidity

In addition to the real bounds, another important preliminary step towards establishing the *smooth rigidity* of multicritical circle maps (to be examined in Sect. 10.1) is to answer the question: When are two topologically conjugate multicritical circle maps *quasisymmetrically* conjugate? This question pertains to the general study of *quasisymmetric rigidity* of one-dimensional systems. Our purpose in this chapter is twofold:

(a) To derive useful geometric criteria that allow us to decide whether a given homeomorphism is quasisymmetric, or perhaps even smooth; and
(b) To use one such criterion to prove a quasisymmetric rigidity theorem for multicritical circle maps.

The main theorem in this chapter is Theorem 7.2, which provides an answer to the question raised above and yields the quasisymmetric rigidity alluded to in item (b). It states that if we are given two (minimal) multicritical circle maps with the same number of critical points, say f and g, and we know that there is a topological conjugacy h between them that maps each critical point of f to a critical point of g, then h must be a quasisymmetric homeomorphism.

In Chap. 9 we will go a bit further and examine the geometric structure of individual orbits—more specifically, we will examine with a reasonable amount of detail the problem of classifying the orbits of such a map up to quasisymmetric equivalence.

7.1 Quasisymmetry and Fine Grids

The concept of quasisymmetry stems from the theory of quasiconformal mappings. Quasisymmetric homeomorphisms arise as boundary values of *quasiconformal* homeomorphisms of the unit disk or the upper half-plane (see Ahlfors 2006, Ch. 4).

E. de Faria, P. Guarino, *Dynamics of Circle Mappings*, IMPA Monographs,
https://doi.org/10.1007/978-3-031-67495-2_7

Roughly speaking, an orientation-preserving self-homeomorphism of the unit circle or the real line is *quasisymmetric* if it maps every triple of equally spaced points onto a triple of almost equally spaced points. Here is the formal definition.

Definition 7.1 An orientation-preserving homeomorphism of $S^1 = \mathbb{R}/\mathbb{Z}$, say $h : S^1 \to S^1$, is said to be *quasisymmetric* if there exists a constant $K \geqslant 1$ such that

$$\frac{1}{K} \leqslant \frac{h(x+t) - h(x)}{h(x) - h(x-t)} \leqslant K\,, \quad \text{for all } x \in S^1 \text{ and all } t > 0\,. \tag{7.1}$$

If K is such that h satisfies (7.1) for this K, then we say that h is K*-quasisymmetric*. The smallest K with this property is called the *quasisymmetric distortion* of h.

The kind of regularity possessed by a quasisymmetric homeomorphism is very weak. Indeed, most quasisymmetric homeomorphisms are purely singular with respect to Lebesgue measure. They are, however, always Hölder continuous. Moreover, the composition of quasisymmetric homeomorphisms is quasisymmetric, and the inverse of a quasisymmetric homeomorphism is also quasisymmetric. These properties are not obvious from the definition given above, but they are easily proved once it is established that quasisymmetric homeomorphisms are precisely the boundary values of *quasiconformal* self-homeomorphisms of the disk (once again, see Ahlfors 2006, Ch. 4; see also Exercise 7.1).

There is a relationship between quasisymmetry and distortion of cross-ratios, but a full discussion of it would constitute a lengthy digression. There are in fact only a couple of places in the present book where a particular instance of this relationship is required. What we need is a simple consequence of the following result, which we state without proof (*cf.* de Faria and de Melo 2008, p. 130). Here, we will be using the b-cross-ratio, i.e., $[M, T] = b(M, T)$, but both the proposition below and its corollary can be easily recast in terms of the a-cross-ratio.

Proposition 7.1 *If $\phi : S^1 \to S^1$ is quasisymmetric, then there exists a non-decreasing function $\sigma : [0, \infty) \to [0, \infty)$ with $\sigma(t) \to 0$ as $t \to 0$ such that $[\phi(M), \phi(T)] \leqslant \sigma([M, T])$ for every pair of intervals $M, T \subset S^1$ with M compactly contained in the interior of T.*

A proof of this result may be found in Astala et al. (2009). In order to state the corollary in simple terms, it is best to introduce a definition. We say that a homeomorphism $\phi : S^1 \to S^1$ *has weakly bounded cross-ratio distortion* if for every pair of constants $0 < \alpha < \beta < 1$ there exists $B_{\alpha,\beta} > 0$ such that $\mathrm{CrD}(\phi, M, T) \leqslant B_{\alpha,\beta}$ for every pair of intervals M, T (with M compactly contained in the interior of T) such that $\alpha \leqslant [M, T] \leqslant \beta$.

Corollary 7.1 *Every quasisymmetric homeomorphism of the circle has weakly bounded cross-ratio distortion.*

Proof Follows easily from Proposition 7.1; the details are left as an exercise. □

This corollary will be used in its contrapositive, as a criterion for *non-quasisymmetry* (see Chap. 9).

7.1.1 A Criterion for Quasisymmetry

Let us now describe a criterion for quasisymmetry that is particularly useful in the study of circle maps. In order to formulate it, we first need to introduce the concept of a *fine grid*. Here is the definition, reproduced almost verbatim from Estevez and de Faria (2018, Def. 5.1).

Definition 7.2 A *fine grid* is a nested sequence $\{\mathscr{Q}_n\}_{n\geqslant 0}$ of finite interval partitions of the circle (modulo endpoints) having the following properties.

(a) Each $\mathscr{Q}_{n+1}$ is a strict refinement of $\mathscr{Q}_n$.
(b) There exists an integer $a \geqslant 2$ such that each atom $\Delta \in \mathscr{Q}_n$ is the disjoint union of at most a atoms of $\mathscr{Q}_{n+1}$.
(c) There exists $\sigma > 1$ such that $\sigma^{-1}|\Delta| \leqslant |\Delta'| \leqslant \sigma|\Delta|$ for each pair of adjacent atoms $\Delta, \Delta' \in \mathscr{Q}_n$.

The numbers a, σ are called *fine grid constants*.

Remark 7.1 Given a fine grid as above, it is not difficult to check that there exist $0 < \lambda_0 < \lambda_1 < 1$ depending only on the fine grid constants a, σ such that, whenever $\Delta \in \mathscr{Q}_n$, $\Delta^* \in \mathscr{Q}_{n-1}$ and $\Delta \subset \Delta^*$, we have

$$\lambda_0|\Delta^*| \leqslant |\Delta| \leqslant \lambda_1|\Delta^*| . \tag{7.2}$$

In fact, one can take $\lambda_0 = (a\sigma^{a-1})^{-1}$ and $\lambda_1 = (1+\sigma^{-1})^{-1}$. The details are left as an exercise for the reader. In particular, there exists a constant $C_0 > 1$ such that

$$C_0^{-1}\lambda_0^n \leqslant |\Delta| \leqslant C_0\lambda_1^n$$

for all n and each $\Delta \in \mathscr{Q}_n$. When called upon, the constants λ_0, λ_1 will also be referred to as fine grid constants.

The notion of fine grid was first introduced in de Faria and de Melo (1999, §4). Its usefulness lies in the fact that one can sometimes tell how regular a homeomorphism is by looking at the effect it has on a suitable fine grid. This will be illustrated by two results we proceed to present, namely Propositions 7.2 and 7.3, the first of which is the criterion for quasisymmetry that we promised above.

First we need the following lemma.

Lemma 7.1 *Given a fine grid $\{\mathscr{Q}_n\}_{n\geqslant 0}$ with fine grid constants a, σ as above, let $I \subset S^1$ be an interval with non-empty interior, and let $n = n(I)$ be the smallest natural number such that $I \supset \Delta$ for some atom $\Delta \in \mathscr{Q}_n$. Then there exists an interval $U \supset I$ with the following properties:*

(i) U is the union of at most $2a$ atoms of $\mathscr{Q}_n$;
(ii) $|U| \leqslant \lambda_0^{-1}(1+\sigma)|I|$, where λ_0 is the constant in (7.2).

Proof Suppose I intersects 3 distinct consecutive atoms of $\mathscr{Q}_{n-1}$, say $\Delta_1, \Delta_2, \Delta_3$, with Δ_2 lying between Δ_1 and Δ_3. Then we necessarily have $\Delta_2 \subseteq I$; but this contradicts the definition of $n = n(I)$. Hence I is contained in the union U of at most two atoms of $\mathscr{Q}_{n-1}$. Since each atom of $\mathscr{Q}_{n-1}$ is the union of at most a atoms of $\mathscr{Q}_n$, part (i) follows. To prove (ii), given that $I \supset \Delta \in \mathscr{Q}_n$, let Δ^* be the unique atom of $\mathscr{Q}_{n-1}$ that contains Δ. By part (i), U contains Δ^* and at most one other atom $\Delta^{**} \in \mathscr{Q}_{n-1}$ adjacent to Δ^*. Therefore, using property (c) in Definition 7.2 and (7.2), we have

$$|U| \leqslant |\Delta^*| + |\Delta^{**}| \leqslant (1+\sigma)|\Delta^*| \leqslant \lambda_0^{-1}(1+\sigma)|\Delta| \leqslant \lambda_0^{-1}(1+\sigma)|I| .$$

This establishes (ii) and finishes the proof. □

Proposition 7.2 *Let $\{\mathscr{Q}_n\}_{n\geqslant 0}$ be a fine grid in $\boldsymbol{S}^1$ whose fine grid constants are a, σ, and let $h : \boldsymbol{S}^1 \to \boldsymbol{S}^1$ be an orientation-preserving homeomorphism such that*

$$\left| \frac{|h(\Delta')|}{|h(\Delta'')|} - \frac{|\Delta'|}{|\Delta''|} \right| \leqslant \lambda , \tag{7.3}$$

for each pair of adjacent atoms $\Delta', \Delta'' \in \mathscr{Q}_n$, for all $n \geqslant 0$, where λ is a positive constant. Then there exists $K = K(a, \sigma, \lambda) > 1$ such that h is K-quasisymmetric.

Proof We will verify the quasisymmetry condition

$$\frac{1}{K} \leqslant \frac{h(x+t) - h(x)}{h(x) - h(x-t)} \leqslant K$$

for all $x \in \boldsymbol{S}^1 = \mathbb{R}/\mathbb{Z}$ and all $t > 0$, with $K > 1$ a constant to be determined in the course of the argument. Let $I = [x-t, x+t]$ be the interval on the circle that contains x, and write $I = I^- \cup I^+$, where $I^- = [x-t, x]$ and $[x, x+t]$. By Lemma 7.1, there exist $n = n(I)$ and an interval $U \supset I$ such that U is the union of at most $2a$ atoms of $\mathscr{Q}_n$ and $|U| \leqslant \sigma_1|I|$, where $\sigma_1 = \lambda_0^{-1}(1+\sigma)$. Let p be the smallest positive integer such that $\lambda_1^p\sigma_1 < \frac{1}{4}$. Write U as the union of atoms of $\mathscr{Q}_{n+p}$, say

$$U = J_1 \cup J_2 \cup \cdots \cup J_s ,$$

where the $J_i \in \mathscr{Q}_{n+p}$, $1 \leqslant i \leqslant s$ are assumed to be in counterclockwise order on the circle. Note that we must have $s \leqslant 2a^{p+1}$. By (7.2) and induction, we have $|J_i| \leqslant \lambda_1^p|J_i^*|$, where $J_i^* \subseteq U$ is the unique atom of $\mathscr{Q}_n$ that contains J_i. Hence we get

$$|J_i| \leqslant \lambda_1^p |J_i^*| \leqslant \lambda_1^p |U| \leqslant \lambda_1^p \sigma_1 |I| < \frac{1}{4}|I| \, .$$

But this means that at least one of the J_i's, say J_{i_0}, is contained in I^-. Thus, we have on the one hand $J_{i_0} \subset I^-$ and on the other hand $I^+ \subseteq J_{i_0+1} \cup J_{i_0+2} \cup \cdots \cup J_s$. Moreover, by the hypothesis (7.3), for all $1 \leqslant i \leqslant s-1$ we have

$$\frac{|h(J_{i+1})|}{|h(J_i)|} \leqslant \lambda + \frac{|J_{i+1}|}{|J_i|} \leqslant \lambda + \sigma \, ,$$

from which it follows by telescoping that

$$\frac{|h(J_{i+\nu})|}{|h(J_i)|} \leqslant (\lambda+\sigma)^\nu \ \text{ for all } \nu = 1, 2, \ldots, s-i \, .$$

Therefore

$$\begin{aligned} \frac{h(x+t)-h(x)}{h(x)-h(x-t)} = \frac{|h(I^+)|}{|h(I^-)|} &\leqslant \frac{\sum_{i=i_0+1}^{s} |h(J_i)|}{|h(J_{i_0})|} \\ &\leqslant \sum_{\nu=1}^{s-i_0} (\lambda+\sigma)^\nu \leqslant \sum_{\nu=1}^{2a^{p+1}} (\lambda+\sigma)^\nu \, . \end{aligned}$$

This proves that h is K-quasisymmetric with $K = \sum_{\nu=1}^{2a^{p+1}} (\lambda+\sigma)^\nu$, a constant that indeed depends only on the constants a, σ, λ. □

Now, let us agree to say that an orientation-preserving homeomorphism $h : S^1 \to S^1$ is a *fine grid isomorphism* if it maps fine grids to fine grids. Then the criterion for quasisymmetry given by Proposition 7.2 has the following consequence.

Corollary 7.2 *Let $h : S^1 \to S^1$ be an orientation-preserving homeomorphism. Then the following are equivalent.*

(i) h is quasisymmetric;
(ii) h maps some *fine grid onto another fine grid;*
(iii) h is a fine grid isomorphism.

Proof The proof is left as an exercise to the reader. □

As we shall see in the sequence (Sect. 7.2 below), the characterization of quasisymmetry provided by Corollary 7.2 is extremely helpful in the study of critical circle maps.

7.1.2 A Criterion for Smoothness

Our next goal is to present a criterion for $C^{1+\alpha}$ *smoothness* involving fine grids. This criterion will be extremely important later, in our study of renormalization convergence and smooth rigidity (see Chap. 10).

Proposition 7.3 *Let $h : S^1 \to S^1$ be a homeomorphism and let $\{\mathscr{Q}_n\}_{n\geqslant 0}$ be a fine grid. If there exist constants $C > 0$ and $0 < \lambda < 1$ such that*

$$\left| \frac{|I|}{|J|} - \frac{|h(I)|}{|h(J)|} \right| \leqslant C\lambda^n ,$$

for each pair of adjacent atoms $I, J \in \mathscr{Q}_n$, for all $n \geqslant 0$, then h is a $C^{1+\alpha}$-diffeomorphism for some for some $\alpha > 0$.

The proof uses the following calculus lemma concerning lateral derivatives. If ϕ is a real-valued function in an interval or oriented arc on the circle, we define the *right derivative* of ϕ at x to be

$$D^+\phi(x) = \lim_{t\searrow 0} \frac{\phi(x+t) - \phi(t)}{t} ,$$

provided the limit exists. When $D^+\phi(x)$ exists for every x, we say that ϕ is *right-differentiable*.

Lemma 7.2 *Let $\phi_n : [0, 1] \to \mathbb{R}$ be a sequence of continuous, right differentiable functions such that the sequence of right derivatives $D^+\phi_n$ converges uniformly to an α-Hölder continuous function $\varphi : [0, 1] \to \mathbb{R}$, and such that each $D^+\phi_n$ is Riemann-integrable. If ϕ_n converges uniformly to ϕ, then ϕ is $C^{1+\alpha}$ and $D\phi = \varphi$.*

Proof Exercise. □

Proof of Proposition 7.3 Let ϕ_n be the piecewise affine C^0-approximations to h determined by the vertices of $\mathscr{Q}_n$. Then ϕ_n is differentiable on the right, and $D^+\phi_n$ is a step function, so in particular it is Riemann integrable. First we show that $\{D^+\phi_n\}_{n\geqslant 0}$ is a uniform Cauchy sequence, and then that the limit is Hölder continuous. Take an atom I of $\mathscr{Q}_n$, and consider the decomposition

$$I = J_1 \cup J_2 \cup \cdots \cup J_p ,$$

with $J_k \in \mathscr{Q}_{n+1}$ consecutive and pairwise disjoint and $p \leqslant a$. Then $D^+\phi_n$ is constant on I and $D^+\phi_{n+1}$ is constant on each J_k, say

$$\begin{cases} D^+\phi_n(t) = s = \dfrac{|\phi_n(I)|}{|I|} & (\forall t \in I) , \\ \\ D^+\phi_{n+1}(t) = s_k = \dfrac{|\phi_{n+1}(J_k)|}{|J_k|} & (\forall t \in J_k) . \end{cases}$$

From this, and the fact that $|\phi_n(I)| = \sum_{k=1}^{p} |\phi_{n+1}(J_k)|$, we get

$$s|I| = \sum_{k=1}^{p} s_k |J_k| ,$$

and in particular $s' = \min s_k \leqslant s \leqslant \max s_k = s''$. Also, $s'/s'' \leqslant s/s_k \leqslant s''/s'$ for all k. Since by assumption $|1 - (s_{k+1}/s_k)| \leqslant C\lambda^{n+1}$, an easy telescoping trick gives us

$$\frac{s''}{s'} \leqslant (1 + C\lambda^{n+1})^a \leqslant 1 + C\lambda^{n+1} .$$

A similar lower bound holds for s'/s''. Therefore we have

$$1 - C\lambda^n \leqslant \frac{s}{s_k} \leqslant 1 + C\lambda^n , \tag{7.4}$$

for all $k = 1, 2, \dots p$. This shows that the sequence $\{D^+\phi_n\}_{n\geq 0}$ is uniformly bounded, and moreover that for all $m \geqslant n \geq 0$ and all $t \in \mathbf{S}^1$, we have

$$\left|D^+\phi_m(t) - D^+\phi_n(t)\right| \leqslant C\sum_{j=n}^{m-1} \lambda^j < \frac{C}{1-\lambda}\lambda^n . \tag{7.5}$$

Hence $\{D^+\phi_n\}_{n\geq 0}$ is a uniform Cauchy sequence as claimed. Let $\varphi = \lim D^+\phi_n$ be its uniform limit, and let $\alpha > 0$ be such that $\lambda_0^\alpha = \lambda$, where λ_0 is the fine grid constant appearing in Remark 7.1. We prove φ is α-Hölder as follows. It suffices to consider points $x, y \in \mathbf{S}^1$ whose distance is smaller than $\inf_{I\in\mathscr{Q}_0} |I|$. Take the smallest n such that x and y belong to distinct elements of $\mathscr{Q}_n$. Then either $n = 0$ or x and y lie in a common element of $\mathscr{Q}_{n-1}$. Either way we have by (7.4)

$$\left|D^+\phi_n(x) - D^+\phi_n(y)\right| \leq C\lambda^n . \tag{7.6}$$

Combining (7.5) and (7.6), we deduce that

$$\begin{aligned}
|\varphi(x) - \varphi(y)| &\leqslant \left|\varphi(x) - D^+\phi_n(x)\right| + \left|D^+\phi_n(x) - D^+\phi_n(y)\right| \\
&\qquad + \left|D^+\phi_n(y) - \varphi(y)\right| \\
&\leqslant \frac{C}{1-\lambda}\lambda^n + C\lambda^n + \frac{C}{1-\lambda}\lambda^n \leqslant C\lambda_0^{n\alpha} \\
&\leqslant C|x-y|^\alpha ,
\end{aligned}$$

and so φ is α-Hölder as claimed. But then, since the sequence $\{\phi_n\}_{n\geqslant 0}$ converges uniformly to h, we deduce from Lemma 7.2 that $Dh = \varphi$, whence h is indeed $C^{1+\alpha}$. This completes the proof. □

Remark 7.2 The reader who happens to be familiar with probability theory will have no difficulty in translating the above result to the language of conditional expectations. Indeed, viewing each $D^+\phi_n \in L^1$ as a random variable, the sequence $\{D^+\phi_n\}_{n\geq 0}$ satisfies $D^+\phi_{n+1} = \mathbb{E}\left(D^+\phi_n \mid \mathscr{B}_n\right)$, where $\mathscr{B}_n$ is the σ-algebra[1] generated by $\mathscr{Q}_n$, and therefore constitutes a *martingale*. Thus, the existence of a pointwise *a.e* limit φ, merely as an integrable function, is a special case of J. Doob's *martingale convergence theorem*, see Billingsley (1986, p. 490).

7.2 Quasisymmetric Conjugacies

What we have done so far already allow us to give a short proof of the following theorem, originally due to Herman (1988). In this section, since we will consider dynamical partitions associated to different maps, we shall use the notation $\mathscr{P}_n(x, f)$, $I_n(x, f)$, instead of $\mathscr{P}_n(x)$, $I_n(x)$, etc. to emphasize the dependency on f.

Theorem 7.1 *A multicritical circle map without periodic points is quasisymmetrically conjugate to a rigid rotation if and only if its rotation number is of bounded type.*

Proof Let us first assume that $f : S^1 \to S^1$ is a multicritical circle map whose rotation number $\rho = [a_0, a_1, a_2, \ldots]$ is an irrational of bounded type, say $a_n \leqslant A$ for all n. Let $c \in S^1$ be a critical point of f. We claim that the dynamical partitions $\mathscr{P}_{2n}(c, f)$, $n \geqslant 0$ constitute a fine grid. Indeed, every atom of $\mathscr{P}_{2n}(c, f)$ is partitioned into at least 2 and at most $(a_{2n+1} + 1)(a_{2n+2} + 1) \leqslant (A + 1)^2$ atoms of $\mathscr{P}_{2n+2}(c, f)$, and these are all comparable by Theorem 6.3. Hence conditions (a), (b) and (c) of Definition 7.2 are met, and the claim is proved. Let $h : S^1 \to S^1$ be a topological conjugacy between f and the rigid rotation R_ρ (say $h \circ f = R_\rho \circ h$), which exists by Yoccoz's Theorem 6.2. Then one can easily check that the dynamical partitions $\mathscr{P}_{2n}(h(c), R_\rho)$, $n \geqslant 0$, also constitute a fine grid (for R_ρ). But then h satisfies property (ii) of corollary 7.2, and therefore it must be quasisymmetric.

For the converse, suppose $h : S^1 \to S^1$ is a homeomorphism satisfying $h \circ f = R_\rho \circ h$, and suppose the rotation number of f is not of bounded type. Then there exists a subsequence (n_i) with $a_{n_i+1} \to \infty$ as $i \to \infty$. Again we take c to be a critical point of f, and let $x = h(c)$. By the real bounds, the scaling ratios

[1] In fact, $\mathscr{B}_n$ is a finite algebra: each one of its elements is a finite union of atoms of $\mathscr{Q}_n$.

$|I_{n_i+1}(c, f)|/|I_{n_i}(c, f)|$ for f remain bounded, whereas for the rigid rotation we have

$$\frac{|h(I_{n_i+1}(c, f))|}{|h(I_{n_i}(c, f))|} = \frac{|I_{n_i+1}(x, R_\rho)|}{|I_{n_i}(x, R_\rho)|} > a_{n_i+1} \to \infty \text{ as } i \to \infty ,$$

and therefore h cannot be quasisymmetric. □

Remark 7.3 In the above proof, the only reason we did not use the full collection of dynamical partitions as our fine grid is that $\mathscr{P}_{n+1}(c, f)$ is *not* a strict refinement of $\mathscr{P}_n(c, f)$ (the *short* atoms of $\mathscr{P}_n(c, f)$ are not decomposed at all in the next step; they become *long* atoms of $\mathscr{P}_{n+1}(c, f)$). This is why we skipped every other level.

Remark 7.4 An interesting application of Theorem 7.1 to holomorphic dynamics goes as follows. In the complex quadratic family $P_\theta : z \mapsto e^{2\pi i\theta}z + z^2$, one knows that for each Diophantine θ the fixed point at the origin is linearizable, so it belongs to the Fatou set of P_θ. The component of the Fatou set containing 0 is a *Siegel disk*; call it Ω_θ. In Douady (1987), Douady proved that if θ is a number of bounded type, then $\partial\Omega_\theta$ is a *quasicircle* that contains the critical point of P_θ. The rough idea is to start with a Blaschke product B from the family introduced in Sect. 6.1.1 (see Eq. (6.1)) whose restriction to $\mathbf{S}^1$ is a critical circle map f with rotation number θ. Then, using Theorem 7.1, one applies quasiconformal surgery to B, cutting out the unit disk and glueing it back in using as sewing map the quasisymmetric conjugacy h between f and the rigid rotation with the same rotation number. Redefining the map in the interior of the unit disk to be that same rotation, and applying the measurable Riemann mapping theorem, the unit circle is mapped onto a quasicircle, and the post-surgery map becomes P_θ, thereby proving Douady's theorem. This result was later generalized by Petersen and Zakeri (2004). Their theorem allows the rotation number θ to belong to a certain class of unbounded type numbers, and the proof is accomplished through the use of *trans-quasiconformal surgery*.

More important for our purposes is the following immediate consequence of Theorem 7.1.

Corollary 7.3 *Any two multicritical circle maps f and g with the same irrational rotation number* of bounded type *are quasisymmetrically conjugate, and in fact* every *topological conjugacy between f and g is a quasisymmetric homeomorphism.*

Note that in Corollary 7.3 the number of critical points of f and the number of critical points of g need not be the same! But the bounded type hypothesis on the rotation number is essential. In full generality, the above statement is most definitely *false* for unbounded combinatorics; see Chap. 9.

What can be said, then, for arbitrary irrational rotation numbers? If f and g have the same number of critical points and there is a conjugacy between f and g that maps each critical point of f to a critical point of g, the first part of the statement of Corollary 7.3 continues to hold. This will be the main result in Sect. 7.4.

7.3 Almost Parabolic Maps

When studying the geometry of dynamical partitions of a multicritical circle map whose rotation number is of *unbounded* type, one has to deal with the fact that, at certain levels, some short atoms can be much smaller than long atoms. For instance, let f be a unicritical circle map with critical point c, and consider the first return map to a small neighborhood of c, say $I_n(c) \cup I_{n+1}(c)$. If the partial quotient[2] a_{n+1} is very large, then the restriction of $f^{q_{n+1}}$ to $I_n(c)$ is very nearly a parabolic map at the center of a saddle-node bifurcation. The consecutive intervals[3] $\Delta_i = f^{iq_{n+1}+q_n}(I_{n+1}(c)) \subset I_n(c)$ with $0 \leqslant i \leqslant a_{n+1} - 1$ work as *fundamental domains* for the dynamics of $f^{q_{n+1}}|_{I_n(c)}$. By the real bounds, the two outermost of these intervals, Δ_0 and $\Delta_{a_{n+1}-1}$, are comparable to $I_n(c)$, but the ones in the middle, i.e., the Δ_i's with i close to $a_{n+1}/2$, are much smaller. The map $f^{q_{n+1}}|_{I_n(c)}$ is an example of what one calls an *almost parabolic map*.

Such maps can be described abstractly as follows (see de Faria and de Melo 1999, p. 354 or Estevez and de Faria 2018, Def. 4.1).

Definition 7.3 An *almost parabolic map* is a C^3 diffeomorphism

$$\phi : \Delta_1 \cup \Delta_2 \cup \cdots \cup \Delta_\ell \;\to\; \Delta_2 \cup \Delta_3 \cup \cdots \cup \Delta_{\ell+1}\,,$$

where $\Delta_1, \Delta_2, \ldots, \Delta_{\ell+1}$ are consecutive intervals on the circle (or on the line), with the following properties.

(i) One has $\phi(\Delta_k) = \Delta_{k+1}$ for all $1 \leqslant k \leqslant \ell$;
(ii) The Schwarzian derivative of ϕ is everywhere negative.

The positive integer ℓ is called the *length* of ϕ, and the positive real number

$$\sigma = \min\left\{\frac{|\Delta_1|}{|\cup_{k=1}^{\ell}\Delta_k|}\,,\;\frac{|\Delta_\ell|}{|\cup_{k=1}^{\ell}\Delta_k|}\right\}$$

is called the *width* of ϕ.

Remark 7.5 Note the negative Schwarzian hypothesis (ii). As we saw in Sect. 6.5, Proposition 6.2, for sufficiently large n we have $Sf^{q_{n+1}}(x) < 0$ for every regular point $x \in I_n(c)$. Thus, in the unicritical case at least, the restriction $f^{q_{n+1}}|_{\Delta_0\cup\Delta_1\cup\cdots\cup\Delta_{a_{n+1}-1}}$ is an almost parabolic map with length $\ell = a_{n+1} - 1$, provided n is sufficiently large.

[2] See Remark 1.2.

[3] By the expression *consecutive intervals* we mean a pair of intervals that share a common endpoint and have no other points in common.

7.3.1 Yoccoz's Inequality

The basic geometric control of an almost parabolic map is provided by the following fundamental inequality due to Yoccoz.

Lemma 7.3 (Yoccoz) *Let $\phi : \bigcup_{k=1}^{\ell} \Delta_k \to \bigcup_{k=2}^{\ell+1} \Delta_k$ be an almost parabolic map with length ℓ and width σ. There exists a constant $C_\sigma > 1$ (depending on σ but not on ℓ) such that, for all $k = 1, 2, \ldots, \ell$, we have*

$$\frac{C_\sigma^{-1}|I|}{[\min\{k, \ell + 1 - k\}]^2} \leqslant |\Delta_k| \leqslant \frac{C_\sigma |I|}{[\min\{k, \ell + 1 - k\}]^2}, \tag{7.7}$$

where $I = \bigcup_{k=1}^{\ell} \Delta_k$ is the domain of ϕ.

Yoccoz never published a proof of this result, but he was kind enough to explain the idea to the authors of de Faria and de Melo (1999), and as a result the first complete proof appeared as an appendix to that paper.

The main geometric idea behind the proof is to use the negative Schwarzian property of f to *squeeze* the graph of f between the graphs of two Möbius transformations. The required estimate for f will then follow from the corresponding estimate for Möbius transformations. Hence the first thing we do is to state and prove the estimate for Möbius transformations.

Consider the fractional linear transformation $T(x) = x/(1+x)$, and given $\varepsilon > 0$, let $T_\varepsilon(x) = T(x) - \varepsilon$. We are interested in certain quantitative aspects of the orbit $x_n = T_\varepsilon^n(x_0)$ for $x_0 = 1$. Observe that this sequence is strictly decreasing.

Lemma 7.4 *Let $N > 0$ be such that $x_{N+1} \leqslant 0 < x_N$. Then we have $N \asymp 1/\sqrt{\varepsilon}$ and moreover $x_n - x_{n+1} \asymp 1/n^2$ for $n = 0, 1, \ldots, N$.*

Proof Writing $\delta_n = T^n(x_0) - T_\varepsilon^n(x_0)$, we have

$$\delta_n = \varepsilon + \frac{\delta_{n-1}}{\left(1 + \frac{1}{n}\right)\left(1 + \frac{1}{n} - \delta_{n-1}\right)} \tag{7.8}$$

for all $n = 1, 2, \ldots, N + 1$. We claim that

$$\frac{n\varepsilon}{6} \leqslant \delta_n \leqslant n\varepsilon . \tag{7.9}$$

The last inequality is clear. To prove the first, we note from (7.8) that

$$\delta_n \geqslant \varepsilon + \left(\frac{n}{n+1}\right)^2 \delta_{n-1} .$$

By induction, this gives us

$$\delta_n \geq \frac{\varepsilon}{(n+1)^2}\left(1^2+2^2+\cdots+n^2\right) = \frac{\varepsilon}{(n+1)^2}\frac{n(n+1)(2n+1)}{6} \geqslant \frac{n\varepsilon}{6},$$

which proves the claim. Now, from the fact that $x_{N+1} \leqslant 0 < x_N$ we have the inequalities

$$\delta_N < \frac{1}{N+1}, \quad \delta_{N+1} \geqslant \frac{1}{N+2}.$$

Then, using (7.9), we get

$$\frac{1}{(N+1)(N+2)} \leqslant \varepsilon < \frac{6}{N(N+1)}, \tag{7.10}$$

which proves the first assertion.

Next, note that since $[x_{N+1}, x_N] \subseteq \left[T_\varepsilon(0), T_\varepsilon^{-1}(0)\right] = [-\varepsilon, \varepsilon/(1-\varepsilon)]$, we have

$$\varepsilon < x_N - x_{N+1} < 3\varepsilon \tag{7.11}$$

Hence, by (7.10), we get $x_N - x_{N+1} \asymp 1/N^2$ and the second assertion is proved when $n = N$. To prove it in general using this information, observe that

$$x_n - x_{n+1} = \frac{x_{n-1}-x_n}{(1+x_{n-1})(1+x_n)} = \frac{x_{n-1}-x_n}{(1+\frac{1}{n}-\delta_{n-1})(1+\frac{1}{n+1}-\delta_n)}$$

implies

$$x_n - x_{n+1} \geq \frac{n}{n+2}(x_{n-1}-x_n).$$

By induction, this gives on one hand

$$x_n - x_{n+1} \geq \frac{2}{(n+1)(n+2)}(x_0 - x_1) \geq \frac{1}{(n+1)(n+2)},$$

and on the other hand, using (7.10) and (7.11),

$$x_n - x_{n+1} \leqslant (x_N - x_{N+1})\prod_{j=1}^{N-n}\left(\frac{n+j+2}{n+j}\right) < \frac{54}{(n+1)(n+2)}.$$

This proves the second assertion in all cases. □

Now recall that $\phi : \Delta_1 \cup \Delta_2 \cup \cdots \cup \Delta_\ell \to \mathbb{R}$ satisfies $\phi(\Delta_j) = \Delta_{j+1}$ for all j. Without loss of generality, we can assume that $\phi(x) < x$ for all x. Thus, if we call

x_0 the right endpoint of Δ_1 and write $x_j = \phi^j(x_0)$, we have $\Delta_j = [x_j, x_{j-1}]$ for all j. Since ϕ is a negative-Schwarzian diffeomorphism, there exists a unique z in the domain of ϕ such that $\varepsilon = |\phi(z) - z| \leqslant |\phi(x) - x|$ for all x. Since the statement we want to prove is invariant under affine changes of coordinates, we may assume also that $z = 0$ and $x_0 = 1$. In this setting, we want to prove that $|\Delta_j| \asymp 1/j^2$ for all j such that $\Delta_j \subseteq [0, 1]$. Note that $\phi'(0) = 1$.

Next, let A be the Möbius transformation on the line such that $A(x_0) = \phi(x_0)$ and $A(0) = \phi(0)$ and $A'(0) = \phi'(0) = 1$. This determines A uniquely, and in fact

$$A(x) = \frac{x}{1 + \lambda x} - \varepsilon ,$$

for some $\lambda > 0$. Since $S\phi < 0$, we see that $A(x) \leqslant \phi(x)$ for all $x \in [0, 1]$.

Likewise, let B be the Möbius transformation such that $B(x_\ell) = \phi(x_\ell)$, $B(0) = \phi(0)$ and $B'(0) = \phi'(0) = 1$. This determines B uniquely, and in fact

$$B(x) = \frac{x}{1 + \mu x} - \varepsilon ,$$

for some $\mu > 0$. This time, since $x_\ell < 0$ and $S\phi < 0$, we have $\phi(x) \leqslant B(x)$ for all $x \in [0, 1]$. In particular, $\lambda > \mu$. It is easy to see that $\lambda/\mu \leqslant c_\sigma$, where c_σ depends only on the constant σ in the statement.

Lemma 7.5 *Let $x \in [0, 1]$ and $k > 0$ be such that $A(x) < B^k(x)$. Then $k \leqslant 1 + \lambda/\mu$.*

Proof By induction we have

$$B^k(x) \leqslant \frac{x}{1 + (k-1)\mu x} - \varepsilon .$$

Therefore $A(x) < B^k(x)$ implies $(k-1)\mu x < \lambda x$. □

Now, let us write $\alpha_n = A^n(x_0)$ and $\beta_n = B^n(x_0)$. By Lemma 7.5, the number of β_j's inside each interval of the form $[\alpha_{n+1}, \alpha_n]$ is bounded independently of n. Moreover, since $\alpha_n < x_n < \beta_n$ for all n, the number of x_j's inside each $[\alpha_{n+1}, \alpha_n]$ is also bounded independently of n. To prove that $|\Delta_j| \asymp 1/j^2$, we proceed as follows. Let $m > 0$ be such that $\beta_{m+1} \leqslant x_j \leqslant \beta_m \leqslant x_{j-1}$. Then Lemma 7.5 says that $m \leqslant Cj$, and we have also

$$|\beta_{m+1} - \beta_m| < |B(x_{j-1}) - x_{j-1}| < |x_j - x_{j-1}| .$$

Since by Lemma 7.4 we have

$$|\beta_{m+1} - \beta_m| \asymp \frac{1}{m^2} \geqslant \frac{1}{Cj^2} ,$$

it follows that $|\Delta_j| = |x_j - x_{j-1}| \geq 1/Cj^2$.

To prove an inequality in the opposite direction, let p be the largest integer such that $\alpha_p > x_{j-1}$. Then, again by Lemma 7.5, we have $j \leqslant Cp$. Since $A(x) < \phi(x) < x$ for all x, we also have $\Delta_j \subseteq [\alpha_{p+2}, \alpha_p]$. Using Lemma 7.4 once more, we deduce that

$$|\Delta_j| \leqslant \frac{C}{p^2} \leqslant \frac{C}{j^2} \, .$$

This completes the proof of Yoccoz's Lemma.

Remark 7.6 Let us define the *order* of a fundamental domain Δ_k as above to be $\mathrm{ord}(\Delta_\nu) = \min\{k, \ell + 1 - k\}$. Then the conclusion of Lemma 7.3 reads: for all $k = 1, 2, \ldots, \ell$, we have $|\Delta_k| \asymp (\mathrm{ord}(\Delta_k))^{-2}|I|$ with comparability constant depending only on σ. This can be expressed in simple words as follows: *the relative size of a fundamental domain in an almost parabolic map is inversely proportional to the square of its order.*

7.3.2 Balanced Decompositions

The following lemma exhibits a special way of grouping together the fundamental domains of an almost parabolic map.

Lemma 7.6 *Let ϕ be an almost parabolic map with domain $I = \bigcup_{\nu=1}^{\ell} \Delta_\nu$, and let $d \in \mathbb{N}$ be largest such that $2^{d+1} \leqslant \ell/2$. There exists a descending chain of (closed) intervals (see Fig. 7.1)*

$$I = M_0 \supset M_1 \supset \cdots \supset M_{d+1}$$

for which, letting L_i, R_i denote the (left and right) connected components of $M_i \setminus M_{i+1}$ for all $0 \leqslant i \leqslant d$, the following properties hold.

(i) Each of the intervals L_i, R_i is the union of exactly 2^i consecutive atoms (fundamental domains) of I.
(ii) We have

$$I = \bigcup_{i=0}^{d} L_i \cup M_{d+1} \cup \bigcup_{i=0}^{d} R_i \, . \tag{7.12}$$

(iii) For each $0 \leqslant i \leqslant d$ we have $|L_i| \asymp |M_{i+1}| \asymp |R_i|$, with comparability constants depending only on the width σ of ϕ.

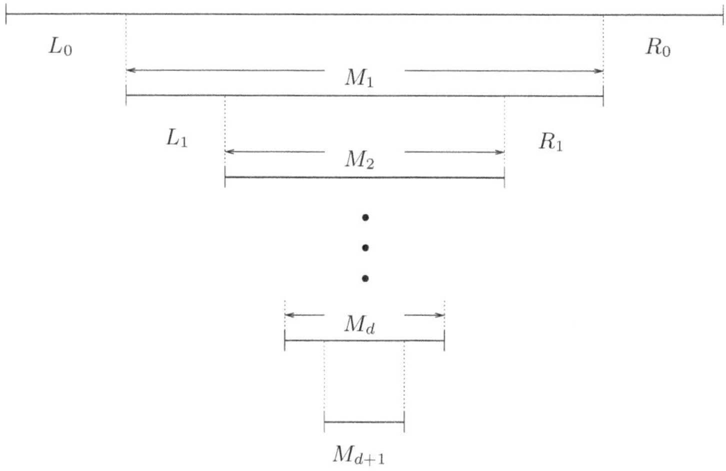

Fig. 7.1 Balanced decomposition of the domain of an almost parabolic map

Proof We define, for each $0 \leqslant i \leqslant d$,

$$L_i = \bigcup_{\nu=2^i}^{2^{i+1}-1} \Delta_\nu \;\; ; \;\; R_i = \bigcup_{\nu=\ell+2-2^{i+1}}^{\ell+1-2^i} \Delta_\nu \, .$$

Also, for each $0 \leqslant i \leqslant d+1$, we let

$$M_i = \bigcup_{\nu=2^i}^{\ell+1-2^i} \Delta_\nu$$

Then we immediately have (i) and (ii). Hence all we have to do is prove (iii). Let us fix $0 \leqslant i \leqslant d$. In all that follows, the implicit comparability constants are either universal or depend on the constant C_σ of Yoccoz's Lemma 7.3. Applying that lemma, we see that

$$|L_i| = \sum_{\nu=2^i}^{2^{i+1}-1} |\Delta_\nu| \asymp \left(\sum_{\nu=2^i}^{2^{i+1}-1} \frac{1}{\nu^2} \right) |I| \asymp 2^{-i}|I| \, . \tag{7.13}$$

Similarly, we have

$$|R_i| \asymp 2^{-i}|I| \, . \tag{7.14}$$

Moreover, we can write

$$|M_{i+1}| = \sum_{\nu=2^{i+1}}^{2^{i+2}-1} |\Delta_\nu| \asymp 2\left(\sum_{2^{i+1}\leqslant \nu \leqslant \frac{\ell}{2}} \frac{1}{\nu^2}\right)|I| = 2A|I|\,, \tag{7.15}$$

where the number A satisfies

$$\sum_{\nu=2^{i+1}}^{2^{i+2}-1} \frac{1}{\nu^2} \leqslant A \leqslant \sum_{\nu=2^{i+1}}^{\infty} \frac{1}{\nu^2}\,. \tag{7.16}$$

Both sums appearing in (7.16) are comparable to 2^{-i-1} (use the integral test). Hence (7.15) and (7.16) put together yield

$$|M_{i+1}| \asymp 2^{-i}|I|\,. \tag{7.17}$$

Combining (7.13), (7.14) and (7.17), we see that (iii) holds as well, and the proof is complete. □

Remark 7.7 Given an interval I partitioned into atoms Δ_ν, $1 \leqslant \nu \leqslant \ell$, as above, a decomposition of the form (7.12) satisfying properties (i), (ii), (iii) of Lemma 7.6 is called a *balanced decomposition* of I (relative to its given partition into atoms). Thus, Lemma 7.6 can be re-stated as saying that the domain of an almost parabolic map always admits a balanced decomposition. In such balanced decomposition, the intervals M_i, $0 \leqslant i \leqslant d+1$, are said to be *central*, whereas the intervals L_i, R_i, $0 \leqslant i \leqslant d$, are said to be *lateral*. The positive integer d is the *depth* of the decomposition.

Remark 7.8 The following fact, more general than what was used in the proof of Lemma 7.6, holds for the fundamental domains Δ_ν ($1 \leqslant \nu \leqslant \ell$) of any almost parabolic map ϕ: For all $1 \leqslant k < l < m \leqslant \ell$, one has

$$\frac{|\Delta_{l+1}| + |\Delta_{l+2}| + \cdots + |\Delta_m|}{|\Delta_{k+1}| + |\Delta_{k+2}| + \cdots + |\Delta_l|} \asymp \frac{k(m-l)}{m(l-k)}\,,$$

with comparability constant depending only on the width σ of ϕ[4] Again, this follows from Yoccoz's Lemma 7.3. This fact will be useful in the proof of Proposition 7.6.

Remark 7.9 Let I, I^* be two closed intervals with I^* contained in the interior of I. Let I^* be partitioned into a finite number ℓ of atoms, consecutively labelled Δ_ν, $1 \leqslant \nu \leqslant \ell$ as before, and suppose such atoms satisfy the inequalities (7.7) (for some choice of the constant C_σ)—so that we have a balanced decomposition of I^* (as in

[4] In fact, the comparability constant can be taken to be equal to (a universal constant times) C_σ^2, where C_σ is the constant in Lemma 7.3.

Lemma 7.6). Then, adding both lateral components of $I \setminus I^*$ to the collection of Δ_ν's and re-labelling these $\ell + 2$ atoms from first to last, one sees that the inequalities (7.7) hold for the new collection also (with a different comparability constant, in general) Thus, we get a balanced decomposition of I as well. This remark will be used in the proof of Corollary 7.4.

Remark 7.10 Note that the comparability bounds given in Lemma 7.6 (iii) depend only on the width σ of ϕ, via the constant C_σ in Lemma 7.3. If σ is small, then C_σ is potentially very bad. However, in the present text we only apply Lemma 7.6 to the cases when $\phi = f^{q_{n+1}}$ for some n and the domain of ϕ is what we call a *bridge* at level n (roughly speaking, a bridge is the interval between two consecutive critical points of a return map $f^{q_{n+1}}|_{I_n(c)}$—see Sect. 7.4.2.1 for the precise definition). In these cases, σ is uniformly bounded from below by a constant that depends only on the real bounds.

7.4 Quasisymmetric Rigidity

In this section we will prove the first major theorem of this chapter, establishing that (minimal, C^3) multicritical circle maps are *quasisymmetrically rigid.* This was informally stated in the introduction to the present chapter. More precisely, we will prove the following theorem, which first appeared in Estevez and de Faria (2018).

Theorem 7.2 *Let $f, g : S^1 \to S^1$ be two C^3 multicritical circle maps with the same irrational rotation number and the same number of (non-flat) critical points, and let $h : S^1 \to S^1$ be a homeomorphism conjugating f to g, i.e., such that $h \circ f = g \circ h$. If h maps each critical point of f to a corresponding critical point of g, then h is quasisymmetric.*

In the special case of maps with a single critical point, this theorem was first proved by Yoccoz (unpublished, but see de Faria and de Melo 1999, Cor. 4.6). Here, the presence of *at least one* critical point is absolutely crucial: the statement is *false* for diffeomorphisms. Indeed, there exist diffeomorphisms of the circle, even analytic ones, that are topologically conjugate to an irrational rotation and yet no conjugacy between them is quasisymmetric—as the reader will be able to check, this is precisely what happens with the Arnold examples given by Theorem 4.6 (see also de Melo and van Strien 1993, p. 75).

The basic idea behind the proof of Theorem 7.2 is to build for each multicritical circle map f an associated fine grid in a *canonical* way, and then apply Corollary 7.2. By canonical here we mean that the partitions making up this fine grid must be defined in a dynamically invariant way, i.e., in purely combinatorial terms. An obvious first attempt is to use the dynamical partitions $\mathscr{P}_n(c, f)$, where c is a critical point of f, all of whose vertices lie in the forward orbit of c. But even if we skip levels (to circumvent the fact that $\mathscr{P}_{n+1}(c, f)$ is not a strict refinement of $\mathscr{P}_n(c, f)$) and look at a subsequence of this sequence of partitions, we are in trouble because, whenever a partial quotient a_{n+1} is very large, there are atoms of $\mathscr{P}_{n+1}(c, f)$ which are much smaller than the atoms of $\mathscr{P}_n(c, f)$ in which they are contained. We need to group some of these small atoms together, but to do that we first need to understand their geometry.

7.4.1 More on the Geometry of Dynamical Partitions

Let us present some further geometric consequences of the real bounds that will be crucial in the proof of Theorem 7.2. The results below refer to the dynamical partitions $\mathscr{P}_n(c_k)$ $(0 \leqslant k \leqslant N-1, n \in \mathbb{N})$ of a multicritical circle map f for which the real bounds of Theorem 6.3 are satisfied. Recall that the atoms of each partition $\mathscr{P}_n(c_k)$ are of two types: the *long* atoms, i.e. those of the form $I_n^i(c_k), 0 \leqslant i < q_{n+1}$, and the *short* atoms, i.e. those of the form $I_{n+1}^j(c_k)$, $0 \leqslant j < q_n$. In what follows, we use the notion (and notation) of comparability introduced in Sect. 6.3.3.

7.4.1.1 Intersecting Atoms Are Comparable

The first result states that any two intersecting atoms belonging to dynamical partitions of two distinct critical points at the same level n are comparable.

Lemma 7.7 *Let c, c' be any two critical points of our map f. If $\Delta \in \mathscr{P}_n(c)$ and $\Delta' \in \mathscr{P}_n(c')$ are two atoms such that $\Delta \cap \Delta' \neq \emptyset$, then $|\Delta| \asymp |\Delta'|$, i.e. they are comparable.*

Proof Let $C = C(f) > 1$ be the constant given by the real bounds (Theorem 6.3). There are three cases to consider, according to the types of atoms we have: long/long, long/short, and short/short. More precisely, we have the following three cases.

(i) We have $\Delta = I_n^i(c)$ and $\Delta' = I_n^j(c')$, where $0 \leqslant i, j < q_{n+1}$. Here we may assume that $f^j(c') \in \Delta = [f^i(c), f^{i+q_n}(c)]$. Then $f^{i+q_n}(c) \in \Delta' = [f^j(c'), f^{j+q_n}(c')]$, and we have the situation shown in Fig. 7.2a. Using the monotonicity of f^{q_n}, we see that $\Delta' \subset \Delta \cup f^{q_n}(\Delta)$. Applying Lemma 6.3 to $x = f^{i+q_n}(c)$, we see that $\Delta = [f^{-q_n}(x), x]$ and $f^{q_n}(\Delta) = [x, f^{q_n}(x)]$ satisfy $|f^{q_n}(\Delta)| \leqslant C|\Delta|$, and from this it follows that $|\Delta'| \leqslant (1 + C)|\Delta|$. Conversely, we also have $\Delta \subset f^{-q_n}(\Delta') \cup \Delta'$. Again applying Lemma 6.3, this time to $x = f^j(c')$, we deduce just as before that $|f^{-q_n}(\Delta')| \leqslant C|\Delta'|$, and therefore $|\Delta| \leqslant (1+C)|\Delta'|$. Hence Δ and Δ' are comparable in this case.
(ii) We have $\Delta = I_n^i(c)$ and $\Delta' = I_{n+1}^j(c')$, where $0 \leqslant i < q_{n+1}$ and $0 \leqslant j < q_n$. Here, we look at the interval $I_{n+1}^{i+q_n}(c) \subset \Delta$. This interval shares an endpoint with Δ (namely $f^{i+q_n}(c)$) and it is also an atom of $\mathscr{P}_{n+1}(c)$. In particular, $|I_{n+1}^{i+q_n}(c)| \asymp |\Delta|$, by the real bounds. There are now two sub-cases.

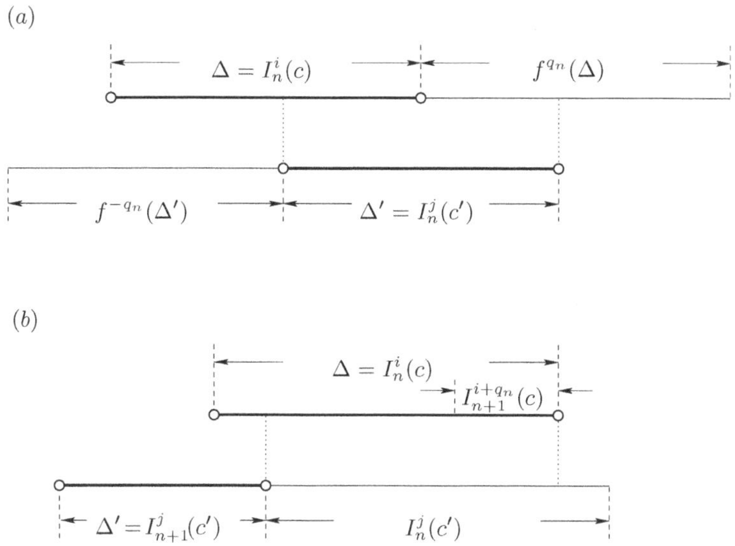

Fig. 7.2 The cases long/long and long/short of Lemma 7.7

If $\Delta' \cap I_{n+1}^{i+q_n}(c) \neq \emptyset$, then, since Δ' also belongs to $\mathscr{P}_{n+1}(c')$, case (i) above tells us that $|\Delta'| \asymp |I_{n+1}^{i+q_n}(c)|$, and therefore Δ' is comparable to Δ in this sub-case. On the other hand, if $\Delta' \cap I_{n+1}^{i+q_n}(c) = \emptyset$, then we must have $f^j(c') \in \Delta$ (see Fig. 7.2b). In this sub-case, we consider the interval $I_n^j(c') \in \mathscr{P}_n(c')$, which also has $f^j(c')$ as an endpoint. Then we have $\Delta \cap I_n^j(c') \neq \emptyset$, and again by case (i) we have $|\Delta| \asymp |I_n^j(c')|$. But by the real bounds we have $|I_n^j(c')| \asymp |I_{n+1}^j(c')| = |\Delta'|$, so Δ' is comparable to Δ also in this sub-case.

(iii) We have $\Delta = I_{n+1}^i(c)$ and $\Delta' = I_{n+1}^j(c')$, where $0 \leqslant i, j < q_n$. This case is entirely analogous to case (i).

□

Remark 7.11 The above lemma still holds if one of the critical points, say c', is replaced by an arbitrary regular point $x_0 \in S^1$, see de Faria and Guarino (2021, Lem. A.4) for the details.

7.4.1.2 Critical Atoms Are Large

Let us now consider the first return map to the interval $I_n(c_0) \cup I_{n+1}(c_0)$, or equivalently the pair of maps $f^{q_n}|_{I_{n+1}(c_0)}$, $f^{q_{n+1}}|_{I_n(c_0)}$. Besides c_0 (which is critical for both maps in the pair), this return map has at most $N - 1$ other critical points: some in $I_n(c_0)$, and some in $I_{n+1}(c_0)$. Our next auxiliary result states that the intervals of the dynamical partition at the next level ($\mathscr{P}_{n+1}(c_0)$) which contain these critical points of the return map at level n must be comparable with their parent atom ($I_n(c_0)$ or $I_{n+1}(c_0)$).

Lemma 7.8 *Let $0 \leqslant k < a_{n+1}$ be such that the interval $f^{q_n+kq_{n+1}}(I_{n+1}(c_0)) \subset I_n(c_0)$ contains a critical point of $f^{q_{n+1}}$. Then*

$$\left|f^{q_n+kq_{n+1}}(I_{n+1}(c_0))\right| \asymp |I_n(c_0)| \ . \tag{7.18}$$

Proof If $k = 0$ there is nothing to prove, since we already know from the real bounds that $|f^{q_n}(I_{n+1}(c_0))| \asymp |I_n(c_0)|$. Hence we assume that $1 \leqslant k \leqslant a_{n+1} - 1$. Let us write $\Delta = f^{q_n+kq_{n+1}}(I_{n+1}(c_0))$ in this proof. Let $0 < j \leqslant q_{n+1}$ be such that $f^j(\Delta) \ni c_1$, where $c_1 \neq c_0$ is another critical point of f. Note that $I_n^j(c_0) = f^j(I_n(c_0)) \supset f^j(\Delta)$. We claim that $|f^j(\Delta)| \asymp |f^j(I_n(c_0))|$. This is a consequence of the following two facts.

(i) We have $|I_n^j(c_0)| \asymp |I_{n+1}(c_1)|$. Indeed, these two intervals have non-empty intersection (they both contain c_1), and since $I_n^j(c_0) \in \mathscr{P}_n(c_0)$ and $I_{n+1}(c_1) \in \mathscr{P}_n(c_1)$, their comparability follows from Lemma 7.7.
(ii) We have $|I_{n+1}(c_1)| \asymp |f^j(\Delta)|$. To see why, first note that

$$j + q_n + kq_{n+1} \leqslant q_n + (k+1)q_{n+1} \leqslant q_n + a_{n+1}q_{n+1} = q_{n+2} \ ,$$

from which it follows that

$$f^j(\Delta) = I_{n+1}^{j+q_n+kq_{n+1}}(c_0) \in \mathscr{P}_{n+1}(c_0) \ .$$

Since $I_{n+1}(c_1) \in \mathscr{P}_{n+1}(c_1)$, and $f^j(\Delta) \cap I_{n+1}(c_1) \supset \{c_1\} \neq \emptyset$, we may again apply Lemma 7.7 to deduce that $I_{n+1}(c_1)$ and $f^j(\Delta)$ are comparable.

We now proceed as follows. Consider the (closure of the) gap between Δ and $I_{n+1}^{q_n}$ inside $I_n(c_0)$, namely the interval $J = \bigcup_{i=1}^{k-1} I_{n+1}^{q_n+iq_{n+1}}(c_0)$. Note that if $k = 1$ then $J = \emptyset$; in this case Δ and $I_{n+1}^{q_n}$ are two adjacent atoms of $\mathscr{P}_{n+1}(c_0)$, hence they are comparable by the real bounds (Theorem 6.3) and there is nothing to prove. Therefore we assume that $k \geqslant 2$, so that $J \neq \emptyset$. We already know from the above claim that $|f^j(\Delta)| \asymp |I_n^j(c_0)|$, and the real bounds also tell us that $|I_n^j(c_0)| \asymp |I_{n+1}^{j+q_n}(c_0)|$. Moreover, we have $I_{n+1}^{j+q_n+q_{n+1}}(c_0) \subseteq f^j(J) \subset I_n^j(c_0)$. Since $|I_{n+1}^{j+q_n+q_{n+1}}(c_0)| \asymp |I_{n+1}^{j+q_n}(c_0)|$, because these two intervals are consecutive atoms of $\mathscr{P}_{n+1}(c_0)$, it follows that $|f^j(J)| \asymp |I_{n+1}^{j+q_n}(c_0)|$. In other words, the consecutive intervals $f^j(\Delta)$, $f^j(J)$ and $I_{n+1}^{j+q_n}(c_0)$ are pairwise comparable. In particular, the b-cross-ratio determined by these three intervals is bounded from above and from below, i.e. there exists a constant $K > 1$ depending only on the constant C of the real bounds such that

$$K^{-1} \leqslant [f^j(J), f^j(T)] \leqslant K \,. \tag{7.19}$$

Here we have written $T = \Delta \cup J \cup I_{n+1}^{q_n}(c_0)$. Note that $T, f(T), \ldots, f^j(T)$ are pairwise disjoint. Therefore, by the Cross-ratio Inequality applied to the homeomorphism f^j (and $m = 1$), we have $\mathrm{CrD}(f^j; J, T) \leqslant C$, or equivalently $[f^j(J), f^j(T)] \leqslant C[J, T]$. Using the lower estimate in (7.19), we see that $[J, T] \geqslant C^{-1}K^{-1}$, that is,

$$\frac{|\Delta|\,|I_{n+1}^{q_n}(c_0)|}{|\Delta \cup J|\,|J \cup I_{n+1}^{q_n}(c_0)|} \geqslant (CK)^{-1} \,. \tag{7.20}$$

But, since $J \supseteq I_{n+1}^{q_n+q_{n+1}}(c_0)$, and since $I_{n+1}^{q_n+q_{n+1}}(c_0)$ and $I_{n+1}^{q_n}(c_0)$ are adjacent atoms of $\mathscr{P}_{n+1}(c_0)$, we have by the real bounds

$$|\Delta \cup J| > |J| \geqslant |I_{n+1}^{q_n+q_{n+1}}(c_0)| \geqslant C^{-1}|I_{n+1}^{q_n}(c_0)| \,.$$

Moreover, $|I_{n+1}^{q_n}(c_0)| \geqslant C^{-1}|I_n(c_0)|$, again by the real bounds. Putting these facts back into (7.20), we deduce that

$$|\Delta| \geqslant C^{-2}K^{-1}|J \cup I_{n+1}^{q_n}(c_0)| > C^{-3}K^{-1}|I_n(c_0)| \,.$$

This shows that Δ and $I_n(c_0)$ are comparable. Hence (7.18) is established, and the proof of Lemma 7.8 is complete. □

Remark 7.12 Similarly to what we observed in Remark 7.11, the statement of Lemma 7.8 is still true if we replace the critical point c_0 by an arbitrary regular point on the circle (see Exercise 7.9).

7.4.2 Building a Suitable Fine Grid

Recall that our aim is to build, for each multicritical circle map f, a fine grid $\mathscr{G}(f) = \{\mathscr{Q}_n(f)\}_{n\geqslant 0}$ which is *adapted* to f in the sense that all of its vertices are dynamically labeled (in a canonical way that depends solely on the combinatorics of $\rho(f)$). The vertices are taken from the forward orbit of one of the critical points of f, say $c_0 \in \mathrm{Crit}(f)$. For each $n \geqslant 0$, the atoms of $\mathscr{Q}_n(f)$ will be built as unions of atoms belonging to the dynamical partitions $\mathscr{P}_m(f)$ with $m \geqslant n$. The construction is subtle, and involves first building certain auxiliary partitions, using what we already know about the geometry of dynamical partitions and Yoccoz's inequality, and then applying a recursive scheme.

7.4.2.1 Auxiliary Partitions

The first step is to construct a suitable refinement of the dynamical partition $\mathscr{P}_n(c_0)$ (for each $n \geqslant 1$). This auxiliary partition, which we denote by $\mathscr{P}_n^*(c_0)$, is finer than $\mathscr{P}_n(c_0)$ but coarser than $\mathscr{P}_{n+1}(c_0)$. Such auxiliary partition will be needed in the construction of the fine grid presented in Proposition 7.6.

From now on we write, for $0 \leqslant k < a_{n+1}$, $\Delta_k = f^{q_n+kq_{n+1}}(I_{n+1}(c_0))$. Note that each Δ_k is an atom of the dynamical partition $\mathscr{P}_{n+1}(c_0)$, and that

$$\bigcup_{k=0}^{a_{n+1}-1} \Delta_k = I_n(c_0) \setminus I_{n+2}(c_0) .$$

We consider the times $0 \leqslant k_1 < k_2 < \cdots < k_r < a_{n+1}$ having the property that Δ_{k_i} contains a critical point of $f^{q_{n+1}}$. These are called the *critical times* at level n. For convenience of notation, we also define $k_0 = 0$. Note that $f^{q_{n+1}}$ has at most N critical points in $I_n(c_0)$, where N is the total number of critical points of f. Since each such critical point belongs to at most two of the Δ_k's, we see that $r \leqslant 2N$. Thus, although the non-negative integer r may depend on n (the level of renormalization), it nevertheless ranges over only finitely many values. The critical times k_i also depend on n. The intervals Δ_{k_i} for $0 \leqslant i \leqslant r$ will be called *critical spots*.

For each $i = 0, 1, \ldots, r-1$, let $G_i \subseteq I_n(c_0) \setminus I_{n+2}(c_0)$ be the gap between the two consecutive critical spots Δ_{k_i} and $\Delta_{k_{i+1}}$ inside $I_n(c_0)$, namely the interval

$$G_i = \bigcup_{k=k_i+1}^{k_{i+1}-1} \Delta_k .$$

We also define, for $i = r$,

$$G_r = \bigcup_{k=k_r+1}^{a_{n+1}-1} \Delta_k .$$

We call G_i the i-th *bridge* of $I_n(c_0)$. See Fig. 7.3. We remark that it may well be the case that $G_i = \emptyset$ for some (or all!) values of i.

Lemma 7.9 *Each non-empty bridge G_i is comparable to $I_n(c_0)$.*

Proof If $G_i \neq \emptyset$, then G_i contains at the very least the atom Δ_{k_i+1}, adjacent to Δ_{k_i}, and so we have $|G_i| \geqslant |\Delta_{k_i+1}| \asymp |\Delta_{k_i}|$, by the real bounds. By Lemma 7.8, we have $|\Delta_{k_i}| \asymp |I_n(c_0)|$. Since we also have $G_i \subset I_n(c_0)$, it follows that $|G_i| \asymp |I_n(c_0)|$. □

Thus, we have the following decomposition of $I_n(c_0) \setminus I_{n+2}(c_0)$ as union of at most $2r + 2 \leqslant 4N + 2$ intervals:

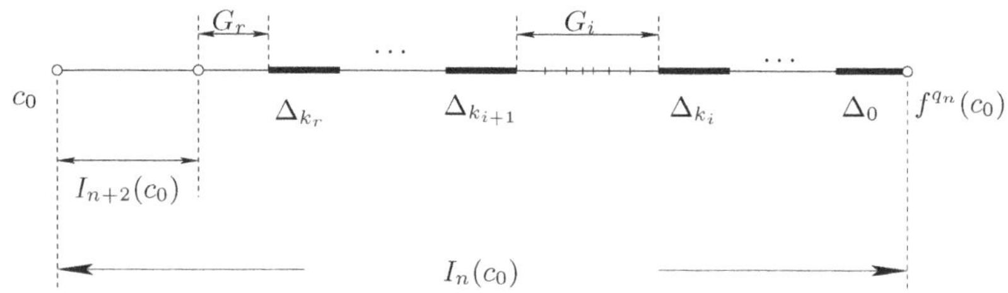

Fig. 7.3 Primary bridges and critical spots

$$I_n(c_0) \setminus I_{n+2}(c_0) \;=\; \bigcup_{i=0}^{r} \Delta_{k_i} \cup \bigcup_{i=0}^{r} G_i \,. \tag{7.21}$$

In view of Lemmas 7.8 and 7.9, as well as the real bounds, each interval in the above decomposition is comparable to $I_n(c_0)$. In particular, they are all pairwise comparable.

Remark 7.13 Note that the image of each critical spot Δ_{k_i} under $f^{q_{n+1}}$ is also comparable to $I_n(c_0)$; this is simply because $f^{q_{n+1}}(\Delta_{k_i}) = \Delta_{k_i+1}$ is adjacent to Δ_{k_i} in $\mathscr{P}_{n+1}(c_0)$. Likewise, the image of each bridge G_i under $f^{q_{n+1}}$ is also comparable to $I_n(c_0)$, because either $i < r$ and $f^{q_{n+1}}(G_i)$ contains the critical spot $\Delta_{k_{i+1}}$, or $i = r$, in which case $f^{q_{n+1}}(G_r)$ contains $I_{n+2}(c_0)$.

Let us now map the decomposition (7.21) forward by f to get corresponding decompositions of all long atoms $I_n^j(c_0) \in \mathscr{P}_n(c_0)$, for $j = 1, 2, \ldots, q_{n+1} - 1$. We get in this fashion a new partition $\mathscr{P}_n^*(c_0)$ of the circle (modulo endpoints). More precisely, let

$$\begin{aligned} \mathscr{P}_n^*(c_0) \;=\; & \Big\{ f^j(\Delta_{k_i}) : 0 \leqslant i \leqslant r \;;\; 0 \leqslant j \leqslant q_{n+1} - 1 \Big\} \\ & \cup \Big\{ f^j(G_i) : 0 \leqslant i \leqslant r \;;\; 0 \leqslant j \leqslant q_{n+1} - 1 \Big\} \\ & \cup \Big\{ f^j(I_{n+2}) : 0 \leqslant j \leqslant q_{n+1} - 1 \Big\} \\ & \cup \Big\{ f^\ell(I_{n+1}) \;:\; 0 \leqslant \ell \leqslant q_n - 1 \Big\} \,. \end{aligned} \tag{7.22}$$

This partition refines $\mathscr{P}_n(c_0)$, although not strictly because each short atom of $\mathscr{P}_n(c_0)$ is left untouched by the above procedure.

Remark 7.14 Generalizing the nomenclature introduced earlier, all atoms of $\mathscr{P}_n^*(c_0)$ of the form $f^j(\Delta_{k_i})$ are called *critical spots*, and all those of the form $f^j(G_i)$ are called *bridges*. We sometimes refer to bridges and critical spots contained in $I_n(c)$ (i.e., those with $j = 0$) as *primary*, and to the remaining ones as *secondary*.

Proposition 7.4 *Any two consecutive atoms of $\mathscr{P}_n^*(c_0)$ are comparable.*

Proof By the real bounds (Theorem 6.3), the partition $\mathscr{P}_n(c_0)$ has the stated comparability property. Hence it suffices to check that all bridges and critical spots of $\mathscr{P}_n^*(c_0)$ inside each long atom $I_n^j(c_0) \in \mathscr{P}_n(c_0)$ are comparable to $I_n^j(c_0)$. We already know this for $j = 0$ (see Lemma 7.9 and the paragraph following its proof). For the other values of j, map $I_n^j(c_0)$ forward by $f^{q_{n+1}-j}$ onto $I_n^{q_{n+1}}(c_0) \subset I_n(c_0) \cup I_{n+1}(c_0)$ and apply the Cross-ratio Inequality, combined with Remark 7.13. □

7.4.2.2 Balanced Decompositions of Bridges

We distinguish two types of atoms belonging to the partition $\mathscr{P}_n^*(c_0)$:

(a) *Regular atoms*: These consist of all short atoms of $\mathscr{P}_n(c_0)$, all of which belong to $\mathscr{P}_n^*(c_0)$, all intervals of the form $f^j(I_{n+2})$ (with $0 \leqslant j \leqslant q_{n+1} - 1$), all critical spots $f^j(\Delta_{k_i})$ (with $0 \leqslant i \leqslant r$, $0 \leqslant j \leqslant q_{n+1} - 1$), together with all those bridges $G_{i,j} = f^j(G_i)$ that have less than 1000 atoms of $\mathscr{P}_{n+1}(c_0)$ in it (i.e., those with $k_{i+1} - k_i \leqslant 1000$).
(b) *Saddle-node atoms*: These are the remaining bridges; to wit, those $G_{i,j}$ whose decomposition as a union of atoms of $\mathscr{P}_{n+1}(c_0)$ has at least 1000 such atoms in it (i.e., those with $k_{i+1} - k_i > 1000$).

Proceeding by analogy with a procedure first described in de Faria and de Melo (1999, §4.3), we will show, with the help of Yoccoz's Lemma 7.3, how to get a *balanced decomposition* of a saddle-node bridge.

The following lemma is the key to showing that, on every primary saddle-node bridge, say $G_i \subset I_n(c_0)$ given by

$$G_i = \bigcup_{k=k_i+1}^{k_{i+1}-1} \Delta_k ,$$

the return map $f^{q_{n+1}}$ acts as an almost parabolic map.

Lemma 7.10 *There exists a positive integer $n_0 = n_0(f)$ such that the following holds for all $n \geqslant n_0$. For each non-empty primary bridge $G_i \subset I_n(c_0)$, the restriction $f^{q_{n+1}}|_{G_i}$ has negative Schwarzian derivative everywhere, i.e., for all $x \in G_i$ we have $Sf^{q_{n+1}}(x) < 0$.*

Proof This is an immediate consequence of Proposition 6.2 (see Sect. 6.5). □

From Lemma 7.10, we deduce the following result concerning the bridges G_i, $0 \leqslant i \leqslant r$, contained in the closest return interval $I_n(c_0)$ (see Fig. 7.4).

Proposition 7.5 *For all $n \geqslant n_0$, where n_0 is as in Lemma 7.10, and each $i = 0, 1, 2, \ldots, r$ for which the bridge $G_i \subset I_n(c_0)$ is non-empty, the restriction*

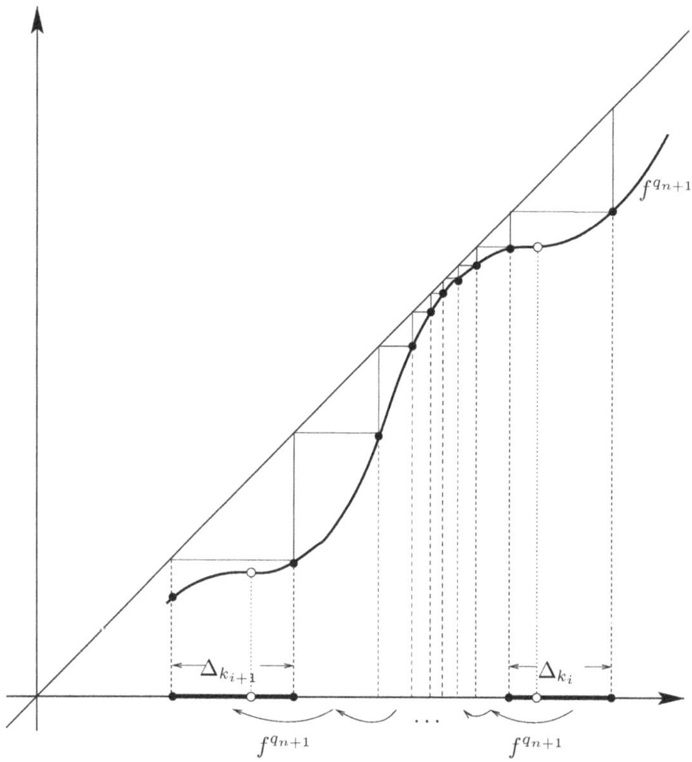

Fig. 7.4 Two consecutive critical spots and the bridge joining them: the dynamical picture

$$f^{q_{n+1}}|_{G_i} : G_i \to f^{q_{n+1}}(G_i)$$

is an almost parabolic map with length $\ell_i = k_{i+1} - k_i - 1$ *and width* $\sigma_i \geqslant \sigma$*, where* $\sigma = \sigma(C) > 0$ *depends only on the constant* C *in the real bounds.*

Proof By construction, the map $\phi = f^{q_{n+1}}|_{G_i}$ has no critical points, hence it is a diffeomorphism onto its image. Since $G_i = \bigcup_{k=k_i+1}^{k_{i+1}-1} \Delta_k$ and $\phi(\Delta_k) = f^{q_{n+1}}(\Delta_k) = \Delta_{k+1}$ for all k, it follows that the length of ϕ is as stated. Moreover, by Lemma 7.10, we have $S\phi = Sf^{q_{n+1}} < 0$ throughout. Finally, since the intervals Δ_{k_i+1} and $\Delta_{k_{i+1}-1}$ are both comparable to G_i (by the real bounds and Lemma 7.9), the last statement concerning the width of ϕ follows as well. □

Combining Poroposition 7.5 with Lemma 7.6 and the Koebe distortion principle, we deduce that every saddle-node bridge admits a *balanced decomposition*. More precisely, we have the following result.

Corollary 7.4 *For all* $n \in \mathbb{N}$*, each non-empty bridge* $G_{i,j} = f^j(G_i) \in \mathscr{P}_n^*(c_0)$ *admits a balanced decomposition (with uniform comparability constants depending only on the real bounds for* f*).*

Proof We may of course assume that $n \geqslant n_0$, where n_0 is as in Lemma 7.10. For *primary* bridges, namely $G_{i,0} = G_i \subset I_n(c_0)$ (i.e., those with $j = 0$), the assertion follows from Proposition 7.5 and Lemma 7.6. For *secondary* bridges, namely $G_{i,j} = f^j(G_i)$, $1 \leqslant j \leqslant q_{n+1} - 1$, use the fact that $f^j : \mathrm{int}(G_i) \to \mathrm{int}(G_{i,j})$ is a diffeomorphism and apply Koebe's distortion principle (the image under f^j of the balanced decomposition of G_i yields a balanced decomposition of $G_{i,j}$, as desired). □

7.4.2.3 The Recursive Scheme

Now we define an auxiliary collection of intervals $\mathscr{P}_n^{**}(c_0)$, for each $n \geq 1$. The intervals belonging to $\mathscr{P}_n^{**}(c_0)$ are all atoms of $\mathscr{P}_n^{*}(c_0)$ which are not saddle-node, together with the atoms of the balanced partitions of all saddle-node atoms of $\mathscr{P}_n^{*}(c_0)$. We warn the reader that, unlike $\mathscr{P}_n^{*}(c_0)$, the collection $\mathscr{P}_n^{**}(c_0)$ is *not* a partition of $\boldsymbol{S}^1$ (modulo endpoints), since it contains, for instance, all central intervals of any given saddle-node atom of $\mathscr{P}_n^{*}(c_0)$, and these are in fact nested. The partition $\mathscr{Q}_n(f)$ that we want is constructed using elements from $\mathscr{P}_m^{**}(c_0)$ and $\mathscr{P}_m^{*}(c_0)$ for various values of $m \leqslant n$. The construction follows a recursive scheme that we proceed to describe.

Proposition 7.6 *There exists a fine grid $\{\mathscr{Q}_n(f)\}$ in $\boldsymbol{S}^1$ with the following properties.*

(a) Every atom of $\mathscr{Q}_n(f)$ is the union of at most $a = 4N + 3$ atoms[5] *of $\mathscr{Q}_{n+1}(f)$.*
(b) Every atom $\Delta \in \mathscr{Q}_n(f)$ is a union of atoms of $\mathscr{P}_m^{*}(c_0)$ for some $m \leqslant n$, and there are three possibilities:

(b_1) Δ is a single atom of $\mathscr{P}_m^{*}(c_0)$;
(b_2) Δ is a central interval of $\mathscr{P}_m^{**}(c_0)$;
(b_3) Δ is the union of at least two atoms of $\mathscr{P}_{m+1}^{*}(c_0)$ contained in a single atom of $\mathscr{P}_m^{**}(c_0)$.

Proof The proof is by induction on n. The first partition $\mathscr{Q}_1(f)$ consists of all atoms of $\mathscr{P}_1^{*}(c_0)$ which are not saddle-node atoms together with the intervals L_0, M_1 and R_0 of each saddle-node interval $I \in \mathscr{P}_1^{*}(c_0)$ ($I = L_0 \cup M_1 \cup R_0$). It is clear that each atom of $\mathscr{Q}_1(f)$ falls within one of the categories (b_1)-(b_3) above.

Assuming $\mathscr{Q}_n(f)$ defined, define $\mathscr{Q}_{n+1}(f)$ as follows. Take an atom $I \in \mathscr{Q}_n(f)$ and consider the four cases below.

(1) If I is a single atom of $\mathscr{P}_m^{*}(c_0)$ then one of two things can happen:

[5] As we saw in Sect. 7.4.2.1, each long interval $I_n^i(c_0) \in \mathscr{P}_n(c_0)$ is decomposed as the union of $2r + 3 \leqslant 4N + 3$ atoms of $\mathscr{P}_n^{*}(c_0)$.

(i) I is a saddle-node atom: In this case write $I = L_0 \cup M_1 \cup R_0$ as above and take L_0, R_0 and M_1 as atoms of $\mathcal{Q}_{n+1}$. Note that the lateral intervals L_0 and R_0 are atoms of type (b_1), while the central interval M_1 is of type (b_2).

(ii) I is not a saddle-node atom: Here, there are two sub-cases to consider. The first possibility is that I is a single (regular) atom of $\mathcal{P}_m(c_0)$, in which case we break it into the union of at most a atoms of $\mathcal{P}^*_{m+1}(c_0)$ and take them as atoms of $\mathcal{Q}_{n+1}(f)$, all of which are of type (b_1). The second possibility is that I is a (short) bridge, in which case we break it up into its $\leqslant 1000$ constituent atoms of $\mathcal{P}_{m+1}(c_0)$ and take them as atoms of $\mathcal{Q}_{n+1}(f)$, again all of type (b_1).

(2) If I is a central interval of $\mathcal{P}^*_m(c_0)$ which is not the final interval, consider the next central interval of (c_0) inside I, say M, and the two corresponding lateral intervals L and R such that $I = L \cup M \cup R$, and declare L, R and M members of $\mathcal{Q}_{n+1}(f)$. Note that L and R are of type (b_3), while M is of type (b_2).

(3) If I is a union of $p \geq 2$ consecutive atoms $J_1, \ldots, J_p$ of $\mathcal{P}_{m+1}(c_0)$ inside a single atom of $\mathcal{P}^*_m(c_0)$ (this happens when I is contained in a lateral interval of the balanced decomposition of a long bridge), divide it up into two approximately equal parts. More precisely, write $p = 2q + r$, where $r = 0$ or 1, and consider $I = L \cup R$ where

$$L = \bigcup_{j=1}^{q} J_j \,, \; R = \bigcup_{j=q+1}^{p} J_j \,.$$

We obtain in this fashion two new atoms of $\mathcal{Q}_{n+1}(f)$ (namely L and R) which are either single atoms of $\mathcal{P}_{m+1}(c_0)$, and therefore of type (b_1), or once again intervals of type (b_3).

This completes the induction. That $\{\mathcal{Q}_n(f)\}_{n\geq 1}$ constitutes a fine grid follows easily from the real bounds, Lemma 7.6, Remark 7.8 and Corollary 7.4. Indeed, it suffices to verify that condition (c) of Definition 7.2 is satisfied (for some constant $\rho > 1$ depending only on the real bounds). Given two adjacent atoms $\Delta, \Delta' \in \mathcal{Q}_n$, there are two cases to consider.

(a) There exist $m, m' \leqslant n$ such that Δ is a single atom of $\mathcal{P}_m(c_0)$ and Δ' is a single atom of $\mathcal{P}_{m'}(c_0)$. In this case, either $m = m'$, or m and m' differ by 1 (this is easily proved by induction on n from the construction of $\mathcal{Q}_n$ given above). But then we have $|\Delta| \asymp |\Delta'|$ by the real bounds (Theorem 6.3).

(b) For some $m \leqslant n$, at least one of the two atoms, say Δ, is the union of $p \geqslant 2$ atoms of $\mathcal{P}_{m+1}(c_0)$ inside a single atom of $\mathcal{P}^*_m(c_0)$, which is necessarily a bridge. This implies that both Δ and Δ' are contained in the same bridge $G \in \mathcal{P}^*_m(c_0)$. Looking at the balanced decomposition of G (given by Corollary 7.4), we see that there are two possibilities. The first possibility is that both Δ and Δ' are contained in the same lateral interval (L_i, R_i) or the same central interval (M_i) of said balanced decomposition. In this case, Δ and Δ' are both unions of the *same* number of fundamental domains of G, and we have $|\Delta| \asymp |\Delta'|$

by Lemma 7.6 and Remark 7.8. The second possibility is that Δ and Δ' are contained in adjacent intervals of the balanced decomposition of G. In this case, one of the two atoms, Δ or Δ', is the union of at most twice as many fundamental domains of G as the other, and we have $|\Delta| \asymp |\Delta'|$, again by Lemma 7.6 and Remark 7.8.

This establishes the desired comparability of adjacent atoms of $\mathscr{Q}_n(f)$ in all cases, with uniform constants depending only on the real bounds, and the proof is complete. □

7.4.3 The Punchline

The proof of Theorem 7.2 is now within reach.

Proof of Theorem 7.2 By hypothesis, the conjugacy h sets a bijective correspondence between the critical points of f and the critical points of g. Let c be a critical point of f, and let $h(c)$ be the corresponding critical point of g. Then h maps each partition $\mathscr{P}_n(c, f)$ onto the corresponding partition $\mathscr{P}_n(h(c), g)$, sending critical spots to critical spots and bridges to bridges. Therefore if $\mathscr{G}_f = \{\mathscr{Q}_n(c, f)\}$ and $\mathscr{G}_g = \{\mathscr{Q}_n(h(c), g)\}$ are the fine grids for f and g, respectively, given by Proposition 7.6, it follows that h maps $\mathscr{G}_f$ bijectively onto $\mathscr{G}_g$. But then, by Corollary 7.2, h is quasisymmetric. This finishes the proof. □

Exercises

Exercise 7.1 Let $\psi : S^1 \to S^1$ be a quasisymmetric homeomorphism. Given an interval $T \subset S^1$, let $m_T \geqslant 1$ be the infimum over all $C > 1$ such that $C^{-1}|\psi(I)| \leqslant |\psi(J)| \leqslant C|\psi(I)|$ for all pairs of adjacent intervals $I, J \subset T$ of equal length.

(i) If $I \subset T \subset S^1$ are intervals sharing an endpoint and satisfying $\theta = |I|/|T| \leqslant 1/2$, show that

$$\theta^{\beta_T(\theta)} \leqslant \frac{|\psi(I)|}{|\psi(T)|} \leqslant \theta^{\gamma_T(\theta)} ,$$

where

$$\beta_T(\theta) = \left(1 + \frac{1}{k-1}\right) \log_2(1 + m_T) , \quad \gamma_T(\theta) = \left(1 - \frac{1}{k}\right) \log_2(1 + m_T^{-1}) ,$$

and where $k \geqslant 2$ is the unique integer such that $2^{-k} < \theta \leqslant 2^{-(k-1)}$. [*Hint.* For each $n \geqslant 1$, let $T_n \subset T$ be the subinterval sharing an endpoint with both I and

T and having length $2^{-n}|T|$. First estimate $|\psi(T_{n+1}|/|\psi(T_n)|$ and then use a telescoping decomposition.]

(ii) Deduce from (i) that every quasisymmetric homeomorphism is bi-Hölder continuous.

[Reference: de Faria 1996.]

Exercise 7.2 Prove Corollary 7.1.

Exercise 7.3 Prove the assertions made in Remark 7.1.

Exercise 7.4 Prove Corollary 7.2.

Exercise 7.5 Prove Lemma 7.2.

Exercise 7.6 Let f be a multicritical circle map with irrational rotation number $\theta = \rho(f)$ of bounded type, and let $h : \boldsymbol{S}^1 \to \boldsymbol{S}^1$ be a quasisymmetric homeomorphism conjugating f to the rotation R_θ (such an h exists by Herman's Theorem 7.1). Show that h is purely singular with respect to Lebesgue measure, i.e., $Dh(x) = 0$ for Lebesgue almost every $x \in \boldsymbol{S}^1$.

Exercise 7.7 Prove the assertion made following the statement of Theorem 7.2, namely, that the Arnold analytic diffeomorphisms given in Theorem 4.6 cannot be quasisymmetrically conjugate to a rotation.

Exercise 7.8 Given a quasisymmetric homeomorphism $\phi : \boldsymbol{S}^1 \to \boldsymbol{S}^1$, consider its scalewise logarithmic quasisymmetric distortion $\epsilon_\phi : \mathbb{R}^+ \to \mathbb{R}^+$ defined by

$$\epsilon_\phi(t) = \sup_{x\in\mathbb{R},\, |\tau|\leqslant t} \log \frac{\phi(x+\tau)-\phi(x)}{\phi(x)-\phi(x-\tau)} .$$

Note that $\epsilon_\phi(t)$ is a non-decreasing function of t. The purpose of this exercise is to guide the reader to a proof of the following theorem due to L. Carleson (1967).

Theorem *If* $\displaystyle\int_0^1 \frac{[\epsilon_\phi(t)]^2}{t}\,dt < \infty$, *then ϕ is absolutely continuous, and in fact its derivative $D\phi$ belongs to $L^2(\boldsymbol{S}^1)$.*

Let $(\phi_n)_{n\geqslant 0}$ be the sequence of dyadic C^0 approximations[6] to ϕ, and for each $n \geqslant 0$ let

$$K_n = \sup_{1\leqslant k\leqslant 2^n-1} \frac{\phi_n((k+1)2^{-n})-\phi_n(k2^{-n})}{\phi_n(k2^{-n})-\phi_n((k-1)2^{-n})} .$$

For each $n \geqslant 0$, let $\varphi_n = D\phi_n$. Note that, since ϕ_n is piecewise affine, φ_n is a step function, and we have

[6] As usual in this book, we think of the circle $\mathbb{R}/\mathbb{Z}$ as $[0, 1]$ with the endpoints 0 and 1 identified.

$$\phi_n(x) = \phi_n(0) + \int_0^x \varphi_n(t)\,dt\,. \tag{7.23}$$

(i) Show that $\log K_n \leqslant \epsilon_\phi(2^{-n})$, and deduce from this that

$$\sum_{n=0}^{\infty}(K_n - 1)^2 < \infty\,.$$

(ii) Show that for each $m > n > 0$ we have

$$\|\varphi_m - \varphi_n\|^2 = \|\varphi_m\|^2 - \|\varphi_n\|^2\,.$$

(iii) Using (i) and (ii), show that $(\varphi_n)_{n\geqslant 0}$ is a Cauchy sequence in $L^2(S^1)$; hence it is also a Cauchy sequence in $L^1(S^1)$ (why?).
(iv) Combining (iii) with (7.23) and the fact that (ϕ_n) converges uniformly to ϕ, deduce that ϕ is absolutely continuous, and that $D\phi = \varphi$ (Lebesgue a.e.), where $\varphi = \lim \varphi_n \in L^2(S^1)$.

[Reference: See the expository note de Faria n.d.]

Remark In the language of probability theory, what we have in Exercise 7.8 is an instance of an L^2 *martingale convergence theorem.*

Exercise 7.9 Let $f : S^1 \to S^1$ be a multicritical circle map with irrational rotation number, and let $x \in S^1$. If $0 \leqslant i < a_{n+1}$ is such that $\Delta = f^{q_n+iq_{n+1}}(I_n(x))$ is a critical spot at level n, show that $|\Delta| \asymp |I_n(x)|$. Do this by working through the following steps:

(i) Let $\xi \in \Delta$ be a critical point of $f^{q_{n+1}}$, say $\xi = f^{-j}(c)$ for some $c \in \text{Crit}(f)$ and some $0 \leqslant j < q_{n+1}$. Show that the interval $L \subset I_n(x)$ with endpoints ξ and $f^{q_{n+1}}(\xi)$ is comparable to Δ.
(ii) Show that the interval M with endpoints ξ and $f^{q_n}(\xi)$ that does not contain x is comparable to $I_n(x)$.
(iii) Let R be the interval with endpoints $f^{q_n}(\xi)$ and $f^{2q_n}(\xi)$ that does not contain x, and let $T = L \cup M \cup R$. Show that the cross-ratio $[f^j(M)\,,\; f^j(T)]$ is bounded from below.
(iv) Using (iii) and the Cross-ratio Inequality, show that $|L| \asymp |M|$, and deduce from this that $|\Delta| \asymp |I_n(x)|$, as desired.

Chapter 8
Ergodic Aspects

In this chapter we examine multicritical circle maps from the point of view of measurable dynamics. We have seen in Theorem 2.3 that every homeomorphism of the circle without periodic points is uniquely ergodic. In particular, every multicritical circle map f with irrational rotation number is uniquely ergodic. If μ denotes the unique Borel probability measure invariant under f, then we also know from Corollary 2.2 that μ is either absolutely continuous or purely singular with respect to Lebesgue measure. Can we resolve this dichotomy?

The answer is *yes*. As we will see in Sect. 8.2, Khanin (1991) proved that the measure μ is always purely singular with respect to Lebesgue measure. After establishing this fact, we will prove in Sect. 8.3 that the Lyapunov exponent of f under μ is equal to zero (compare with Theorem 3.11 in Sect. 3.4.2). In Sect. 8.4 we will revisit the notions of automorphic measures and invariant distributions, introduced in Sect. 3.4.3. We will close this chapter with the statements of some results on the Hausdorff dimension of the invariant measure μ (see Sect. 8.5).

8.1 The Integrability of log Df

As before, let f be a C^3 multicritical circle map with finitely many non-flat critical points and with irrational rotation number $\rho(f)$, and let μ be its unique invariant Borel probability measure. By Yoccoz's Theorem 6.2, there exists a circle homeomorphism $h : S^1 \to S^1$ which is a topological conjugacy between f and the rigid rotation by angle $\rho(f)$, namely $R_{\rho(f)}$. More precisely, the following diagram commutes.

E. de Faria, P. Guarino, *Dynamics of Circle Mappings*, IMPA Monographs,
https://doi.org/10.1007/978-3-031-67495-2_8

$$\begin{array}{ccc} (S^1, \mu) & \xrightarrow{\ f\ } & (S^1, \mu) \\ {\scriptstyle h}\big\downarrow & & \big\downarrow{\scriptstyle h} \\ (S^1, m) & \xrightarrow[R_{\rho(f)}]{} & (S^1, m) \end{array}$$

where m denotes the normalized Lebesgue measure in the unit circle (the Haar measure for the multiplicative group of complex numbers of modulus 1). Therefore, μ is just the push-forward of Lebesgue measure under h^{-1}, that is, $\mu(A) = \left(h_*^{-1}m\right)(A) = m\left(h(A)\right)$ for any Borel set A in the unit circle (recall from Exercise 2.3 that the conjugacy h is unique up to post-composition with rotations, so the measure μ is well-defined).

In this section we prove that $\log Df$ belongs to $L^1(\mu)$. Let us denote by $c_1, c_2, \ldots, c_N$ the critical points of f. Let $\varphi : S^1 \to \mathbb{R}$ be given by $\varphi = |\log Df|$. For each $1 \leqslant j \leqslant N$ and each $n \geqslant 1$, let $J_n(c_j) = I_n(c_j) \cup I_{n+1}(c_j)$. We will use the following four facts:

F1. From the real bounds (Theorem 6.3) there exists $0 < \lambda < 1$ such that $|I_k(c_j)| \geqslant \lambda^k$ for all $k \geqslant 1$ and each $1 \leqslant j \leqslant N$.

F2. As explained above, the measure μ is the pullback of the Lebesgue measure under any topological conjugacy between f and the corresponding rigid rotation. In particular, for each $1 \leqslant j \leqslant N$ and for all $k \geqslant 1$, we have $\mu(I_k(c_j)) = |q_k\theta - p_k|$ and by Theorem 1.2(i):

$$\frac{1}{q_k + q_{k+1}} < \mu(I_k(c_j)) \leqslant \frac{1}{q_{k+1}} \quad \text{for all } k \geqslant 1 \text{ and each } 1 \leqslant j \leqslant N.$$

F3. By combinatorics, we have $\mu(I_k(c_j) \setminus I_{k+2}(c_j)) = a_{k+1}\mu(I_{k+1}(c_j))$, for all $k \geqslant 0$ and for each $1 \leqslant j \leqslant N$.

F4. Since each c_j is a non-flat critical point, there exist $C_0 > 0$ and a neighborhood V_j of c_j such that for all $x \in V_j$ we have

$$\varphi(x) \leqslant C_0 \log \frac{1}{|x - c_j|}\,. \tag{8.1}$$

We may assume, of course, that the V_j's are pairwise disjoint.

With all these facts at hand we are ready to prove the desired integrability result. This result was first obtained by Przytycki (1993, Th. B), but the proof presented here is taken from de Faria and Guarino (2016).

Proposition 8.1 *The function* $\log Df$ *is* μ*-integrable, i.e.,* $\log Df \in L^1(\mu)$.

Proof For each $1 \leqslant j \leqslant N$ and each $n \geqslant 1$, we define $E_n = \bigcup_{j=1}^{N} J_n(c_j)$ and consider $\varphi_n : S^1 \to \mathbb{R}$ given by

$$\varphi_n = \mathbb{1}_{S^1 \setminus E_n} \cdot \varphi\,,$$

that is, $\varphi_n = 0$ on each $J_n(c_j)$ and $\varphi_n = \varphi$ on the complement of their union. Note that the sequence $\{\varphi_n\}$ converges monotonically to $\varphi = |\log Df|$. Let n_0 be the smallest positive integer such that $J_{n_0}(c_j) \subseteq V_j$ for all $1 \leqslant j \leqslant N$. We only look at values of n greater than n_0. Then, since φ_n is identically zero on E_n and agrees with φ everywhere else, we can write

$$\int_{S^1} \varphi_n \, d\mu = \int_{S^1 \setminus E_{n_0}} \varphi \, d\mu + \sum_{j=1}^{N} \sum_{k=n_0}^{n-1} \int_{I_k(c_j)\setminus I_{k+2}(c_j)} \varphi \, d\mu \tag{8.2}$$

The first integral on the right-hand side is a fixed number independent of n. Hence it suffices to bound the last double sum. Using (8.1) and the fact that in $I_k(c_j) \setminus I_{k+2}(c_j)$ the closest point to c_j is $f^{q_{k+2}}(c_j)$, we see that (see Fig. 8.1)

$$\sum_{k=n_0}^{n-1} \int_{I_k(c_j)\setminus I_{k+2}(c_j)} \varphi \, d\mu \leqslant$$

$$\leqslant C_0 \sum_{k=n_0}^{n-1} \mu(I_k(c_j) \setminus I_{k+2}(c_j)) \log \frac{1}{|I_{k+2}(c_j)|} \tag{8.3}$$

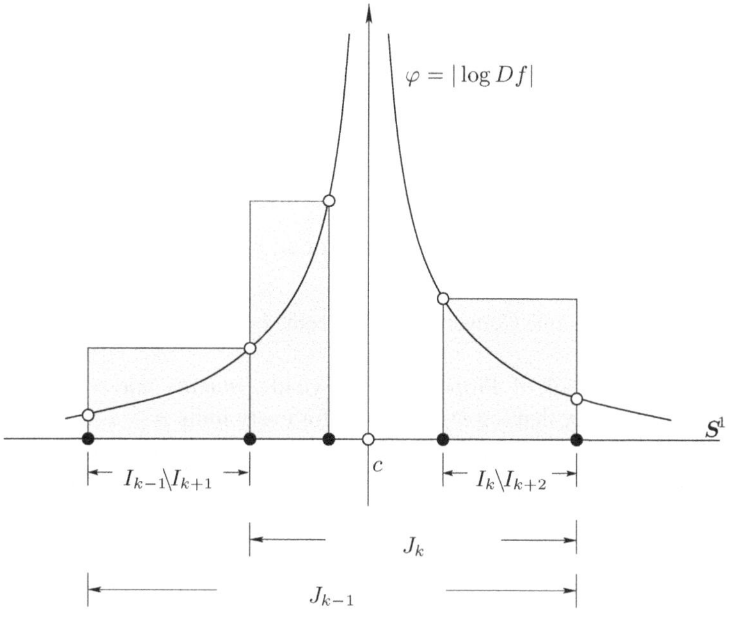

Fig. 8.1 Bounding the integral of $\varphi = |\log Df|$ near a critical point c

Applying facts F1, F2 and F3 to this last sum, we see that

$$\sum_{k=n_0}^{n-1} \mu(I_k(c_j) \setminus I_{k+2}(c_j)) \log \frac{1}{|I_{k+2}(c_j)|} \leqslant$$

$$\leqslant C_1 \sum_{k=n_0}^{n-1} (k+2)\, a_{k+1}\, |q_{k+1}\theta - p_{k+1}| \qquad (8.4)$$

However we know from Theorem 1.2 that

$$|q_{k+1}\theta - p_{k+1}| \leqslant \frac{1}{q_{k+2}} = \frac{1}{a_{k+1}q_{k+1} + q_k} < \frac{1}{a_{k+1}q_{k+1}} \qquad (8.5)$$

Putting (8.5) into (8.4) we get

$$\sum_{k=n_0}^{n-1} \mu(I_k(c_j) \setminus I_{k+2}(c_j)) \log \frac{1}{|I_{k+2}(c_j)|} \leqslant C_1 \sum_{k=n_0}^{n-1} \frac{(k+2)}{q_{k+1}} . \qquad (8.6)$$

Since the q_k's grow exponentially fast (at least as fast as the Fibonacci numbers), we have

$$\sum_{k=0}^{\infty} \frac{(k+2)}{q_{k+1}} < \infty .$$

Hence the left-hand side of (8.6) is uniformly bounded. Taking this information back to (8.3) and then to (8.2), we deduce that there exists a constant $C_2 > 0$ such that

$$\int_{S^1} \varphi_n \, d\mu \leqslant C_2 \quad \text{for all } n \geqslant 1.$$

But then, by the Monotone Convergence Theorem, φ is μ-integrable, as desired. □

Remark 8.1 The proof of Proposition 8.1 yields, *mutatis mutandis*, a slightly stronger result, namely that $\log Df \in L^p(\mu)$ for every finite $p \geqslant 1$.

8.2 No Invariant σ-Finite Measures

As mentioned before, the unique Borel probability measure which is invariant under a minimal multicritical circle map is *purely singular* with respect to Lebesgue measure. This result was first proved by Khanin (1991, Th. 4) in the late eighties, with the help of a certain thermodynamic formalism (see also Graczyk and Świątek

1993, Prop. 1). We will follow a very different approach from the one used by Khanin. We will in fact prove a stronger result, namely the following theorem.

Theorem 8.1 *Let $f : S^1 \to S^1$ be a C^3 multicritical circle map with irrational rotation number. Then f does not admit a σ-finite invariant measure which is absolutely continuous with respect to Lebesgue measure.*

This theorem was recently proved by the authors, see de Faria and Guarino (2021). Our entire exposition here is extracted almost *ipsis verbis* from that paper.

The fact that a circle map f is uniquely ergodic does not eliminate the possibility that f leaves invariant an *infinite*, *σ-finite* measure which is absolutely continuous with respect to Lebesgue measure. For instance, if f happens to be a Denjoy counterexample, it is easy to construct a plethora of such measures (see Exercise 8.1). In fact, examples of *minimal C^∞ diffeomorphisms* of the circle possessing infinite, σ-finite invariant measures have been shown to exist by Katznelson (1977) (see Exercise 8.2 for a non-smooth example).

Theorem 8.1 is saying that the above phenomenon cannot occur in the realm of multicritical circle maps. For its proof, one can argue by contradiction. If f is a minimal multicritical circle map and μ is an infinite, σ-finite invariant measure, let us denote by $\psi = d\mu/dm$ the Radon–Nikodým derivative of μ with respect to Lebesgue measure m. Then ψ is a Borel function such that $0 < \psi < \infty$ Lebesgue-a.e., and the following cocycle identity is satisfied:

$$\psi(x) = \psi \circ f(x) \cdot Df(x) \quad \text{for Lebesgue a.e. } x \in S^1 . \tag{8.7}$$

The rough idea will be to show that, due to the presence of (non-flat) critical points, f has the following property. Near every point x on the circle, and at every scale, one can find two intervals of very different lengths, say I and J, and an iterate of f mapping one of them onto the other diffeomorphically, say $J = f^k(I)$, with bounded distortion. However, if E denotes a positive Lebesgue measure set of points on the circle where ψ is approximately constant, we can take x to be a Lebesgue density point of E, and choose I and J so close to x that they are both almost filled-in by points of E. The cocycle identity (8.7) and a bounded distortion argument then imply that Df^k is approximately equal to 1 inside I. But this implies that I and J have approximately the same length, a contradiction.

Let us turn this rough idea into a formal criterion.

8.2.1 *The Katznelson Criterion*

The proof of Theorem 8.1, given originally in de Faria and Guarino (2021) and reproduced in Sect. 8.2.2 below, is based on a criterion for non-existence of σ-finite measures which is a generalization of a criterion given by Katznelson (1977, Th. 1.1).

Recall from Chap. 7 that a *nested sequence of partitions* $\{\mathscr{Q}_n\}_{n\geqslant 0}$ is a sequence of finite interval partitions of S^1 (modulo endpoints) with the property that each atom of $\mathscr{Q}_n$ is a union of atoms of $\mathscr{Q}_{n+1}$, for all $n \geqslant 0$, and such that $\mathrm{mesh}(\mathscr{Q}_n) \to 0$ as $n \to \infty$.[1]

Definition 8.1 A C^1 circle homeomorphism f has the *Katznelson property* if there exist a nested sequence of partitions $\{\mathscr{Q}_n\}_{n\geqslant 0}$ and constants $1 < b_0 < b_1$ and $0 < \theta < 1$ such that the following holds. For each $\Delta \in \mathscr{Q}_n$, the collection $\mathscr{A}^\Delta = \{J \in \mathscr{Q}_{n+1} : J \subset \Delta\}$ can be decomposed as a disjoint union $\mathscr{A}^\Delta = \mathscr{A}_1^\Delta \cup \mathscr{A}_2^\Delta \cup \mathscr{A}_3^\Delta$ with the following properties:

(i) For each $J_1 \in \mathscr{A}_1^\Delta$ and each $J_2 \in \mathscr{A}_2^\Delta$ we have $|J_1| \geqslant b_0|J_2|$;
(ii) For each $J_1 \in \mathscr{A}_1^\Delta$ and each $J_2 \in \mathscr{A}_2^\Delta$ there exists $k \in \mathbb{N}$ such that $f^k|_{J_1}$ is a diffeomorphism mapping J_1 onto J_2, and we have $Df^k(x) \geqslant b_1^{-1}$ for all $x \in J_1$.
(iii) We have $\lambda(\Omega) \geqslant \theta|\Delta|$, where

$$\Omega = \bigcup_{J\in\mathscr{A}_1^\Delta\cup\mathscr{A}_2^\Delta} J\,.$$

(iv) The sub-collections $\mathscr{A}_1^\Delta$ and $\mathscr{A}_2^\Delta$ have the same number of elements.

Remark 8.2 The sub-collection $\mathscr{A}_3^\Delta$, about which nothing is said in the above definition, plays no role in the arguments to come. Only $\mathscr{A}_1^\Delta$ and $\mathscr{A}_2^\Delta$ matter.

Theorem 8.2 *Let $f : S^1 \to S^1$ be a C^1 minimal homeomorphism, and suppose that f has the Katznelson property. Then f does not admit a σ-finite invariant measure which is absolutely continuous with respect to Lebesgue measure.*

We refer to this theorem as the *Katznelson criterion.*

Proof Assume by contradiction that there exists a σ-finite measure μ which is invariant under f and is absolutely continuous with respect to Lebesgue measure. Let $\psi = d\mu/dm$ be the corresponding Radon–Nikodým derivative. This is a Borel measurable function that is positive and finite Lebesgue a.e., and it satisfies the cocycle identity (8.7). By an easy induction, that cocycle identity can be written more generally as

$$\psi(x) = \psi \circ f^k(x) \cdot Df^k(x) \quad \text{for Lebesgue a.e. } x \in S^1\,,\ \text{for all } k \in \mathbb{Z}\,. \tag{8.8}$$

Fix a small number $0 < \delta < 1$; we will need it small enough that $(1+\delta)^{-1}b_0 > 1$. For each real number c consider the Borel set $E_c = \{x \in S^1 : c \leqslant \psi(x) \leqslant c(1+\delta)\}$. Then we have $m(E_c) > 0$ for some choice of c. We choose such c and from now on write $E = E_c$.

[1] As customary, the *mesh* of a partition is the maximum length of its atoms.

By the Lebesgue density theorem, m-a.e. $x \in E$ is such that the density of E at x is 1. Hence for each $\epsilon > 0$ we can find a suitable level $n \in \mathbb{N}$ and an atom $\Delta \in \mathcal{Q}_n$ such that

$$\frac{m(E \cap \Delta)}{|\Delta|} \geqslant 1 - \epsilon\,. \tag{8.9}$$

We will show that the assumption at the start of this proof contradicts our standing hypothesis on f if we take ϵ sufficiently small. How small ϵ has to be will be determined in the course of the argument to follow.

Let $\mathscr{A}^\Delta$ and $\mathscr{A}_i^\Delta$, $i = 1, 2, 3$ be as defined before, and for $i = 1, 2$ let $\Omega_i = \bigcup_{J \in \mathscr{A}_i^\Delta} J$. Then (iii) in our standing hypothesis tells us that $\Omega = \Omega_1 \cup \Omega_2$ satisfies $m(\Omega) \geqslant \theta|\Delta|$. Hence from (8.9) we have

$$\frac{m(E \cap \Omega)}{m(\Omega)} \geqslant 1 - \epsilon\theta^{-1}\,, \tag{8.10}$$

provided ϵ is so small that $\epsilon\theta^{-1} < 1$. Note that our standing hypothesis also tells us that $b_0 m(\Omega_2) \leqslant m(\Omega_1) \leqslant b_1 m(\Omega_2)$. These inequalities imply that

$$m(\Omega) \leqslant (1 + b_0^{-1}) m(\Omega_1) \;\text{ and }\; m(\Omega) \leqslant (1 + b_1) m(\Omega_2)\,. \tag{8.11}$$

Using (8.10) and the first inequality in (8.11), we get

$$\begin{aligned} m(\Omega_1) &\leqslant m(E \cap \Omega_1) + m(\Omega \setminus E) \\ &\leqslant m(E \cap \Omega_1) + \epsilon\theta^{-1} m(\Omega) \\ &\leqslant m(E \cap \Omega_1) + \epsilon\theta^{-1}(1 + b_0^{-1}) m(\Omega_1)\,. \end{aligned}$$

Hence we have

$$\frac{m(E \cap \Omega_1)}{m(\Omega_1)} \geqslant 1 - \epsilon\theta^{-1}(1 + b_0^{-1})\,,$$

and this lower bound will be positive (in fact close to one) provided ϵ is sufficiently small. Similarly, using (8.10) and the second inequality in (8.11), we deduce that

$$\frac{m(E \cap \Omega_2)}{m(\Omega_2)} \geqslant 1 - \epsilon\theta^{-1}(1 + b_1)\,.$$

Thus, writing $\eta = \epsilon\theta^{-1} \max\{1 + b_0^{-1},\, 1 + b_1\} = \epsilon\theta^{-1}(1 + b_1)$, we have

$$\frac{m(E \cap \Omega_i)}{m(\Omega_i)} \geqslant 1 - \eta\,, \quad \text{for } i = 1, 2\,. \tag{8.12}$$

Note that $\eta \to 0$ when $\epsilon \to 0$. Now, since both Ω_1 and Ω_2 are disjoint unions of atoms in $\mathscr{Q}_{n+1}$, it follows from (8.12) that there exist atoms $J_1 \in \mathscr{A}_1^\Delta$ and $J_2 \in \mathscr{A}_2^\Delta$ such that

$$m(J_i \cap E) \geqslant (1-\eta)|J_i|\,, \quad \text{for } i = 1, 2\,. \tag{8.13}$$

Let $k \in \mathbb{N}$ be such that f^k maps J_1 diffeomorphically onto J_2, and let us estimate the Lebesgue measure of $f^{-k}(J_2 \setminus E)$. By (ii) in our standing hypothesis and the chain rule we have $Df^{-k}(y) \leqslant b_1$ for all $y \in J_2$. Since by (8.13) we have $m(J_2 \setminus E) \leqslant \eta|J_2|$, we get

$$m(f^{-k}(J_2 \setminus E)) = \int_{J_2 \setminus E} Df^{-k}\, dm \leqslant b_1\eta|J_2|\,. \tag{8.14}$$

Letting $J_1^* = \{x \in J_1 \cap E : f^k(x) \in E\}$, it follows from (8.13) and (8.14) that

$$m(J_1^*) = m(J_1 \cap E) - m(f^{-k}(J_2 \setminus E)) \geqslant [(1-\eta)b_0 - \eta b_1]\,|J_2|\,. \tag{8.15}$$

But now observe that the equality $\psi = (\psi \circ f^k)Df^k$ holds Lebesgue almost everywhere: this is simply the cocycle identity (8.8). Since for every $x \in J_1^*$ we have both $x \in E$ and $f^k(x) \in E$, it follows from this equality and the definition of E that for Lebesgue a.e. $x \in J_1^*$ we have $Df^k(x) \geqslant (1+\delta)^{-1}$. Therefore

$$|J_2| > m(f^k(J_1^*)) = \int_{J_1^*} Df^k\, dm \geqslant (1+\delta)^{-1} m(J_1^*)\,. \tag{8.16}$$

Combining (8.15) and (8.16) and cancelling out $|J_2|$ from both sides of the resulting inequality, we deduce at last that

$$(1+\delta)^{-1}[(1-\eta)b_0 - \eta b_1] < 1\,. \tag{8.17}$$

But since $(1+\delta)^{-1}b_0 > 1$, the inequality (8.17) is clearly violated if η is sufficiently small, which is certainly the case if we choose ϵ sufficiently small. We have reached the desired contradiction, and the proof is complete. □

8.2.2 *Proof of Theorem 8.1*

The proof of Theorem 8.1 entails two separate arguments, presented in separate sections below as *first step* and *second step*, respectively. Which argument applies for a given map f depends on the nature of its rotation number—more precisely, on the behavior of the partial quotients of the continued fraction development of $\rho(f)$.

The first argument deals with all irrational rotation numbers *except* those numbers (of bounded type) whose partial quotients are bounded by a certain constant B that depends only on the real bounds (Theorem 6.3). The second argument takes care of the bounded type case. They are presented as two separate theorems, namely Theorems 8.3 and 8.4, respectively.

The arguments presented in both proofs have different flavors, exploiting different aspects of the geometry of multicritical circle maps. In particular, while the proof of Theorem 8.3 uses the real bounds *and* Yoccoz's inequality (Lemma 7.3), the proof of Theorem 8.4 uses only the real bounds.

8.2.2.1 First Step

The precise result we shall prove here is the following weaker version of Theorem 8.1.

Theorem 8.3 *Given $N \geqslant 1$ in $\mathbb{N}$ and $d > 1$ there exists a universal constant $B = B(N, d) \in \mathbb{N}$ such that the following holds. If f is a multicritical circle map with at most N critical points whose criticalities are bounded by d, and if the rotation number of f is irrational and its partial quotients a_n satisfy $\limsup a_n \geqslant B$, then f does not admit an invariant σ-finite measure which is absolutely continuous with respect to Lebesgue measure.*

In the proof of Theorem 8.3, we will make extensive use of the following fact, which is an immediate consequence of Lemma 7.8.

Lemma 8.1 *Let c_0 be a critical point of f, and let $0 \leqslant k < a_{n+1}$ be such that the interval $f^{q_n+kq_{n+1}}\big(I_{n+1}(c_0)\big) \subset I_n(c_0)$ contains a critical point of $f^{q_{n+1}}$. Then*

$$\left|f^i\big(f^{q_n+kq_{n+1}}(I_{n+1}(c_0))\big)\right| \asymp \left|f^i\big(I_n(c_0)\big)\right| \quad \textit{for all } i \in \{0, 1, \ldots, q_{n+1}\}.$$

Proof We only sketch the proof. For $i = 0$ the statement is just Lemma 7.8. Moreover, by Theorem 6.3, the image of each critical spot under $f^{q_{n+1}}$ is also comparable to $I_n(c_0)$; this is simply because

$$f^{q_{n+1}}\big(f^{q_n+kq_{n+1}}(I_{n+1}(c_0))\big) = f^{q_n+(k+1)q_{n+1}}\big(I_{n+1}(c_0)\big)$$

is adjacent to $f^{q_n+kq_{n+1}}\,(I_{n+1}(c_0))$ in $\mathscr{P}_{n+1}(c_0)$. So the statement of our lemma also holds for $i = q_{n+1}$. Now, for each $i \in \{1, \ldots, q_{n+1} - 1\}$ consider the iterate $f^{q_{n+1}-i}$, and apply the Cross-ratio Inequality (Theorem 5.1). □

In keeping with the terminology introduced in Sect. 7.4.2.1, an interval such as $f^{q_n+kq_{n+1}}\big(I_{n+1}(c_0)\big)$ appearing in the statement above, containing some critical point of $f^{q_{n+1}}$, is called a *critical spot*. Thus, Lemma 8.1 is saying that every critical spot is large, i.e., is comparable to the atom of $\mathscr{P}_n(c_0)$ in which it is contained, and the same happens to all its images up to time $i = q_{n+1}$.

Proof of Theorem 8.3 By Theorem 8.2, it suffices to show that an f as in the statement satisfies the *standing hypothesis* previously formulated, provided $\limsup a_n$ is sufficiently large. This will be proved with the help of the real bounds (Theorem 6.3), Yoccoz's inequality (Lemma 7.3) and Lemma 8.1 above.

Let c_0 be a critical point of f and consider the associated dynamical partitions $\mathscr{P}_n(c_0)$ for $n \geqslant n_0(f)$, where $n_0(f)$ is as in Theorem 6.3. We are also assuming that such n is large enough that the iterates f^{q_n} and $f^{q_{n+1}}$ have negative Schwarzian derivative at all points in $I_{n+1}(c_0)$ ($I_n(c_0)$ respectively) where their derivatives do not vanish (this is possible by Proposition 6.2). We will only consider in the proof long atoms of $\mathscr{P}_n(c_0)$, the proof for the short ones being the same. Moreover, we will decompose first the collection $\{J \in \mathscr{P}_{n+1}(c_0) : J \subset I_n(c_0)\}$, and then we will spread this decomposition iterating by f. So let $\Delta = I_n(c_0)$, and consider the following consecutive atoms of $\mathscr{P}_{n+1}(c_0)$ inside Δ: $\Delta_0 = f^{q_n}(I_{n+1})$ and $\Delta_j = f^{jq_{n+1}}(\Delta_0)$ for $j = 1, 2, \ldots, a_{n+1}-1$; note that $\Delta_j = f^{q_{n+1}}(\Delta_{j-1})$ for all $1 \leqslant j \leqslant a_{n+1}-1$. Some of these intervals may be critical spots (which are always comparable in size with $|\Delta|$, by Lemma 8.1). We look at the *bridges* between such critical spots, and pick the longest one. More precisely, let $0 \leqslant j_1 \leqslant j_2 \leqslant a_{n+1} - 1$ with $j_2 - j_1$ maximal with the property that $\phi = f^{q_{n+1}}|_{\Delta_{j_1} \cup \cdots \cup \Delta_{j_2}}$ is a diffeomorphism onto its image. Let $T_n = \Delta_{j_1} \cup \cdots \cup \Delta_{j_2}$, $R_n = \Delta_{j_1}$, $L_n = \Delta_{j_2}$ and $M_n = T_n \setminus (L_n \cup R_n) = \Delta_{j_1+1} \cup \cdots \cup \Delta_{j_2-1}$. Note that $\phi|_{M_n}$ is an *almost parabolic map* (Definition 7.3) with length $\ell = j_2 - j_1 - 1$, and note that $\ell \geqslant a_{n+1}/(N+1)$, where N is the number of critical points of f. Let us write $J_1 = \Delta_{j_1+1}$, $J_2 = \Delta_{j_1+2}$, $\ldots$, $J_\ell = \Delta_{j_1+\ell} = \Delta_{j_2-1}$. From the real bounds (Theorem 6.3), we have $|J_1| \asymp |\Delta| \asymp |J_\ell|$, with *beau* comparability constants. Therefore, by Yoccoz's inequality (Lemma 7.3), there exists a constant $C_0 > 1$, depending only on f, such that, for all $1 \leqslant j \leqslant \ell$,

$$\frac{C_0^{-1}}{\min\{j\,,\,\ell - j\}^2} \leqslant \frac{|J_j|}{|\Delta|} \leqslant \frac{C_0}{\min\{j\,,\,\ell - j\}^2} \tag{8.18}$$

Now we claim that there exists a constant $\tau > 0$ (depending only on f) such that

$$\left|f^i(L_n)\right| > \tau\left|f^i(M_n)\right| \quad \text{and} \quad \left|f^i(R_n)\right| > \tau\left|f^i(M_n)\right|$$

for all $i \in \{0, \ldots, q_{n+1}\}$. Indeed, again by combining Theorem 6.3 with Lemma 8.1 we obtain the claim for both $i = 0$ and $i = q_{n+1}$. By the Cross-ratio Inequality (note that the intervals $T_n, f(T_n), \ldots, f^{q_{n+1}-1}(T_n)$ are pairwise disjoint), we deduce the claim for any $i \in \{1, \ldots, q_{n+1} - 1\}$. With this at hand, and since $f^i|_{T_n}$ is a diffeomorphism for any $i \in \{0, \ldots, q_{n+1}\}$, we can apply Koebe distortion principle (Lemma 5.2) in order to obtain a constant $K = K(f) > 1$ such that $f^i|_{M_n}$ has distortion bounded by K for each $i \in \{0, \ldots, q_{n+1}\}$. Let us now define $B = 2(N+1)\left\lceil\sqrt{2KC_0}\right\rceil + 1$. We are assuming from now on that n is one of infinitely many natural numbers such that $a_{n+1} \geqslant B$. Let m be the smallest natural number such that $KC_0^2 m^{-2} \leqslant \frac{1}{2}$; in other words, let $m = \left\lceil\sqrt{2KC_0}\right\rceil$. Since $a_{n+1} \geqslant B$, we have

$$\frac{\ell}{2} \geqslant \frac{a_{n+1}}{2(N+1)} \geqslant \frac{B}{2(N+1)} > \left\lceil \sqrt{2K}C_0 \right\rceil = m\,.$$

Thus, setting $J' = J_1$ and $J'' = \phi^{m-1}(J') = J_m$, it follows from (8.18) that

$$\frac{1}{C_0^2 m^2} \leqslant \frac{|J''|}{|J'|} \leqslant \frac{C_0^2}{m^2} \leqslant \frac{1}{2K} < \frac{1}{2}\,. \tag{8.19}$$

We are now ready to define the desired decomposition of $\mathscr{A}^\Delta$, the collection of all atoms of $\mathscr{P}_{n+1}(c_0)$ that are contained in $\Delta = I_n(c_0)$. Let $\mathscr{A}_1^\Delta = \{J'\}$, let $\mathscr{A}_2^\Delta = \{J''\}$ and let $\mathscr{A}_3^\Delta = \mathscr{A}^\Delta \setminus (\mathscr{A}_1^\Delta \cup \mathscr{A}_2^\Delta)$. We claim that this decomposition satisfies all conditions (i)-(iv) in the standing hypothesis. From (8.19), we have $|J'| \geqslant 2|J''|$, so (i) is satisfied with $b_0 = 2$. By the mean value theorem, there exists $\xi \in J'$ such that

$$D\phi^{m-1}(\xi) = \frac{|J''|}{|J'|} \geqslant \frac{1}{C_0^2 m^2}\,,$$

where we have again used (8.19). By Koebe distortion principle, there exists $C_1 > 1$ (depending only on f) such that

$$C_1^{-1} \leqslant \frac{D\phi^{m-1}(x)}{D\phi^{m-1}(\xi)} \leqslant C_1\,, \quad \text{for all } x \in J'\,.$$

Combining these facts we deduce that $D\phi^{m-1}(x) \geqslant (C_0^2 C_1 m^2)^{-1}$, and so (ii) is certainly satisfied if we take $k = q_{n+1}(m-1)$ and $b_1 = KC_0^2 C_1 m^2 = KC_0^2 C_1 \left\lceil \sqrt{2K}C_0 \right\rceil^2$. Note that $b_1 > 2 = b_0$. For $\Omega = J' \cup J''$, we now have, using (8.18), the simple bound $m(\Omega) = |J'| + |J''| \geqslant |J'| \geqslant C_0^{-1}|\Delta|$. This shows that (iii) is satisfied if we choose $\theta = C_0^{-1} < 1$. Finally, condition (iv) is trivially satisfied because both $\mathscr{A}_1^\Delta$ and $\mathscr{A}_2^\Delta$ have a single element.

Now we spread the previous decomposition along the whole family of long intervals of $\mathscr{P}_n(c_0)$. More precisely, for each $i \in \{1, \ldots, q_{n+1} - 1\}$ we define a decomposition of $\mathscr{A}^\Delta$, the collection of all atoms of $\mathscr{P}_{n+1}(c_0)$ that are contained in $\Delta = f^i\big(I_n(c_0)\big)$, as follows: let $\mathscr{A}_1^\Delta = \{f^i(J')\}$, let $\mathscr{A}_2^\Delta = \{f^i(J'')\}$ and let $\mathscr{A}_3^\Delta = \mathscr{A}^\Delta \setminus (\mathscr{A}_1^\Delta \cup \mathscr{A}_2^\Delta)$. Again, we claim that this decomposition satisfies all conditions (i)-(iv) in the standing hypothesis. Indeed, for each $i \in \{1, \ldots, q_{n+1} - 1\}$ let $x_i' \in J'$ and $x_i'' \in J''$ be given by the mean value theorem:

$$\frac{\big|f^i(J'')\big|}{\big|f^i(J')\big|} = \frac{Df^i(x_i'')}{Df^i(x_i')}\,\frac{|J''|}{|J'|}\,.$$

By bounded distortion and (8.19) we obtain

$$\frac{\left|f^i(J'')\right|}{\left|f^i(J')\right|}=\frac{Df^i(x_i'')}{Df^i(x_i')}\,\frac{|J''|}{|J'|}\leqslant K\,\frac{|J''|}{|J'|}\leqslant\frac{K\,C_0^2}{m^2}\leqslant\frac{1}{2}\,.$$

So (i) is again satisfied with $b_0=2$. Now if we conjugate $\phi^{m-1}:J'\to J''$ with the iterate f^i, we obtain a diffeomorphism $f^i\circ\phi^{m-1}\circ f^{-i}:f^i(J')\to f^i(J'')$ which satisfies the following for all $x\in f^i(J')$:

$$\begin{aligned}D\big(f^i\circ\phi^{m-1}\circ f^{-i}\big)(x)&=D\phi^{m-1}\big(f^{-i}(x)\big)\,Df^i\big(\phi^{m-1}\circ f^{-i}(x)\big)\,Df^{-i}(x)\\&=D\phi^{m-1}\big(f^{-i}(x)\big)\,\frac{Df^i\big(\phi^{m-1}\circ f^{-i}(x)\big)}{Df^i\big(f^{-i}(x)\big)}\,.\end{aligned}$$

Since $f^{-i}(x)$ belongs to J', $\phi^{m-1}\big(f^{-i}(x)\big)$ belongs to J'' and then

$$D\big(f^i\circ\phi^{m-1}\circ f^{-i}\big)(x)\geqslant\frac{1}{K}\,D\phi^{m-1}\big(f^{-i}(x)\big)\geqslant\frac{1}{K}\,(C_0^2C_1m^2)^{-1}\,.$$

Therefore, just as before, (ii) is again satisfied with $k=q_{n+1}(m-1)$ and $b_1=KC_0^2C_1m^2=KC_0^2C_1\left\lceil\sqrt{2K}C_0\right\rceil^2$. By Lemma 8.1, the i-th iterate of a critical spot, contained in $I_n(c_0)$, is comparable to $f^i\big(I_n(c_0)\big)$ for all $i\in\{0,1,\ldots,q_{n+1}\}$ and then, by Theorem 6.3, the interval $f^i(J')$ is comparable to $f^i\big(I_n(c_0)\big)$ as well, which implies (iii). Again, condition (iv) is trivially satisfied. Summarizing, we have shown that, for infinitely many values of n, the partitions $\mathscr{P}_n(c_0)$ satisfy conditions (i) through (iv) of the standing hypothesis. Therefore, by Theorem 8.2, f does not admit a σ-finite invariant measure equivalent to Lebesgue measure. This finishes the proof. □

8.2.2.2 Second Step

We now move to the bounded type case. Here our goal will be to prove the following result.

Theorem 8.4 *If f is a multicritical circle map with an irrational rotation number of bounded type, then f does not admit an invariant σ-finite measure which is absolutely continuous with respect to Lebesgue measure.*

In the proof of Theorem 8.4 we will make use of the following two auxiliary results.

Proposition 8.2 *Given a multicritical circle map f with an irrational rotation number of bounded type, there exist constants $C_0>1$ and $0<\lambda_0<\lambda_1<1$*

with the following property. For each $x \in \mathbf{S}^1$, each $n, k \geqslant 0$ and every pair of atoms $I \in \mathscr{P}_n(x)$ and $J \in \mathscr{P}_{n+k}(x)$ with $J \subseteq I$, we have

$$C_0^{-1}\lambda_0^k \leqslant \frac{|J|}{|I|} \leqslant C_0\lambda_1^k .$$

Proof Exercise. □

Proposition 8.3 *Given a multicritical circle map f with an irrational rotation number of bounded type, there exists $n_0 = n_0(f) \in \mathbb{N}$ such that for all $n \geqslant n_0$ we have*

$$Sf^{q_{n+1}}(x) < 0 \quad \textit{for all } x \in \mathbf{S}^1 \textit{ regular point of } f^{q_{n+1}}.$$

Likewise, we have

$$Sf^{q_n}(x) < 0 \quad \textit{for all } x \in \mathbf{S}^1 \textit{ regular point of } f^{q_n}.$$

We postpone the proof of Proposition 8.3 until the end of this section (see Sect. 8.2.3).

Remark 8.3 We emphasize that the statement of Proposition 8.2 is obviously *false* for unbounded combinatorics. On the other hand, Proposition 8.3 is most likely true for *any* irrational rotation number.

Our proof of Theorem 8.4 will be based on the following lemma. Recall that we are fixing our attention on a critical point c of f. Below, we use the following notation: for all $i \geqslant 0$, let $c_{-i} = f^{-i}(c)$; we write accordingly $I_n(c_{-i}) = f^{-i}(I_n(c))$ for all $n \geqslant 0$ and all $i \geqslant 0$.

Lemma 8.2 *There exist constants $K > 1$ and $0 < \theta < 1$ such that the following holds for all n sufficiently large and each $0 \leqslant i < q_n$. There exist subintervals $\Delta'_{i,n} \subset I_{n+1}(c_{-i})$ and $\Delta''_{i,n} \subset I_n(c_{-i})$ such that*

(i) $\Delta'_{i,n} \cap \Delta''_{i,n} = \emptyset$;
(ii) $|\Delta'_{i,n}| \geqslant 2|\Delta''_{i,n}|$;
(iii) $|\Delta''_{i,n}| \geqslant \theta |I_n(c_{-i})|$;
(iv) $\Delta''_{i,n} = f^{q_n}(\Delta'_{i,n})$, *and* $f^{q_n}|_{\Delta'_{i,n}} : \Delta'_{i,n} \to \Delta''_{i,n}$ *is a diffeomorphism whose distortion is bounded by K.*

Proof We assume from the start that n is so large that $f^{q_n}|_{I_{n+1}(c_{-i})}$ has negative Schwarzian derivative for all $0 \leqslant i < q_n$. This is possible by Proposition 8.3. *Note that each c_{-i} for $0 \leqslant i < q_{n+1}$ is a critical point of f^{q_n}.* In what follows, we keep n and $0 \leqslant i < q_n$ fixed.

Note that for all $k \geqslant 0$ *even* we have $I_{n+k+1}(c_{-i}) \subseteq I_{n+1}(c_{-i})$. By Proposition 8.2, there exist constants $0 < \lambda_0 < \lambda_1 < 1$ and $C_0 > 1$ such that

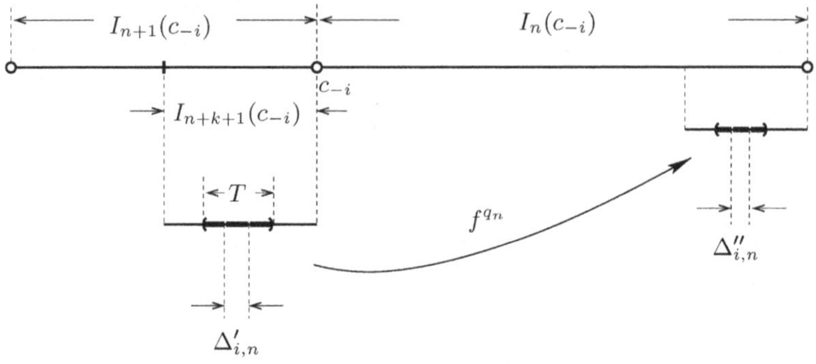

Fig. 8.2 The iterate f^{q_n} maps $\Delta'_{i,n}$ diffeomorphically onto $\Delta''_{i,n}$ with bounded distortion

$$C_0^{-1}\lambda_0^k \leqslant \frac{|I_{n+k+1}(c_{-i})|}{|I_n(c_{-i})|} \leqslant C_0\lambda_1^k . \tag{8.20}$$

Moreover, if we denote by $d = d(i,n) > 1$ the power-law at the critical point c_{-i} of f^{q_n}, then we have[2]

$$\frac{|f^{q_n}(I_{n+k+1}(c_{-i}))|}{|I_n(c_{-i})|} \asymp \left(\frac{|I_{n+k+1}(c_{-i})|}{|I_n(c_{-i})|}\right)^d . \tag{8.21}$$

Let us write $I = I_{n+k+1}(c_{-i})$ and $J = f^{q_n}(I)$; these are obviously disjoint intervals (see Fig. 8.2), and they are both atoms of $\mathscr{P}_{n+k}(c_{-i})$. Combining (8.20) with (8.21), we deduce that there exists a constant $C_1 > 1$ (independent of n and k) such that

$$C_1^{-1}\lambda_0^{k(d-1)}|I| \leqslant |J| \leqslant C_1\lambda_1^{k(d-1)}|I| \tag{8.22}$$

Note that $f^{q_n}|_I : I \to J$ has at most N critical points,[3] and has negative Schwarzian at all regular points. Note that, by choosing k sufficiently large, we can make $|J|$ definitely smaller than $|I|$. The meaning of "definitely smaller", and thus how large k has to be, will be clear in a moment.

For $p \geqslant 0$, let us denote the number of atoms of $\mathscr{P}_{n+k+p}(c_{-i})$ inside I (or J) by $a = a(n,k,p)$. Then we have $2^p \leqslant a \leqslant (A+1)^p$ (where $A = \sup a_n < \infty$ is the least upper bound on the convergents of the rotation number of f). Choose $p = p(N)$ smallest with the property that $2^p > 3N+2$. Since $f^{q_n}|_I$ has at most N critical points, and since $a > 3N+2$, it follows from the pigeonhole principle that there exist 3 consecutive atoms of $\mathscr{P}_{n+k+p}(c_{-i})$ inside I, say L, M, R, such that the open interval $T = \text{int}(L \cup M \cup R)$ contains no critical point of f^{q_n}. Hence

[2] One can easily check that $d_{\min} \leqslant d(i,n) \leqslant d_{\max}^N$, where $d_{\min}$ and $d_{\max}$ are the smallest and largest power-law exponents of the critical points of f, and N is the number of such critical points.

[3] Again, N is the total number of critical points of f.

$f^{q_n}|_T : T \to f^{q_n}(T)$ is a diffeomorphism with negative Schwarzian derivative. Applying Koebe's nonlinearity principle, we see that

$$|D \log Df^{q_n}(x)| \leqslant \frac{2}{\tau} \quad \text{for all } x \in M . \tag{8.23}$$

where τ is the *space* of M inside T, namely

$$\tau = \min\left\{\frac{|L|}{|M|}, \frac{|R|}{|M|}\right\} .$$

From the real bounds, we know that $\tau \geqslant C_2$, for some constant $C_2 > 0$. Using this fact in (8.23) and integrating the resulting inequality, we deduce that

$$e^{-2/C_2} \leqslant \frac{Df^{q_n}(x)}{Df^{q_n}(y)} \leqslant e^{2/C_2} , \quad \text{for all } x, y \in M . \tag{8.24}$$

Now, applying once again Proposition 8.2 (note that we are using the bounded type hypothesis!), it follows that there exists a constant $C_3 > 1$ depending on A such that

$$C_3^{-1}\lambda_0^p \leqslant \frac{|M|}{|I|} \leqslant C_3\lambda_1^p , \tag{8.25}$$

as well as

$$C_3^{-1}\lambda_0^p \leqslant \frac{|f^{q_n}(M)|}{|J|} \leqslant C_3\lambda_1^p , \tag{8.26}$$

Putting together (8.22), (8.25) and (8.26), we deduce that

$$|M| \geqslant C_1^{-1}C_3^{-2}\lambda_0^p\lambda_1^{-k(d-1)-p}\,|f^{q_n}(M)| . \tag{8.27}$$

Likewise, putting together (8.20), (8.22) and (8.27), we get

$$|f^{q_n}(M)| \geqslant (C_0C_1C_3)^{-1}\lambda_0^{kd+p}|I_n(c_{-i})| . \tag{8.28}$$

Now let us choose $k \geqslant 1$ smallest with the property that

$$C_1^{-1}C_3^{-2}\lambda_0^p\lambda_1^{-k(d_0-1)-p} \geqslant 2 , \tag{8.29}$$

where $d_0 = \min_{i,n} d(i,n) > 1$ Such k exists (and is independent of n) because $\lambda_1 < 1$.

To finish the proof, we define $\Delta'_{i,n} = M$ and $\Delta''_{i,n} = f^{q_n}(M)$. These, we claim, are the intervals satisfying properties (i)-(iv) in the statement. Indeed, property (i) is clear. Property (iv) follows directly from (8.24) if we take $K = e^{2/C_2}$. Property

(ii) follows from inequalities (8.27) and (8.29). Finally, property (iii) follows from (8.28), provided we take $\theta = (C_0C_1C_3)^{-1}\lambda_0^{kd+p}$. The proof is complete. □

Proof of Theorem 8.4 The proof will based on the generalized Katznelson criterion given by Theorem 8.2. Our argument combines Lemma 8.2 with the Cross Ratio Inequality.

It is enough to show that f possesses the Katznelson property with respect to (a subsequence of) the sequence of dynamical partitions $\mathscr{P}_n(c)$ for some choice of critical point c. For this purpose, as we have seen in the proof of that theorem, and also taking into account the result of Exercise 8.4, it suffices to prove the following statement.

Claim For every sufficiently large n, every atom $I \in \mathscr{P}_n(c)$ contains two disjoint subintervals Δ', Δ'' such that: (a) $|\Delta'| \geqslant 2|\Delta''|$; (b) $|\Delta'| \asymp |I| \asymp |\Delta''|$; (c) there exists $q \geqslant 1$ such that $\Delta'' = f^q(\Delta')$ and $f^q|_{\Delta'} : \Delta' \to \Delta''$ is a diffeomorphism with bounded distortion.[4]

The comparability constants and bounds implicit in this statement depend only on the real bounds for f and the bound on the combinatorics. To simplify the notation a bit, let us write $J_k = I_k(c) \cup I_{k+1}(c)$ for all $k \geqslant 0$. In order to prove the claim, we proceed through the following steps.

(i) We may assume that I is a *long* atom of $\mathscr{P}_n(c)$, say $I = f^{q_{n+1}-i}(I_n(c))$, where $1 \leqslant i \leqslant q_{n+1} - 1$. If I happens to be a *short* atom, all we have to do is recall that every short atom of $\mathscr{P}_n(c)$ is a long atom of $\mathscr{P}_{n+1}(c)$.

(ii) The interval $T = f^{q_{n+1}}(I_n(c))$ contains the interval J_{n+4} in its interior, with definite space on both sides (see Fig. 8.3). To see why this is true, first note that, by the real bounds, the interval J_{n+4} is comparable to $|I_n(c)|$, i.e., $|J_{n+4}| \asymp |I_n(c)|$. Consider the following two atoms of $\mathscr{P}_{n+1}(c)$, which also lie inside T:

$$L^* = f^{q_{n+1}}(I_{n+2}) \subset I_{n+1}(c) \quad \text{and} \quad R^* = f^{q_n+q_{n+1}}(I_{n+1}(c)) \subset I_n(c)\ .$$

Both these intervals share an endpoint with T (one on the left, the other on the right). By simple combinatorics, we see that $J_{n+4} \subset T$ is disjoint from both L^* and R^*. But by the real bounds, we have $|L^*| \asymp |I_{n+1}(c)|$ and $|R^*| \asymp |I_n(c)|$. If we denote by L and R the two connected components of $T \setminus J_{n+4}$, then one of them contains L^* and the other contains R^*. For definiteness, we assume that $L \supseteq L^*$ and $R^* \supseteq R$. Hence we have $|L| \asymp |I_{n+1}(c)| \asymp |T|$ and $|R^*| \asymp |I_n(c)| \asymp |T|$.

(iii) In particular, (ii) tells us that the cross-ratio $[J_{n+4}, f^{q_{n+1}}(I_n(c))]$ is bounded away from 0 and ∞.

(iv) Now look at the interval

[4] The claim's proof will show that $q = q_n$ or $q = q_{n+1}$, depending on whether I is a *long* or *short* atom of $\mathscr{P}_n(c)$, respectively.

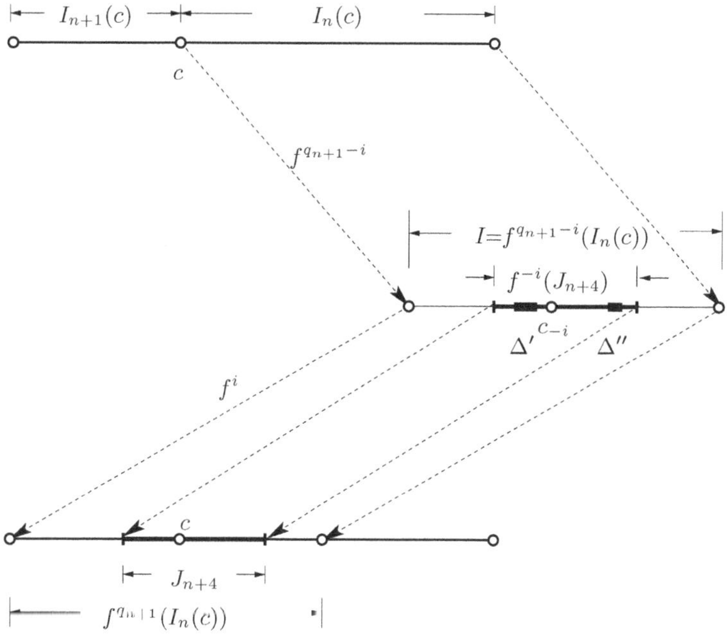

Fig. 8.3 Finding two intervals, long and short, inside an atom $I \in \mathscr{P}_n(c)$

$$f^{-i}(J_{n+4}) \subset f^{-i}(f^{q_{n+1}}(I_n(c))) = f^{q_{n+1}-i}(I_n(c)) = I \,.$$

Observe that $f^{-i}(J_{n+4}) = I_{n+4}(c_{-i}) \cup I_{n+5}(c_{-i})$ (in the notation introduced prior to Lemma 8.2). Hence we can apply Lemma 8.2 (with n replaced by $n+4$) and deduce that there exist intervals

$$\Delta' = \Delta'_{i,n+4} \subset I_{n+5}(c_{-i}) \quad \text{and} \quad \Delta'' = \Delta''_{i,n+4} \subset I_{n+4}(c_{-i})$$

satisfying properties (i)-(iv) of that lemma. In particular, we have

$$|\Delta'| \asymp |f^{-i}(J_{n+4})| \asymp |\Delta''| \,. \tag{8.30}$$

(v) The intervals Δ' and Δ'' already satisfy properties (a) and (c) in the claim. Therefore, all we have to do is to verify that (b) holds as well. For this, it suffices to show that the intervals $f^{-i}(J_{n+4})$ and $I = f^{q_{n+1}}(I_n(c_{-i}))$ have comparable lengths. Let $L_i = f^{-i}(L)$ and $R_i = f^{-i}(R)$ be the two connected components of $I \setminus f^{-i}(J_{n+4})$. Since $L_i \supset f^{-i}(L^*)$ and $R_i \supset f^{-i}(R^*)$, and since

$$f^{-i}(L^*) = f^{q_{n+1}-i}(I_{n+2}) \quad \text{and} \quad f^{-i}(R^*) = f^{q_n+q_{n+1}-i}(I_{n+1}(c))$$

are both atoms of $\mathscr{P}_{n+1}(c)$ contained in the same atom $I \in \mathscr{P}_n(c)$, we deduce from the real bounds that $|L_i| \asymp |I| \asymp |R_i|$. By the cross-ratio inequality, the cross-ratio distortion $\mathrm{CrD}(f^i; f^{-i}(J_{n+4}), I)$ is bounded above. Combining this fact with (iii), we deduce that the cross-ratio $[f^{-i}(J_{n+4}), I]$ is bounded below. Since the two lateral intervals $L_i, R_i \subset I$ and the total interval I have comparable lengths, it follows that the middle interval $f^{-i}(J_{n+4}) \subset I$ also has length comparable to $|I|$. Together with (8.30), this shows at last that $|\Delta'| \asymp |I| \asymp |\Delta''|$.

This completes the proof of our claim. And as we had already observed, the claim implies that f satisfies the hypotheses of Theorem 8.2. Therefore it satisfies the conclusion as well: f does not admit a σ-finite absolutely continuous invariant measure. This finishes the proof of Theorem 8.4. □

8.2.2.3 The Punchline

Our main theorem, namely Theorem 8.1, is now an immediate consequence of steps 1 and 2, or more precisely, of Theorem 8.3 and 8.4.

8.2.3 Negative Schwarzian Redux

As promised, we offer a proof of Proposition 8.3, which we rephrase as follows.

Proposition 8.4 (The Negative Schwarzian Property) *For any given multicritical circle map f with bounded combinatorics there exists $n_0 = n_0(f) \in \mathbb{N}$ such that for all $x_0 \in \mathbf{S}^1$ and all $n \geqslant n_0$ we have*

$$Sf^{q_{n+1}}(x) < 0 \quad \textit{for all } x \in I_n(x_0) \textit{ regular point of } f^{q_{n+1}}.$$

Likewise, we have

$$Sf^{q_n}(x) < 0 \quad \textit{for all } x \in I_{n+1}(x_0) \textit{ regular point of } f^{q_n}.$$

Both the statement above and the proof below are extracted from our paper de Faria and Guarino (2021). It should be clear to the reader that Propostion 8.3 is indeed an immediate consequence of Proposition 8.4. As for the latter, we already know its statement to be true in the case when x_0 is a critical point of f—in which case it holds in fact for any irrational rotation number: this is precisely what we did in Sect. 6.5, Proposition 6.2. Hence all we need is to extend the proof to the case when x_0 is a regular point of a multicritical circle map with bounded combinatorics. This requires, by way of preparation, a couple of auxiliary results.

8.2.3.1 Bounded Geometry

We say that a minimal circle homeomorphism f has *bounded geometry* at $x \in \mathbf{S}^1$ if there exists $K > 1$ such that for all $n \in \mathbb{N}$ and for every pair I, J of adjacent atoms of $\mathscr{P}_n(x)$ we have

$$K^{-1}\,|I| \leqslant |J| \leqslant K\,|I|\,.$$

Obviously, every irrational rotation by an angle of bounded type has bounded geometry, and so does every homeomorphism smoothly or even quasisymmetrically conjugate to such a rotation. Thus, multicritical circle maps with rotation number of bounded type have bounded geometry *at every point*. Here is the precise statement.

Theorem 8.5 *For any given multicritical circle map f with bounded combinatorics, there exists a constant $C > 1$ depending only on f, such that for any given point $x \in \mathbf{S}^1$, for all $n \in \mathbb{N}$, and for every pair I, J of adjacent atoms of $\mathscr{P}_n(x)$ we have:*

$$C^{-1}\,|I| \leqslant |J| \leqslant C\,|I|\,.$$

Proof By Herman's Theorem 7.1, f is quasisymmetrically conjugate to an irrational rotation. □

It is also possible to prove this result without using Herman's theorem (see Exercise 8.7). Theorem 8.5 is most definitely *false* for maps with rotation number of *unbounded* type. We will have a lot more to say about bounded geometry in Chap. 9.

If the rotation number $\rho(f) = [a_0, a_1, \ldots]$ satisfies $\sup_{n\in\mathbb{N}}\{a_n\} \leqslant B$, we say that f has *combinatorics bounded by B*. With this terminology, we can state the following simple consequence of the beau bounds (Theorem 6.4).

Lemma 8.3 *Given $B > 1$, $N \geqslant 1$ in $\mathbb{N}$ and $d > 1$ there exists $C = C(B, N, d) > 1$ with the following property: for any given multicritical circle map f with combinatorics bounded by B, and with at most N critical points whose criticalities are bounded by d, there exists $n_0 = n_0(f) \in \mathbb{N}$ such that for each critical point c of f, for all $n \geqslant n_0$ and for every pair of intervals $I \in \mathscr{P}_n(c)$ and $J \in \mathscr{P}_{n+1}(c)$ satisfying $J \subseteq I$, we have that $|I| \leqslant C\,|J|$.*

Proof Exercise. □

The next auxiliary result we need is the analogue of Fact 6.2 in Sect. 6.4. For each $n \geqslant 0$, we consider the intervals $L_n(x_0) = I_{n+1}(x_0)$ and $R_n(x_0) = f^{q_n}\big(I_n(x_0)\big)$. As usual, we write $L_n^j(x_0) = f^j(L_n(x_0))$, etc., for the images of these intervals under the iterates of f.

Lemma 8.4 *There exists a constant $\tau > 0$ (depending only on f) such that*

$$|L_n^j(x_0)| > \tau |I_n^j(x_0)| \quad \text{and} \quad |R_n^j(x_0)| > \tau |I_n^j(x_0)|$$

for each $j \in \{0, \dots, q_{n+1}\}$ and all $n \in \mathbb{N}$.

Proof The proof of Fact 6.2 given in Sect. 6.4 applies here, *mutatis mutandis*. The only difference occurs at the moment when we need to claim that $|I_n^{q_n}(x_0)| \asymp |I_n(x_0)|$. If x_0 is a critical point of f, this fact is immediate from the real bounds, and holds under no restriction on the rotation number. But if x_0 is a regular point, then we need to use Theorem 8.5 instead, and this is the reason for the bounded type hypothesis. □

This, in turn, can be used to prove the following analogue of Lemma 6.5.

Proposition 8.5 (The C^1 Bounds) *For any given multicritical circle map f with bounded combinatorics there exists a constant $K = K(f) > 1$ such that the following holds. For any given $x_0 \in S^1$ and $n \in \mathbb{N}$ let $I_n = I_n(x_0)$ and $I_{n+1} = I_{n+1}(x_0)$. Then we have*

(i) $Df^k(x) \leqslant K \dfrac{|f^k(I_n)|}{|I_n|}$ *for all $x \in I_n$ and all $k \in \{0, 1, \dots, q_{n+1}\}$;*

(iii) $Df^k(x) \leqslant K \dfrac{|f^k(I_{n+1})|}{|I_{n+1}|}$ *for all $x \in I_{n+1}$ and all $k \in \{0, 1, \dots, q_n\}$;*

(iv) $\|f^{q_n}\|_{C^1(I_{n+1})} \leqslant K$ *and* $\|f^{q_{n+1}}\|_{C^1(I_n)} \leqslant K$;.

Proof Again, the proof of Lemma 6.5 given in Sect. 6.4 can be repeated here, *mutatis mutandis*. The only difference, of course, is that Fact 6.2 is replaced by Lemma 8.4, which only holds in the bounded type case. □

8.2.3.2 Proving that the Schwarzian Is Negative

We are finally ready for the proof of Proposition 8.4.

Proof of Proposition 8.4 Let us fix $x_0 \in S^1$ and $n \in \mathbb{N}$. We give the proof only for the case $x \in I_n(x_0)$ regular point of $f^{q_{n+1}}$ (the other case being entirely analogous). Let $j \in \{0, \dots, q_{n+1} - 1\}$ be the minimum positive integer such that

$$f^j\big(I_n(x_0)\big) \cap J_n(c_i) \neq \emptyset$$

for some $i \in \{0, \dots, N-1\}$. Without loss of generality, we may assume that $i = 0$. By Lemma 7.7 (and Remark 7.11), the intervals $f^j(I_n(x_0))$ and $J_n(c_0)$ have comparable lengths, In other words, there exists $C_0 > 1$, depending only on f, such that

$$\big|f^j(x) - c_0\big| \leqslant C_0 \big|f^j(I_n(x_0))\big| \quad \text{for all } x \in I_n(x_0).$$

Moreover, by Koebe distortion principle there exists $C_1 > 1$ (also depending only on f) such that $f^j|_{I_n(x_0)}$ has distortion bounded by C_1, that is:

$$\frac{1}{C_1} \leqslant \frac{Df^j(x)}{Df^j(y)} \leqslant C_1 \quad \text{for all } x, y \in I_n(x_0).$$

Recall that, by the non-flatness condition, for each critical point c_i there exist a neighborhood $U_i \subseteq \mathbf{S}^1$ of c_i and a positive constant K_i such that for all $x \in U_i \setminus \{c_i\}$ we have

$$Sf(x) < -\frac{K_i}{(x - c_i)^2} < 0. \tag{8.31}$$

Let $\mathscr{U} = \bigcup_{i=0}^{N-1} U_i$, and let $\mathscr{V} \subset \mathbf{S}^1$ be an open set whose closure contains no critical point of f and such that $\mathscr{U} \cup \mathscr{V} = \mathbf{S}^1$. Since f is of class C^3, we know that $M = \sup_{y \in \mathscr{V}} |Sf(y)|$ is finite. Let $\delta_n = \max_{x_0 \in \mathbf{S}^1} \max_{0 \leqslant k < q_{n+1}} |f^k(I_n(x_0))|$. Since f is minimal, $\delta_n \to 0$ as $n \to \infty$. We choose $n_0 = n_0(f)$ so large that δ_n is smaller than the Lebesgue number of the covering $\{\mathscr{U}, \mathscr{V}\}$ of the circle for all $n \geqslant n_0$. Moreover, we also require that $\delta_n < K_0/M\,K^2\,C_0^2\,C_1^2$ for all $n \geqslant n_0$, where $K = K(f) > 1$ is given by Proposition 8.5. Using the chain rule for the Schwarzian derivative, we have for all $\ell \in \{j+1, \ldots, q_{n+1}\}$ and for all $x \in I_n(x_0)$ regular point of f^ℓ the following identity:

$$Sf^\ell(x) = \sum_{k=0}^{\ell-1} Sf(f^k(x)) \left[Df^k(x)\right]^2.$$

We decompose this expression as $\Sigma_1^{(n)}(x) + \Sigma_2^{(n)}(x)$, where

$$\Sigma_1^{(n)}(x) = \sum_{k: f^k(I_n(x_0)) \subset \mathscr{U}} Sf(f^k(x)) \left[Df^k(x)\right]^2, \tag{8.32}$$

and $\Sigma_2^{(n)}(x)$ is the sum over the remaining terms, and we treat both cases separately.

(i) Since $f^j(I_n(x_0)) \cap J_n(c_0) \neq \emptyset$, we have $f^j(I_n(x_0)) \subset \mathscr{U}$ and then the sum in the right-hand side of (8.32) includes the term with $k = j$, namely $Sf(f^j(x)) \left[Df^j(x)\right]^2$. Since all the other terms in (8.32) are negative as well, and since $|f^j(x) - c_0| \leqslant C_0 |f^j(I_n(x_0))|$, we deduce from (8.31) that:

$$\Sigma_1^{(n)}(x) < -\frac{K_0}{C_0^2 |f^j(I_n(x_0))|^2} \left[Df^j(x)\right]^2.$$

Let $y \in I_n(x_0)$ be such that $\left|f^j(I_n(x_0))\right| = Df^j(y)\,|I_n(x_0)|$. By bounded distortion, we obtain:

$$\Sigma_1^{(n)}(x) < -\frac{K_0}{C_0^2}\frac{1}{|I_n(x_0)|^2}\left[\frac{Df^j(x)}{Df^j(y)}\right]^2 < -\frac{K_0}{C_0^2\,C_1^2}\frac{1}{|I_n(x_0)|^2}. \tag{8.33}$$

(ii) Observe that

$$\left|\Sigma_2^{(n)}(x)\right| \leqslant \sum_{k:f^k(I_n(x_0))\subset\mathscr{V}} \left|Sf(f^k(x))\right|\left[Df^k(x)\right]^2.$$

By Proposition 8.5, there exists $K > 1$ such that

$$\begin{aligned}
\left|\Sigma_2^{(n)}(x)\right| &\leqslant \sum_{k:f^k(I_n(x_0))\subset\mathscr{V}} \left|Sf(f^k(x))\right| K^2\frac{|f^k(I_n(x_0))|^2}{|I_n(x_0)|^2}\\
&\leqslant \frac{MK^2}{|I_n(x_0)|^2}\sum_{k:f^k(I_n(x_0))\subset\mathscr{V}} \left|f^k(I_n(x_0))\right|^2\\
&\leqslant \frac{MK^2}{|I_n(x_0)|^2}\max_{0\leqslant k\leqslant \ell-1}\left|f^k(I_n(x_0))\right| \sum_{k:f^k(I_n(x_0))\subset\mathscr{V}}\left|f^k(I_n(x_0))\right|\\
&\leqslant \frac{MK^2}{|I_n(x_0)|^2}\,\delta_n.
\end{aligned} \tag{8.34}$$

By our choice of n_0, we know that $\delta_n < K_0/M\,K^2\,C_0^2\,C_1^2$ for all $n \geqslant n_0$, and then we deduce from (8.33) and (8.34) that, indeed, $Sf^\ell(x) < 0$ for all $\ell \in \{j+1,\ldots,q_{n+1}\}$ and all $x \in I_n(x_0)$ regular point of f^ℓ. □

8.3 Lyapunov Exponents

Recall from Theorem 3.11 (Sect. 3.4.2), that every diffeomorphism of the circle without periodic points has zero Lyapunov exponents everywhere. We will see in this section that an analogous result holds for multicritical circle maps (Theorem 8.6). The proof of this result is considerably more difficult than the one of Theorem 3.11, since in this case $\log Df$ is not a continuous function: it is defined only in the complement of the critical set of f, and it is unbounded (recall Fig. 6.3).

8.3.1 The Collet–Eckmann Condition

The result we wish to present is taken from de Faria and Guarino (2016). For its proper formulation, it is best to introduce the notion of *Collet–Eckmann* condition. We do this in the restricted context of homeomorphisms of the circle, but of course a much more general definition is possible.

Definition 8.2 We say that a multicritical circle map f satisfies the *Collet–Eckmann* condition at a critical point $c \in \mathrm{Crit}(f)$ if there exist $C > 0$ and $\lambda > 1$ such that $Df^n\big(f(c)\big) \geqslant C\lambda^n$ for all $n \in \mathbb{N}$, or equivalently

$$\liminf_{n\to\infty} \frac{1}{n} \log Df^n(f(c)) \;\geqslant\; \log\lambda \;>\; 0\,.$$

Theorem 8.6 (Zero Lyapunov Exponent) *Let $f : S^1 \to S^1$ be a C^3 multicritical circle map with irrational rotation number, and let μ be its unique invariant Borel probability measure. Then* $\log Df$ *belongs to $L^1(\mu)$ and it has zero μ-mean, in other words*

$$\int_{S^1} \log Df \, d\mu = 0\,.$$

Moreover, no critical point of f satisfies the Collet–Eckmann condition.

Remark 8.4 In a recent note by Ji (2022), it has been established that

$$\lim_{n\to\infty} \frac{1}{n} \log Df^n(f(c)) \;=\; 0\,,$$

for each critical point c of f.

Remark 8.5 The Collet–Eckmann condition has a long history in one-dimensional dynamics for it provides, under mild conditions, absolutely continuous invariant measures for smooth multimodal maps of the interval (see de Melo and van Strien 1993, Ch. V and references therein, and see also Bruin et al. (2008) for recent developments).

8.3.2 The Key Step

Our proof of Theorem 8.6 relies on Proposition 8.6 below. As before, let $\{q_n\}_{n\in\mathbb{N}}$ be the sequence of return times given by the irrational rotation number of f. Let us denote by $c_1, c_2, \ldots, c_N$ the critical points of f ($N \geqslant 1$) and let $d_i > 1$ denote the criticality of each c_i. Conjugating f by a suitable C^3-diffeomorphism (which does not affect its Lyapunov exponent—see Exercise 3.5 of Chap. 3) we may assume that each c_i has an open neighborhood $V(c_i)$ where f is a *power-law* of the form:

$$f(x) = f(c_i) + (x - c_i)|x - c_i|^{d_i - 1} \quad \text{for all } x \in V(c_i). \tag{8.35}$$

We also assume, of course, that $V(c_i) \cap V(c_j) = \emptyset$ whenever $i \neq j$.

Recall from the real bounds (Theorem 6.3) that, for each $c \in \{c_1, c_2, \ldots, c_N\}$, the dynamical partitions $\{\mathscr{P}_n(c)\}_{n\in\mathbb{N}}$ have the *comparability* property: any two consecutive atoms of $\mathscr{P}_n(c)$ have comparable lengths. We will need a couple of further consequences of the real bounds, that we proceed to state and prove.

Lemma 8.5 *There exists a constant $C = C(f) > 1$ with the following property. For each $n \geqslant 0$, let*

$$\mathscr{C}_n = \left\{I_n^j\right\}_{j=0}^{2q_{n+1}} \bigcup \left\{I_{n+1}^k\right\}_{k=0}^{q_n+q_{n+1}}$$

be the set of all atoms of $\mathscr{P}_n$ together with their forward images under f up to iterate $q_{n+1} + 1$. Then, for any $J_1, J_2 \in \mathscr{C}_n$ that share a common endpoint we have

$$C^{-1}\,|J_1| \leqslant |J_2| \leqslant C\,|J_1|\ .$$

Proof Recall first Proposition 6.1: the six intervals I_n, I_{n+1}, $I_n^{q_n}$, $I_n^{q_{n+1}}$, $I_{n+1}^{q_n}$ and $I_n^{q_{n+1}-q_n}$ are pairwise comparable (see Fig. 6.5). The idea of the proof of Lemma 8.5 is to prove that: (i) if $J \in \mathscr{C}_n$, then there is $\Delta \in \mathscr{P}_n$ such that $J \subset \Delta^*$ (in fact, there are at most three such Δ); and that (ii) if J and Δ are as above, then $J \asymp \Delta^*$. This is enough to finish the proof: if $J_1, J_2 \in \mathscr{C}_n$ share a common endpoint and $\Delta_1, \Delta_2 \in \mathscr{P}_n$ are such that $J_1 \subset \Delta_1^*$ and $J_2 \subset \Delta_2^*$, then Δ_1^* and Δ_2^* must contain at least one atom of $\mathscr{P}_n$ in common. Thus, by the real bounds, they must be comparable, which implies that $J_1 \asymp J_2$.

To prove (i) is quite simple: if $J = I_{n+1}^j$, $0 \leqslant j < q_n$, or $J = I_n^k$, $0 \leqslant k < q_{n+1}$, then J is itself an atom of $\mathscr{P}_n$, so we can take $\Delta = J$. If $J = I_{n+1}^{q_n+j}$, $0 \leqslant j < q_{n+1}$, then $J \subset I_n^j$, so we can take $\Delta = I_n^j$. If $J = I_{n+1}^{q_n+q_{n+1}}$, then $J \subset I_n \cup I_{n+1}$, so we can take either of these as Δ. If $J = I_n^{q_{n+1}+k}$, $0 \leqslant k < q_n$, then $J \subset I_{n+1}^k \cup I_n^k$, so we can take either of these as Δ. If $J = I_n^{q_{n+1}+q_n+k}$, $0 \leqslant k < q_{n+1} - q_n$, then $J \subset I_{n+1}^{q_n+k} \cup I_n^{q_n+k} \subset I_n^k \cup I_n^{q_n+k}$, so we can take $\Delta = I_n^k$ or $\Delta = I_n^{q_n+k}$. Finally, if $J = I_n^{2q_{n+1}}$, then $J \subset I_{n+1}^{q_{n+1}} \cup I_n^{q_{n+1}} \subset I_n^{q_{n+1}-q_n} \cup I_{n+1} \cup I_n = (I_{n+1})^*$, so we can take $\Delta = I_{n+1}$. This proves (i).

By the real bounds, we only need to prove (ii) for intervals of $\mathscr{C}_n$ that are not themselves atoms of $\mathscr{P}_n$. With this in mind, we first note that given $J \in \mathscr{C}_n$, though there may be (at most) three different choices of Δ such that $J \subset \Delta^*$, the triples of atoms obtained in this way are all comparable; thus, it suffices to prove (ii) for one such choice of Δ. We claim that for $0 \leqslant j \leqslant q_{n+1}$ we have

$$I_{n+1}^{q_n+j} \asymp I_n^j .$$

Proof of the Claim For one, $I_{n+1}^{q_n+j} \subset I_n^j$, so we immediately get $I_n^j \geqslant I_{n+1}^{q_n+j}$. We must now prove that $I_{n+1}^{q_n+j} \geqslant I_n^j$ as well. □

We split into two cases: $a_{n+1} = 1$ and $a_{n+1} \geqslant 2$. If $a_{n+1} = 1$, then

$$I_n^j = I_{n+1}^{q_n+j} \cup I_{n+2}^j , \tag{8.36}$$

and since $I_{n+1}^{q_n+j}, I_{n+2}^j \in \mathscr{P}_n \cup \{I_{n+1}^{q_{n+2}}, I_{n+2}^{q_{n+1}}\}$, it follows from Proposition 6.1 (applied to the partition at level $n+1$) that

$$I_{n+1}^{q_n+j} \asymp I_{n+2}^j . \tag{8.37}$$

Combining (8.36) and (8.37), we conclude that $I_{n+1}^{q_n+j} \asymp I_n^j$ in this case.

We now address the case $a_{n+1} \geqslant 2$, and we start by analyzing the case $j = q_{n+1}$. In this case, we have $I_{n+1}^{q_n+j} = I_{n+1}^{q_n+q_{n+1}}$, which is a long atom of $\mathscr{P}_{n+1}$ (since $a_{n+1} \geqslant 2$) adjacent to $I_{n+1}^{q_n}$. Thus,

$$I_{n+1}^{q_n+q_{n+1}} \asymp I_{n+1}^{q_n} \asymp I_n^{q_{n+1}} , \tag{8.38}$$

which proves the claim in this case.

Now, let $T = I_{n+1}^{q_{n+1}} \cup I_n^{q_{n+1}}$, $M = I_n^{q_{n+1}} \setminus I_{n+1}^{q_n+q_{n+1}}$, and observe that M is compactly contained in the interior of T. Let L, R be the connected components of $T \setminus M$, i.e., $I_{n+1}^{q_{n+1}}$ and $I_{n+1}^{q_n+q_{n+1}}$. Assume, without loss of generality, that $L = I_{n+1}^{q_{n+1}}$ and $R = I_{n+1}^{q_n+q_{n+1}}$. We claim that the cross-ratio $[M, T]$ is bounded from below by a constant depending only on f. Indeed, since $I_{n+1}^{q_{n+1}}, I_{n+1}$ are adjacent atoms of $\mathscr{P}_{n+1}$,

$$L \asymp I_{n+1} \asymp I_n^{q_{n+1}} ,$$

so we conclude that L and T are comparable. From (8.38),

$$R \asymp I_n^{q_{n+1}} \asymp T .$$

It now follows from $|T| = |L| + |M| + |R|$ that M must be comparable to the three other intervals as well. As a consequence, we conclude that there exists a constant $C_0 = C_0(f) > 0$, depending only on f, such that

$$[M, T] = [I_n^{q_{n+1}} \setminus I_{n+1}^{q_n+q_{n+1}}, I_{n+1}^{q_{n+1}} \cup I_n^{q_{n+1}}] \geqslant C_0 .$$

We now turn to the case $0 \leqslant j < q_{n+1}$. Since the family

$$\left\{f^i(I_{n+1}^j \cup I_n^j)\right\}_{i=0}^{q_{n+1}-j}$$

has multiplicity of intersection at most 3, the Cross-Ratio Inequality (Theorem 5.1) implies that the cross-ratio distortion of $f^{q_{n+1}-j}$ on the pair $I_n^j \setminus I_{n+1}^{q_n+j}, I_{n+1}^j \cup I_n^j$ is bounded by some constant $C_1 = C_1(f) > 0$:

$$\frac{[M, T]}{[I_n^j \setminus I_{n+1}^{q_n+j}, I_{n+1}^j \cup I_n^j]} \leqslant C_1 .$$

Manipulating this inequality, we get

$$[I_n^j \setminus I_{n+1}^{q_n+j}, I_{n+1}^j \cup I_n^j] \geqslant C_0/C_1 .$$

Thus, the cross-ratio $[I_n^j \setminus I_{n+1}^{q_n+j}, I_{n+1}^j \cup I_n^j]$ is bounded from below by a constant depending only on f. Consequently, the space of $I_n^j \setminus I_{n+1}^{q_n+j}$ inside $I_{n+1}^j \cup I_n^j$ is bounded from below by a constant depending only on f; this is easily seen to imply that

$$I_{n+1}^{q_n+j} \succcurlyeq I_n^j \setminus I_{n+1}^{q_n+j} . \tag{8.39}$$

Combining (8.39) with the fact that $\left|I_n^j\right| = \left|I_{n+1}^{q_n+j}\right| + \left|I_n^j \setminus I_{n+1}^{q_n+j}\right|$, we get $I_{n+1}^{q_n+j} \succcurlyeq I_n^j$. This finishes the proof of the claim.

With the claim at hand, we proceed to finish the proof of Lemma 8.5. Since I_n^j is comparable to $(I_n^j)^*$ if $0 \leqslant j < q_{n+1}$, or to I_n^* if $j = q_{n+1}$, (ii) holds if $J \in \mathscr{C}_n$ is the image of a short atom of $\mathscr{P}_n$, i.e., $J = I_{n+1}^i$ for some $q_n \leqslant i \leqslant q_n + q_{n+1}$. That (ii) also holds for images of long atoms of $\mathscr{P}_n$ is a simple consequence of the above claim. Indeed, it suffices to show that, for $0 \leqslant k \leqslant q_{n+1}$,

$$I_n^{q_{n+1}+k} \asymp I_n^k .$$

We first show that $I_n^k \succcurlyeq I_n^{q_{n+1}+k}$. If $0 \leqslant k < q_{n+1}$, then $I_n^{q_{n+1}+k} \subset (I_n^k)^* \asymp I_n^k$, so $I_n^k \succcurlyeq I_n^{q_{n+1}+k}$; and in the case $k = q_{n+1}$, $I_n^{2q_{n+1}} \subset (I_n)^* \asymp I_n \asymp I_n^{q_{n+1}}$, so $I_n^{q_{n+1}} \succcurlyeq I_n^{2q_{n+1}}$ as well. It remains to prove that $I_n^{q_{n+1}+k} \succcurlyeq I_n^k$, but this follows immediately from the claim, since we have

$$I_n^{q_{n+1}+k} \supset I_{n+1}^k$$

and

$$I_{n+1}^k \asymp I_{n+1}^{q_n+k} \asymp I_n^k .$$

The first comparability in the previous equation follows from the real bounds, while the second follows from the claim. This finishes the proof of Lemma 8.5. □

Lemma 8.5 implies the following two results.

Lemma 8.6 *There exists $B_0 = B_0(f) > 1$ such that for each $c \in \{c_1, c_2, \dots, c_N\}$, for each $n \in \mathbb{N}$ and for each atom $\Delta \in \mathscr{P}_n(c)$ we have*

$$\frac{|\Delta|}{B_0} \leqslant \left|f^{q_n}(\Delta)\right| \leqslant B_0|\Delta| .$$

Lemma 8.7 *There exists $B_1 = B_1(f) > 1$ with the following property. Let $\Delta \in \mathscr{P}_n(c)$ and denote by Δ^* the union of Δ with its two immediate neighbours in $\mathscr{P}_n(c)$. If $0 \leqslant j < k \leqslant q_n$ are such that the intervals $f^j(\Delta^*)$, $f^{j+1}(\Delta^*)$,..., $f^{k-1}(\Delta^*)$ do not contain any critical point of f, then the map $f^{k-j} : f^j(\Delta) \to f^k(\Delta)$ has distortion bounded by B_1, that is*

$$\frac{1}{B_1} \leqslant \frac{Df^{k-j}(x)}{Df^{k-j}(y)} \leqslant B_1 \quad \textit{for all } x, y \in f^j(\Delta).$$

Proof Since the iterate $f^{k-j} : f^j(\Delta^*) \to f^k(\Delta^*)$ is a diffeomorphism, we would like to apply Koebe's distortion principle (Lemma 5.2). By combinatorics, we already know that the family $f^j(\Delta^*), f^{j+1}(\Delta^*), \dots, f^{k-1}(\Delta^*)$ has multiplicity of intersection equal to 3. Thus, we only need to prove that $f^k(\Delta)$ has definite space inside $f^k(\Delta^*)$, which follows at once from Lemma 8.5. □

Yet another consequence of the real bounds:

Lemma 8.8 *There exists $B_2 = B_2(f) > 1$ with the following property: if $c \neq c'$ are critical points of f and $\Delta \in \mathscr{P}_n(c)$, $\Delta' \in \mathscr{P}_n(c')$ for some $n \in \mathbb{N}$ are such that $\Delta \cap \Delta' \neq \emptyset$, then $B_2^{-1}|\Delta'| \leqslant |\Delta| \leqslant B_2|\Delta'|$.*

Proof This follows from the combinatorial fact that Δ is contained in the union of two adjacent atoms of $\mathscr{P}_n(c')$, one of which is Δ', and likewise for Δ'. □

For each $k \geqslant 0$ and each critical point c we will use the notation $J_k(c) = I_k(c) \cup I_{k+1}(c) = \left[f^{q_{k+1}}(c), f^{q_k}(c)\right] \ni c$. The key step in the proof of Theorem 8.6 is the following fact.

Proposition 8.6 *There exists $C = C(f) > 0$ with the following properties:*

1. *For each $x \in \mathbf{S}^1$ and all $n \geqslant 0$ we have $\log Df^{q_n}(x) \leqslant C$.*
2. *For all $n \geqslant 0$, if $x \in \mathbf{S}^1$ is such that $f^i(x) \notin \bigcup_{j=1}^{j=N} J_{2n}(c_j)$ for all $0 \leqslant i \leqslant q_n$, then $\log Df^{q_n}(x) \geqslant -Cn$.*

In what follows we denote by $C_0, C_1, C_2, C_3, \dots$ positive constants (greater than 1, in fact) depending only on f. Moreover, for any two positive numbers a and b

we use the notation $a \asymp b$ to mean that $C^{-1}a \leqslant b \leqslant Ca$ for some constant $C > 1$ depending only on f.

Proof Let us fix once and for all a critical point $c \in \mathrm{Crit}(f)$. We assume that $n \geqslant 0$ is large enough so that each atom of $\mathscr{P}_n(c)$ contains at most one critical point of f. Let $x \in \mathbf{S}^1$ and let $\Delta \in \mathscr{P}_n(c)$ be such that $x \in \Delta$. Let $\Delta^* \supseteq \Delta$ be as in Lemma 8.7. Just by taking n larger still, we may assume that, for $0 \leqslant k < q_n$, each $f^k(\Delta^*)$ contains at most one critical point of f. We say that $0 \leqslant k < q_n$ is a *critical time* for x if $f^k(\Delta^*)$ contains a critical point of f. Let us write $0 \leqslant k_1 < k_2 < \cdots < k_m < q_n$ for the sequence of all critical times for x. Note that $m \leqslant 3N$ since the family $\{f^k(\Delta^*)\}_{0 \leqslant k < q_n}$ has intersection multiplicity equal to 3. Using these critical times and the chain rule we can write:

$$Df^{q_n}(x) = Df^{k_1}(x)\left[\prod_{j=1}^{m-1} Df^{k_{j+1}-k_j-1}\big(f^{k_j+1}(x)\big)Df\big(f^{k_j}(x)\big)\right] \times Df^{q_n-k_m-1}\big(f^{k_m+1}(x)\big)Df\big(f^{k_m}(x)\big). \tag{8.40}$$

We proceed to estimate each term in the product (8.40) above. From Lemma 8.7 (with $j = 0$ and $k = k_1$) we have:

$$Df^{k_1}(x) \asymp \frac{|f^{k_1}(\Delta)|}{|\Delta|}. \tag{8.41}$$

Again from Lemma 8.7 (with $k_j + 1$ and k_{j+1} replacing j and k respectively) we have for all $j \in \{1, \ldots, m-1\}$:

$$Df^{k_{j+1}-k_j-1}\big(f^{k_j+1}(x)\big) \asymp \frac{|f^{k_{j+1}}(\Delta)|}{|f^{k_j+1}(\Delta)|}.$$

For each $j \in \{1, \ldots, m\}$ let $\beta_j \in \mathrm{Crit}(f)$ be the (unique) critical point of f in $f^{k_j}(\Delta^*)$, and let d_j be its criticality. Since we are assuming that n is sufficiently large, we may suppose that $f^{k_j}(\Delta^*) \subseteq V(\beta_j)$ for all $j \in \{1, \ldots, m\}$. Then, from the power-law expression (8.35) we have:

$$Df\big(f^{k_j}(x)\big) \asymp \big|f^{k_j}(x) - \beta_j\big|^{d_j-1}, \tag{8.42}$$

and recall that $d_j - 1 > 1$ for all $j \in \{1, \ldots, m\}$. Still using the power-law expression we see that:

$$\big|f^{k_j+1}(\Delta)\big| \asymp \big|f^{k_j}(\Delta)\big|^{d_j} \quad \text{for all } j \in \{1, \ldots, m\}. \tag{8.43}$$

Using Lemma 8.7 yet again, we also see that:

$$Df^{q_n-k_m-1}\big(f^{k_m+1}(x)\big) \asymp \frac{\big|f^{q_n}(\Delta)\big|}{\big|f^{k_m+1}(\Delta)\big|}\,.$$

Let us now prove assertions (1) and (2) in the statement of the proposition. Note that (8.42) yields:

$$Df\big(f^{k_j}(x)\big) \leqslant C_0\big|f^{k_j}(\Delta)\big|^{d_j-1} \quad \text{for all } j \in \{1,\ldots,m\}, \tag{8.44}$$

where $C_0 = C_0(f) > 0$. Combining all these facts, namely (8.41)–(8.44), we deduce the following (upper) telescoping estimate:

$$\begin{aligned} Df^{q_n}(x) \leqslant C_1\frac{\big|f^{k_1}(\Delta)\big|}{|\Delta|}\left[\prod_{j=1}^{m-1}\frac{\big|f^{k_{j+1}}(\Delta)\big|}{\big|f^{k_j+1}(\Delta)\big|}\big|f^{k_j}(\Delta)\big|^{d_j-1}\right]\times \\ \times\big|f^{k_m}(\Delta)\big|^{d_m-1}\frac{\big|f^{q_n}(\Delta)\big|}{\big|f^{k_m+1}(\Delta)\big|} \\ \asymp \frac{\big|f^{k_1}(\Delta)\big|}{|\Delta|}\left[\prod_{j=1}^{m-1}\frac{\big|f^{k_{j+1}}(\Delta)\big|}{\big|f^{k_j}(\Delta)\big|}\right]\frac{\big|f^{q_n}(\Delta)\big|}{\big|f^{k_m}(\Delta)\big|} = \frac{\big|f^{q_n}(\Delta)\big|}{|\Delta|} \leqslant C_2\,, \end{aligned} \tag{8.45}$$

where in the last line we have used (8.43) and finally Lemma 8.6. This proves item (1). In order to prove item (2) note first that all estimates provided above are two-sided, except (8.44). In order to get a lower bound for the left side of (8.44) we use the hypothesis in (2). Since $f^{k_j}(x) \notin J_{2n}(\beta_j)$ we have:

$$\big|f^{k_j}(x) - \beta_j\big| \geqslant C_3\big|I_{2n}(\beta_j)\big|. \tag{8.46}$$

From the real bounds we know that there exists $\lambda \in (0,1)$ depending only on f such that $C_4^{-1}\lambda^n\big|I_n(\beta_j)\big| \leqslant \big|I_{2n}(\beta_j)\big| \leqslant C_4\lambda^n\big|I_n(\beta_j)\big|$. Moreover, we claim that $\big|I_n(\beta_j)\big|$ is comparable to $\big|f^{k_j}(\Delta)\big|$. Indeed, this follows from Lemma 8.8 because $I_n(\beta_j) \in \mathscr{P}_n(\beta_j)$ intersects an atom of $\mathscr{P}_n(c)$ in $f^{k_j}(\Delta^*)$, and this atom has length comparable to $\big|f^{k_j}(\Delta)\big|$ (such atom is either $f^{k_j}(\Delta)$ itself, or one of its neighbours). Using these facts in (8.46) we deduce that:

$$Df\big(f^{k_j}(x)\big) \geqslant C_5\,\lambda^{n(d_j-1)}\big|f^{k_j}(\Delta)\big|^{d_j-1}.$$

Using this lower estimate in place of the upper estimate (8.44) and proceeding as in (8.45) we arrive at the estimate

$$Df^{q_n}(x) \geqslant C_6\,\lambda^{n(d_1+d_2+\cdots+d_m-m)}\,, \tag{8.47}$$

where again $C_6 = C_6(f) > 1$. Note that $0 < d_1 + d_2 + \cdots + d_m - m < 3(d_1 + d_2 + \cdots + d_N)$, and since $\alpha = 3(d_1 + d_2 + \cdots + d_N)$ is a positive constant depending only on f we get:

$$Df^{q_n}(x) \geqslant C_6\, \lambda^{n\alpha}\,,$$

and then:

$$\log Df^{q_n}(x) \geqslant -n\alpha \log \frac{1}{\lambda} + \log C_6 \geqslant -C_7\, n\,.$$

□

With Proposition 8.6 at hand we are ready to prove Theorem 8.6.

Proof of Theorem 8.6 The fact that no critical point of f satisfies the Collet–Eckmann condition follows at once from item (1) of Proposition 8.6. By Proposition 8.1 we know that $\log Df \in L^1(\mu)$, and then we know from Birkhoff's ergodic theorem that

$$\lim_{n \to +\infty} \left\{ \frac{\log Df^n(x)}{n} \right\} = \int_{S^1} \log Df\, d\mu\,,$$

for μ-almost every $x \in S^1$. For each $n \geqslant 0$ let

$$\begin{aligned} A_n &= S^1 \setminus \bigcup_{j=1}^{j=N} \bigcup_{i=0}^{q_n-1} f^{-i}\big(J_{2n}(c_j)\big) \\ &= \left\{ x \in S^1 : \forall\ 0 \leqslant i \leqslant q_n - 1 : f^i(x) \in S^1 \setminus \bigcup_{j=1}^{j=N} J_{2n}(c_j) \right\}, \end{aligned}$$

and consider

$$A = \limsup_{n \in \mathbb{N}} A_n = \bigcap_{k=1}^{+\infty} \bigcup_{n=k}^{+\infty} A_n\,.$$

We claim that A has full μ-measure. Indeed, since

$$\mu\big(J_{2n}(c_j)\big) = \mu\big(I_{2n}(c_j)\big) + \mu\big(I_{2n+1}(c_j)\big) \leqslant \frac{1}{q_{2n+1}} + \frac{1}{q_{2n+2}}\,,$$

we deduce that $q_n\, \mu\big(J_{2n}(c_j)\big) \to 0$ (exponentially fast in n, in fact) and since $\mu(A_n) \geqslant 1 - N q_n\, \mu\big(J_{2n}(c_j)\big)$ we see that $\mu(A_n) \to 1$ as $n \to +\infty$. This implies the claim that $\mu(A) = 1$. Now for each $x \in A$ we have from Proposition 8.6 that there exists a sequence $n_k \to +\infty$ such that:

$$\frac{-Cn_k}{q_{n_k}} \leqslant \frac{\log Df^{q_{n_k}}(x)}{q_{n_k}} \leqslant \frac{C}{q_{n_k}},$$

and letting $k \to +\infty$ we get that:

$$\lim_{k\to+\infty} \frac{\log Df^{q_{n_k}}(x)}{q_{n_k}} = 0.$$

Therefore:

$$\lim_{n\to+\infty} \left\{\frac{\log Df^{n}(x)}{n}\right\} = 0$$

for μ-almost every $x \in A$, and then we are done since A has full μ-measure. □

8.4 Further Ergodic Properties

In Sect. 3.4.3 we introduced the notion of automorphic measure for a circle diffeomorphism, and discussed an important result by Douady and Yoccoz (1999), namely Theorem 3.12, which states both existence and uniqueness of automorphic measures of any given exponent for $C^{1+\mathrm{BV}}$ diffeomorphisms. It turns out that, at least for positive exponents, the same holds for multicritical circle maps. More precisely, we have the following result.

Theorem 8.7 *Let $f : S^1 \to S^1$ be a multicritical circle map. For any given $s \geqslant 0$ there exists a unique automorphic measure of exponent s for f. This measure has no atoms, is supported on the whole circle and it is ergodic under f.*

For a proof of the *existence* part of this theorem, see Exercise 8.9. We would like to mention here that, in a recent preprint by Goncharuk and Yampolsky (2023), both existence and uniqueness of s-automorphic measures have been established for the particular exponent $s = -1$.

As we have also seen in Chap. 3, any $C^{1+\mathrm{BV}}$ circle diffeomorphism without periodic points is ergodic with respect to Lebesgue measure (recall Theorem 3.10). Therefore, it is worth mentioning here, for future reference, the following special case of Theorem 8.7.

Theorem 8.8 *If $f : S^1 \to S^1$ is a multicritical circle map, then f is ergodic with respect to Lebesgue measure.*

Just as in the proof of Theorem 3.10, one can use a Lebesgue density argument to prove Theorem 8.8 (see also the proof of Lemma 1.5 in Chap. 1). The needed distortion estimates, however, are considerably more involved than those in the diffeomorphism case.

Another important ergodic object introduced in Sect. 3.4.3 was the notion of invariant distribution. Recall that, as proved by Avila and Kocsard (2011), every C^∞ diffeomorphism of the circle with irrational rotation number is distributionally uniquely ergodic, i.e., each linear space $\mathscr{D}'_k(f)$ is one-dimensional, spanned by the unique f-invariant probability measure (Theorem 3.13, see also Theorem 3.14). For multicritical circle maps and distributions of order 1, the following is known.

Theorem 8.9 *If $f : S^1 \to S^1$ is a multicritical circle map, then the linear space $\mathscr{D}'_1(f)$ is one-dimensional (spanned by the unique f-invariant probability measure).*

Both Theorem 8.7 and Theorem 8.9 were proved in the recent manuscript by de Faria et al. (2024).

Remark 8.6 It is reasonable to conjecture that every sufficiently smooth multicritical circle map with irrational rotation number is distributionally uniquely ergodic. This conjecture is certainly made plausible by Theorem 8.9. As we write these lines, however, the problem is still open.

8.5 Hausdorff Dimension

We close this chapter with an elegant recent result due to Trujillo (2020). Only the statement will be given, however—discussing the proof would constitute a digression into the realm of geometric measure theory.

As we have seen in Sect. 8.2, the unique invariant probability measure μ under a multicritical circle map f is purely singular with respect to Lebesgue measure. The *Hausdorff dimension* of μ, denoted $\dim_H(\mu)$, is by definition the smallest of the Hausdorff dimensions of all measurable sets having full μ-measure. More precisely,

$$\dim_H(\mu) = \inf\{\dim_H(E) \; : \; E \subset S^1 \text{ is measurable and } \mu(E) = 1\} .$$

A natural question to ask is: how does the Hausdorff dimension of μ vary with f? Obviously, a priori it should only depend on the bi-Lipschitz conjugacy class of f. In his recent paper, Trujillo (2020) establishes lower and upper bounds for $\dim_H(\mu)$ that depend only on the Diophantine nature of the rotation number of f. In order to state his result, let us first recall from Chap. 4 that an irrational number α is said to be *Diophantine of order* $\delta \geqslant 0$ if there exists a positive constant $C = C(\alpha) > 0$ such that

$$\left|\alpha - \frac{p}{q}\right| \geqslant \frac{C}{q^{2+\delta}} , \quad \text{for all } p, q \in \mathbb{Z},\ q \neq 0 .$$

We denote by $\mathscr{D}_\delta$ the set of all Diophantine numbers of order δ. Recall from Exercise 4.11 that $\mathscr{D}_0$ is precisely the set of numbers of *bounded type*. As we will see in Appendix A, the set $\mathscr{D}_0$ has zero Lebesgue measure (see Lemma A.1), whereas for each $\delta > 0$ the set $\mathscr{D}_\delta$ has full Lebesgue measure (see Lemma A.4).

Theorem 8.10 *If f is a C^3 multicritical circle map with irrational rotation number $\rho = \rho(f)$ and μ is its unique invariant probability measure, then the following holds.*

(i) If $\rho \in \mathscr{D}_\delta$ for some $\delta \geqslant 0$, then there exists $\nu > 0$ such that

$$\dim_H(\mu) \geqslant \frac{1}{2\delta + \nu} .$$

(ii) If $\rho \notin \mathscr{D}_\delta$ for some $\delta > 0$, then

$$\dim_H(\mu) \leqslant \frac{1}{\delta + 1} .$$

Note that the above theorem does not provide an upper bound for $\dim_H(\mu)$ in the case when the rotation number $\rho(f)$ is of bounded type (i.e., lies in $\mathscr{D}_0$). Nevertheless, it had already been known since the work of Graczyk and Świątek (1993) that in the bounded type case the Hausdorff dimension of μ lies strictly between 0 and 1.

Exercises

Exercise 8.1 Let $f : S^1 \to S^1$ be a C^1 non-minimal circle homeomorphism with irrational rotation number (*e.g.*, a Denjoy counterexample). Show that there exist uncountably many σ-finite f-invariant measures which are non-atomic and have pairwise disjoint supports.

Exercise 8.2 Work through the following steps to construct a *minimal* homeomorphism of S^1, with arbitrary irrational rotation number, having an infinite, σ-finite invariant measure which is absolutely continuous with respect to Lebesgue measure. Fix an irrational $\alpha \in (0, 1)$, and consider the rotation R_α.

(i) Show that there exists a closed, perfect and totally disconnected set $E_0 \subset S^1$ whose rotated copies $E_n = R_\alpha^n(E_0)$, $n \in \mathbb{Z}$ are pairwise disjoint.
(ii) Let ν_0 be a non-atomic Borel probability measure on S^1 supported by E_0, and let $\nu_n = (R_\alpha^n)_*\nu_0$, $n \in \mathbb{Z}$, be its rotated copies under R_α. Set $\nu = \sum_{n\in\mathbb{Z}} \nu_n$ and

$$\mu = \frac{1}{3}\sum_{n\in\mathbb{Z}} 2^{-|n|}\nu_n .$$

Show that μ is a probability measure on S^1 whose support is the whole circle, and that ν is an infinite, σ-finite measure which is invariant under the rotation R_α.

(iii) Show that the measures μ and ν are equivalent, i.e., they are mutually absolutely continuous.

(iv) Let $h : S^1 \to S^1$ be the *primitive* of μ given by $h(x) = \int_0^x d\mu(t) = \mu[0, x]$, and verify that h is a homeomorphism. Show that $h_*\mu = m$.

(v) Check that the measure $\tilde{\nu} = h_*\nu$ is infinite but σ-finite, and prove that it is absolutely continuous with respect to Lebesgue measure.

(vi) Finally, letting $f_\alpha = h \circ R_\alpha \circ h^{-1}$, show that f_α leaves $\tilde{\nu}$ invariant, i.e., $(f_\alpha)_*\tilde{\nu} = \tilde{\nu}$.

This example is due to Katznelson (1977, p. 11).

Exercise 8.3 Show that the homeomorphism $f_\alpha : S^1 \to S^1$ constructed in the previous exercise is not ergodic with respect to Lebesgue measure. Deduce that f_α cannot be C^2, or even $C^{1+\mathrm{BV}}$.

Exercise 8.4 Let $f : S^1 \to S^1$ be a minimal C^1 homeomorphism satisfying the following hypothesis. There exist constants $\alpha, K > 1$ and $\beta > 0$ such that for any given interval I on the circle, we can find (a) two disjoint intervals $J', J'' \subset I$ with $|J'| \geqslant \alpha|J''|$ and $|J''| \geqslant \beta|I|$; and (b) an iterate of f mapping J' diffeomorphically onto J'' with distortion bounded by K. Show that f satisfies the conclusion of Theorem 8.2.

Exercise 8.5 Fill in the missing details of the proof of Lemma 8.1.

Exercise 8.6 Prove Proposition 8.2.

Exercise 8.7 Give a proof of Theorem 8.5 that is independent of Herman's Theorem 7.1.

Exercise 8.8 *Divergent Poincaré series.* Let $f : S^1 \to S^1$ be a multicritical circle map with irrational rotation number, and let μ be its unique invariant Borel probability measure. For each $s > 0$, we define the *Poincaré series* of f of exponent s to be the function $P_{s,f} : S^1 \to [0, \infty]$ given by

$$P_{s,f}(x) = \sum_{n=0}^{\infty} \left(Df^n(x)\right)^s .$$

The main purpose of this exercise is to show that $P_{s,f}$ diverges μ-a.e.

(i) Show that there exists a countable dense set on the circle on which $P_{s,f}$ is finite for all $s > 0$.

(ii) Show that the identity

$$P_{s,f}(x) = 1 + \left(Df^n(x)\right)^s P_{s,f}(f(x))$$

holds for each $x \in S^1$.

(iii) Using Theorem 8.6 and the fact that μ is f-invariant, show with the help of Jensen's inequality that

$$\log\left(\int_{S^1} (Df^n)^s \, d\mu\right) \geqslant 0 ,$$

for each $n \geqslant 0$.

(iv) Deduce from (iii) that

$$\int_{S^1} \sum_{k=0}^{n-1} (Df^{q_k})^s \, d\mu \geqslant n ,$$

for each $n \geqslant 1$ (where, as usual, $\{q_k\}_{k\geqslant 0}$ is the sequence of return times associated with the rotation number $\rho(f)$).

(v) Now let $A = \{x \in S^1 : P_{s,f}(x) = \infty\}$. Show that A is Borel measurable and, using (ii), that A is f-invariant.

(vi) For each $m \geqslant 1$, let

$$X_m = \left\{x \in S^1 : P_{s,f}(x) \leqslant m\right\} .$$

Show that if $\mu(A) = 0$, then $\lim_{m\to\infty} \mu(X_m) = 1$. Combine this fact with (iv) to arrive at a contradiction.

(vii) From (v), (vi) and the fact that μ is ergodic, deduce that $\mu(A) = 1$.

Exercise 8.9 *Existence of automorphic measures.* Let $f : S^1 \to S^1$ be a multicritical circle map with irrational rotation number. Fix $s > 0$, let $P_{s,f}$ be the corresponding Poincaré series, and consider the set A defined in Exercise 8.8. Fix $x \in A$, and for each $n \geqslant 1$, let

$$S_n(x) = \sum_{i=0}^{q_n-1} (Df^i(x))^s .$$

Consider the atomic probability measure

$$\mu_{s,x,n} = \frac{1}{S_n(x)} \sum_{i=0}^{q_n-1} (Df^i(x))^s \, \delta_{f^i(x)} ,$$

(i) Show that there exist a probability measure $\mu_{s,x} \in \mathscr{P}(S^1)$ and a monotone sequence $(n_k) \subseteq \mathbb{N}$ such that, for all $\varphi \in C^0(S^1)$,

$$\int_{S^1} \varphi \, d\mu_{s,x,n_k} \longrightarrow \int_{S^1} \varphi \, d\mu_{s,x} \, .$$

(ii) Let $K = \sup \|Df^{q_n}\|_{C^0(S^1)}$ and recall from the real bounds that $K < \infty$. Show that for each $k \geqslant 1$ we have

$$\left| \int_{S^1} [\varphi - (\varphi \circ f)(Df)^s] \, d\mu_{s,x,n_k} \right| \leqslant \frac{1}{S_{n_k}(x)} \|\varphi\|_{C^0(S^1)} (1 + K^s) \, .$$

(iii) Deduce from (i) and (ii) that $\mu_{s,x}$ is an s-automorphic measure for f.

Chapter 9
Orbit Flexibility

In this chapter we study the geometric structure of individual orbits of a multicritical circle map with irrational rotation number.

From the dynamical standpoint, a minimal circle homeomorphism $f : S^1 \to S^1$ is topologically very homogeneous: all orbits look topologically the same. But are such orbits geometrically the same? In order to turn this somewhat vague question into a mathematically meaningful one, we need to properly define the underlying concept of geometric equivalence. We also need to assume that f has some reasonable degree of smoothness.

Let us agree to say that the orbits $\mathscr{O}_f(x)$ and $\mathscr{O}_f(y)$ of two points $x, y \in S^1$ are *geometrically equivalent* if there exists a self-conjugacy $h : S^1 \to S^1$ ($h \circ f = f \circ h$) which is a quasisymmetric homeomorphism carrying $\mathscr{O}_f(x)$ to $\mathscr{O}_f(y)$. So let us ask that question again: are two given orbits $\mathscr{O}_f(x)$ and $\mathscr{O}_f(y)$ geometrically equivalent?

The answer is easily seen to be *yes* if f is smoothly conjugate to a rotation. For instance, when f is a smooth diffeomorphism with Diophantine rotation number, the rigidity results discussed in Sect. 4.5 imply that f is smoothly conjugate to a rotation—and all orbits under a rigid rotation are not just merely geometrically equivalent, they are *isometric*. By contrast, our purpose in the present chapter is to explain that the answer to the above question is *almost always no* when f is a critical circle map. Precise statements will be given in due time. The entire exposition here is extracted from our own recent work, de Faria and Guarino (2022b).

9.1 Geometric Equivalence of Orbits

Ideally, we would like to classify orbits of a (reasonably smooth) minimal homeomorphism of the circle up to *quasisymmetric equivalence*.

E. de Faria, P. Guarino, *Dynamics of Circle Mappings*, IMPA Monographs,
https://doi.org/10.1007/978-3-031-67495-2_9

As we have seen in Chap. 7, quasisymmetry can be regarded as a very weak form of geometric regularity. Indeed, it is so weak that one might guess, for instance, that *every* conjugacy between two critical circle maps is quasisymmetric. This guess is reinforced by Theorem 7.1, which, we recall, states that every multicritical circle map whose rotation number is an irrational of *bounded type* is quasisymmetrically conjugate to the corresponding rotation.

However, the above guess is unfortunately wrong. Our purpose in the present chapter is to explain that a conjugacy between two critical circle maps is *almost never quasisymmetric*. We will first identify a *mechanism* which forces the breakdown of quasisymmetry for a topological conjugacy (see Lemma 9.4 in Sect. 9.6), and then we will show that this mechanism is *abundant*, both from the topological and measure-theoretical viewpoints (see Theorem 9.6 in Sect. 9.2). The precise statements will be given below—see Theorem 9.1–9.3.

9.1.1 Orbit-Flexibility

Some of these results can be stated in the light of the complementary concepts of *orbit-rigidity* and *orbit-flexibility*, which we presently describe.

Definition 9.1 We say that a minimal circle homeomorphism f is *quasisymmetrically orbit-rigid* if for any pair of points x, y on the circle there exists a quasisymmetric homeomorphism $h_{x,y}$ which conjugates f to itself and maps x to y. If f is *not* quasisymmetrically orbit-rigid, we say that f is *quasisymmetrically orbit-flexible*.

Example 9.1 Every rotation is quasisymmetrically orbit-rigid, and the equivalence between orbits is in fact given by an isometry.

Example 9.2 Every sufficiently smooth circle diffeomorphism with Diophantine rotation number is quasisymmetrically orbit-rigid. This follows from the Herman–Yoccoz theorems of Sect. 4.5.

For multicritical circle maps, we have the following simple consequence of Theorem 7.1.

Proposition 9.1 *Every multicritical circle map f with rotation number $\alpha = \rho(f)$ of* bounded *type is quasisymmetrically orbit-rigid.*

Proof Let $h : S^1 \to S^1$ be a quasisymmetric conjugacy between f and the rotation R_α, whose existence is guaranteed by Theorem 7.1. Given $x, y \in S^1$, let R be the circle rotation with $R(x) = h^{-1}(y)$. Then $\varphi = h \circ R$ is a quasisymmetric homeomorphism, and since any two rotations commute, we have

$$\varphi^{-1} \circ f \circ \varphi = (h \circ R)^{-1} \circ f \circ (h \circ R) = R^{-1} \circ R_\alpha \circ R = R_\alpha .$$

Therefore φ conjugates f to R_α, and since $\varphi(x) = y$, it maps the orbit of x onto the orbit of y. □

By contrast, we will show in Theorem 9.1 that (uni)critical circle maps whose rotation numbers belong to a certain full-measure set are quasisymmetrically orbit-flexible (see also Proposition 9.2). In particular, the centralizers of such maps in the group of all homeomorphisms of the circle contain non-quasisymmetric elements (see Sect. 9.1.4 below).

9.1.2 Statement for Unicritical Maps

In the *unicritical* case we have the following *coexistence* phenomenon.

Theorem 9.1 *There exists a full Lebesgue measure set* $\boldsymbol{R}_A \subset [0, 1]$ *of irrational numbers with the following property. Let* f *and* g *be two* C^3 *circle homeomorphisms with a single (non-flat) critical point (say,* c_f *and* c_g *respectively) and with* $\rho(f) = \rho(g) \in \boldsymbol{R}_A$*. For any given* $x \in \mathbf{S}^1$ *let* $h_x \in \mathrm{Diff}^0_+(\mathbf{S}^1)$ *be the topological conjugacy between* f *and* g *determined by* $h_x(x) = c_g$*. Let* $\mathscr{A}$ *be the set of points* $x \in \mathbf{S}^1$ *such that the homeomorphism* h_x *is quasisymmetric, and let* $\mathscr{B} = \mathbf{S}^1 \setminus \mathscr{A}$ *be its complement in the unit circle (that is,* $\mathscr{B}$ *is the set of points* $x \in \mathbf{S}^1$ *such that the homeomorphism* h_x *is* not *quasisymmetric). Then* $\mathscr{A}$ *is dense in* $\mathbf{S}^1$*, while* $\mathscr{B}$ *contains a residual set (in the sense of Baire) and it has full* μ_f*-measure, where* μ_f *denotes the unique* f*-invariant probability measure.*

Remark 9.1 A somewhat related coexistence phenomenon occurs in the context of Lorenz maps, and also in the context of circle maps with flat intervals (see Martens et al. 2018 and references therein).

Remark 9.2 The proof of Theorem 9.1, to be given in Sect. 9.6, still works if *one of the two* maps has more than one critical point.

Let us pose two questions that arise from Theorem 9.1.

Question 9.1 Denote by BT $\subset (0, 1)$ the set of irrational numbers of *bounded type*. Corollary 7.3 implies that $\boldsymbol{R}_A$ is disjoint from BT (since, in this case, *all* conjugacies are quasisymmetric). Is it true that $\boldsymbol{R}_A = [0, 1] \setminus (\mathbb{Q} \cup \mathrm{BT})$? Is it true, at least, that $\boldsymbol{R}_A$ contains a residual subset of $[0, 1]$?

Question 9.2 Note that both sets $\mathscr{A}$ and $\mathscr{B}$ defined in Theorem 9.1 are f-invariant. Indeed, this follows from the identity $h_x = h_{f(x)} \circ f$ and the fact that f itself (hence f^n for all $n \in \mathbb{Z}$) is a quasisymmetric homeomorphism. By Theorem 7.2, the critical point of f belongs to $\mathscr{A}$ (and then its whole orbit), since h_{c_f} is always a quasisymmetric homeomorphism. It could be the case that $\mathscr{A} = \{f^n(c_f) : n \in \mathbb{Z}\}$. Is it true, at least, that $\mathscr{A}$ is a countable set?

In Sect. 9.1.4 below we describe more precisely the notion of orbit-flexibility, and state some straightforward consequences of Theorem 9.1. In Sect. 9.1.5 we state some further consequences of Theorem 9.1, this time involving geometric bounds for dynamical partitions (see Theorem 9.5).

9.1.3 Statements for Multicritical Maps

As usual, we denote by $a_n = a_n(\rho)$, $n \in \mathbb{N}$, the infinite sequence of partial quotients of the continued fraction development of a given irrational number ρ. Let us consider the set $\mathbb{E}_\infty$ consisting of all numbers $\rho \in (0, 1)$ for which the corresponding a_n's are *even* and $\lim_{n\to\infty} a_n = \infty$. It is not difficult to see that $\mathbb{E}_\infty$ is a meager set whose Lebesgue measure is equal to zero (see Exercise 9.3). Despite being both topologically and measure-theoretically negligible, this set does contain some interesting Diophantine, Liouville and transcendental numbers, see Sect. 9.5. Our second goal in this chapter is to prove the following result.

Theorem 9.2 *There exists a set $\mathscr{G} \subset [0, 1]^2$, which contains a residual set (in the Baire sense) and has full Lebesgue measure, for which the following holds. Let f and g be two C^3 multicritical circle maps with the same irrational rotation number ρ and such that the map f has exactly one critical point c_0, whereas the map g has exactly two critical points c_1 and c_2. Denote by α and $1-\alpha$ the μ_g-measures of the two connected components of $\boldsymbol{S}^1 \setminus \{c_1, c_2\}$, where μ_g denotes the unique invariant probability measure of g. If (ρ, α) belongs to $\mathscr{G}$, then the topological conjugacy between f and g that takes c_0 to c_1 is not quasisymmetric. Moreover, the set of rotation numbers $\boldsymbol{R}_B = \{\rho : (\rho, \alpha) \in \mathscr{G} \text{ for some } \alpha\}$ contains the set $\mathbb{E}_\infty$ defined above.*

The proofs of both Theorems 9.1 and 9.2 will be given in Sect. 9.6. The following auxiliary result, a complete proof of which can be found in de Faria and Guarino (2022b), will be discussed in Sect. 9.7.

The C^∞ Realization Lemma *For any given $(\rho, \alpha) \in \big([0, 1] \setminus \mathbb{Q}\big) \times (0, 1)$ there exists a C^∞ bi-critical circle map with rotation number ρ, a unique invariant Borel probability measure μ and with exactly two critical points c_1 and c_2 such that the two connected components of $\boldsymbol{S}^1 \setminus \{c_1, c_2\}$ have μ-measures equal to α and $1-\alpha$ respectively.*

Remark 9.3 It is possible to prove a similar *Analytic Realization Lemma* using the results of Zakeri (1999, Section 7).

Together with Theorem 9.2, the C^∞ Realization Lemma implies the following result.

Theorem 9.3 *There exists a set $\boldsymbol{R}_C \subset [0, 1]$ of irrational numbers, which contains a residual set (in the Baire sense), has full Lebesgue measure and contains $\mathbb{E}_\infty$, for* *which the following holds. For each $\rho \in \boldsymbol{R}_C$, there exist two C^∞ multicritical circle maps $f, g : \boldsymbol{S}^1 \to \boldsymbol{S}^1$ with the following properties:*

1. *Both maps have the same rotation number ρ;*
2. *The map f has exactly one critical point c_0, whereas the map g has exactly two critical points c_1 and c_2;*
3. *The topological conjugacy between f and g that takes c_0 to c_1 is not quasisymmetric.*

9.1.4 Centralizers

Following Yoccoz (1984a, 1995a), we denote by $Z_0(f) = \{h \in \mathrm{Diff}^0_+(\boldsymbol{S}^1) : h \circ f = f \circ h\}$ the *centralizer* of f in $\mathrm{Diff}^0_+(\boldsymbol{S}^1)$. We also denote by $\mathrm{QS}(\boldsymbol{S}^1)$ the subgroup of $\mathrm{Diff}^0_+(\boldsymbol{S}^1)$ consisting of those homeomorphisms of the circle that are quasisymmetric. In this language, Theorem 9.1 has the following immediate consequence.

Theorem 9.4 *If $f : \boldsymbol{S}^1 \to \boldsymbol{S}^1$ is a unicritical circle map with $\rho(f) \in \boldsymbol{R}_A$, then f is quasisymmetrically orbit-flexible. In particular, $Z_0(f) \setminus \mathrm{QS}(\boldsymbol{S}^1) \neq \emptyset$.*

See also Avila et al. (2022, Section 4) for recent results on the centralizers of some analytic circle maps. In fact, much more can be obtained from Theorem 9.1. First, we need a definition. Let $f : \boldsymbol{S}^1 \to \boldsymbol{S}^1$ be a minimal circle homeomorphism.

Definition 9.2 If $x, y \in \boldsymbol{S}^1$, we say that x is f-equivalent to y, and write $x \sim_f y$, if there exists a quasisymmetric homeomorphism $h \in Z_0(f)$ such that $h(x) = y$.

It is clear that $\sim_f$ is an equivalence relation, so we can consider the set of equivalence classes $X_f = \boldsymbol{S}^1/\sim_f$. Below, in the proof of Proposition 9.2, we will use the following observation.

Lemma 9.1 *All equivalence classes are homeomorphic to each other.*

Proof Let us mark some point $c \in \boldsymbol{S}^1$. For any given $x \in \boldsymbol{S}^1$ consider $F_x : \boldsymbol{S}^1 \to \boldsymbol{S}^1$ defined as follows: given $y \in \boldsymbol{S}^1$ let $h_{x,y} \in Z_0(f)$ be determined by $h_{x,y}(x) = y$, and then let $F_x(y)$ be defined by $h_{x,y}\big(F_x(y)\big) = c$. It not difficult to prove that F_x is a circle homeomorphism which identifies the class of x with the class of c. In particular, given $x, y \in \boldsymbol{S}^1$, the homeomorphism $F_y^{-1} \circ F_x$ identifies the class of x with the class of y. □

Note that if f is either a diffeomorphism or a (C^3) multicritical circle map, then points in the same f-orbit are f-equivalent. More generally, for such f's, if $x \sim_f y$ then for each $x' \in \mathcal{O}_f(x)$ and each $y' \in \mathcal{O}_f(y)$ we have $x' \sim_f y'$. This happens because, in the cases considered, f itself (hence f^n for all $n \in \mathbb{Z}$) is a quasisymmetric homeomorphism. Note that, being f-invariant, all equivalence classes are dense in the unit circle.

In the language introduced before, if X_f reduces to a single point, then f is quasisymmetrically orbit-rigid, whereas if X_f has more than one point, then f is quasisymmetrically orbit-flexible. Now we can state the following simple consequence of Theorem 9.1.

Proposition 9.2 *If $f : \boldsymbol{S}^1 \to \boldsymbol{S}^1$ is a unicritical circle map whose rotation number belongs to the set $\boldsymbol{R}_A$ of Theorem 9.1, then all its equivalence classes are meagre (in the sense of Baire). In particular X_f is uncountable.*

Proof By definition, the set $\mathscr{A}$ given by Theorem 9.1 (applied to the particular case $g = f$) is the equivalence class of c_f, the critical point of f. Being disjoint from the residual set $\mathscr{B}$, the set $\mathscr{A}$ is meagre. By Lemma 9.1, all classes are meagre, and therefore, by Baire's theorem, their number is uncountable. □

As already explained, if $f : \boldsymbol{S}^1 \to \boldsymbol{S}^1$ is an irrational rotation or a smooth diffeomorphism whose rotation number is Diophantine, then X_f is a single point. By Proposition 9.1, the same happens with any multicritical circle map with rotation number of bounded type.

9.1.5 Unbounded Geometry

Let f be a C^3 multicritical circle map with irrational rotation number. Just as in Sect. 8.2.3, we say that f has *bounded geometry* at $x \in \boldsymbol{S}^1$ if there exists $K > 1$ such that for all $n \in \mathbb{N}$ and for every pair I, J of adjacent atoms of $\mathscr{P}_n(x)$ we have

$$K^{-1}\,|I| \leqslant |J| \leqslant K\,|I|\,,$$

where $\left\{\mathscr{P}_n(x)\right\}_{n\in\mathbb{N}}$ is the standard sequence of dynamical partitions of the circle associated to $x \in \boldsymbol{S}^1$ (see Sect. 6.3.1). With this at hand, consider the set

$$\mathscr{A} = \mathscr{A}(f) = \{x \in \boldsymbol{S}^1 : f \text{ has bounded geometry at } x\}\,.$$

The relation between bounded geometry and quasisymmetric homeomorphisms is given by the following result.

Proposition 9.3 *Let f be a multicritical circle map with irrational rotation number, and let $x \in \mathscr{A}(f)$. As before, for any given $y \in \boldsymbol{S}^1$ let $h_{x,y} \in \mathrm{Z}_0(f)$ be determined by $h_{x,y}(x) = y$. Then*

$$y \in \mathscr{A}(f) \Leftrightarrow h_{x,y} \in \mathrm{QS}(\boldsymbol{S}^1)\,.$$

Proof For the "if" implication suppose, by contradiction, that $y \notin \mathscr{A}$. This means that there exists a sequence $\{n_k\}_{k\in\mathbb{N}} \subset \mathbb{N}$ such that for each $k \in \mathbb{N}$ we can find a pair I_k, J_k of adjacent atoms of $\mathscr{P}_{n_k}(y)$ satisfying $\lim_k |I_k|/|J_k| = +\infty$. However, both intervals $h_{x,y}^{-1}(I_k)$ and $h_{x,y}^{-1}(J_k)$ are adjacent and belong to $\mathscr{P}_{n_k}(x)$, and since $x \in \mathscr{A}$, the ratios $|h_{x,y}^{-1}(I_k)|/|h_{x,y}^{-1}(J_k)|$ are bounded. But this implies that the quasisymmetric homeomorphism $h_{x,y}$ does *not* have weakly bounded cross-ratio distortion, which is impossible by Corollary 7.1. For the "only if" implication we refer the reader to Exercise 9.2. □

An immediate consequence of Proposition 9.3 is that the set $\mathscr{A}$ is f-invariant, since f itself (hence f^n for all $n \in \mathbb{Z}$) is a quasisymmetric homeomorphism. As it follows from the real bounds (Theorem 6.3), all critical points of f belong to $\mathscr{A}$. Being f-invariant and non-empty, the set $\mathscr{A}$ is dense in the unit circle. However, the following consequence of Theorem 9.1 shows that $\mathscr{A}$ can be rather small.

Theorem 9.5 *Let $\boldsymbol{R}_A \subset (0, 1)$ be the full Lebesgue measure set given by Theorem 9.1, and let f be a C^3 critical circle map with a single (non-flat) critical point and rotation number $\rho \in \boldsymbol{R}_A$. Then the set $\mathscr{A}(f)$ is meagre (in the sense of Baire) and it has zero μ_f-measure.*

To prove Theorem 9.5, note first that Proposition 9.3 is saying that the set $\mathscr{A}(f)$ is an equivalence class for the $\sim_f$ relation and then, by Proposition 9.2, we already know that it is meagre. Moreover, since the critical point of f belongs to $\mathscr{A}$ (by Theorem 6.3), we deduce that $\mathscr{A}(f)$ is precisely the equivalence class of the critical point. With this observation at hand, Theorem 9.5 follows at once from Theorem 9.1, just by considering the particular case $g = f$.

By contrast, recall that if f has bounded combinatorics, then the set $\mathscr{A}(f)$ is the whole circle: f has bounded geometry at any point in the unit circle.

9.2 Renormalization Trails and Ancestors

Let $f : S^1 \to S^1$ be an orientation preserving circle homeomorphism with irrational rotation number $\rho = [a_0, a_1, a_2, \dots]$. Let us fix some point x in the unit circle, and for each $n \geqslant 0$ let us consider the interval $I_n = I_n(x)$ having x and $f^{q_n}(x)$ as its endpoints.[1] Suppose we are given another point on the circle, say y. Looking at the *past* of y, i.e., at its negative orbit $\mathscr{O}_f^-(y)$, we see that for each $n \geq 0$ there exists in $\mathscr{O}_f^-(y)$ a most recent visitor to $I_n \cup I_{n+1}$; this point is called the *n-th generation ancestor* of y (with respect to x and f).

Let us be a bit more formal. Consider the rectangle $R = [0, 1] \times [-1, 1]$ in $\mathbb{R}^2$, and let $M = \big([0, 1] \setminus \mathbb{Q}\big) \times [-1, 1] \subset R$. For any given y in S^1, we will define/construct in what follows a sequence of pairs $(\rho_n, \alpha_n) \in M$, called renormalization trail (see Definition 9.4 below) of y with respect to x and f. Let us define simultaneously the initial cases $n = 0$ and $n = 1$. First, let $\rho_0 = \rho =$

[1] The interval $I_n \cup I_{n+1}$, whose interior contains x, is sometimes called the *n-th renormalization domain* of f around x. The meaning of the word "renormalization" will be explained in Chap. 10.

$[a_0, a_1, a_2, \dots] \in [0,1] \setminus \mathbb{Q}$ and $\rho_1 = G(\rho) = [a_1, a_2, \dots] \in [0,1] \setminus \mathbb{Q}$, where $G : [0,1] \to [0,1]$ is the Gauss map (recall Chap. 1, see also Appendix A). To define α_0 and α_1 consider both intervals

$$I_0 = \big(x, f(x)\big] \quad \text{and} \quad I_1 = \big(f^{a_0}(x), x\big]$$

If y belongs to the short interval I_1 we define

$$\alpha_0 = \mu\big((x,y)\big) \in [0, 1 - a_0\,\rho_0] \quad \text{and} \quad \alpha_1 = -\frac{\mu\big((x,y)\big)}{\mu(I_1)} \in [-1, 0]\,.$$

Otherwise, there exist y_0 in the long interval I_0 and $i \in \{0, 1, \dots, a_0 - 1\}$ such that $f^i(y_0) = y$, in which case we define

$$\alpha_0 = 1 - \big[\mu\big((x, y_0)\big) + i\,\rho_0\big] = 1 - \mu\big((x,y)\big) \quad \text{and} \quad \alpha_1 = \frac{\mu\big((x, y_0)\big)}{\mu(I_0)}\,.$$

Note that $\alpha_0 \in [1 - a_0\,\rho_0, 1]$, whereas $\alpha_1 \in [0,1]$. It should be noted also that, in the definition of α_0, we are measuring arcs in the *counterclockwise* sense: in the first case, we measure $\mu\big((x,y)\big)$ considering the arc determined by x and y which is contained in I_1, while in the second case we measure $\mu\big((x,y_0)\big)$ considering the arc determined by x and y_0 which is contained in I_0. In this way we obtain the first two terms of the sequence of pairs $(\rho_n, \alpha_n) \in M = \big([0,1] \setminus \mathbb{Q}\big) \times [-1,1]$. After the first n terms are defined, let $\rho_{n+1} \in [0,1] \setminus \mathbb{Q}$ be given by

$$\rho_{n+1} = G^{n+1}(\rho) = G^{n+1}\big([a_0, a_1, \dots]\big) = [a_{n+1}, a_{n+2}, \dots]\,.$$

If y belongs to the long interval $f^i(I_n)$ for some $i \in \{0, 1, \dots, q_{n+1} - 1\}$, let $y_n \in I_n$ be such that $f^i(y_n) = y$. Otherwise, y belongs to the short interval $f^j(I_{n+1})$ for some $j \in \{0, 1, \dots, q_n - 1\}$, and then let $y_n \in I_{n+1}$ be given by $f^j(y_n) = y$. In the first case, see Fig. 9.1, we define

$$\alpha_{n+1} = \frac{\mu\big((x, y_n)\big)}{\mu(I_n)} \in [0,1],$$

while in the second case we define

$$\alpha_{n+1} = -\frac{\mu\big((y_n, x)\big)}{\mu(I_{n+1})} \in [-1, 0].$$

We can now formally define the notion of ancestor.

Definition 9.3 The points y_n, $n \geq 0$, defined above are called the *renormalization ancestors* (or simply the *ancestors*) of y with respect to x and f.

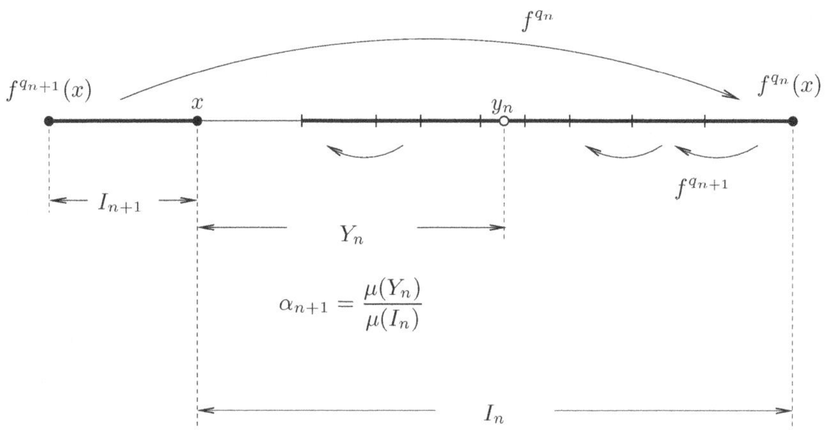

Fig. 9.1 Calculating renormalization trails

We are in fact more interested in the sequence of pairs $(\rho_n, \alpha_n) \in M = \big([0, 1] \setminus \mathbb{Q}\big) \times [-1, 1]$. Accordingly, we formulate the following definition.

Definition 9.4 The sequence $\big\{(\rho_n, \alpha_n)\big\}_{n\geq 0} \subset M$ is called the *renormalization trail*, or simply the *trail*, of the point y with respect to x and f.

In Sect. 9.4, we will prove the following result.

Theorem 9.6 *There exists a full Lebesgue measure set $\boldsymbol{R} \subset [0, 1]$ of irrational numbers with the following property: given a minimal circle homeomorphism f with $\rho(f) \in \boldsymbol{R}$ and given any point $x \in \mathbf{S}^1$ there exists a set $\mathscr{B}_x \subset \mathbf{S}^1$ which is residual (in the Baire sense) and has full μ_f-measure such that for all $y \in \mathscr{B}_x$ the renormalization trail $\big\{(\rho_n, \alpha_n)\big\}$ of y (with respect to x and f) is dense in the rectangle* $[0, 1] \times [-1, 1]$.

Being dense in $[0, 1]$, the orbit under the Gauss map of any element of $\boldsymbol{R}$ accumulates at the origin. In particular, $\boldsymbol{R}$ is disjoint from the set BT $\subset [0, 1]$ of *bounded type* numbers. Note also that $\mathscr{B}_x$ is disjoint from $\mathscr{O}_f^+(x) = \big\{x, f(x), f^2(x), \ldots\big\}$, since for $n \geq 0$ the second coordinate of the renormalization trail of $f^n(x)$ with respect to x and f eventually becomes constant equal to 0.

9.3 The Skew Product

In this section we construct a *skew product* (see Sect. 9.3.2 below) that will be crucial in order to prove Theorem 9.6 (its proof will be given in Sect. 9.4) and also to prove Theorem 9.2 (see Sect. 9.6).

9.3.1 The Fiber Maps

For any given $\rho \in [0,1] \setminus \mathbb{Q}$ consider the piecewise affine dynamical system $T_\rho : [-1,1] \to [-1,1]$ given by

$$T_\rho(\alpha) = \begin{cases} -\alpha & \text{for } \alpha \in \big[-1, 0\big] \\[2ex] -\dfrac{\alpha}{\rho\, G(\rho)} & \text{for } \alpha \in \big[0, \rho\, G(\rho)\big] \\[2ex] \left\{\dfrac{1-\alpha}{\rho}\right\} & \text{for } \alpha \in \big(\rho\, G(\rho), 1\big], \end{cases}$$

where G is, as before, the Gauss map. Each T_ρ is a *Markov* map, its graph is depicted in Fig. 9.2.

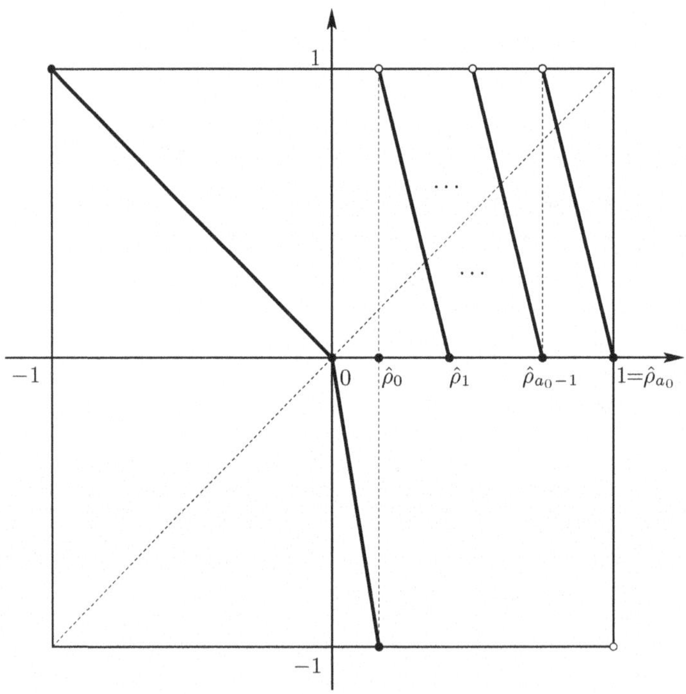

Fig. 9.2 The fiber map T_ρ; here, $\hat\rho_j = (G(\rho)+j)\rho$ for each $0 \le j \le a_0$, where $a_0 = \lfloor \frac{1}{\rho} \rfloor$

9.3.2 The Skew Product

As before (see Sect. 9.2) we consider the rectangle $R = [0, 1] \times [-1, 1]$ in $\mathbb{R}^2$, and let $M = \big([0, 1] \setminus \mathbb{Q}\big) \times [-1, 1] \subset R$. Consider the skew product $T : M \to M$ given by

$$T(\rho, \alpha) = \big(G(\rho)\,,\ T_\rho(\alpha)\big)\,,$$

where G is the Gauss map, and where the fiber maps T_ρ were introduced in the previous section (Sect. 9.3.1). The main dynamical property of the skew product T that we will need here is the following.

Proposition 9.4 *There exists a set $\mathscr{G}_0 \subset [0, 1] \times [-1, 1]$, which is residual (in the Baire sense) and has full Lebesgue measure, such that any initial condition in $\mathscr{G}_0$ has a positive orbit under T which is dense in $[0, 1] \times [-1, 1]$.*

The set $\mathscr{G}_0$ given by Proposition 9.4 will be crucial in the proof of Theorem 9.6 (which will be given in Sect. 9.4 below), and also in the proof of Theorem 9.2 (see Sect. 9.6). In Sect. 9.7 we will also need the following fact.

Lemma 9.2 (Topologically Exactness) *Let U be a subset of the rectangle R with non-empty interior. Then there exists $n \in \mathbb{N}$ such that $T^n(U \cap M) = M$.*

We postpone the proofs of Proposition 9.4 and Lemma 9.2 until Appendix C.

9.4 Proof of Theorem 9.6

Just as before, let $f : S^1 \to S^1$ be an orientation preserving circle homeomorphism with irrational rotation number ρ. With Proposition 9.4 at hand, Theorem 9.6 will be a straightforward consequence of the following fact:

Lemma 9.3 *Given x and y in S^1 we have*

$$(\rho_n, \alpha_n) = T^n(\rho_0, \alpha_0) \quad \textit{for all } n \in \mathbb{N},$$

where $\{(\rho_n, \alpha_n)\}$ is the renormalization trail of y with respect to x and f, as defined in Sect. 9.2, and $T : M \to M$ is the skew product constructed in Sect. 9.3.2.

Throughout the proof of Lemma 9.3 we will make repeated use of the formula $\mu(I_n) = \prod_{j=0}^{n} G^j(\rho)$ (recall Exercise 6.2).

Proof By our definition of renormalization trails, $\rho_n = G^n(\rho)$ for all $n \in \mathbb{N}$, which coincides with the first coordinate $T^n(\rho_0, \alpha_0)$, as we can see directly from the definition of our skew product T. So we only need to deal with the second coordinate of the trails.

Let us treat first the cases $n = 0$ and $n = 1$. If on one hand y belongs to the short interval $I_1 = \big(f^{a_0}(x), x\big]$, we have $\alpha_0 \in \big[0, \rho_0\, G(\rho_0)\big]$ and then

$$T_{\rho_0}(\alpha_0) = T_{\rho_0}\big(\mu((x, y))\big) = -\frac{\mu((x, y))}{\rho_0\, G(\rho_0)} = -\frac{\mu\big((x, y)\big)}{\mu(I_1)} = \alpha_1 \,.$$

On the other hand, if $y \notin I_1$ then there exist y_0 in the long interval $I_0 = \big(x, f(x)\big]$ and $i \in \{0, 1, \ldots, a_0 - 1\}$ such that $f^i(y_0) = y$, in which case we have $\alpha_0 \in \big[\rho_0\, G(\rho_0), 1\big]$ and then

$$\begin{aligned} T_{\rho_0}(\alpha_0) = T_{\rho_0}\big(1 - \mu((x, y_0)) - i\,\rho_0\big) &= \left\{\frac{\mu((x, y_0)) + i\,\rho_0}{\rho_0}\right\} \\ &= \left\{\frac{\mu((x, y_0))}{\rho_0}\right\} = \frac{\mu((x, y_0))}{\rho_0} = \frac{\mu((x, y_0))}{\mu(I_0)} = \alpha_1 \,. \end{aligned}$$

In any case, $\alpha_1 = T_{\rho_0}(\alpha_0)$ and then $(\rho_1, \alpha_1) = T(\rho_0, \alpha_0)$, as desired.

In order to prove the desired result for the remaining values of n, we have three cases to consider.

(i) If $y_n \in I_{n+2}$, we have

$$0 \le \alpha_{n+1} = \frac{\mu((x, y_n))}{\mu(I_n)} \le \frac{\mu(I_{n+2})}{\mu(I_n)} = \rho_{n+1}\, G(\rho_{n+1}) \,,$$

and then

$$\begin{aligned} T_{\rho_{n+1}}(\alpha_{n+1}) &= -\frac{\alpha_{n+1}}{\rho_{n+1}\, G(\rho_{n+1})} \\ &= -\frac{\alpha_{n+1}\, \mu(I_n)}{\mu(I_{n+2})} = -\frac{\mu((x, y_n))}{\mu(I_{n+2})} = \alpha_{n+2} \,. \end{aligned} \tag{9.1}$$

(ii) If $y_n \in I_n \setminus I_{n+2}$, we have

$$\frac{\mu(I_{n+2})}{\mu(I_n)} < \alpha_{n+1} \le 1 \,,$$

which implies $\alpha_{n+1} \in \big(\rho_{n+1}\, G(\rho_{n+1}), 1\big]$, and then

$$T_{\rho_{n+1}}(\alpha_{n+1}) = \left\{\frac{1 - \alpha_{n+1}}{\rho_{n+1}}\right\} .$$

Consider the fundamental domains $\Delta_{j,n} \subset I_n$ for $f^{q_{n+1}}$ given by

$$\Delta_{j,n} = f^{j\, q_{n+1} + q_n}(I_{n+1}) = \big(f^{(j+1)\, q_{n+1} + q_n}(x), f^{j\, q_{n+1} + q_n}(x)\big]$$

for $j \in \{0, 1, \ldots, a_{n+1} - 1\}$, and let $\ell_n \in \{0, 1, \ldots, a_{n+1} - 1\}$ be defined by $y_n \in \Delta_{\ell_n,n}$. We claim that $\ell_n = \left\lfloor \dfrac{1 - \alpha_{n+1}}{\rho_{n+1}} \right\rfloor$. Indeed, since $\mu(\Delta_{j,n}) = \mu(I_{n+1})$ for all $j \in \{0, 1, \ldots, a_{n+1} - 1\}$, it follows that

$$\ell_n\, \mu(I_{n+1}) \leq (1 - \alpha_{n+1})\, \mu(I_n) < (\ell_n + 1)\, \mu(I_{n+1}).$$

Equivalently,

$$\ell_n \leq (1 - \alpha_{n+1})\, \frac{\mu(I_n)}{\mu(I_{n+1})} < \ell_n + 1\,.$$

Finally, from

$$\frac{\mu(I_n)}{\mu(I_{n+1})} = \frac{\prod_{j=0}^{n} G^j(\rho)}{\prod_{j=0}^{n+1} G^j(\rho)} = \frac{1}{G^{n+1}(\rho)} = \frac{1}{\rho_{n+1}}\,,$$

we deduce that $\ell_n \leq \dfrac{1 - \alpha_{n+1}}{\rho_{n+1}} < \ell_n + 1$, which implies the claim. With this at hand we deduce that

$$\begin{aligned}
T_{\rho_{n+1}}(\alpha_{n+1}) &= \left\{\frac{1 - \alpha_{n+1}}{\rho_{n+1}}\right\} \\
&= \frac{1 - \alpha_{n+1}}{\rho_{n+1}} - \ell_n \\
&= \frac{\mu(I_n) - \alpha_{n+1}\, \mu(I_n)}{\mu(I_{n+1})} - \ell_n \\
&= \frac{\mu(I_n) - \big[\mu((x, y_n)) + \ell_n\, \mu(I_{n+1})\big]}{\mu(I_{n+1})} = \alpha_{n+2}\,.
\end{aligned}$$

(iii) Whenever y_n belongs to the *short* interval I_{n+1}, we have $\alpha_{n+1} \in [-1, 0)$ and then $T_{\rho_{n+1}}(\alpha_{n+1}) = -\alpha_{n+1} = \alpha_{n+2}$, since $y_{n+1} = y_n$ belongs now to the *long* interval I_{n+1}.

This finishes the proof of Lemma 9.3. □

Proof of Theorem 9.6 Let $\mathcal{G}_0 \subset \boldsymbol{R}$ be given by Proposition 9.4. By Fubini's theorem, there exists a full Lebesgue measure set $\boldsymbol{R} \subset [0, 1]$ such that for each $\rho \in \boldsymbol{R}$, the set $\boldsymbol{R}_\rho = \big\{\alpha \in [-1, 1] : (\rho, \alpha) \in \mathcal{G}_0\big\}$ has full Lebesgue measure in

$[-1, 1]$. In particular, $\boldsymbol{R}_\rho$ is also residual[2] in $[-1, 1]$ for all $\rho \in \boldsymbol{R}$. Given a minimal circle homeomorphism f with $\rho(f) \in \boldsymbol{R}$ and given any point $x \in \boldsymbol{S}^1$, the map that sends $\alpha \in (0, 1)$ to the point $y \in \boldsymbol{S}^1 \setminus \{x\}$ which satisfies $\mu_f([x, y]) = \alpha$ (and note that such point is unique if we fix, say, the counterclockwise orientation) is a homeomorphism that, by definition, identifies the Lebesgue measure in $(0, 1)$ with the probability measure μ_f in $\boldsymbol{S}^1 \setminus \{x\}$. By combining Proposition 9.4 with Lemma 9.3, we deduce that it is enough to take $\mathscr{B}_x$ as the image (under the homeomorphism described above) of $\boldsymbol{R}_\rho \cap (0, 1)$. □

9.5 Even-Type Rotation Numbers

Let us now present a result concerning trails for maps whose rotation number belongs to the special class appearing in the statements of Theorems 9.2 and 9.3. We denote by $\mathbb{E}$ the set of those irrationals $0 < \theta < 1$ all of whose partial quotients $a_n(\theta)$ are *even* (in particular $a_n(\theta) \geq 2$ for all n). We also consider the subset $\mathbb{E}_\infty = \{\theta \in \mathbb{E} : \lim_{n\to\infty} a_n(\theta) = \infty\}$.

Remark 9.4 We note *en-passant* that $\mathbb{E}_\infty$ contains some Diophantine numbers: for example, the number $\theta = [a_1, a_2, \ldots, a_n, \ldots]$ with $a_n = 2^n$ is Diophantine, and it clearly belongs to $\mathbb{E}_\infty$. The set $\mathbb{E}_\infty$ also contains many Liouville numbers: for instance, any $\theta = [a_1, a_2, \ldots, a_n, \ldots]$ with a_n even and $a_n > e^{n^n}$ for all $n \in \mathbb{N}$ belongs to $\mathbb{E}_\infty$. Finally, note that the transcendental number $\lambda = (e-1)/(e+1)$ also belongs to $\mathbb{E}_\infty$; indeed, its continued fraction expansion has $a_n = 4n - 2$ for all $n \geq 1$, i.e., $\lambda = [2, 6, 10, 14, \ldots]$—this is a special case of an old identity due to Euler and Lambert.[3]

Proposition 9.5 *Let $f : \boldsymbol{S}^1 \to \boldsymbol{S}^1$ be a minimal circle homeomorphism with $\rho(f) = \rho$. Given $x, y \in \boldsymbol{S}^1$ distinct, let $\{(\rho_n, \alpha_n)\}_{n\geq 0}$ be the renormalization trail of y with respect to x and f. If $\rho \in \mathbb{E}$ and $\alpha_0 = \frac{1}{2}$, then for all $n \geq 1$ we have $\rho_n < \frac{1}{2}$, and*

$$\alpha_n = \begin{cases} \dfrac{\rho_n}{2} & \text{if } n \text{ is odd,} \\[2ex] \dfrac{1}{2} + \rho_n & \text{if } n \text{ is even.} \end{cases} \tag{9.2}$$

[2] Indeed, let $\{A_n\}$ be a sequence of open and dense sets in R such that $\cap A_n = \mathscr{G}_0$. For each $\rho \in \boldsymbol{R}$ and each n we have that $(\{\rho\} \times [-1, 1]) \cap A_n$ is open and has full Lebesgue measure in $\{\rho\} \times [-1, 1]$, and in particular it is also dense in $\{\rho\} \times [-1, 1]$.

[3] Which states that $\tanh(x^{-1}) = [x, 3x, 5x, 7x, \ldots]$ for all $x \in \mathbb{N}$; see Lang (1995, p. 71).

In particular, if $\rho \in \mathbb{E}_\infty$, then there exists a subsequence $n_i \to \infty$ such that $\alpha_{n_i} \to \frac{1}{2}$.

Proof First note that, if $a_0, a_1, a_2, \ldots$ are the partial quotients of the continued fraction expansion of ρ_0, then by hypothesis $a_n \geq 2$ for all n, and this already implies that $\rho_n < \frac{1}{a_n} \leq \frac{1}{2}$ for all $n \geq 1$. This takes care of the first assertion in the statement. In order to prove the second assertion, we will use Lemma 9.3 and induction on n.

(1) Base of Induction We have $\alpha_0 = \frac{1}{2}$, and since $\alpha_0 > \rho_0 G(\rho_0) = \rho_0\rho_1$, Lemma 9.3 tells us that

$$\alpha_1 = T_{\rho_0}(\alpha_0) = \left\{\frac{1-\alpha_0}{\rho_0}\right\} = \left\{\frac{1}{2\rho_0}\right\} .$$

But $\rho_0^{-1} = a_0 + \rho_1$, where $a_0 \geq 2$ is *even*. Therefore

$$\alpha_1 = \left\{\frac{1}{2}(a_0 + \rho_1)\right\} = \frac{\rho_1}{2} .$$

This verifies (9.2) for $n = 1$. Let us now look at α_2. We have $\alpha_1 > \rho_1 G(\rho_1) = \rho_1\rho_2$. Hence, using Lemma 9.3 and the fact that $\rho_1^{-1} = a_1 + \rho_2$, we see that

$$\begin{aligned}
\alpha_2 = T_{\rho_1}(\alpha_1) &= \left\{\frac{1-\alpha_1}{\rho_1}\right\} \\
&= \left\{\frac{1}{\rho_1} - \frac{1}{2}\right\} \\
&= \left\{a_1 + \rho_2 - \frac{1}{2}\right\} \\
&= \left\{\rho_2 - \frac{1}{2}\right\} \\
&= \frac{1}{2} + \rho_2 .
\end{aligned}$$

This verifies (9.2) for $n = 2$. Summarizing, we have established the base of the induction.

(2) Induction Step Suppose (9.2) holds for n. In order to show that this assertion holds for $n + 1$, there are two cases to consider, according to whether n is odd or even.

(i) If n is odd, then we are assuming that $\alpha_n = \frac{1}{2}\rho_n$. In particular, we have $\alpha_n > \rho_n\rho_{n+1} = \rho_n G(\rho_n)$, so Lemma 9.3 tells us that

$$\alpha_{n+1} = T_{\rho_n}(\alpha_n) = \left\{\frac{1-\alpha_n}{\rho_n}\right\} = \left\{\frac{1}{\rho_n} - \frac{1}{2}\right\}$$

Using here that $\rho_n^{-1} = a_n + \rho_{n+1}$, we get

$$\alpha_{n+1} = \left\{a_n + \rho_{n+1} - \frac{1}{2}\right\} = \frac{1}{2} + \rho_{n+1} .$$

This establishes the induction step when n is odd.

(ii) If n is even, then we are assuming that $\alpha_n = \frac{1}{2}+\rho_n$, by the induction hypothesis. Hence we have $\alpha_n > \frac{1}{2} > \rho_n\rho_{n+1} = \rho_n G(\rho_n)$, and therefore from Lemma 9.3 we deduce that

$$\begin{aligned}\alpha_{n+1} = T_{\rho_n}(\alpha_n) &= \left\{\frac{1-\alpha_n}{\rho_n}\right\}\\ &= \left\{\frac{1}{2\rho_n} - 1\right\}\\ &= \left\{\frac{1}{2\rho_n}\right\} .\end{aligned} \tag{9.3}$$

Again, using that $\rho_n^{-1} = a_n + \rho_{n+1}$, we see that

$$\alpha_{n+1} = \left\{\frac{1}{2}a_{n+1} + \frac{1}{2}\rho_{n+1}\right\} = \frac{\rho_{n+1}}{2} ,$$

where in the last equality we have at last used the fact that a_n is an even integer! This establishes the induction step when n is even, and completes the proof of the second assertion.

Finally, the last assertion in the statement is easily proved: if $\rho \in \mathbb{E}_\infty$, then $\rho_n \to 0$ as $n \to \infty$. Hence by (9.2) we see that $\alpha_{2i} \to \frac{1}{2}$ as $i \to \infty$. This concludes the proof.

□

Remark 9.5 The above proof still works if only the odd partial quotients a_{2k+1} are required to be even (but still requiring $a_n \neq 1$ for all n). The resulting class of numbers with this property is a bit larger than $\mathbb{E}$, but still has zero Lebesgue measure.

9.6 Proofs of Theorems 9.1 and 9.2

In this section we prove our first two main results, namely Theorem 9.1 and Theorem 9.2. We first recall the setup for both theorems, and fix some notation.

Let $f, g : S^1 \to S^1$ be two C^3 (multi)critical circle maps with the same irrational rotation number $\rho = [a_0, a_1, \ldots, a_n, \ldots]$. Let $h : S^1 \to S^1$ be a topological conjugacy between f and g mapping orbits of f to orbits of g (i.e.,, such that $h \circ f = g \circ h$). Let $x, z \in S^1$ be such that $h(x) = z$. Suppose also that $w \in S^1$, $w \neq z$, is a critical point for g. Assume one of the following two scenarios (which correspond to the situations in Theorems 9.1 and 9.2, respectively).

Scenario A. Both f and g are *uni-critical* circle maps, with critical points at x and w, respectively.
Scenario B. The map f is *uni-critical* with critical point at x, whereas the map g is *bi-critical* with critical points at z and w.

In either scenario, let $y = h^{-1}(w)$ and let y_n, $n \geq 0$, be the renormalization ancestors of y (with respect to x and f). Likewise, let $w_n = h(y_n)$, $n \geq 0$, denote the renormalization ancestors of $w = h(y)$ (with respect to z and g). Finally, let (ρ_n, α_n), $n \geq 0$, be the renormalization trail of y (with respect to x and f) – which is also the renormalization trail of w (with respect to z and g).

Both Theorems 9.1 and 9.2 will be straightforward consequences of the following result (Fig. 9.3).

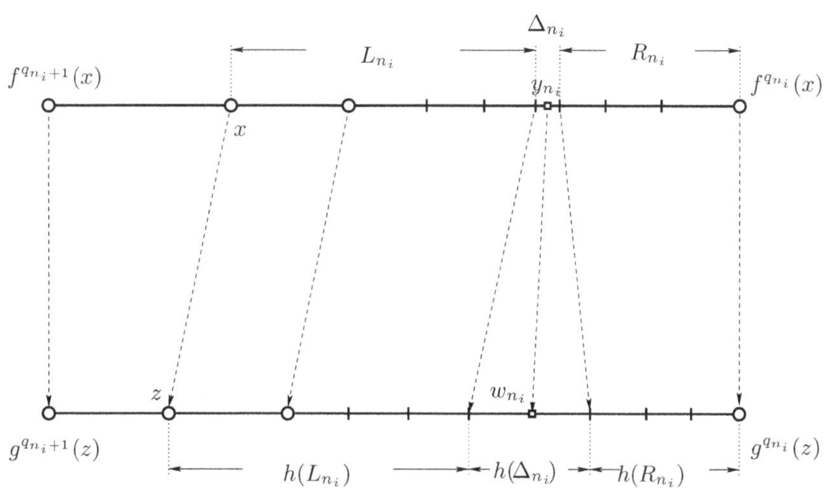

Fig. 9.3 The distortion of cross-ratios is large

Lemma 9.4 *In either of the two scenarios above, suppose that there exists a subsequence $n_i \to \infty$ such that $\rho_{n_i+1} \to 0$ as $i \to \infty$, and $\left|\alpha_{n_i+1} - \frac{1}{2}\right| < \frac{1}{4}$ for all i. Then the conjugacy h is* not *quasisymmetric.*

The proof of this lemma, in turn, depends on the fact that, inside every interval of the form $I_n(x)$, *critical spots are large*. We have seen this already (in Chap. 7), but for convenience we repeat the statement here.

Lemma 9.5 *Let $0 \le k < a_{n+1}$ be such that the interval*

$$\Delta_{k,n} = f^{q_n + kq_{n+1}}(I_{n+1}(x)) \subset I_n(x)$$

contains a critical point of $f^{q_{n+1}}$. Then $|\Delta_{k,n}| \asymp |I_n(x)|$.

Proof The proof is outlined in Sect. 7.4.3, Exercise 7.9; the reader is invited to fill in the details. □

Proof of Lemma 9.4 The idea is to show that h has unbounded distortion of cross-ratios: once this is proved, then Corollary 7.1 implies that the homeomorphism h is not quasisymmetric. Passing to a subsequence if necessary, we may assume that either (a) $y_{n_i} \in I_{n_i}$ for all i; or (b) $y_{n_i} \in I_{n_i+1}$ for all i. We give the proof assuming that case (a) holds. The proof in case (b) is the same, *mutatis mutandis*.

By restricting our attention to sufficiently large i, we may assume that $\rho_{n_i+1} < \frac{1}{9}$, which implies that $a_{n_i+1} > 8$. Then we must have $y_{n_i} \in I_{n_i} \setminus I_{n_i+2}$. Indeed, if $y_{n_i} \in I_{n_i+2}$, then $\alpha_{n_i+1} \le \mu(I_{n_i+2})/\mu(I_{n_i}) < \frac{1}{9}$, which contradicts the hypothesis. Since the intervals

$$\Delta^{(j)} = f^{q_{n_i} + jq_{n_i+1}}(I_{n_i+1}), \quad 0 \le j \le a_{n_i+1} - 1,$$

constitute a partition of $I_{n_i} \setminus I_{n_i+2}$ (modulo endpoints), it follows that there exists $0 \le k_{n_i} \le a_{n_i+1} - 1$ such that $y_{n_i} \in \Delta_{n_i} = \Delta^{(k_{n_i})}$.

Claim We have $k_{n_i} \asymp a_{n_i+1} \asymp a_{n_i+1} - k_{n_i}$.

In order to prove this claim, we first recall that

$$1 - \alpha_{n_i+1} = \frac{\mu([y_{n_i}, f^{q_{n_i}}(x)])}{\mu(I_{n_i})}, \tag{9.4}$$

where as before μ is the unique Borel probability measure invariant under f. Moreover, we have

$$\bigcup_{j=0}^{k_{n_i}-1} \Delta^{(j)} \subseteq [y_{n_i}, f^{q_{n_i}}(x)] \subseteq \Delta_{n_i} \cup \bigcup_{j=0}^{k_{n_i}-1} \Delta^{(j)}. \tag{9.5}$$

Since $\mu(\Delta^{(j)}) = \mu(I_{n_i+1})$ for all j, from (9.4) and (9.5) we get

$$k_{n_i}\,\frac{\mu(I_{n_i+1})}{\mu(I_{n_i})} \;\le\; 1-\alpha_{n_i+1} \;\le\; (k_{n_i}+1)\,\frac{\mu(I_{n_i+1})}{\mu(I_{n_i})}\,. \tag{9.6}$$

Taking into account that

$$\rho_{n_i+1} \;=\; \frac{\mu(I_{n_i+1})}{\mu(I_{n_i})}$$

and that, by hypothesis, $\frac{1}{4} < 1-\alpha_{n_i+1} < \frac{3}{4}$, we deduce from (9.6) that

$$\frac{1}{4\rho_{n_i+1}} - 1 \;<\; k_{n_i} \;<\; \frac{3}{4\rho_{n_i+1}}\,.$$

But $\rho_{n_i+1}^{-1} = a_{n_i+1} + \rho_{n_i+2}$, and $0 < \rho_{n_i+2} < 1$, so

$$\frac{1}{4} - \frac{1}{a_{n_i+1}} \;<\; \frac{k_{n_i}}{a_{n_i+1}} \;<\; \frac{3}{4}\left(1+\frac{1}{a_{n_i+1}}\right),$$

and since $\rho_{n_i+1} < \frac{1}{9}$ implies $a_{n_i+1} > 8$, we deduce that

$$\frac{1}{8} \;<\; \frac{k_{n_i}}{a_{n_i+1}} \;<\; \frac{27}{32}\,.$$

This proves the claim.

Now, provided n_i is sufficiently large, the map $f^{q_{n_i+1}}$ restricted to the interval $I_{n_i} \setminus I_{n_i+2}$ is an almost parabolic map (see Definition 7.3 in Sect. 7.3). Here we need n_i large enough so that, restricted to the interval in question, the map $f^{q_{n_i+1}}$ is a diffeomorphism with negative Schwarzian derivative, and this is possible by Proposition 8.3. By Yoccoz's inequality (Lemma 7.3) and the above claim, we have

$$\frac{|\Delta_{k_{n_i}}|}{|I_{n_i}|} \;\asymp\; \frac{1}{\min\{k_{n_i}^2\,,\,(a_{n_i+1}-k_{n_i})^2\}} \;\asymp\; \frac{1}{a_{n_i+1}^2}\,.$$

Letting L_{n_i} and R_{n_i} denote the left and right components of $I_{n_i}\setminus\Delta_{n_i}$, we know from the real bounds (Theorem 6.3) that $|L_{n_i}| \asymp |I_{n_i}| \asymp |R_{n_i}|$. Therefore we see that

$$[\Delta_{n_i}, I_{n_i}] \;=\; \frac{|\Delta_{n_i}||I_{n_i}|}{|L_{n_i}||R_{n_i}|} \;\asymp\; \frac{1}{a_{n_i+1}^2}\,. \tag{9.7}$$

The next step is to estimate the cross-ratio determined by the pair of intervals $h(\Delta_{n_i})$ and $h(I_{n_i})$. Here, we first note that $w_{n_i} = h(y_{n_i}) \in h(\Delta_{n_i})$ is a critical point for the map $g^{q_{n_i+1}}$; in the terminology of Estevez and de Faria (2018), $h(\Delta_{n_i})$ is therefore a *critical spot* of $g^{q_{n_i+1}}|_{h(I_{n_i})}$. As we saw in Lemma 9.5, every critical spot of a renormalization return map is comparable to the interval domain of said return map.

Hence we have $|h(\Delta_{n_i})| \asymp |h(I_{n_i})|$. Moreover, by the real bounds for g, we have $|h(L_{n_i})| \asymp |h(I_{n_i})| \asymp |h(R_{n_i})|$. These facts show that

$$[h(\Delta_{n_i}), h(I_{n_i})] = \frac{|h(\Delta_{n_i})||h(I_{n_i})|}{|h(L_{n_i})||h(R_{n_i})|} \asymp 1 . \tag{9.8}$$

Combining (9.7) with (9.8), we finally get an estimate on the cross-ratio distortion of the pair of intervals $\Delta_{n_i} \subset I_{n_i}$ under h, to wit

$$\mathrm{CrD}(h; \Delta_{n_i}, I_{n_i}) = \frac{[h(\Delta_{n_i}), h(I_{n_i})]}{[\Delta_{n_i}, I_{n_i}]} \asymp a_{n_i+1}^2 .$$

But since $\rho_{n_i+1} \to 0$, we have $a_{n_i+1} \to \infty$. This shows that the cross-ratio distortion of h blows up, and so h cannot be quasisymmetric (again, recall Corollary 7.1). The proof of Lemma 9.4 is complete. □

Proof of Theorem 9.1 Consider the sets $\boldsymbol{R}$ and $\mathscr{B}_{c_f}$ given by Theorem 9.6 (applied to f and $x = c_f$), and define $\boldsymbol{R}_A = \boldsymbol{R}$. Then Lemma 9.4 (applied in the *Scenario* A case) implies that $\mathscr{B}_{c_f} \subset \mathscr{B}$, which proves Theorem 9.1. Remember also that, as explained in Sect. 9.1.2, the fact that the complement of $\mathscr{B}$ is dense follows from the fact that it is non-empty and invariant under the minimal homeomorphism f. □

Proof of Theorem 9.2 By Lemma 9.4 (applied in the *Scenario B* case), it is enough to consider

$$\mathscr{G} = \mathscr{G}_0 \cup \left(\mathbb{E}_\infty \times \left\{\frac{1}{2}\right\}\right) \subset R ,$$

where $\mathscr{G}_0$ is given by Proposition 9.4, and $\mathbb{E}_\infty$ is given by Proposition 9.5. □

9.7 The C^∞ Realization Lemma

9.7.1 Admissible Pairs

We start Sect. 9.7 with a definition. Remember that R denotes the rectangle $[0, 1] \times [-1, 1]$ in $\mathbb{R}^2$, and $M = \big([0, 1] \setminus \mathbb{Q}\big) \times [-1, 1] \subset R$.

Definition 9.5 A pair $(\rho, \alpha) \in M$ is said to be *admissible* if there exists a C^∞ multicritical circle map g with irrational rotation number ρ, a unique invariant measure μ and with exactly two critical points c_1 and c_2 such that the two connected components of $\boldsymbol{S}^1 \setminus \{c_1, c_2\}$ have μ-measures equal to α and $1 - \alpha$ respectively.

The set of admissible pairs is denoted by $\mathbb{A}$. Let us examine some of its properties.

Lemma 9.6 *Any pair* $(\rho, \alpha) \in (0, 1)^2$ *such that* $\rho \notin \mathbb{Q}$ *and* $\rho - 2\alpha = 0$ *belongs to* $\mathbb{A}$.

Proof Let f_0 be a C^∞ critical circle map with a single critical point $c(f_0)$ and such that $\rho(f_0) = \alpha$ (note that f_0 can be chosen to be real-analytic, say from the Arnold's family). Let us denote by μ the unique invariant Borel probability measure of f_0. Define $g = f_0^2 = f_0 \circ f_0$, and note that g is a real-analytic bi-critical circle map, with irrational rotation number $\rho(g) = 2\rho(f_0) = 2\alpha = \rho$ and with two critical points $c_1(g) = c(f_0)$ and $c_2(g) = f_0^{-1}\big(c(f_0)\big)$. Moreover, the unique invariant Borel probability measure of g is μ, and the two connected components of $\mathbf{S}^1 \setminus \{c_1, c_2\}$ have μ-measures equal to α and $1 - \alpha$ respectively, since $c_1 = f_0(c_2)$. □

Lemma 9.7 *The set* $\mathbb{A}$ *of admissible pairs is forward invariant under* T, *where* $T : M \to M$ *is the skew product constructed in Sect. 9.3.*

Proof Let $(\rho, \alpha) \in \mathbb{A}$ and let f be a C^∞ bi-critical circle map, with critical points c_1 and c_2, such that (ρ, α) is the initial term of the renormalization trail of c_2 with respect to c_1 and f. For some fixed $n \in \mathbb{N}$, we want to prove that $T^{n+1}(\rho, \alpha) \in \mathbb{A}$. By Lemma 9.3, $T^{n+1}(\rho, \alpha)$ coincides with the $(n + 1)$-th term $(\rho_{n+1}, \alpha_{n+1})$ of the renormalization trail of c_2 (with respect to c_1 and f). Recall, from Sect. 9.2, that $\rho_{n+1} = G^{n+1}(\rho)$ and that if c_2 belongs to the long interval $f^i\big(I_n(c_1)\big)$ for some $i \in \{0, 1, \ldots, q_{n+1} - 1\}$, we have that

$$\alpha_{n+1} = \frac{\mu\big((c_1, y_n)\big)}{\mu(I_n)},$$

where $y_n \in I_n(c_1)$ is given by $f^i(y_n) = c_2$. Otherwise, c_2 belongs to the short interval $f^j\big(I_{n+1}(c_1)\big)$ for some $j \in \{0, 1, \ldots, q_n - 1\}$, and then

$$\alpha_{n+1} = -\frac{\mu\big((y_n, c_1)\big)}{\mu(I_{n+1})},$$

where $y_n \in I_{n+1}(c_1)$ is given by $f^j(y_n) = c_2$. Let us assume that we are in the first case (the proof for the second one being the same), and note that the iterate f^{q_n} restricts to a C^∞ homeomorphism (with a critical point at c_1) between the intervals

$$I_{n+1}(c_1) \cup f^{-q_{n+1}}\big(I_{n+1}(c_1)\big) = \big[f^{q_{n+1}}(c_1), f^{-q_{n+1}}(c_1)\big] \quad \text{and}$$

$$\Delta_{0,n} \cup f^{-q_{n+1}}\big(\Delta_{0,n}\big) = \big[f^{q_{n+1}+q_n}(c_1), f^{-q_{n+1}+q_n}(c_1)\big],$$

where $\Delta_{0,n} = f^{q_n}\big(I_{n+1}(c_1)\big) = \big(f^{q_{n+1}+q_n}(c_1), f^{q_n}(c_1)\big]$, as defined in the course of the proof of Lemma 9.3. Identifying points in this way we obtain from the interval

$$I_{n+1}(c_1) \cup I_n(c_1) \cup f^{-q_{n+1}}\big(\Delta_{0,n}\big) = \big[f^{q_{n+1}}(c_1), f^{-q_{n+1}+q_n}(c_1)\big],$$

a compact boundaryless one-dimensional topological manifold N. Denote by $\pi : I_{n+1}(c_1) \cup I_n(c_1) \cup f^{-q_{n+1}}(\Delta_{0,n}) \to N$ the quotient map, and let $\phi : N \to S^1$ be any homeomorphism which is a C^∞ diffeomorphism between $N \setminus \{\pi(c_1)\}$ and $S^1 \setminus \{\phi(\pi(c_1))\}$. Note that $\phi \circ \pi$ maps the interior of $I_n(c_1)$ C^∞-diffeomorphically onto $S^1 \setminus \{\phi(\pi(c_1))\}$. Let $g : S^1 \to S^1$ be given by the identity

$$g \circ \phi \circ \pi = \phi \circ \pi \circ f^{q_{n+1}} \text{ in } I_n(c_1),$$

and note that g is a well-defined C^∞ circle homeomorphism, with irrational rotation number equal to $\rho_{n+1} = G^{n+1}(\rho)$. Moreover, g has exactly two critical points in S^1, given by $\hat{c}_1 = \phi \circ \pi(c_1)$ and $\hat{c}_2 = \phi \circ \pi(y_n)$. Finally, note that the unique invariant Borel probability measure μ_g of g in S^1 is given by

$$\mu_g(\phi \circ \pi(A)) = \mu(A) \,/\, \mu(I_n(c_1)) = \mu(A) \,/\, \prod_{j=0}^{n} G^j(\rho),$$

for any Borel set $A \subset I_n(c_1)$. In particular, the two connected components of $S^1 \setminus \{\hat{c}_1, \hat{c}_2\}$ have μ_g - measures equal to α_{n+1} and $1 - \alpha_{n+1}$ respectively. This finishes the proof of Lemma 9.7. □

We remark that the *glueing procedure* described in the proof of Lemma 9.7 was introduced by Lanford in the eighties, see Sect. 10.2 and Lanford (1987, 1988) for much more.

Lemma 9.8 *The set $\mathbb{A}$ of admissible pairs has non-empty interior in M.*

We will not prove this result here. For a proof, see the original paper de Faria and Guarino (2022b, Prop. 7.5). We are now in a position to give a quick proof of the C^∞ Realization Lemma, which we restate as follows.

Theorem 9.7 (The C^∞ Realization Lemma) *Every pair in M is admissible; in other words, $\mathbb{A} = M$.*

Proof Since the set $\mathbb{A}$ of admissible pairs is obviously non-empty (see for instance Lemma 9.6 above), Theorem 9.7 follows by combining Lemmas 9.2 and 9.7 with Lemma 9.8. □

Finally, when combined with Theorem 9.2, the C^∞ Realization Lemma implies Theorem 9.3. This is left as an exercise to the reader.

Exercises

Exercise 9.1 Show that if $f : S^1 \to S^1$ is quasisymmetrically conjugate to a rigid rotation R_α, then *every* conjugacy between f and R_α is quasisymmetric.

Exercise 9.2 Let $f : S^1 \to S^1$ be a multicritical circle map, and let $x \in S^1$ be such that f has bounded geometry at x.

(i) Show that there exists a fine grid $\mathscr{G}_x = \{\mathscr{Q}_n(x)\}_{n \geq 0}$ with the property that, for every $n \geq 0$, each atom $\Delta \in \mathscr{Q}_n(x, f)$ is a union of atoms of $\mathscr{P}_m(x)$ for some $m \leq n + 1$. [*Hint*: imitate the recursive construction used in the proof of Proposition 7.6].

(ii) Let $y \in S^1$ be another point such that f has bounded geometry at y, and let $h_{x,y} \in Z_0(f)$ be the self-conjugacy such that $h_{x,y}(x) = y$. Using (i), prove that $h_{x,y}$ is quasisymmetric.

Exercise 9.3 Prove that the set $\mathbb{E}_\infty$ defined in Sect. 9.1.3 is uncountable. Prove also that $\mathbb{E}_\infty$ is a set of first category of Baire (i.e., it is meager) and that its Lebesgue measure is equal to zero. [*Hint*: see Appendix A.]

Exercise 9.4 Let f be a smooth bi-critical circle map with irrational rotation number ρ_f, unique invariant measure μ_f and critical points c_1 and c_2. Say that the two connected components of $S^1 \setminus \{c_1, c_2\}$ have μ_f-measures equal to α_f and $1 - \alpha_f$ respectively. As we know from Corollary 2.1, the rotation number ρ_f is continuous under C^0 perturbations. Show that α_f is continuous under smooth perturbations. More precisely, prove the following statement, borrowed from de Faria and Guarino (2022b, Lem. 7.6): given $\varepsilon > 0$ there exists $\delta = \delta(\varepsilon, f) > 0$ such that if g is a smooth bi-critical circle map with irrational rotation number ρ_g satisfying $d_{C^1}(f, g) < \delta$, then $|\rho_f - \rho_g| < \varepsilon$ and $|\alpha_f - \alpha_g| < \varepsilon$ [*Hint*: recall Theorem 3.3, the Denjoy–Koksma inequality].

Exercise 9.5 Deduce Theorem 9.3 from Theorem 9.2 and the C^∞ Realization Lemma.

Part IV
Renormalization Theory

Chapter 10
Smooth Rigidity and Renormalization

In recent years, the main new tool introduced in dynamics to understand the fine-scale structure of a low-dimensional system is *renormalization*. The notion of renormalization stems from statistical mechanics and field theory, and was introduced in the context of one-dimensional dynamics—more precisely, in the study of bifurcations of one-parameter families of unimodal maps—more than four decades ago, through the numerical observations and conjectures formulated by Coullet and Tresser (1978) and independently by Feigenbaum (1978).

In a nutshell, to renormalize a (smooth) dynamical system around some point of interest (usually a critical point) means to consider a small, dynamically defined neighborhood of that point in phase space, and to take the *first return map* to that neighborhood, linearly rescaling it to unit size. If this can be done for a sequence of smaller and smaller neighborhoods of the special point, then we say that the system is *infinitely renormalizable* at that point. Now, if two systems are topologically equivalent, and are infinitely renormalizable, it makes sense to compare their successive renormalizations around corresponding special points. If these corresponding successive renormalizations get closer and closer together (say in the C^0 sense), this points to both systems having the same asymptotic geometric structure near their special points—and we expect this to happen at all points in the forward orbits of the special points. In other words, the general *ansatz* is that the convergence of successive renormalizations implies a form of *geometric rigidity*. This is the rosy picture, but the reality is rather thorny, as we shall see.

Our goal in the present chapter is to explain the interplay between rigidity and renormalization convergence in the specific context of (multi)critical circle maps. We shall see that, when correctly interpreted, the above *ansatz* is true most of the time (Sect. 10.3), but not always (Sect. 10.5).

E. de Faria, P. Guarino, *Dynamics of Circle Mappings*, IMPA Monographs,
https://doi.org/10.1007/978-3-031-67495-2_10

10.1 Smooth Rigidity

The notion of *smooth rigidity* first appeared in hyperbolic geometry in the sixties, through the seminal work of Mostow, who showed that the fundamental group (the *topology*) of a complete, finite-volume hyperbolic manifold of dimension greater than two completely determines its *geometry*. In dynamical systems, smooth rigidity means that a finite number of dynamical invariants determines the fine scale structure of orbits. More precisely, maps that are topologically conjugate and share these invariants are in fact smoothly conjugate. Numerical observations by Feigenbaum et al. (1982), Ostlund et al. (1983), and Shenker (1982) suggested in the early eighties that this was the case for C^3 critical circle maps with a single critical point and with irrational rotation number of bounded type. This was posed as a conjecture in several works by Lanford (1987, 1988), Ostlund et al. (1983), Rand (1987, 1988, 1992), Feigenbaum et al. (1982), and Shenker (1982) among others. We proceed to state the most recent results in this area, namely Theorems 10.1 and 10.2 below.

Theorem 10.1 *Let f and g be two C^4 circle homeomorphisms with the same irrational rotation number and with a unique critical point of the same odd integer criticality. Let h be the unique topological conjugacy between f and g that maps the critical point of f to the critical point of g. Then:*

1. *h is a C^1 diffeomorphism.*
2. *h is $C^{1+\alpha}$ at the critical point of f for a universal $\alpha > 0$.*
3. *There exists a full Lebesgue measure set of rotation numbers (containing those of bounded type) for which the conjugacy h is a global $C^{1+\alpha}$ diffeomorphism.*

Recall that an irrational number $\rho = [a_0, a_1, \dots]$ is of *bounded type* if $\sup\{a_n\}$ is finite (see Chap. 4 and Appendix A). By Theorem 6.2, the rotation number is the unique invariant of the C^0 conjugacy classes of critical circle maps with no periodic orbits. Theorem 10.1 is saying that, inside each topological class, the order of the critical point is the unique invariant of the C^1 conjugacy classes! This is what we call *rigidity*.

A delicate problem is to precisely determine "how smooth" the conjugacy h is. By comparing with the material presented in Chap. 4 we see that, on the one hand, the presence of the critical point gives us more rigidity than in the case of diffeomorphisms: a smooth conjugacy is obtained for all irrational rotation numbers, with no need of a Diophantine condition. On the other hand, in Sect. 10.5 we will construct examples where the conjugacy h is not globally $C^{1+\alpha}$. It might be possible, but probably quite difficult, to obtain a sharp arithmetical condition on the rotation number that would allow us to decide whether the conjugacy is "better than C^1".

Theorem 10.2 *Any two C^3 critical circle maps with a single critical point, with the same irrational rotation number of bounded type and with the same odd integer criticality are conjugate to each other by a $C^{1+\alpha}$ circle diffeomorphism, for some $\alpha > 0$.*

This fourth and last part of the present book is entirely devoted to explaining the proof of these two fundamental results; deep tools coming from Renormalization Theory and Holomorphic Dynamics will be introduced along the way. As it will be clear to the reader (specially in Chap. 14), the ideas of Dennis Sullivan, Curtis McMullen and Mikhail Lyubich (conceived in the context of unimodal maps of the interval) were crucial to develop the rigidity theory of critical circle maps (for an overview of Sullivan's major contributions to the area of Dynamical Systems, we refer the reader to the recent expository paper by de Faria and van Strien 2023).

We remark that the statement of Theorem 10.2 is the precise statement of the *rigidity conjecture* mentioned above. Together, Theorems 10.1 and 10.2 can be regarded as the *state of the art* concerning rigidity of critical circle maps with a single critical point of integer criticality (see also the recent paper by Gorbovickis and Yampolsky (2020), where rigidity is obtained for real-analytic unicritical circle maps of bounded combinatorics, with non-integer criticalities which are close enough to an odd integer). As it will be explained in the next chapters, both results were first proved for *real-analytic* unicritical circle maps, mainly as a combination of works by the first named author together with de Melo, and by Yampolsky (see de Faria 1999; de Faria and de Melo 1999, 2000; Yampolsky 1999, 2001, 2002, 2003; Khmelev and Yampolsky 2006; Khanin and Teplinsky 2007). In the current form (i.e., for C^r maps), Theorem 10.1 was proved by Guarino et al. (2018), while Theorem 10.2 was proved by Guarino and de Melo (2017). In both cases, the main task is to reduce the rigidity problem for C^r unicritical circle maps to the real-analytic case. This reduction will be discussed in detail in Chap. 13.

What about dynamics with more critical points? Let f be a C^3 multicritical circle map with irrational rotation number $\rho \in (0, 1)$, unique invariant Borel probability measure μ and $N \geq 1$ critical points c_i, for $0 \leq i \leq N-1$. As before, all critical points are assumed to be *non-flat*: in C^3 local coordinates around c_i, the map f can be written as $t \mapsto t\,|t|^{d_i-1}$ for some $d_i > 1$ (Definition 5.1). Moreover, just as in Chap. 6 (recall Definition 6.2), we define the *signature* of f to be the $(2N+2)$-tuple

$$(\rho\,;\,N;\; d_0, d_1, \ldots, d_{N-1};\; \delta_0, \delta_1, \ldots, \delta_{N-1}),$$

where d_i is the criticality of the critical point c_i, and $\delta_i = \mu[c_i, c_{i+1})$ (with the convention that $c_N = c_0$).

Now consider two multicritical circle maps, say f and g, with the same irrational rotation number. By Theorem 6.2, they are topologically conjugate to each other. By elementary reasons, if f and g have the same signature there exists a circle homeomorphism h, which is a topological conjugacy between f and g, identifying each critical point of f with a corresponding critical point of g having the same criticality. As proved in Chap. 7, such conjugacy h is a quasisymmetric homeomorphism (Theorem 7.2).

Question 10.1 Is this conjugacy a smooth diffeomorphism?

Of course, only such an h conjugating f and g has the chance of being smooth (in fact, as explained in Chap. 9, for Lebesgue almost every rotation number most conjugacies between f and g fail to be even quasisymmetric). The following result follows by combining the recent papers Estevez and Guarino (2023), Estevez et al. (2022), and Yampolsky (2019).

Theorem 10.3 *Let f and g be real-analytic bi-critical circle maps with the same irrational rotation number, both critical points of cubic type and with the same signature. If their common rotation number is of bounded type, then the topological conjugacy h is a $C^{1+\alpha}$ diffeomorphism.*

To the best of our knowledge, Theorem 10.3 is the first rigidity statement available for maps with more than one critical point. In other words, Question 10.1 remains wide open (but see the recent preprint Gorbovickis and Yampolsky 2021).

As stated in the introduction to this chapter, the main tool to study rigidity problems in low dimensional dynamics is renormalization theory. Renormalization of a dynamical system with a marked point (usually a critical point) means a (suitably rescaled) first return map to a neighborhood of such point. Thus, renormalization can be thought as a *supra* dynamical system, acting on an infinite dimensional phase space made up by the original dynamics (see Sect. 10.2 for the precise definition of renormalization of multicritical circle maps). In the context of one dimensional dynamics, the renormalization program was initiated by Dennis Sullivan in the eighties (Sullivan 1986, 1992), and then carried out by mathematicians such as Yoccoz, Douady, Hubbard, Shishikura, McMullen, Lyubich, Martens, de Melo, Yampolsky, van Strien and Avila among others.

A fundamental principle in this theory states that exponential convergence of renormalization orbits implies rigidity: topological conjugacies are actually smooth (when restricted to the attractors of the original systems). We refer the reader to de Melo and van Strien (1993, Section VI.9) for the seminal case of unimodal maps with bounded combinatorics (specifically, see Theorem 9.4 on page 552).

Let us be more precise: by Yoccoz's Theorem 6.2, two multicritical circle maps f and g with the same irrational rotation number are topologically conjugate to the corresponding rigid rotation, and in particular to each other. To obtain a smooth conjugacy between f and g, we need to assume the existence of a topological conjugacy h that identifies their critical sets, while preserving corresponding criticalities. In other words, f and g need to have the same signature (recall Definition 6.2). It turns out that for Lebesgue almost every rotation number, such conjugacy h is a $C^{1+\alpha}$ diffeomorphism, provided the successive renormalizations of f and g (around critical points identified under h) converge together exponentially fast in the C^1 topology (see the recent paper by Estevez and Guarino 2023). For unicritical circle maps, it is sufficient to have exponential convergence of renormalizations in the C^0 topology, and this is the main theorem that we will prove in this chapter (see Theorem 10.4 in Sect. 10.3 below). Our proof will follow very closely the original source, de Faria and de Melo (1999).

Thus, the main step to obtain rigidity, as in Sect. 10.1 above, is to establish geometric contraction of the successive renormalizations of multicritical circle maps with the same signature. The dynamics of renormalization, however, is usually difficult to understand. To begin with, its phase space is neither bounded nor locally compact. Therefore, no recurrence is given a priori. This makes some basic dynamical questions, such as existence of attractors and periodic orbits, quite difficult to solve.[1] In particular, proving exponential contraction is a challenging problem. In the case of a single critical point and real-analytic dynamics, exponential contraction was obtained in de Faria and de Melo (2000) for rotation numbers of bounded type, and extended in Khmelev and Yampolsky (2006) to cover all irrational rotation numbers (Theorem 13.1). Both papers lean heavily on complex dynamics techniques (to be discussed in Chaps. 11 and 14), and therefore an additional hypothesis is required: the criticality at both critical points has to be an odd integer. These results have been recently extended in at least two directions: in Gorbovickis and Yampolsky (2020), exponential contraction is obtained for real-analytic critical circle maps of bounded combinatorics, with non-integer criticalities which are close enough to an odd integer, while in Guarino et al. (2018) and Guarino and de Melo (2017), exponential contraction is established for critical circle maps with a *finite degree of smoothness*, (but still with odd integer criticalities, see Theorem 13.2 and 13.3). Finally, in the case of two critical points, it is proved in Yampolsky (2019) both the existence of periodic orbits and the hyperbolicity (under renormalization) of those periodic orbits, for real-analytic bi-critical circle maps (with both critical points of cubic type). These results have been recently extended to bounded combinatorics in Estevez et al. (2022). See Chap. 13 for more details.

10.2 Renormalization of Commuting Pairs

As mentioned before, to renormalize a dynamical system means to consider a first return map around some interesting point, and then to rescale this return map. In the context of circle maps, the first return map to a small neighborhood of a point is always discontinuous. Hence it was already clear from the start (see Feigenbaum et al. 1982; Ostlund et al. 1983) that the natural thing to do is to construct a renormalization operator (see Definition 10.3) acting not on the space of critical circle maps but on a suitable space of *critical commuting pairs*, whose precise definition is the following.

Definition 10.1 A C^r *critical commuting pair* $\zeta = (\eta, \xi)$ consists of two C^r orientation preserving homeomorphisms $\eta : I_\eta \to \eta(I_\eta)$ and $\xi : I_\xi \to \xi(I_\xi)$ where:

1. $I_\eta = [0, \xi(0)]$ and $I_\xi = [\eta(0), 0]$ are compact intervals in the real line;
2. $\big(\eta \circ \xi\big)(0) = \big(\xi \circ \eta\big)(0) \neq 0$;

[1] As a first step, the real bounds (Theorem 6.3) can be used to establish C^r bounds for return maps, as in Sect. 6.4 (see also de Faria and de Melo 1999, App. A). A standard Arzelà–Ascoli argument gives then pre-compactness of renormalization orbits.

3. $D\eta(x) > 0$ for all $x \in I_\eta \setminus \{0\}$ and $D\xi(x) > 0$ for all $x \in I_\xi \setminus \{0\}$;
4. The origin is a non-flat critical point for both η and ξ, with the same criticality.
5. The left-derivatives of the composition $\eta \circ \xi$ at the origin coincide with the corresponding right-derivatives of $\xi \circ \eta$: for each $j \in \{1, 2, \dots, r\}$ we have $D_-^j\big(\eta \circ \xi\big)(0) = D_+^j\big(\xi \circ \eta\big)(0)$.

For a commuting pair as above, both η and ξ extend to C^r homeomorphisms, defined on interval neighborhoods of their respective domains, which commute around the origin. In other words, the commuting condition (2) in Definition 10.1 actually holds on an open interval. Let us be more precise.

Lemma 10.1 *There exist open intervals $V_- \supseteq I_\xi$ and $V_+ \supseteq I_\eta$ and C^r homeomorphic extensions $\widehat{\xi} : V_- \to \widehat{\xi}(V_-) \subset \mathbb{R}$ and $\widehat{\eta} : V_+ \to \widehat{\eta}(V_+) \subset \mathbb{R}$ of ξ and η respectively, satisfying $\big(\widehat{\eta} \circ \widehat{\xi}\big)(x) = \big(\widehat{\xi} \circ \widehat{\eta}\big)(x)$ for all x in the open interval C around the origin given by $C = \big\{x \in V_- \cap V_+ : \widehat{\eta}(x) \in V_- \text{ and } \widehat{\xi}(x) \in V_+\big\}$.*

Proof Since the origin is a non-flat critical point of odd criticality, there exists an open interval C around it on which we can extend both η and ξ to C^r homeomorphisms $\widehat{\eta} : C \to A$ and $\widehat{\xi} : C \to B$, where A is an open interval around $\eta(0)$ and B is an open interval around $\xi(0)$ (we may suppose that A, B and C are pairwise disjoint). Moreover, since the criticality of both $\widehat{\eta}$ and $\widehat{\xi}$ at the origin is the same odd integer, the composition $\widehat{\xi} \circ \widehat{\eta}^{-1} : A \to B$ is actually a C^r diffeomorphism.

Let $V_- = A \cup I_\xi \cup C$, which is an open interval where I_ξ is compactly contained, and in the same way let $V_+ = C \cup I_\eta \cup B$.

Since the composition $\eta \circ \xi$ is already defined at the left part of C, the extension of η defined above (given by the non-flatness of the critical point) allows us to extend ξ to the left part of A in the following way: for any $y \in A$ there exists a unique $x \in C$ such that $\widehat{\eta}(x) = y$ (since $A = \widehat{\eta}(C)$ and $\widehat{\eta} : C \to A$ is invertible) and then we define $\widehat{\xi} : A \to \mathbb{R}$ as $\widehat{\xi}(y) = \eta\big(\xi(x)\big) = \big(\eta \circ \xi \circ \widehat{\eta}^{-1}\big)(y)$ if $y < \eta(0)$ and $\widehat{\xi}(y) = \xi(y)$ if $y \geq \eta(0)$.

By Condition (5) in Definition 10.1, the left-derivatives of the composition $\eta \circ \xi \circ \widehat{\eta}^{-1}$ at the point $\eta(0)$ coincide with the corresponding right-derivatives of ξ at $\eta(0)$, that is, $\widehat{\xi}$ is of class C^r at the point $\eta(0)$ (and therefore on the whole domain V_-). Note also that $\widehat{\xi}$ has no critical points on $V_- \setminus \{0\}$ since $\widehat{\xi} \circ \widehat{\eta}^{-1} : A \to B$ is a C^r diffeomorphism and η has no critical points in $B \cap I_\eta$ by Condition (3).

In the same way, since the composition $\xi \circ \eta$ is already defined at the right part of C and since ξ is also defined on C, we extend η to the right part of B by imposing the commuting condition $\widehat{\eta} \circ \widehat{\xi} = \widehat{\xi} \circ \widehat{\eta}$ on C as before. □

The following construction was introduced by Lanford (see Lanford 1987, 1988), and it is known as the *glueing procedure* (recall also the proof of Lemma 9.7 in the previous chapter). It follows from Lemma 10.1 that the map ξ extends to a diffeomorphism from a neighborhood of $\eta(0)$ onto a neighborhood of $\xi \circ \eta(0)$. Identifying those points in this way, we obtain from the interval $\big[\eta(0), \xi \circ \eta(0)\big]$ a smooth, compact one-dimensional manifold M without boundary. The discontinuous piecewise smooth map

$$f_\zeta(t) = \begin{cases} \xi \circ \eta(t) \text{ for } t \in \big[\eta(0), 0\big) \\ \\ \eta(t) \qquad \text{for } t \in \big[0, \xi \circ \eta(0)\big] \end{cases}$$

projects to a smooth homeomorphism on the quotient manifold M. Choosing any diffeomorphism $\psi : M \to S^1$, we obtain a multicritical circle map in S^1 simply by conjugating with ψ. Although there is no canonical choice for the diffeomorphism ψ, any two different choices give rise to smoothly-conjugate multicritical circle maps in S^1. Therefore any critical commuting pair represents a whole *smooth conjugacy class* of multicritical circle maps. In particular, this procedure allows us to define the rotation number of a commuting pair.

On the other hand, any critical circle map f with irrational rotation number ρ gives rise to a sequence of critical commuting pairs in a natural way: let F be the lift of f to the real line (for the canonical covering $t \mapsto e^{2\pi i t}$) satisfying $DF(0) = 0$ and $0 < F(0) < 1$. For each $n \geq 1$ let $\widehat{I_n}$ be the closed interval in the real line, adjacent to the origin, that projects under $t \mapsto e^{2\pi i t}$ to I_n. Let $T : \mathbb{R} \to \mathbb{R}$ be the translation $x \mapsto x + 1$, and define $\eta : \widehat{I_n} \to \mathbb{R}$ and $\xi : \widehat{I_{n+1}} \to \mathbb{R}$ as:

$$\eta = T^{-p_{n+1}} \circ F^{q_{n+1}} \quad \text{and} \quad \xi = T^{-p_n} \circ F^{q_n} ,$$

where $\{p_n/q_n\}$ is the sequence of convergents associated to ρ, as defined in Chap. 1. It is not difficult to check that $(\eta|_{\widehat{I_n}}, \xi|_{\widehat{I_{n+1}}})$ is a critical commuting pair, usually denoted by $(f^{q_{n+1}}|_{I_n}, f^{q_n}|_{I_{n+1}})$.

For a commuting pair $\zeta = (\eta, \xi)$ we denote by $\widetilde{\zeta}$ the pair $(\widetilde{\eta}|_{\widetilde{I_\eta}}, \widetilde{\xi}|_{\widetilde{I_\xi}})$, where tilde means *rescaling* by the linear factor $1/|I_\xi|$. In other words, $|\widetilde{I_\xi}| = 1$ and the length of $\widetilde{I_\eta}$ equals the ratio between those of I_η and I_ξ.

Given two critical commuting pairs $\zeta_1 = (\eta_1, \xi_1)$ and $\zeta_2 = (\eta_2, \xi_2)$ let A_1 and A_2 be the Möbius transformations such that for $i = 1, 2$:

$$A_i\big(\eta_i(0)\big) = -1, \quad A_i(0) = 0 \quad \text{and} \quad A_i\big(\xi_i(0)\big) = 1 .$$

Definition 10.2 For any $0 \leq r < \infty$ define the C^r metric on the space of C^r critical commuting pairs in the following way:

$$d_r(\zeta_1, \zeta_2) = \max\left\{\left|\frac{\xi_1(0)}{\eta_1(0)} - \frac{\xi_2(0)}{\eta_2(0)}\right|, \left\|A_1 \circ \zeta_1 \circ A_1^{-1} - A_2 \circ \zeta_2 \circ A_2^{-1}\right\|_r\right\}$$

where $\|\cdot\|_r$ is the C^r-norm for maps in $[-1, 1]$ with one discontinuity at the origin, and ζ_i is the piecewise map defined by η_i and ξ_i:

$$\zeta_i : I_{\xi_i} \cup I_{\eta_i} \to I_{\xi_i} \cup I_{\eta_i} \quad \text{such that} \quad \zeta_i|_{I_{\xi_i}} = \xi_i \quad \text{and} \quad \zeta_i|_{I_{\eta_i}} = \eta_i$$

When we are dealing with real analytic critical commuting pairs, we consider the C^{ω}-topology defined in the usual way: we say that $(\eta_n, \xi_n) \to (\eta, \xi)$ if there exist two open sets $U_\eta \supset I_\eta$ and $U_\xi \supset I_\xi$ in the complex plane and $n_0 \in \mathbb{N}$ such that η and η_n for $n \geq n_0$ extend continuously to $\overline{U_\eta}$, are holomorphic in U_η and we have $\|\eta_n - \eta\|_{C^0(\overline{U_\eta})} \to 0$, and such that ξ and ξ_n for $n \geq n_0$ extend continuously to $\overline{U_\xi}$, are holomorphic in U_ξ and we have $\|\xi_n - \xi\|_{C^0(\overline{U_\xi})} \to 0$. We say that a set $\mathscr{C}$ of real analytic critical commuting pairs is closed if every time we have $\{\zeta_n\} \subset \mathscr{C}$ and $\{\zeta_n\} \to \zeta$, we have $\zeta \in \mathscr{C}$. This defines a Hausdorff topology, stronger than the C^r-topology for any $0 \leq r \leq \infty$ (in particular any C^{ω}-compact set of real analytic critical commuting pairs is certainly C^r-compact also, for any $0 \leq r \leq \infty$).

Note that d_r is not a metric but rather a *pseudo-metric*, since it assigns distance zero to any pair of commuting pairs that are conjugate by a homothety: if α is a positive real number, $H_\alpha(t) = \alpha t$ and $\zeta_1 = H_\alpha \circ \zeta_2 \circ H_\alpha^{-1}$, then $d_r(\zeta_1, \zeta_2) = 0$. In order to have a metric, we simply need to restrict to *normalized* critical commuting pairs, as defined above.

Let $\zeta = (\eta, \xi)$ be a critical commuting pair according to Definition 10.1, and recall that $(\eta \circ \xi)(0) = (\xi \circ \eta)(0) \neq 0$. Let us suppose that $(\xi \circ \eta)(0) \in I_\eta$ (see Fig. 10.1) and define the *height* $\chi(\zeta)$ of ζ as $a \in \mathbb{N}$ if

$$\eta^{a+1}\big(\xi(0)\big) < 0 \leq \eta^a\big(\xi(0)\big),$$

and $\chi(\zeta) = \infty$ if no such a exists. Thus, the height of the commuting pair $(f^{q_{n+1}}|_{I_n}, f^{q_n}|_{I_{n+1}})$ induced by a critical circle map f is exactly a_{n+1}, where $\rho(f) = [a_0, a_1, \ldots]$. Now, for $\zeta = (\eta, \xi)$ with $(\xi \circ \eta)(0) \in I_\eta$ and $\chi(\zeta) = a < \infty$, the pair

$$\big(\eta|_{[0, \eta^a(\xi(0))]}, \ \eta^a \circ \xi|_{I_\xi}\big)$$

is again a commuting pair, and if ζ is induced by a critical circle map, i.e.,

$$\zeta = \big(f^{q_{n+1}}|_{I_n}, \ f^{q_n}|_{I_{n+1}}\big),$$

then we have

$$\big(\eta|_{[0, \eta^a(\xi(0))]}, \ \eta^a \circ \xi|_{I_\xi}\big) = \big(f^{q_{n+1}}|_{I_{n+2}}, \ f^{q_{n+2}}|_{I_{n+1}}\big).$$

This motivates the following definition.

Definition 10.3 Let $\zeta = (\eta, \xi)$ be a critical commuting pair with $(\xi \circ \eta)(0) \in I_\eta$. We say that ζ is *renormalizable* if $\chi(\zeta) = a < \infty$. In this case, we define the *pre-renormalization* of ζ as the critical commuting pair

$$p\mathscr{R}(\zeta) = \big(\eta|_{[0, \eta^a(\xi(0))]}, \ \eta^a \circ \xi|_{I_\xi}\big),$$

and we define the *renormalization* of ζ as the normalization of $p\mathscr{R}(\zeta)$; that is,

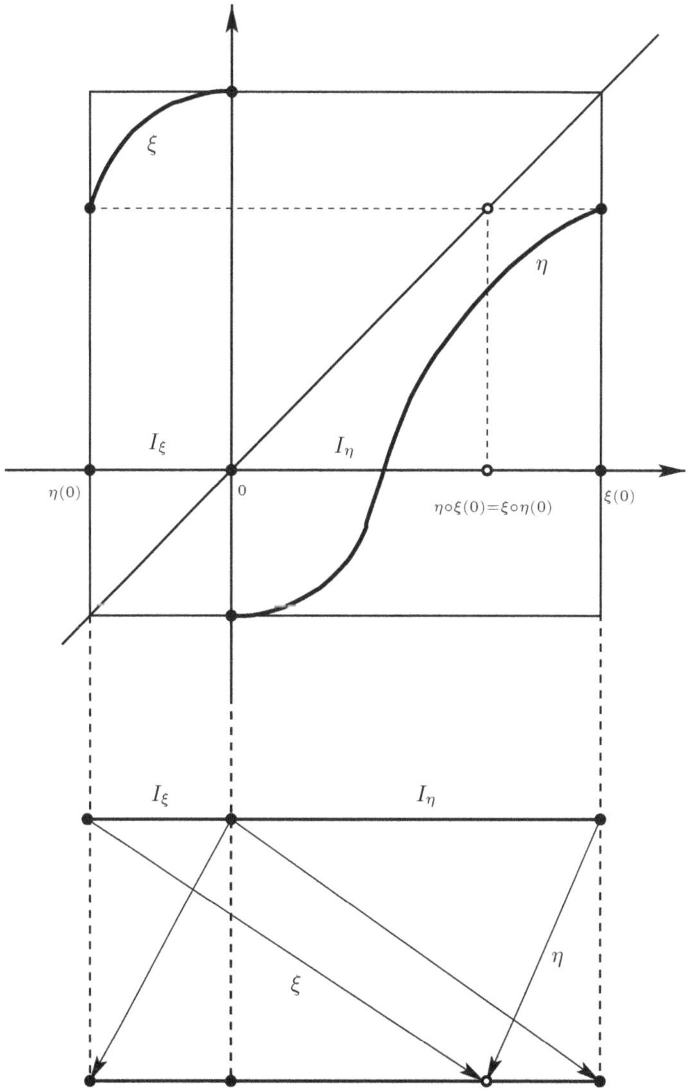

Fig. 10.1 A critical commuting pair and its underlying interval exchange

$$\mathscr{R}(\zeta) = \widetilde{p\mathscr{R}(\zeta)} = \left(\widetilde{\eta}|_{[0,\eta^a(\tilde{\xi}(0))]}, \widetilde{\eta^a \circ \xi}|_{\tilde{I}_\xi}\right).$$

A critical commuting pair is a special case of a *generalized interval exchange map* of two intervals, and the renormalization operator defined above is just the restriction of the *Zorich accelerated version* of the *Rauzy–Veech renormalization* for interval exchange maps (see for instance Yoccoz 2006). However, we keep in this book the classical terminology for critical commuting pairs.

If $\chi(\mathscr{R}^j(\zeta)) < \infty$ for $j \in \{0, 1, \dots, n-1\}$ we say that ζ is *n-times renormalizable*, and if $\chi(\mathscr{R}^j(\zeta)) < \infty$ for all $j \in \mathbb{N}$ we say that ζ is *infinitely renormalizable*. The space of all infinitely renormalizable commuting pairs is the natural phase-space for renormalization. For such a pair, the irrational number whose continued fraction expansion equals

$$\left[\chi(\zeta), \chi\left(\mathscr{R}(\zeta)\right), \dots, \chi\left(\mathscr{R}^n(\zeta)\right), \chi\left(\mathscr{R}^{n+1}(\zeta)\right), \dots\right]$$

is, by definition, the *rotation number* of the critical commuting pair ζ (note that if ζ is induced by a critical circle map with irrational rotation number, then it is infinitely renormalizable and both definitions of rotation number coincide).

To understand the action of renormalization on the rotation number of a commuting pair, recall that the *Gauss map* $G : [0, 1] \to [0, 1]$ is given by

$$G(\rho) = \left\{\frac{1}{\rho}\right\} \text{ for } \rho \neq 0, \text{ and } G(0) = 0.$$

If $\rho = [a_0, a_1, a_2, \dots]$ belongs to $\left(1/(k+1), 1/k\right)$, then $1/\rho = a_0 + [a_1, a_2, \dots]$ and then $a_0 = \left\lfloor \frac{1}{\rho} \right\rfloor = k$ and $G(\rho) = [a_1, a_2, \dots]$. This shows that the Gauss map acts as a *left shift* on the continued fraction expansion of ρ, and therefore the action of the renormalization operator on the rotation number is given by

$$\rho\left(\mathscr{R}(\zeta)\right) = G\left(\rho(\zeta)\right) = \sigma\left([a_0, a_1, a_2, \dots]\right) = [a_1, a_2 \dots]. \tag{10.1}$$

In particular, the way the renormalization operator $\mathscr{R}$ acts on (infinitely renormalizable) critical commuting pairs is by sending topological classes to topological classes.

10.3 A Fundamental Principle

Recall that, by Yoccoz's Theorem 6.2, two C^3 multicritical circle maps, say f and g, with the same irrational rotation number are topologically conjugate to each other. If f and g have the same signature (Definition 6.2) there exists a homeomorphism $h : S^1 \to S^1$, which is a topological conjugacy between f and g, identifying each critical point of f with a critical point of g having the same criticality. By Theorem 7.2, h is a quasisymmetric homeomorphism. Such an h, mapping critical points to critical points and preserving criticalities, is the only hope of a smooth conjugacy between f and g (as explained in Chap. 9, it turns out that for almost every rotation number most conjugacies between f and g fail to be quasisymmetric).

10.3.1 Main Theorem

The following result, originally proved by de Faria and de Melo (1999, First Main Theorem), is the main result of this chapter.

Theorem 10.4 *There exists a set $\mathbb{A}$ of rotation numbers, having full Lebesgue measure and containing all numbers of bounded type, for which the following holds. Let f and g be topologically conjugate C^3 critical circle maps, and let h be the conjugacy between f and g that maps the critical point of f to the critical point of g. If their common rotation number belongs to $\mathbb{A}$, and if their renormalizations converge together exponentially fast in the C^0-topology, then h is $C^{1+\alpha}$ for some $\alpha > 0$.*

This theorem has been recently extended by Estevez and Guarino (2023) to cover the multicritical case. Here, one needs to assume, of course, that both maps have the same signature, and the hypothesis of exponential convergence in the C^0-topology has to be replaced by exponential convergence in the C^1-topology. Indeed, contraction of the first derivatives is needed in order to control the relative position of the various critical points for the return maps. As mentioned in the introduction of this chapter, proving exponential contraction of renormalization is a challenging problem, to be discussed in Chap. 13.

The set $\mathbb{A} \subset (0, 1)$ of rotation numbers considered in the statement of Theorem 10.4 was introduced in de Faria and de Melo (1999, Section 4.4). Its precise definition is the following.

Definition 10.4 Let $\mathbb{A} \subset (0, 1)$ be the set of irrational numbers $\rho = [a_0, a_1, \ldots]$ satisfying:

1. $\displaystyle \limsup_{n\to\infty} \frac{1}{n} \sum_{j=0}^{n-1} \log a_j < \infty\,,$
2. $\displaystyle \lim_{n\to\infty} \frac{1}{n} \log a_n = 0\,,$
3. $\displaystyle \frac{1}{n} \sum_{j=k+1}^{k+n} \log a_j \leq \omega\left(\frac{n}{k}\right),$

for all $0 < n \leq k$, where ω is a monotone function (that depends on ρ) such that $\omega(t) > 0$ for all $t > 0$, and such that $t\,\omega(t) \to 0$ as $t \to 0$.

The set $\mathbb{A}$ has full Lebesgue measure in $(0, 1)$, and a proof of this fact will be given in Appendix A (see Corollary A.1 and Lemma A.3). Obviously, all bounded type numbers satisfy the three conditions above. The number whose partial quotients are given by $a_n = k$ if $n = 2^k$ with $k \geq 1$ and $a_n = 1$ otherwise is an explicit element of $\mathbb{A}$ that is not of bounded type. This number satisfies (3) with $\omega(t) = 1/\sqrt{t}$.

Still in the unicritical case, if one asks for the conjugacy to be only C^1, rather than $C^{1+\alpha}$, we have the following result obtained by Khanin and Teplinsky (2007, Th. 2).

Theorem 10.5 *Let f and g be C^3 unicritical circle maps with the same irrational rotation number. If the renormalizations of f and g converge together exponentially fast in the C^2 topology, then f and g are conjugate to each other by a C^1 diffeomorphism.*

This theorem will not be proved here; we refer the reader to their original paper. Let us mention that it would be important, for the rigidity problem for multicritical circle maps discussed in Sect. 10.1 (recall Question 10.1), to adapt their approach to the multicritical case. In other words, to prove that exponential convergence of renormalization orbits implies C^1 rigidity for multicritical circle maps with arbitrary irrational rotation numbers. To the best of our knowledge, this has not yet been established.

The proof of Theorem 10.4 to be given here is the same proof given in de Faria and de Melo (1999). In addition to the real bounds from Chap. 6 (Theorem 6.3 and 6.4), several tools from Chap. 7, such as the notion of fine grids (Definition 7.2), the criterion for smoothness given by Proposition 7.3, and Yoccoz's lemma on almost parabolic maps (Definition 7.3 and Lemma 7.3), will be used in the proof.

Remark 10.1 As pointed out in de Faria and de Melo (1999, Prop. 2.2), the real bounds imply that exponential convergence of renormalizations is preserved under conjugacy by a smooth diffeomorphism. In other words, if two C^r maps are C^r conjugate, then the C^{r-1} distance between their successive renormalizations goes to zero exponentially fast. This is true even in the general multicritical case: see Exercise 10.1.

10.3.2 Comparing Orbits of Two Almost Parabolic Maps

The following consequence of Yoccoz's inequality will be need in the proof of Theorem 10.4.

Proposition 10.1 *Let ϕ and ψ be two almost parabolic maps with the same length ℓ defined on the same interval. Then for all $x \in J_1(\phi) \cap J_1(\psi)$ and for all $0 \le k \le \ell/2$, we have*

$$|\phi^k(x) - \psi^k(x)| \le C\, k^3\, \|\phi - \psi\|_{C^0}. \tag{10.2}$$

Proof First note, using the mean-value theorem, that

$$
\begin{aligned}
|\phi^k(x) - \psi^k(x)| &= \left|\sum_{j=0}^{k-1} \left(\phi^{k-j-1}(\phi(\psi^j(x))) - \phi^{k-j-1}(\psi^{j+1}(x))\right)\right| \\
&\leq \sum_{j=0}^{k-1} \left|D\phi^{k-j-1}(\xi_j)\right| \left|\phi(\psi^j(x)) - \psi(\psi^j(x))\right| ,
\end{aligned}
$$

where ξ_j lies between $\phi(\psi^j(x))$ and $\psi^{j+1}(x)$. Hence we have

$$
|\phi^k(x) - \psi^k(x)| \leq \|\phi - \psi\|_0 \sum_{j=0}^{k-1} \left|D\phi^{k-j-1}(\xi_j)\right| . \tag{10.3}
$$

Let us estimate each summand in the right-hand side of (10.3). Let $m = m(j)$ be such that $\xi_j \in \Delta_{j+m}(\phi)$, and assume also that $j + m \leq a/2$. This last condition is always satisfied if the central fundamental domain of ψ lies to the left of the central fundamental domain of ϕ (if this is not the case, then reverse the roles of ϕ and ψ in (10.3) and throughout). Using Yoccoz's Lemma 7.3, we see that

$$
|D\phi^{k-j-1}(\xi_j)| \asymp \frac{(j+m)^2}{(a-k-m+1)^2} \leq \left(\frac{j+m}{j+1}\right)^2 . \tag{10.4}
$$

Hence, it suffices to estimate m as a function of j. For this purpose, let $n = n(j)$ be such that $\psi^{j+1}(x) \in [\phi^{j+n-1}(x), \phi^{j+n}(x)]$. We claim that $m \leq n + 1$. There are two possibilities. The first is that $\phi(\psi^j(x)) \geq \psi^{j+1}(x)$: in this case we see easily that

$$
\xi_j \in [\psi^{j+1}(x), \phi(\psi^j(x))] \subseteq [\phi^{j+n-1}(x), \phi^{j+n+1}(x)]
$$

and so $m \leq n + 1$. The second is that $\phi(\psi^j(x)) < \psi^{j+1}(x)$. In this case we have $\xi_j < \psi^{j+1}(x) < \phi^{j+n}(x) \in \Delta_{j+n+1}(\phi)$, so once again $m \leq n+1$. This proves our claim.

So now we must bound n as a function of j. Again, there are two cases to consider.

(a) We have $[\psi^{j+1}(x), \psi^{j+2}(x)] \subseteq [\phi^{j+n-1}(x), \phi^{j+n}(x)]$ (as depicted in Fig. 10.2a). In this case, Yoccoz's Lemma gives us

$$
\frac{1}{j^2} \leq \frac{C}{(j+n)^2},
$$

which implies $n \leq Cj$.

(b) We have $\psi^{j+2}(x) > \phi^{j+n}(x)$. In this case, $\phi^{j+n}(x)$ is the first point in the ϕ-orbit of x that lands inside the interval $\Delta = [\psi^{j+1}(x), \psi^{j+2}(x)]$ (see Fig. 10.2b). Let p be such that $\phi^{j+n+i}(x) \in \Delta$ for $i = 0, 1, \ldots, p-1$ but

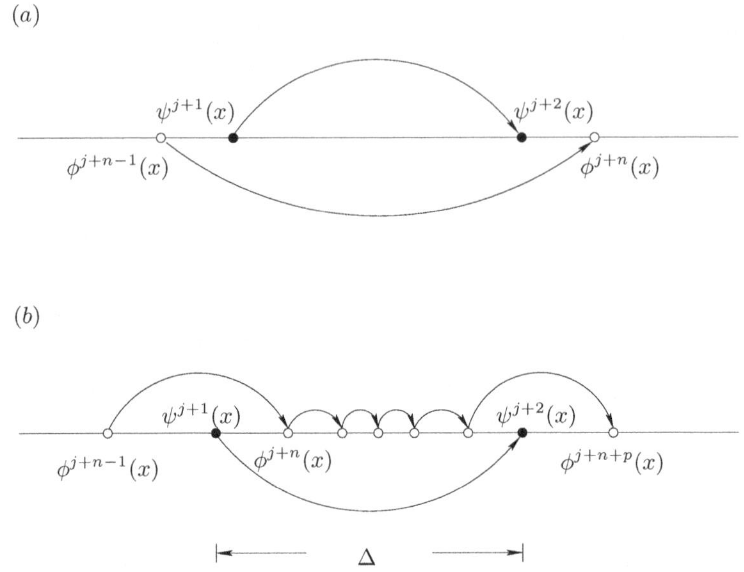

Fig. 10.2 Bounding n in terms of j

$\phi^{j+n+p}(x) \notin \Delta$. Then we have $\Delta \subseteq [\phi^{j+n-1}(x), \phi^{j+n+p}(x)]$, and this time Yoccoz's Lemma gives us

$$\frac{1}{j^2} \leq C \left(\frac{1}{(j+n)^2} + \frac{1}{(j+n+1)^2} + \cdots + \frac{1}{(j+n+p)^2} \right) \leq \frac{C}{j+n}$$

Therefore $n \leq Cj^2$ in this case.

In either case we see that $m \leq Cj^2$. Carrying this information back to (10.4), we deduce that

$$|D\phi^{k-j-1}(\xi_j)| \leq C\, j^2 . \tag{10.5}$$

Substituting (10.5) into (10.3), we arrive at (10.2), and the proof is complete. □

Remark 10.2 It is worth pointing out that Proposition 10.1, which as we saw is based on the geometric inequalities given by Yoccoz's Lemma 7.3, will be significantly improved in Chap. 12 (see for instance Lemma 12.11 and Proposition 12.3). Such sharper estimates, although not needed in the present chapter, will be crucial in Chap. 13.

10.3.3 Proof of Theorem 10.4

Recall that we are dealing here with *unicritical* circle maps. There is no loss of generality in assuming that the critical point c is the same for both maps. Let $\{\mathcal{Q}_n(f)\}_{n\geq 0}$ be the fine grid for f constructed in Proposition 7.6. The idea of the proof is to show that the conjugacy h and this fine grid satisfy, at each level n, the *coherence condition*

$$\left|\frac{|I|}{|J|} - \frac{|h(I)|}{|h(J)|}\right| \leq C\lambda^n , \tag{10.6}$$

for each pair of adjacent atoms $I, J \in \mathcal{Q}_n(f)$ and some constants $C > 0$ and $0 < \beta < 1$, and then invoke Proposition 7.3.

First we introduce some notation, to be used throughout the proof. We write $x_n = x_n(f) = f^{q_n}(c)$. Accordingly, we write $I_n(f)$ instead of $I_n(c, f)$, so that the endpoints of $I_n(f)$ are c and $x_n(f)$. We denote by $J_n(f)$ the interval $I_n(f)\cup I_{n+1}(f)$ and by $f_n : J_n(f) \to J_n(f)$ the first return map to this interval. Finally, we write $\boldsymbol{f}_n = \mathcal{R}^n f$ for the n-th renormalization of f around c (this is just the return map f_n linearly rescaled so that $I_n(f)$ becomes the unit interval).

Now, the first thing to observe is that, if the renormalizations $\boldsymbol{f}_n$ and $\boldsymbol{g}_n$ converge together exponentially fast, then $|x_n(f) - c|/|x_n(g) - c|$ converges to a limit exponentially fast also. More precisely, we have the following lemma.

Lemma 10.2 *If* $\|\boldsymbol{f}_n - \boldsymbol{g}_n\|_0 \leq C\mu^k$ *for some* $0 < \mu < 1$ *and all* $n \geq 0$, *then the ratio* $|x_n(f) - c|/|x_n(g) - c|$ *converges to a limit exponentially fast. Moreover, for all* $m, k \geq 1$ *we have*

$$\left|\frac{|I_m(f)|}{|I_k(f)|} - \frac{|I_m(g)|}{|I_k(g)|}\right| \leq C\mu^{\min\{m,k\}}\frac{|I_m(f)|}{|I_k(f)|} . \tag{10.7}$$

Proof The hypothesis tells us that

$$\left|\frac{|I_{n+1}(f)|}{|I_n(f)|} - \frac{|I_{n+1}(g)|}{|I_n(g)|}\right| \leq C_1\mu^n ,$$

for all $n \geq 1$. Writing $\alpha_n = |x_n(f) - c|/|x_n(g) - c| = |I_n(f)|/|I_n(g)|$, and taking into account that $C_2^{-1}|I_n(g)| \leq |I_{n+1}(g)| \leq C_2|I_n(g)|$ by the real bounds (for some $C_2 > 1$), we see that the above inequality is equivalent to

$$\left|\frac{\alpha_{n+1}}{\alpha_n} - 1\right| \leq C_3\mu^n .$$

This is the same as $\alpha_{n+1} = (1 + \epsilon_n)\alpha_n$ where $|\epsilon_n| \leq C_3\mu^n$. Therefore $\alpha_n = \alpha_1\prod_{j=1}^{n-1}(1 + \epsilon_j)$, and this shows that $\lim \alpha_n$ exists. Finally, note that if $m > k \geq 1$ then

$$\left|\frac{\alpha_m}{\alpha_k} - 1\right| \leq \left|\prod_{j=k}^{m-1}(1+\epsilon_j) - 1\right| \leq C_4 \sum_{j=k}^{m-1} \epsilon_j < C_5 \mu^k ,$$

and similarly for $|1 - \alpha_k/\alpha_m|$, and these facts clearly imply (10.7). □

Remark 10.3 Having established this lemma, we may assume, after conjugating one of the maps (say g) by a suitable smooth diffeomorphism, that the limit of the ratios $|I_n(f)|/|I_n(g)|$ is in fact equal to one. This will be our standing hypothesis from now on (used at the end of the proof of Lemma 10.5 below).

Definition 10.5 Let $f_m : J_m(f) \to J_m(f)$ be the m-th first return map of f and let $k \neq 0$ be an integer such that $|k| \leq \lceil a_{m+1}/2 \rceil$ (where $\lceil x \rceil$ denotes the smallest integer $\geq x$). The *restricted domain* of f_m^k, denoted $D_{m,k}$, is defined as follows.

$$D_{m,k} = \begin{cases} I_{m+1} \cup \left[f_m^{\lceil \frac{a_{m+1}}{2} \rceil - k}(x_m), x_m \right], & \text{when } k > 0 \\ \\ \left[f_m(x_{m+2}), f_m^{\lceil \frac{a_{m+1}}{2} \rceil - k}(x_m) \right], & \text{when } k \leq -1 \end{cases}$$

In informal terms, the restricted domain $D_{m,k}$ is the set of points in J_m which can be iterated k times by f_m without ever going *across* the central fundamental domain of f_m in $J_m(f) \setminus J_{m+1}(f)$.

Lemma 10.3 *For all $x \in D_{m,k}$ we have $|Df_m^k(x)| \leq K$, where $K \geq 1$ depends only on the real bounds.*

Proof Follows easily from the real bounds and Yoccoz's Lemma 7.3. The details are left to the reader as an exercise. □

Lemma 10.4 *Let v be a vertex of $\mathscr{P}_{k+p}(f)$ such that $v \in J_k(f)$. Then there exist $k \leq m \leq k+p$ and $1 \leq N \leq p$ such that v can be represented in the form*

$$v = \phi_1 \circ \phi_2 \circ \cdots \circ \phi_N(x_m) ,$$

where $\phi_j = f_{m_j}^{k_j}$ for some $k \leq m_j \leq k+p$ and $|k_j| \leq \lceil a_{m_j+1}/2 \rceil$, and where the point $\phi_{j+1} \circ \cdots \circ \phi_N(x_m)$ belongs to the restricted domain of ϕ_j for each j.

Proof For simplicity of notation, we write $J_i = J_i(f)$ in this proof. Let $k \leq m_1 \leq k+p$ be largest with the property that $v \in J_{m_1} \setminus J_{m_1+1}$, and let $0 < i \leq a_{m_1+1}$ be such that $f_{m_1}^i(v) \in J_{m_1+1}$. If $i \leq \lceil a_{m_1}/2 \rceil$ then let $k_1 = -i$; otherwise let $k_1 = a_{m_1+1} - i$. We get $\phi_1 = f_{m_1}^{k_1}$ and a new vertex $v_1 = f_{m_1}^{-k_1}(v) \in J_{m_1+1}$. If $v_1 \in J_{k+p}$ then $v_1 = f_{k+p}(x_{k+p})$ necessarily, and we can stop. On the other hand, if $v_1 \notin J_{k+p}$, then once again there exists m_2 in the range $m_1 < m_2 < k+p$ such that $v_1 \in J_{m_2} \setminus J_{m_2+1}$, and we can proceed inductively. At the end of this process we get sequences $m_1 < m_2 < \cdots < m_N \leq k+p$ (so $N \leq p$) and $v_1, v_2, \ldots, v_N$

with $v_j \in J_{m_j} \setminus J_{m_j+1}$, and for each j an integer k_j with $|k_j| \le \lceil a_{m_j+1}/2 \rceil$ such that $v_{j+1} = f_{m_j}^{-k_j}(v_j)$. The last vertex v_N is necessarily x_m for some $m \le k+p$. Hence it suffices to take $\phi_j = f_{m_j}^{k_j}$ to get the desired representation. □

From now on, we assume that the corresponding successive renormalizations of f and g approach each other exponentially, in other words $\|\boldsymbol{f}_n - \boldsymbol{g}_n\|_0 \le C\mu^n$ for some $0 < \mu < 1$ and all $n \ge 0$, just as stated in the hypothesis of Lemma 10.2.

Lemma 10.5 *There exists a constant $0 < \mu_* < 1$ for which the following holds. Let $v \in J_k(f)$ be a vertex of $\mathscr{P}_{k+p}(f)$ and let $w = h(v) \in J_k(g)$ be the corresponding vertex of $\mathscr{P}_{k+p}(g)$. If $\rho(f)$ satisfies condition* (2), *then we have*

$$|v - w| \le C|J_k(f)|K^p\mu_*^k , \tag{10.8}$$

where $K \ge 1$ is the constant of Lemma 4.8.

Proof By Lemma 10.4 above, there exist points $x_m = x_m(f)$, $y_m = x_m(g)$ and a number $N \le p$ such that

$$|v - w| = |\phi_1 \circ \phi_2 \circ \cdots \circ \phi_N(x_m) - \psi_1 \circ \psi_2 \circ \cdots \circ \psi_N(y_m)| ,$$

where $\phi_j = f_{m_j}^{k_j}$ and $\psi_j = g_{m_j}^{k_j}$, with $k \le m_j \le k+p$ and $|k_j| \le \lceil a_{m_j+1}/2 \rceil$. For each $i \ge 1$, let $\Lambda_{i,f}$ be the affine map $x \mapsto c + |I_i(f)|x$, and define $\Lambda_{i,g}$ in the same way. For each $i \ge k$, let $A_{i,f} = \Lambda_{k,f}^{-1} \circ \Lambda_{i,f}$ and $A_{i,g} = \Lambda_{k,g}^{-1} \circ \Lambda_{i,g}$. In order to estimate $|v - w|$, we shall estimate $|v^* - w^*|$, where $v^* = \Lambda_{k,f}^{-1}(v)$ and $w^* = \Lambda_{k,g}^{-1}(w)$. To do this, for each $i \ge k$ consider the map $f_i^* : \Lambda_{k,f}^{-1}(J_i(f)) \to \Lambda_{k,f}^{-1}(J_i(f))$ given by

$$f_i^* = \Lambda_{k,f}^{-1} \circ f_i \circ \Lambda_{k,f} = A_{i,f} \circ \boldsymbol{f}_i \circ A_{i,f}^{-1} ,$$

and let g_i^* be similarly defined.

First we claim that for all $x \in \Lambda_{k,f}^{-1}(J_i(f)) \cap \Lambda_{k,g}^{-1}(J_i(g))$ we have

$$|f_i^*(x) - g_i^*(x)| \le C_1\mu^k \frac{|I_i(f)|}{|I_k(f)|} . \tag{10.9}$$

To see why, note that by inequality (10.7) of Lemma 10.2 we have, for all z in the domain of both renormalizations $\boldsymbol{f}_i$ and $\boldsymbol{g}_i$,

$$\left|A_{i,f}(z) - A_{i,g}(z)\right| = \left|\frac{|I_i(f)|}{|I_k(f)|} - \frac{|I_i(g)|}{|I_k(g)|}\right| |z| \le C_2\mu^k \frac{|I_i(f)|}{|I_k(f)|} .$$

Similarly, for all $x \in \Lambda_{k,f}^{-1}(J_i(f)) \cap \Lambda_{k,g}^{-1}(J_i(g))$ we have, again by (10.7),

$$\left|A_{i,f}^{-1}(x) - A_{i,g}^{-1}(x)\right| = \left|\frac{|I_k(f)|}{|I_i(f)|} - \frac{|I_k(g)|}{|I_i(g)|}\right| |x| \leq C_3\mu^k \frac{|I_k(f)|}{|I_i(f)|} |x| \leq C_4\mu^k .$$

Here we have used that $|x| \leq |J_i(f)|/|I_k(f)| \leq C_5|I_i(f)|/|I_k(f)|$ (recall from the real bounds that $|J_i(f)| \asymp |I_i(f)|$). Also, by hypothesis we have $\|\boldsymbol{f}_i - \boldsymbol{g}_i\|_0 \leq C_6\mu^k$. Combining these three estimates with a standard telescoping trick, we get (10.9), and the claim is proved.

Now let $\phi_j^* = A_{m_j,f} \circ \phi_j \circ A_{m_j,f}^{-1}$ and $\psi_j^* = A_{m_j,g} \circ \psi_j \circ A_{m_j,g}^{-1}$. Applying (10.9) with $i = m_j$ and using Proposition 10.1, we have

$$\left|\phi_j^*(x) - \psi_j^*(x)\right| \;\leq\; C_7\, |k_j|^3 \mu^k \frac{|I_{m_j}(f)|}{|I_k(f)|} . \tag{10.10}$$

By the real bounds, there exists $0 < \lambda_1 < 1$ such that $|I_{m_j}(f)|/|I_k(f)| \leq C_8\lambda_1^{m_j-k}$. Taking $\lambda = \max\{\mu, \lambda_1\}$, we deduce from (10.10) that

$$\left|\phi_j^*(x) - \psi_j^*(x)\right| \;\leq\; C_9\, a_{m_j+1}^3 \lambda^{m_j} . \tag{10.11}$$

We can at last start our estimate of $|v^* - w^*|$. First, note that $x_m = \Lambda_{m,f}(1)$ and $y_m = \Lambda_{m,g}(1)$. Writing $x_m^* = \Lambda_{k,f}^{-1}(x_m)$ and $y_m^* = \Lambda_{k,g}^{-1}(y_m)$, we see after a simple computation that $|x_m^* - y_m^*| \leq C_{10}\lambda^m$. Combining this fact with (10.11) and using Lemma 10.3, we get

$$\begin{aligned} |\phi_N^*(x_m^*) - \psi_N^*(y_m^*)| \;&\leq\; |\phi_N^*(x_m^*) - \psi_N^*(x_m^*)| + |\psi_N^*(x_m^*) - \psi_N^*(y_m^*)| \\ &\leq\; C_9 a_{m_N+1}^3 \lambda^{m_N} + C_{10} K \lambda^m . \end{aligned}$$

From this, and since

$$\begin{aligned} &|\phi_{N-1}^*(\phi_N^*(x_m^*)) - \psi_{N-1}^*(\psi_N^*(y_m^*))| \leq \\ &\quad |\phi_{N-1}^*(\phi_N^*(x_m)) - \psi_{N-1}^*(\phi_N^*(x_m^*))| + |\psi_{N-1}^*(\phi_N^*(x_m^*)) - \psi_{N-1}^*(\psi_N^*(y_m^*))| , \end{aligned}$$

we deduce that

$$\begin{aligned} &|\phi_{N-1}^*(\phi_N^*(x_m^*)) - \psi_{N-1}^*(\psi_N^*(y_m^*))| \leq \\ &\qquad C_9\left(a_{m_{N-1}+1}^3 \lambda^{m_{N-1}} + K a_{m_N+1}^3 \lambda^{m_N}\right) + C_{10} K^2 \lambda^m . \end{aligned}$$

Proceeding inductively in this fashion, we get in the end

$$|v^* - w^*| \;\leq\; C_9 \sum_{j=1}^{N} K^{j-1} a_{m_j+1}^3 \lambda^{m_j} + C_{10} K^N \lambda^m .$$

Using that $N \leq p$ and taking $C_{11} = \max\{C_9, C_{10}\}$, we arrive at

$$|v^* - w^*| \leq C_{11}K^p \left(\lambda^m + \sum_{j=1}^{N} a_{m_j+1}^3 \lambda^{m_j} \right) . \tag{10.12}$$

We have of course $\lambda^m \leq \lambda^k$. Moreover, since $k \leq m_j < m_{j+1}$ for all j, we have

$$\sum_{j=1}^{N} a_{m_j+1}^3 \lambda^{m_j} < \sum_{n=k}^{\infty} a_n^3 \lambda^n .$$

But since (a_n) satisfies condition (2), we know that $\lim (a_n^3)^{1/n} = 1$. In particular, if $\varepsilon > 0$ is such that $(1+\varepsilon)\sqrt{\lambda} = 1$, there exists $C_{12} = C_{12}(\varepsilon) > 0$ such that $a_n^3 < C_{12}(1+\varepsilon)^n$ for all n. Therefore

$$\sum_{n=k}^{\infty} a_n^3 \lambda^n < C_{12} \sum_{n=k}^{\infty} \left(\sqrt{\lambda}\right)^n = \frac{C_{12}}{1-\sqrt{\lambda}} \left(\sqrt{\lambda}\right)^k .$$

Taking this back to (10.12) yields $|v^* - w^*| \leq C_{13}K^p] \left(\sqrt{\lambda}\right)^k$. Therefore, noting that under the assumption given in the remark after Lemma 10.2 we have

$$|v - w| = |\Lambda_{k,f}(v^*) - \Lambda_{k,g}(w^*)| \leq |I_k(f)| \left(|v^* - w^*| + C_{14}\mu^k \right) ,$$

and taking $\mu_* = \sqrt{\lambda}$, we get (10.8) as desired. □

Lemma 10.6 *There exists a constant $M > 0$ depending only on the real bounds such that if $\Delta^* \in \mathscr{P}_k(f)$ and $\Delta \in \mathscr{P}_{k+p}(f)$ is contained in Δ^*, then*

$$|\Delta| \geq \frac{M^p}{\left(a_{k+1}a_{k+2}\cdots a_{k+p}\right)^2} |\Delta^*| .$$

Proof This again follows from Yoccoz's Lemma 7.3 and a simple inductive argument. □

Let us now consider the fine grid $\{\mathscr{Q}_n(f)\}_{n \geq 0}$ constructed before. It will be convenient to use the following terminology.

Definition 10.6 The *level* of an atom $\Delta \in \mathscr{Q}_n(f)$, denoted $\ell(\Delta)$, is the largest $m \leq n$ such that Δ is contained in an atom of $\mathscr{P}_m(f)$.

Lemma 10.7 *If $\mathscr{Q}_n(f)$ contains an atom of level m, then*

$$n \leq c_0 \sum_{j=1}^{m} \log\left(1 + a_{j+1}\right) \tag{10.13}$$

for some absolute constant $c_0 > 0$*. In particular, if the partial quotients of* $\rho(f)$ *satisfy* (1)*, then* $m \geq c_1 n$ *for some constant* $0 < c_1 < 1$ *that depends only on* $\rho(f)$*.*

Proof Let $\Delta \in \mathcal{Q}_n(f)$ be an atom of level m. Let $\Delta_1 \supseteq \Delta_2 \supseteq \cdots \supseteq \Delta_n = \Delta$ be such that $\Delta_k \in \mathcal{Q}_k(f)$, and note that $1 = \ell(\Delta_1) \leq \ell(\Delta_2) \leq \cdots \leq \ell(\Delta_n) = m$. Given $1 \leq l \leq m$, let i and s (maximal) be such that

$$\ell(\Delta_{i+1}) = \ell(\Delta_{i+2}) = \cdots = \ell(\Delta_{i+s}) = l\,.$$

Then there exists $I \in \mathcal{P}_l(f)$ such that each Δ_j with $i+1 \leq j \leq i+s$ is a union of atoms of $\mathcal{P}_{l+1}(f)$ inside I. From the very construction of the partitions $\mathcal{Q}_j(f)$ (Proposition 4.5), we see that the number of atoms of $\mathcal{P}_{l+1}(f)$ inside Δ_j is at least *twice* the number of such atoms inside Δ_{j+1}, for each $i+1 \leq j \leq i+s-1$. Moreover, Δ_{i+s} contains at least two such atoms, otherwise its level would be $l+1$. Since the total number of atoms of $\mathcal{P}_{l+1}(f)$ that lie inside I is at most $1 + a_{l+1}$, it follows that $2^s \leq 1 + a_{l+1}$, whence $s \leq \log_2(1 + a_{l+1})$. This proves (10.13) with $c_0 = 1/\log 2$.

Now, if $\rho(f)$ satisfies (1), then there exists $B > 0$ depending on $\rho(f)$ such that $\sum_{j=1}^{m} \log a_{j+1} \leq Bm$. Therefore

$$n \leq c_0 \sum_{j=1}^{m} \log\left(1 + a_{j+1}\right) \leq c_0\,(B + \log 2)\, m\,,$$

which proves the last assertion, with $c_1 = c_0^{-1}(B + \log 2)^{-1}$. □

Lemma 10.8 *If* $\rho(f)$ *satisfies* (2) *and* (3) *then there exists* $0 < \beta < 1$ *with the following property. If* L *and* R *are adjacent atoms of* $\mathcal{Q}_n(f)$ *and we have* $\ell(L) \geq m$ *and* $\ell(R) \geq m$*, then*

$$\left| \frac{|L|}{|R|} - \frac{|h(L)|}{|h(R)|} \right| \leq C\beta^m\,. \tag{10.14}$$

Proof Write $m = k + p$ with $p = \lceil \sigma k \rceil$ where $\sigma > 0$ is a small constant (its size will be determined in the course of the argument). We may assume that $L \cup R$ is contained in a single atom Δ of $\mathcal{P}_k(f)$. There are two cases to consider.

(a) If $L \cup R \subseteq J_k(f)$, then the required coherence estimate (10.14) follows from Lemmas 10.5 and 10.6. To see this, let $v_1, v_2, v_3 \in \mathcal{P}_{k+p}(f)$ be the endpoints of L and R, v_2 being their common endpoint. Let w_1, w_2, w_3 be the corresponding endpoints of $h(L)$ and $h(R)$. Then by Lemma 10.5 we have $|v_i - w_i| \leq C_0 |J_k(f)| \theta^k$, where $\theta = K^\sigma \mu_* < 1$ if σ is small enough. On the other hand, condition (3) tells us that

$$a_{k+1}a_{k+2}\cdots a_{k+p} \le \exp\{p\omega(p/k)\} \le \exp\{p\omega(\sigma)\}\,.$$

Combining this fact with Lemma 10.6, we get

$$|v_1 - v_2| \ge \frac{M^p}{\left(a_{k+1}a_{k+2}\cdots a_{k+p}\right)^2}\,|J_k(f)| \ge \frac{M^p}{e^{2p\omega(\sigma)}}\,|J_k(f)|\,.$$

The same lower bound holds for $|v_2 - v_3|$. From these facts, we deduce after some simple computations that

$$\begin{aligned}\left|\frac{|L|}{|R|} - \frac{|h(L)|}{|h(R)|}\right| &= \left|\frac{|v_1 - v_2|}{|v_2 - v_3|} - \frac{|w_1 - w_2|}{|w_2 - w_3|}\right| \\ &\le C_1\frac{\theta^k e^{2p\omega(\sigma)}}{M^p} \le C_2\left(\frac{\theta e^{2\sigma\omega(\sigma)}}{M^\sigma}\right)^k \le C_3\beta_1^m\,,\end{aligned}$$

where $\beta_1 = \left(\theta e^{2\sigma\omega(\sigma)}/M^\sigma\right)^{1/(1+\sigma)}$. Since $\theta < 1$ and $\sigma\omega(\sigma) \to 0$ as $\sigma \to 0$, we see that $\beta_1 < 1$ if σ is small enough.

(b) If $L \cup R$ is not contained in $J_k(f)$, there exists $j < q_{k+1}$ such that f^j is a diffeomorphism on an interval containing Δ and its two neighbors in $\mathscr{P}_k(f)$ and such that $f^j(\Delta) \subseteq J_k(f)$. By the Koebe principle and the real bounds, the distortion of f^j on $L \cup R$ is bounded by $\exp(C_4\mu_0^p)$ (where $0 < \mu_0 < 1$ is the beau constant of Theorem 3.1). Therefore we have

$$\left|\frac{|L|}{|R|} - \frac{|f^j(L)|}{|f^j(R)|}\right| \le C_5\mu_0^p \le C_6\mu_1^m\,, \tag{10.15}$$

where $\mu_1 = \mu_0^{\sigma/(1+\sigma)}$. Working similarly with $h(L), h(R) \in \mathscr{Q}_n(g)$, we get also

$$\left|\frac{|h(L)|}{|h(R)|} - \frac{|g^j(h(L))|}{|g^j(h(R))|}\right| \le C_7\mu_1^m\,. \tag{10.16}$$

Putting (10.15) and (10.16) together and using (a) we get inequality (10.14) with the constant $\beta = \max\{\mu_1, \beta_1\}$.

Hence in both cases (10.14) is established, and we are done. □

The proof of Theorem 10.4 is now almost complete. If L and R are adjacent atoms of $\mathscr{Q}_n(f)$ as above, then combining Lemma 10.7 with Lemma 10.8 we deduce that the coherence condition (10.6) is satisfied with $\lambda = \beta^{c_1}$. Therefore by Proposition 7.3 the conjugacy h is indeed $C^{1+\alpha}$ for some $\alpha > 0$.

10.4 The C^m-Approximation Lemma

Our purpose in this section is to present a technical lemma extracted from de Faria and de Melo (1999, App. A). This lemma will be used in the proof of Proposition 10.2, but it can be applied to many other one-dimensional situations, so it is of some independent interest.

We will use the following notation. Let $m \geq 1$ be a fixed integer and let $I, J \subseteq \mathbb{R}$ be fixed closed intervals. We denote by $C^m(I)$ the Banach space of C^m-mappings $f : I \to \mathbb{R}$ with the norm $\|f\|_m = \max\{\|D^i f\|_0 : 0 \leq i \leq m\}$, where $\|\phi\|_0 = \sup_{x \in I} |\phi(x)|$. If the need arises to emphasize the domain of f, we sometimes write $\|f\|_{I,m}$ instead of $\|f\|_m$. We consider also the closed, convex subset $C^m(I, J) \subseteq C^m(I)$ consisting of those f's such that $f(I) \subseteq J$.

The reader will undoubtedly be familiar with Leibnitz's formula for the k-th derivative of a product of two functions, to wit

$$D^k(uv) = \sum_{j=0}^{k} \binom{k}{j} D^j u \, D^{k-j} v \ ,$$

from which it is clear that

$$\|uv\|_m \leq 2^m \|u\|_m \|v\|_m \tag{10.17}$$

whenever $u, v \in C^m(I)$. Perhaps less familiar to the reader is the fact that something similar holds for the *composition* of two C^m mappings. Namely, we have Faa-di-Bruno's formula (cf. Herman 1979, p. 42), which reads

$$D^k(f \circ g) = \sum_{j=1}^{k} B_{j,k}(D^1 g, D^2 g, \ldots, D^j g) \, D^{k-j+1} f \circ g \ ,$$

where each $B_{j,k}$ is a homogeneous polynomial of degree $k - j + 1$ on j variables whose coefficients are non-negative numbers depending only on k and j. It readily follows from this formula that if $\psi \in C^m(I, J)$ and $\phi \in C^m(J)$ then

$$\|\phi \circ \psi\|_m \leq A(m) \|\phi\|_m \sum_{k=1}^{m} \|\psi\|_m^k \ , \tag{10.18}$$

where $A(m) = \max_{1 \leq k \leq m} \max_{1 \leq j \leq k} B_{j,k}(1, 1, \ldots, 1)$.

Another well-known fact we will need below is the following. Suppose $m > 1$ and consider the composition operator $(f, g) \mapsto f \circ g$ as a map $\Theta : C^m(J) \times C^{m-1}(I, J) \to C^{m-1}(I)$. Then Θ is C^1 and its Fréchet derivative is given by

$$D\Theta(f, g)(u, v) = u \circ g + v \, Df \circ g \ . \tag{10.19}$$

Note that $C^m(J) \times C^{m-1}(I, J) \subseteq C^m(J) \times C^{m-1}(I)$; we consider this last product endowed with the norm

$$\big|(f, g)\big|_{I,J,m} = \max\{\|f\|_{J,m}, \|g\|_{I,m-1}\}\,.$$

Lemma 10.9 *For each $M > 0$, there exists $c(M) > 0$ with the following property. If $f_1, g_1 \in C^m(J)$ and $f_2, g_2 \in C^{m-1}(I, J)$ and if $|(f_1, f_2)|_{I,J,m} < M$ and $|(g_1, g_2)|_{I,J,m} < M$, then*

$$\|f_1 \circ f_2 - g_1 \circ g_2\|_{m-1} \;\leq\; c(M)\,|(f_1 - g_1, f_2 - g_2)|_{I,J,m}\;.$$

Proof By the mean value theorem,

$$\|f_1 \circ f_2 - g_1 \circ g_2\|_{m-1} \;\leq\; \sup_{(\phi,\psi)} \|D\Theta(\phi, \psi)\|\;|(f_1 - g_1, f_2 - g_2)|_{I,J,m}\;,$$

where the supremum is taken over all (ϕ, ψ) in the line segment joining (f_1, f_2) to (g_1, g_2) inside $C^m(J) \times C^{m-1}(I, J)$, and where

$$\|D\Theta(\phi, \psi)\| \;=\; \sup\big\{\|D\Theta(\phi, \psi)(u, v)\|_{m-1} \,:\, |(u, v)|_{I,J,m} \leq 1\big\}$$

is the operator-norm of $D\Theta(\phi, \psi)$. Using (10.19), and then (10.17) and (10.18), we have

$$\|D\Theta(\phi, \psi)(u, v)\|_{m-1} \;\leq\; \|u \circ \psi\|_{m-1} + \|v\, D\phi \circ \psi\|_{m-1} \leq$$

$$\leq\; A(m-1)\left(\|u\|_{m-1} + 2^{m-1}\|v\|_{m-1}\|D\phi\|_{m-1}\right)\sum_{k=1}^{m-1}\|\psi\|_{m-1}^k$$

From this, and taking into account that $\|u\|_{m-1} \leq \|u\|_m \leq |(u, v)|_{I,J,m}$ as well as $\|v\|_{m-1} \leq |(u, v)|_{I,J,m}$, we deduce that

$$\|D\Theta(\phi, \psi)\| \;\leq\; A(m-1)\left(1 + 2^{m-1}\|D\phi\|_{m-1}\right)\sum_{k=1}^{m-1}\|\psi\|_{m-1}^k\,.$$

Finally, since $\|D\phi\|_{m-1} \leq \|\phi\|_m$ and $|(\phi, \psi)|_{I,J,m} < M$, we get

$$\sup_{(\phi,\psi)} \|D\Theta(\phi, \psi)\| \;\leq\; A(m-1)\left(1 + 2^{m-1}M\right)\sum_{k=1}^{m-1} M^k \;=\; c(M)\,.$$

□

Let us denote by $\boldsymbol{B}^m(I; M)$ the ball of radius M centered at the origin in $C^m(I)$.

Lemma 10.10 (The C^m-Approximation Lemma) *For each $M > 0$, there exist constants $\varepsilon_M > 0$ and $C_M > 0$ such that the following holds for all $\varepsilon \leq \varepsilon_M$. Let $\Delta_1, \Delta_2, \ldots, \Delta_{n+1}$ be closed intervals on the line or on the circle, and for each $1 \leq i \leq n$ let $f_i, g_i \in C^m(\Delta_i, \Delta_{i+1})$ be such that*

(a) For all $1 \leq j \leq k \leq n$, we have $f_k \circ f_{k-1} \circ \cdots \circ f_j \in \boldsymbol{B}^m(\Delta_j; M)$;
(b) We have $\sum_{i=1}^n \|f_i - g_i\|_m < \varepsilon$.

Then for all $k \leq n$ we have $g_k \circ g_{k-1} \circ \cdots \circ g_1 \in \boldsymbol{B}^{m-1}(\Delta_1; 2M)$, and moreover

$$\|f_k \circ f_{k-1} \circ \cdots \circ f_1 - g_k \circ g_{k-1} \circ \cdots \circ g_1\|_{m-1} \leq C_M \sum_{j=1}^{k} \|f_j - g_j\|_m .$$

Proof Using the notation of Lemma 10.9, let us write

$$C_M = \max\{1, c(2M), c(2M)c(3M)\}$$

and $\varepsilon_M = M/C_M$. We proceed by induction on k. When $k = 1$, we have $\|f_1 - g_1\|_m \leq \varepsilon$ and there is nothing to prove. Suppose the assertion is valid for all $j < k$, and write (omitting the composition symbols)

$$\|f_k f_{k-1} \cdots f_1 - g_k g_{k-1} \cdots g_1\|_{m-1} \leq$$
$$\leq \sum_{j=1}^{k} \|f_k \cdots f_{j+1} g_j g_{j-1} \cdots g_1 - f_k \cdots f_{j+1} f_j g_{j-1} \cdots g_1\|_{m-1} . \tag{10.20}$$

Since $|(f_j, g_{j-1} \circ \cdots \circ g_1)|_{\Delta_1, \Delta_j, m} < 2M$ and also $|(g_j, g_{j-1} \circ \cdots \circ g_1)|_{\Delta_1, \Delta_j, m} < 2M$, it follows from Lemma 10.9 that

$$\|f_j g_{j-1} \cdots g_1 - g_j g_{j-1} \cdots g_1\|_{m-1} \leq c(2M)\|f_j - g_j\|_m ,$$

for $j = 1, \ldots, k$. In particular, by the induction hypothesis, we have for all $1 \leq j \leq k-1$

$$\|f_j g_{j-1} \cdots g_1\|_{m-1} \leq \|g_j g_{j-1} \cdots g_1\|_{m-1} + \varepsilon_M c(2M) < 3M .$$

Taking this back to (10.20) and applying once again Lemma 10.9, we get

$$\begin{aligned} &\|f_k f_{k-1} \cdots f_1 - g_k g_{k-1} \cdots g_1\|_{m-1} \\ &\qquad \leq c(2M)\|f_k - g_k\|_m + c(2M)c(3M) \sum_{j=1}^{k-1} \|f_j - g_j\|_m \\ &\qquad \leq C_M \sum_{j=1}^{k} \|f_j - g_j\|_m , \end{aligned}$$

and this shows also that $\|g_k g_{k-1} \cdots g_1\|_{m-1} \leq M + \varepsilon_M C_M < 2M$, thereby completing the induction. □

10.5 Counterexamples to $C^{1+\alpha}$ Rigidity

As explained in Sect. 10.1, two C^4 critical circle maps with the same irrational rotation number and with a single critical point of the same odd integer criticality are conjugate to each other by a C^1 diffeomorphism. Moreover, this conjugacy is in fact a $C^{1+\alpha}$ diffeomorphism for Lebesgue almost every rotation number (Theorem 10.1). These results immediately raise the question of whether such conjugacy is *always* $C^{1+\alpha}$. The following result, obtained by Avila (2013), says that the above conjecture is not true, even if we restrict ourselves to the analytic category.

Theorem 10.6 *There exist real-analytic critical circle maps f and g with the same irrational rotation number and with a single critical point (of the same criticality) such that if h is the topological conjugacy between f and g identifying critical points, then h is not $C^{1+\alpha}$ for any $\alpha > 0$.*

The first examples of this kind were obtained by de Faria and de Melo (1999, Second Main Theorem) in the C^∞ category. Our goal in this section is to present a detailed construction of such C^∞ examples (see Theorem 10.7 below). To achieve this goal, we will consider critical circle maps whose rotation number $\rho(f) = [a_0, a_1, \ldots, a_n, \ldots]$ satisfies

$$\begin{cases} \limsup \dfrac{1}{n} \log a_n = \infty \\ \\ a_n \geq 2 \text{ for all } n. \end{cases} \tag{10.21}$$

The class of all rotation numbers satisfying (10.21) will be denoted by $\mathbb{B}$. It can be shown that the Hausdorff dimension of $\mathbb{B}$ is less than or equal to $1/2$, see Good (1941).

Theorem 10.7 *For every $\rho \in \mathbb{B}$ there exist C^∞ critical circle maps f, g with $\rho(f) = \rho(g) = \rho$ such that f and g are not $C^{1+\alpha}$ conjugate for any $\alpha > 0$.*

The proof will make use of a C^∞ surgery procedure that we explain below. These counterexamples have one additional feature: their successive renormalizations do converge together at an exponential rate. This follows from general results, such as Theorem 13.3 below, but it will also be clear from the construction.

10.5.1 Saddle-Node Surgery

Given f as above and a fixed $n \geq 1$, let $J_n = J_n(f) = [f^{q_{n+1}}(c), f^{q_n}(c)] \subseteq \mathbf{S}^1$ be the n-th renormalization interval of f. When n is very large, the first return map $f_n : J_n \to J_n$ is an almost parabolic map of length a_{n+1}.

Let $\Delta_1^{(n)}$ be the fundamental domain of this almost parabolic map which is adjacent to $x_n = f^{q_n}(c)$, and let $\Delta_j^{(n)} = f_n^{j-1}(\Delta_1^{(n)})$, for all $j \leq a_{n+1}$. Let $z_n \in \Delta_1^{(n)}$ be the point such that $f_n^{a_{n+1}}(z_n) = x_{n+3} = f^{q_{n+3}}(c)$, that is, $z_n = f^{q_{n+3}-a_{n+1}q_{n+1}}(c)$. Note that since $a_{n+1} \geq 2$, x_{n+3} is not an endpoint of $f_n^{a_{n+1}}(\Delta_1^{(n)})$, and so by the real bounds it splits $f_n^{a_{n+1}}(\Delta_1^{(n)})$ into two intervals of comparable lengths. Hence the same holds for z_n. Namely, z_n splits $\Delta_1^{(n)}$ into two intervals L_n, R_n with $|L_n| \asymp |R_n|$. In particular we have $\tau|\Delta_1^{(n)}| \leq |L_n| \leq (1-\tau)|\Delta_1^{(n)}|$ (and similarly for R_n) for some constant τ depending on the real bounds. We use this fact in the proof of Proposition 10.2 below.

Consider now another critical circle map $\tilde{f}$ with the same rotation number as f, the interval $\tilde{J}_n = J_n(\tilde{f})$, the first return map $\tilde{f}_n : \tilde{J}_n \to \tilde{J}_n$, the point $\tilde{z}_n = \tilde{f}^{q_{n+3}-a_{n+1}q_{n+1}}(\tilde{c})$ and the corresponding intervals $\tilde{L}_n$, $\tilde{R}_n$. Also, let $N_n = \lceil a_{n+1}/2 \rceil$.

Definition 10.7 The number

$$\left| \frac{|f_n^{N_n-1}(L_n)|}{|f_n^{N_n-1}(R_n)|} - \frac{|\tilde{f}_n^{N_n-1}(\tilde{L}_n)|}{|\tilde{f}_n^{N_n-1}(\tilde{R}_n)|} \right|$$

is called the n-th order discrepancy between f and $\tilde{f}$.

Proposition 10.2 *Given a C^∞ critical circle map f with $\rho(f) \in \mathbb{B}$, consider a function $\sigma(n) \to \infty$ such that*

$$\limsup \frac{1}{n\sigma(n)} \log a_{n+1} = \infty .$$

Then for all $n \geq 1$, there exists a critical circle map $\tilde{f} = F(n; f)$ with the same rotation number and critical point as f and having the following properties.

(a) We have $\tilde{f}^j(c) = f^j(c)$ for $0 \leq j \leq q_{n+1}$; in particular, $J_n(\tilde{f}) = J_n = J_n(f)$.
(b) We have $\tilde{f} = \Phi \circ f$, where Φ is a C^∞ diffeomorphism such that

$$\|\Phi^{\pm 1} - \mathrm{Id}_{\mathbf{S}^1}\|_{C^k} \leq B_k |J_n|^{\sigma(n)-k+1}$$

for all k, where $B_k > 0$ is constant depending only on k.
(c) The n-th order discrepancy between f and $\tilde{f}$ is $\geq C|J_n|^{2\sigma(n)}$.
(d) We have $J_{n+1}(\tilde{f}) = J_{n+1}(f)$ and $\tilde{f}_{n+1} = f_{n+1}$; in particular, m-th order discrepancy between f and $\tilde{f}$ is equal to zero for all $m > n$.

Proof We modify f inside $f^{-1}(\Delta_1^{(n)})$ using a C^∞ bump function so as to move z_n by a distance $\geq C|\Delta_1^{(n)}|^{1+\sigma(n)}$ inside $\Delta_1^{(n)}$. This we do as follows.

Let $\varphi : [0, 1] \to [0, 1]$ be a C^∞ perturbation of the identity such that $|\varphi(x)-x| \geq |\Delta_1^{(n)}|^{\sigma(n)}$ for all $\tau \leq x \leq 1-\tau$ (and τ as above), and such that $|D^k\varphi(x)| \leq B_k|\Delta_1^{(n)}|^{\sigma(n)}$ for all $0 \leq x \leq 1$ and all $k \geq 2$. Define $\phi_n : \Delta_1^{(n)} \to \Delta_1^{(n)}$ by $\phi_n = A_n \circ \varphi \circ A_n^{-1}$ where A_n is the affine orientation-preserving map that carries $[0, 1]$ onto $\Delta_1^{(n)}$. Note that $|\phi_n(z_n) - z_n| \geq |\Delta_1^{(n)}|^{1+\sigma(n)}$. Moreover, since $D^k\phi_n = |\Delta_1^{(n)}|^{1-k}D^k\varphi$, we have

$$\|\phi_n^{\pm 1} - \mathrm{Id}_{\Delta_1^{(n)}}\|_{C^k} \leq B_k|\Delta_1^{(n)}|^{\sigma(n)-k+1}$$

for all k. Define $\psi_n : \Delta_{a_{n+1}}^{(n)} \to \Delta_{a_{n+1}}^{(n)}$ as the conjugate of ϕ_n^{-1} by the diffeomorphism $f_n^{a_{n+1}-1} : \Delta_1^{(n)} \to \Delta_{a_{n+1}}^{(n)}$, namely

$$\psi_n = f_n^{a_{n+1}-1} \circ \phi_n^{-1} \circ (f_n^{a_{n+1}-1})^{-1} . \tag{10.22}$$

Using the C^m Approximation Lemma 10.10, we see from (10.22) that

$$\|\psi_n^{\pm 1} - \mathrm{Id}_{\Delta_{a_n}^{(n)}}\|_{C^{k-1}} \leq C\|\phi_n^{\pm 1} - \mathrm{Id}_{\Delta_1^{(n)}}\|_{C^k} \leq B_k|\Delta_1^{(n)}|^{\sigma(n)-k+1} .$$

Define $\Phi : S^1 \to S^1$ to be equal to ϕ_n on $\Delta_1^{(n)}$, to ψ_n on $\Delta_{a_n}^{(n)}$ and to the identity everywhere else. The critical circle map we look for is $\tilde{f} = \Phi \circ f$. Note that $\|\Phi^{\pm 1} - \mathrm{Id}_{S^1}\|_{C^k} \leq B_k|\Delta_1^{(n)}|^{\sigma(n)-k+1}$ for all k; since $|\Delta_1^{(n)}| \asymp |J_n|$ by the real bounds, this proves (b). It is also clear from the construction that property (a) holds too. It follows in particular that the first $n+1$ partial quotients of the rotation number of $\tilde{f}$ agree with those of f. More remarkable is that, because what ϕ_n does is undone by ψ_n, we have

$$\begin{cases} \tilde{f}^{q_n}|I_{n+1} = f^{q_n}|I_{n+1} \\ \\ \tilde{f}^{q_{n+1}}|I_n = f^{q_{n+1}}|I_n . \end{cases}$$

In other words, $\tilde{f}_n = f_n$, the n-th renormalizations agree. Therefore all subsequent renormalizations agree as well. This shows that $\rho(\tilde{f}) = \rho(f)$ and also proves (d).

It remains to prove (c), so we estimate the n-th order discrepancy between f and $\tilde{f}$ from below. Since $|z_n - \tilde{z}_n| \geq |\Delta_1^{(n)}|^{1+\sigma(n)}$, a simple calculation yields

$$\left|\frac{|L_n|}{|R_n|} - \frac{|\tilde{L}_n|}{|\tilde{R}_n|}\right| \geq C|\Delta_1^{(n)}|^{\sigma(n)} \geq C|J_n|^{2\sigma(n)} , \tag{10.23}$$

provided n is sufficiently large. Since, by the real bounds, the map $f_n^{N-1} : \Delta_1^{(n)} \to \Delta_{N_n}^{(n)}$ has bounded distortion, and since $\tilde{f}_n = f_n$, inequality (10.23) gives us

$$\left| \frac{|f_n^{N_n-1}(L_n)|}{|f_n^{N_n-1}(R_n)|} - \frac{|\tilde{f}_n^{N_n-1}(\tilde{L}_n)|}{|\tilde{f}_n^{N_n-1}(\tilde{R}_n)|} \right| \geq C|J_n|^{2\sigma(n)} ,$$

and this proves (c). □

10.5.2 The Counterexamples

We now iterate the procedure given by Proposition 10.2 to prove Theorem 10.7.

Proof of Theorem 10.7 We start with a C^∞ map f with $\rho(f) \in \mathbb{B}$ as before and select $n_1 < n_2 < \cdots$ such that

$$\lim_{i\to\infty} \frac{1}{n_i\sigma(n_i)} \log a_{n_i+1} = \infty , \tag{10.24}$$

where $\sigma(n)$ is as in Proposition 10.2. Now generate a sequence $g_0, g_1, \ldots, g_i, \ldots$ recursively, starting with $g_0 = f$, and taking, for all $i \geq 0$, $g_{i+1} = F(n_{i+1}, g_i)$, where $F(\cdot, \cdot)$ is as given in Proposition 10.2. Each g_i is a C^∞ critical circle map with $\rho(g_i) = \rho(f)$, and $g_{i+1} = \Phi_{i+1} \circ g_i$, where Φ_{i+1} is a C^∞ diffeomorphism with

$$\|\Phi_{k+1}^{\pm 1} - \mathrm{Id}_{S^1}\|_{C^k} \leq B_k\theta^{n_i(\sigma(n_i)-k+1)} , \tag{10.25}$$

for all k, where $0 < \theta < 1$ is a constant depending only on the real bounds. From (10.25) it follows that $\Phi = \lim \Phi_i \circ \cdots \circ \Phi_1$ exists as a C^∞ diffeomorphism, and therefore so does $g = \lim g_i = \Phi \circ f$ as a critical circle map.

Using properties (c) and (d) of Proposition 10.2 for each g_i, we deduce that the n_i-th order discrepancy between f and g satisfies

$$\left| \frac{|f_{n_i}^{N_i-1}(L_{n_i})|}{|f_{n_i}^{N_i-1}(R_{n_i})|} - \frac{|g_{n_i}^{N_i-1}(\tilde{L}_{n_i})|}{|g_{n_i}^{N_i-1}(\tilde{R}_{n_i})|} \right| \geq C|J_{n_i}|^{2\sigma(n_i)} , \tag{10.26}$$

where $N_i = \lceil a_{n_i+1}/2 \rceil$, etc.

Now, let $h : \boldsymbol{S}^1 \to \boldsymbol{S}^1$ be the conjugacy between f and g mapping the critical point c to itself. Suppose h were $C^{1+\beta}$ for some $\beta > 0$. Then the left-hand side of (10.26) would be $\leq C|f_{n_i}^{N_i-1}(\Delta_1^{(n_i)})|^\beta$, where $\Delta_1^{(n_i)} = L_{n_i} \cup R_{n_i}$. But by Yoccoz's Lemma 7.3, we have

$$|f_{n_i}^{N_i-1}(\Delta_1^{(n_i)})| \asymp \frac{1}{N_i^2}|J_{n_i}| \asymp \frac{1}{a_{n_i+1}^2}|J_{n_i}| \,. \tag{10.27}$$

Combining the above with (10.26) and (10.27), we would get the inequality

$$a_{n_i+1}^{2\beta}|J_{n_i}|^{2\sigma(n_i)-\beta} \leq C \,.$$

But by the real bounds $|J_n| \geq C\mu^n$ for all n, where $0 < \mu < 1$. Therefore, taking logarithms, we would have

$$\limsup \frac{\log a_{n_i+1}}{n_i \sigma(n_i)} \leq \frac{1}{\beta} \log \frac{1}{\mu} \,, \tag{10.28}$$

but this clearly contradicts (10.24). □

Remark 10.4 A closer look at the construction performed above, especially at expressions (10.25) and (10.28), reveals that if

$$\limsup \frac{1}{n} \log a_n > \frac{k}{\beta_0} \log \frac{1}{\mu}$$

then one can construct a pair of C^k critical circle maps (whose renormalizations converge exponentially fast) that are not $C^{1+\beta}$ conjugate for any $\beta \geq \beta_0$.

Exercises

Exercise 10.1 Let f be a C^r multicritical circle map with critical points labeled $c_1, c_2, \ldots, c_N$, and let ϕ a C^r circle diffeomorphism. Prove that there exist constants $C = C(f, \phi) > 0$ and $0 < \mu = \mu(f) < 1$ such that, for all $n \in \mathbb{N}$ and each $1 \leq i \leq N$, we have

$$d_{r-1}\left(\mathscr{R}_i^n f \,,\, \mathscr{R}_i^n(\phi \circ f \circ \phi^{-1})\right) \leq C\,\mu^n \,,$$

where $\mathscr{R}_i^n$ denotes the n-th renormalization around the i-th critical point (i.e., around c_i for f and around $\phi(c_i)$ for the conjugated map).

Exercise 10.2 Give a detailed proof of Lemma 10.6

Exercise 10.3 Show that the number ρ whose partial quotients are $a_n = 2^{2^n}$ is Diophantine (recall Chap. 4) and belongs to the set $\mathbb{B}$ defined by (10.21).

Exercise 10.4 Give a detailed proof of the assertion made in Remark 10.4.

Chapter 11
Quasiconformal Deformations

This chapter should be regarded as a second intermezzo (after Chap. 5). Here we briefly review some standard facts about the theory of quasiconformal mappings in the complex plane and the Riemann sphere. In such a short exposition we can hardly do justice to this beautiful and powerful theory. We refer the reader to the books of Ahlfors (2006) and Lehto and Virtanen (1973), which are classical references for the subject. Modern treatments, highlighting connections with Dynamical Systems and Teichmüller theory, can be found in the books of Gardiner (1987), Carleson and Gamelin (1993), de Melo and van Strien (1993), McMullen (1994, 1996), Gardiner and Lakic (2000), Hubbard (2006), de Faria and de Melo (2008), and Farb and Margalit (2012). Here we limit ourselves to stating some fundamental facts about quasiconformal mappings, and to establishing an approximation result, namely Theorem 11.4 (borrowed from Guarino and de Melo 2017), that will be a useful tool in the discussions of Chap. 13. Some of the ideas mentioned in this chapter will reappear in Chap. 14, which is fully focused on holomorphic methods.

The use of quasiconformal theory in holomorphic dynamics was initiated by Sullivan (1985). He applied one of the cornerstones of the theory—the *measurable Riemann mapping theorem with parameters* or *Ahlfors–Bers theorem* (see Sect. 11.1)—to solve a long-standing conjecture by Fatou, stating that every component of the complement of the Julia set of a rational map of the Riemann sphere is eventually periodic.

What makes quasiconformal maps so useful in the study of holomorphic dynamical systems is the fact that, unlike analytic maps, they are very flexible. In many arguments in dynamics, say in the study of structural stability, it is sometimes necessary to be able to deform a given system into another nearby, within the same topological class, preserving its smoothness. Deformations using conjugation by C^1 diffeomorphisms (or better) are usually inadequate, because they preserve the eigenvalues at all periodic points. In the case of holomorphic dynamics, there is an abundance of periodic points in the non-wandering set of the map (which is essentially its Julia set), and the situation is simply too rigid

E. de Faria, P. Guarino, *Dynamics of Circle Mappings*, IMPA Monographs,
https://doi.org/10.1007/978-3-031-67495-2_11

to allow this type of deformation. By contrast, using conjugation by quasiconformal homeomorphisms, one can deform a holomorphic system into another system which is still holomorphic, but has different multipliers at corresponding periodic points. Moreover, the Ahlfors–Bers theorem yields a continuous path of holomorphic systems of the same topological type (known as *Beltrami paths*, see Chap. 14 below) joining the original system to the deformed one.

In this book, we are interested in the theory of quasiconformal maps only to the extent that it can be applied to the study of critical circle maps. In what follows, we make no attempt at a systematic exposition of this beautiful theory, but simply take stock of the relevant facts that will be needed later.

11.1 Quasiconformal Homeomorphisms

The notion of quasiconformal homeomorphism was born of the necessity to solve a geometric *extremal problem* that can be formulated as follows: Given two rectangles in the plane, what is the *most nearly conformal* homeomorphism mapping one rectangle to the other, sending vertices to vertices? The answer turns out to be the obvious affine map that carries vertices to vertices as specified, but a proof of this fact depends on an inequality established by Grötzsch in 1928. Such affine map will be conformal if and only if the ratios between the "vertical" and "horizontal" sides are the same for both rectangles.

11.1.1 The Geometric Definition

The above extremal problem can be similarly formulated replacing rectangles with round annuli having concentric boundaries. The answer by Grötzsch reveals in particular that a conformal homeomorphism exists between both annuli if and only if the ratios of inner to outer radius are the same for both annuli, i.e., if and only if the have the same *modulus*, as we proceed to define.

Given $0 < r < R \leq \infty$, the *conformal modulus*, or simply *modulus*, of the round annulus $A_{r,R} = \{z \in \mathbb{C} : r < |z| < R\}$ is defined to be $\text{mod}(A_{r,R}) = \log(R/r)$. Now, given any topological annulus in the plane, i.e., any doubly connected region $\Omega \subset \mathbb{C}$ not equal to a punctured disk or plane, it can be shown (as a special case of the famous *uniformization theorem*, see Exercise 11.9) that there exists a conformal equivalence between Ω and some round annulus $A_{r,R}$; hence we define $\text{mod}(\Omega) = \text{mod}(A_{r,R})$.

Thus, the above discussion motivates the following geometric definition of quasiconformality.

Definition 11.1 An orientation-preserving homeomorphism $f : U \to V$ between two regions U, V in the complex plane (or Riemann sphere) is said to be K-

quasiconformal, where $K \geqslant 1$ is a given constant, if for every topological annulus $\Omega \subset U$ we have $K^{-1}\operatorname{mod}(\Omega) \leq \operatorname{mod}(f(\Omega)) \leq K\operatorname{mod}(\Omega)$.

This definition makes it obvious that a composition of a K_1-quasiconformal homeomorphism with a K_2-quasiconformal homeomorphism is K_1K_2-quasiconformal, and that the inverse of a K-quasiconformal homeomorphism is also K-quasiconformal. However, it is not of much practical value when we want to examine quasiconformal maps at the infinitesimal level. For instance, it is far from obvious from this definition that a 1-quasiconformal homeomorphism is in fact conformal (this is known as Weyl's lemma).

11.1.2 The Analytic Definition

Let us first recall the two basic differential operators of complex calculus:

$$\frac{\partial}{\partial z} = \frac{1}{2}\left(\frac{\partial}{\partial x} - i\,\frac{\partial}{\partial y}\right) \quad \text{and} \quad \frac{\partial}{\partial \overline{z}} = \frac{1}{2}\left(\frac{\partial}{\partial x} + i\,\frac{\partial}{\partial y}\right).$$

Instead of $\frac{\partial f}{\partial z}$ and $\frac{\partial f}{\partial \overline{z}}$, we will use the more compact notation ∂f and $\overline{\partial} f$ respectively. In other words, if Ω is a domain in $\mathbb{C}$ and $f : \Omega \to \mathbb{C}$ is differentiable at $w \in \Omega$ (in the real sense), then

$$\big(Df(w)\big)(z) = \partial f(w)\, z + \overline{\partial} f(w)\, \overline{z} \quad \text{for any } z \in \mathbb{C}.$$

Recall also that a continuous function $h : \mathbb{R} \to \mathbb{R}$ is *absolutely continuous* if it is differentiable at Lebesgue almost every point, its derivative is integrable and $h(b) - h(a) = \int_a^b Dh(t)\, dt$, for any a and b in $\mathbb{R}$. A continuous function $f : \Omega \subset \mathbb{C} \to \mathbb{C}$ is *absolutely continuous on lines* in Ω if its real and imaginary parts are absolutely continuous on Lebesgue almost every horizontal line, and Lebesgue almost every vertical line.

Definition 11.2 Let $\Omega \subset \mathbb{C}$ be a domain and let $K \geqslant 1$. An orientation-preserving homeomorphism $f : \Omega \to f(\Omega)$ is K*-quasiconformal* if it is absolutely continuous on lines and

$$\left|\overline{\partial} f(z)\right| \leq \left(\frac{K-1}{K+1}\right) \left|\partial f(z)\right| \quad \text{for a.e. } z \in \Omega.$$

A proof that Definitions 11.1 and 11.2 are equivalent can be found in Ahlfors (2006, Ch. II).

11.1.3 Measurable Riemann Mapping Theorem

Given a K-quasiconformal homeomorphism $f : \Omega \to f(\Omega)$ we define its *Beltrami coefficient* as the measurable function $\mu_f : \Omega \to \mathbb{D}$ given by

$$\mu_f(z) = \frac{\overline{\partial} f(z)}{\partial f(z)} \quad \text{for a.e. } z \in \Omega.$$

Note that μ_f belongs to $L^\infty(\Omega)$ and satisfies $\|\mu_f\|_\infty \le (K-1)/(K+1) < 1$. Conversely, any measurable function from Ω to $\mathbb{C}$ with L^∞ norm less than one is the Beltrami coefficient of a quasiconformal homeomorphism. More precisely, we have the following result, which is known as *Morrey's theorem* or *measurable Riemann mapping theorem.*

Theorem 11.1 *Given any measurable function $\mu : \Omega \to \mathbb{D}$ such that $|\mu(z)| \le (K-1)/(K+1) < 1$ almost everywhere in Ω for some $K \geqslant 1$, there exists a K-quasiconformal homeomorphism $f^\mu : \Omega \to f^\mu(\Omega)$ which is a solution of the* Beltrami equation

$$\overline{\partial} f^\mu(z) = \partial f^\mu(z)\,\mu(z) \quad \textit{for a.e. } z \in \Omega. \tag{11.1}$$

This solution is unique up to post-composition with biholomorphisms. In particular, if Ω is the entire Riemann sphere, there is a unique solution (called the normalized *solution) that fixes* 0, 1 *and* ∞.

See Ahlfors (2006, Ch. V, Section B) or Lehto and Virtanen (1973, Ch. V) for the proof. Note that Theorem 11.1 not only assures the existence of a solution of the Beltrami equation, but also the fact that such a solution is a homeomorphism, i.e., injective in Ω.

Remark 11.1 Theorem 11.1 yields a solution to the classical problem of finding local *isothermal coordinates* on a given Riemannian surface. This problem goes back to Gauss, and in modern language his solution amounts to solving the Beltrami equation in the case when the Beltrami coefficient is a function that can be written as a convergent power series in z and $\overline{z}$. In Exercise 11.10, the reader is invited to find a solution to (11.1) when μ is a polynomial in z and $\overline{z}$.

Later in this chapter (in the proof of Theorem 11.4) we will need the following fact, whose proof can be found in Ahlfors (2006, Ch. V, Section C).

Proposition 11.1 *If $\mu_n \to 0$ in the unit ball of $L^\infty(\widehat{\mathbb{C}})$, then the normalized quasiconformal homeomorphisms f^{μ_n} converge to the identity uniformly on compact sets of $\mathbb{C}$. In general, if $\mu_n \to \mu$ almost everywhere in $\widehat{\mathbb{C}}$ and $\|\mu_n\|_\infty \le k < 1$ for all $n \in \mathbb{N}$, then the normalized quasiconformal homeomorphisms f^{μ_n} converge to f^μ uniformly on compact sets of $\mathbb{C}$.*

The Beltrami equation induces therefore a one-to-one correspondence between the space of quasiconformal homeomorphisms of $\widehat{\mathbb{C}}$ that fix 0, 1 and ∞, and the space of (equivalence classes of) measurable complex-valued functions μ on $\widehat{\mathbb{C}}$ for which $\|\mu\|_\infty < 1$. The following deep result expresses the *analytic dependence* of the solution of the Beltrami equation with respect to μ, and it is known as the *Ahlfors–Bers theorem.*

Theorem 11.2 *Let $\mathscr{U}$ be an open subset of some complex Banach space and consider a map $\mathscr{U} \times \mathbb{C} \to \mathbb{D}$, denoted by $(\lambda, z) \mapsto \mu_\lambda(z)$, satisfying the following properties.*

1. *For every λ the function $\mathbb{C} \to \mathbb{D}$ given by $z \mapsto \mu_\lambda(z)$ is measurable, and $\|\mu_\lambda\|_\infty \leq k$ for some fixed $k < 1$.*
2. *For Lebesgue almost every $z \in \mathbb{C}$, the function $\mathscr{U} \to \mathbb{D}$ given by $\lambda \mapsto \mu_\lambda(z)$ is holomorphic.*

For each $\lambda \in \mathscr{U}$, let f^{μ_λ} be the unique quasiconformal homeomorphism of the Riemann sphere that fixes 0, 1 and ∞, and whose Beltrami coefficient is μ_λ (f^{μ_λ} is given by Theorem 11.1). Then $\lambda \mapsto f^{\mu_\lambda}(z)$ is holomorphic for all $z \in \mathbb{C}$.

Again, we refer the reader to Ahlfors (2006, Ch. V, Section C) for a proof of Theorem 11.2.

11.2 A Simple Dynamical Application

The measurable Riemann mapping theorem and its version with parameters, the Ahlfors–Bers theorem, have countless striking applications to many different areas, such as holomorphic dynamics, Kleinian groups, Riemann surface theory, Teichmüller theory. See the books we mentioned in the introduction to this chapter and references therein.

Here, we would like to discuss a simple application which is more specifically related to critical circle maps. It concerns our old friend, the Arnold family $f_\alpha : \mathbb{C}^* \to \mathbb{C}^*$, with corresponding lifts $F_\alpha : \mathbb{C} \to \mathbb{C}$ given by

$$F_\alpha(z) = z + \alpha - \frac{1}{2\pi} \sin 2\pi z \,.$$

The maps $f_\alpha : \mathbb{C}^* \to \mathbb{C}^*$ are holomorphic branched covering of the cylinder $\mathbb{C}^* = \mathbb{C}/\mathbb{Z}$, branched at $z = 1$, and the restrictions $f_\alpha|_{S^1}$ are critical circle maps with a unique cubic critical point at $z = 1$. We will show here that the elements of this family whose restrictions to S^1 have irrational rotation number are *quasiconformally rigid*. This fact will be relevant in our discussion of *holomorphic commuting pairs* in Chap. 14.

But first, some terminology and general facts. If $f : \mathbb{C}^* \to \mathbb{C}^*$ is holomorphic, we denote by S_f the set of singular values of f, i.e., points in $\mathbb{C}^*$ all neighborhoods

U of which are such that $f^{-1}(U) \xrightarrow{f} U$ fails to be a covering map. We also write $X_f = \mathbb{C}^* \setminus S_f$ for the set of regular values, so that $f^{-1}(X_f) \xrightarrow{f} X_f$ is always a covering map. For example, since $1 \in \partial\mathbb{D}$ is the unique critical point of f_α, it is easy to see that $S_{f_\alpha} = \{f_\alpha(1)\}$; in this case $f_\alpha^{-1}(X_{f_\alpha})$ has an infinite discrete complement in $\mathbb{C}^*$. We let J_f be the Julia set of f (the closure of the set of repelling periodic points). A theorem due to Keen (1988) asserts that, if S_f is finite, then f has no wandering domains, i.e., no connected component of the complement of J_f is wandering. This is certainly the case with the maps in the Arnold family.

We will need the following lemma.

Lemma 11.1 *The family $\{f_\alpha\}$ is topologically complete, i.e., every symmetric, normalized holomorphic self-map of $\mathbb{C}^*$ which is topologically conjugate to a member of the family is a member also.*

Proof Let $f : \mathbb{C}^* \to \mathbb{C}^*$ be holomorphic and suppose $h : \widehat{\mathbb{C}} \to \widehat{\mathbb{C}}$ is an orientation preserving homeomorphism fixing 0 and ∞ and satisfying $h \circ f_\alpha = f \circ h$. Let $A(z) = \lambda z$, where $\lambda = h \circ f_\alpha(1)/f_\alpha(1)$. This A is homotopic to h relative to $S_{f_\alpha} \cup \{0, \infty\}$, so the covering homotopy theorem yields a holomorphic lift $\widehat{A} : f_\alpha^{-1}(X_{f_\alpha}) \to f^{-1}(X_f)$, which is then homotopic to h relative to $f_\alpha^{-1}(S_{f_\alpha}) \cup \{0, \infty\}$. Some easy topology and the removable singularity theorem show that $\widehat{A}$ is Möbius and fixes 0 and ∞. In particular, if f is symmetric about $\partial\mathbb{D}$ and is normalized so that its critical point lies at $1 \in \partial\mathbb{D}$, then $\widehat{A}$ is the identity and $|\lambda| = 1$, say $\lambda = e^{2\pi i\theta}$. Therefore $f = A \circ f_\alpha \circ \widehat{A}^{-1} = f_{\alpha+\theta}$. □

Theorem 11.3 *If $\rho(f_\alpha)$ is irrational then f_α admits no non-trivial, symmetric, invariant Beltrami differentials entirely supported in its Julia set.*

Proof Now suppose μ is an f_α-invariant Beltrami differential in $\widehat{\mathbb{C}}$ with support in J_{f_α}; assume also that μ is symmetric about $\partial\mathbb{D}$. For all sufficiently small real t, let $h_t : \widehat{\mathbb{C}} \to \widehat{\mathbb{C}}$ be the unique solution to $\overline{\partial} h_t = (t\mu)\, \partial h_t$ fixing $\{0, 1, \infty\}$ pointwise, and let $f_t = h_t \circ f_\alpha \circ h_t^{-1}$. Since $t\mu$ is symmetric and f_α-invariant, each f_t is symmetric and holomorphic, and has a single critical point at $1 \in \partial\mathbb{D}$. Using Lemma 11.1, we have $f_t = f_{\alpha_t}$ for some α_t. But then $\rho(f_{\alpha_t}) = \rho(f_\alpha)$ is irrational, so $\alpha_t = \alpha$ for all t (because the function $\alpha \mapsto \rho(f_\alpha)$ is a devil staircase; see Lemma 4.7). Therefore, h_t commutes with f_α for all t; in particular h_t must permute the elements of $Y_n = f_\alpha^{-n}(1)$, which is discrete in $\mathbb{C}^*$, for each $n \geqslant 0$. Since $h_0 = \mathrm{Id}_{\widehat{\mathbb{C}}}$ and for each $z \in \widehat{\mathbb{C}}$ the path $t \to h_t(z)$ is continuous by the Ahlfors–Bers theorem, we deduce that h_t fixes Y_n pointwise for all $n \geqslant 0$, for all t. But by Montel's theorem,

$$J_{f_\alpha} \subseteq \overline{\bigcup_{n \geqslant 0} Y_n}\,,$$

so h_t agrees with the identity over J_{f_α} for all t. Since h_t is conformal outside J_{f_α}, it follows that $h_t \equiv \mathrm{Id}_{\widehat{\mathbb{C}}}$ for all t, and so $\mu \equiv 0$ almost everywhere. □

11.3 Holomorphic Approximation Lemma

As already mentioned, our goal in this brief chapter is to prove the following consequence of Theorem 11.2, borrowed from Guarino and de Melo (2017, Prop. 5.5) and Guarino (2012, Prop. 3.3.2).

Theorem 11.4 *For any bounded domain U in the complex plane there exists a number $C(U) > 0$, with $C(U) \leq C(W)$ if $U \subseteq W$, such that the following holds. Let $\{G_n : U \to G_n(U)\}_{n \in \mathbb{N}}$ be a sequence of quasiconformal homeomorphisms satisfying:*

- *The images $G_n(U)$ are uniformly bounded: there exists $R > 0$ such that $G_n(U) \subset B(0, R)$ for all $n \in \mathbb{N}$;*
- *$\mu_n \to 0$ in L^∞, where μ_n is the Beltrami coefficient of G_n in U.*

Then for any given domain V, compactly contained in U, there exist $n_0 \in \mathbb{N}$ and a sequence $\{H_n : V \to H_n(V)\}_{n \geqslant n_0}$ of bi-holomorphisms such that

$$\|H_n - G_n\|_{C^0(V)} \leq C(U) \left(\frac{R}{d\big(\partial V, \partial U\big)} \right) \|\mu_n\|_\infty \quad \text{for all } n \geqslant n_0,$$

where $d\big(\partial V, \partial U\big)$ denotes the Euclidean distance between the boundaries of U and V.

Proof For each $n \in \mathbb{N}$ we first extend μ_n to the complement of U in the trivial way:

$$\mu_n(z)\, \partial G_n(z) = \overline{\partial} G_n(z) \text{ for a.e. } z \in U, \text{ and } \mu_n(z) = 0 \text{ for all } z \in \widehat{\mathbb{C}} \setminus U.$$

Of course if $\mu_n \equiv 0$ we just take $H_n = G_n|_V$, so we may assume that $\|\mu_n\|_\infty > 0$. Fix some small $\varepsilon \in \big(0, 1 - \|\mu_n\|_\infty\big)$ and denote by $\mathscr{B}_n$ the open disk $B\big(0, (1 - \varepsilon)/\|\mu_n\|_\infty\big)$ centred at the origin with radius $(1 - \varepsilon)/\|\mu_n\|_\infty$ in the complex plane (note that $\overline{\mathbb{D}} \subset \mathscr{B}_n$). Consider the one-parameter family of Beltrami coefficients $\{\mu_n(t)\}_{t \in \mathscr{B}_n}$ defined by

$$\mu_n(t) = t \cdot \mu_n\,,$$

and note that for all $t \in \mathscr{B}_n$ we have $\big\|\mu_n(t)\big\|_\infty < 1 - \varepsilon < 1$. Denote by $f^{\mu_n(t)}$ the solution of the Beltrami equation with coefficient $\mu_n(t)$, given by Theorem 11.1, normalized to fix 0, 1 and ∞. Note that $f^{\mu_n(0)}$ is the identity for all $n \in \mathbb{N}$ and that, by uniqueness, there exists a biholomorphism $H_n : f^{\mu_n(1)}(U) \to G_n(U)$ such that

$$G_n = H_n \circ f^{\mu_n(1)} \text{ in } U.$$

In order to estimate the uniform distance between G_n and H_n, we need to first estimate the distance between $f^{\mu_n(1)}$ and the identity. To be more precise, we will prove now that the ratio $\left\| f^{\mu_n(1)} - \mathrm{Id}\right\|_{C^0(U)} / \|\mu_n\|_\infty$ is bounded by a constant only depending on U (thus, independent of n). Indeed, by Theorem 11.2, we know that for any $z \in \mathbb{C}$ the curve $\left\{ f^{\mu_n(t)}(z) : t \in [0, 1]\right\}$ is smooth. Following Ahlfors (2006, Ch. V, Section C), we use the notation

$$\dot{f}_n(z, s) = \lim_{t\to 0} \frac{f^{\mu_n(s+t)}(z) - f^{\mu_n(s)}(z)}{t}.$$

The limit exists for every $z \in \mathbb{C}$ and every $s \in [0, 1]$, and the convergence is uniform on compact sets of $\mathbb{C}$. Then we have

$$\left\| f^{\mu_n(1)} - \mathrm{Id}\right\|_{C^0(U)} = \sup_{z\in U} \left\{ \left| f^{\mu_n(1)}(z) - z\right| \right\} \le \sup_{z\in U} \left\{ \int_0^1 \left|\dot{f}_n(z, s)\right| ds \right\}.$$

Moreover, $\dot{f}_n$ has the following integral representation, borrowed from Ahlfors (2006, Ch. V, Section C, Theorem 5):

$$\dot{f}_n(z, s) = -\frac{1}{\pi} \iint_U \mu_n(w)\, S\big(f^{\mu_n(s)}(w), f^{\mu_n(s)}(z)\big) \left(\partial f^{\mu_n(s)}(w)\right)^2 dxdy,$$

for every $z \in \mathbb{C}$ and every $s \in [0, 1]$, where $w = x + iy$ and

$$S(w, z) = \frac{1}{w - z} - \frac{z}{w - 1} + \frac{z - 1}{w} = \frac{z(z - 1)}{w(w - 1)(w - z)}.$$

From the well-known formula

$$\det\left(Df^{\mu_n(s)}(w)\right) = \left|\partial f^{\mu_n(s)}(w)\right|^2 - \left|\bar{\partial} f^{\mu_n(s)}(w)\right|^2 \tag{11.2}$$

we obtain

$$\left|\partial f^{\mu_n(s)}(w)\right|^2 = \frac{1}{1 - |s|^2|\mu_n(w)|^2} \det\left(Df^{\mu_n(s)}(w)\right),$$

and then we deduce that $\left|\dot{f}_n(z, s)\right|$ is bounded by

$$\begin{aligned}
&\frac{1}{\pi} \iint_U \frac{|\mu_n(w)|}{1 - |s|^2|\mu_n(w)|^2} \det\left(Df^{\mu_n(s)}(w)\right) \left|S\big(f^{\mu_n(s)}(w), f^{\mu_n(s)}(z)\big)\right| dxdy \\
&\le \frac{1}{\pi} \frac{\|\mu_n\|_\infty}{1 - |s|^2\|\mu_n\|_\infty^2} \iint_U \det\left(Df^{\mu_n(s)}(w)\right) \left|S\big(f^{\mu_n(s)}(w), f^{\mu_n(s)}(z)\big)\right| dxdy \\
&\qquad = \frac{1}{\pi} \frac{\|\mu_n\|_\infty}{1 - |s|^2\|\mu_n\|_\infty^2} \iint_{f^{\mu_n(s)}(U)} \left|S\big(w, f^{\mu_n(s)}(z)\big)\right| dxdy.
\end{aligned}$$

Therefore, the length of the curve $\{f^{\mu_n(t)}(z) : t \in [0, 1]\}$ is bounded by

$$\frac{1}{\pi}\int_0^1 \left[\frac{\|\mu_n\|_\infty}{1-|s|^2\|\mu_n\|_\infty^2}\iint_{f^{\mu_n(s)}(U)} \left|S\big(w, f^{\mu_n(s)}(z)\big)\right| dxdy\right] ds$$
$$\leq \frac{1}{\pi}\frac{\|\mu_n\|_\infty}{1-\|\mu_n\|_\infty^2}\int_0^1 \left[\iint_{f^{\mu_n(s)}(U)} \left|S\big(w, f^{\mu_n(s)}(z)\big)\right| dxdy\right] ds\,.$$

Considering

$$M_n(U) = \frac{1}{\pi}\sup_{z\in U}\left\{\int_0^1 \left[\iint_{f^{\mu_n(s)}(U)} \left|S\big(w, f^{\mu_n(s)}(z)\big)\right| dxdy\right] ds\right\},$$

we get

$$\left\| f^{\mu_n(1)} - \mathrm{Id}\right\|_{C^0(U)} \leq \frac{\|\mu_n\|_\infty}{1-\|\mu_n\|_\infty^2}\, M_n(U).$$

Recall that, by hypothesis, $\mu_n \to 0$ in $L^\infty(U)$. With this at hand, we deduce from Proposition 11.1 that, for any $s \in [0, 1]$, the sequence $\{f^{\mu_n(s)}\}$ converges uniformly to the identity in $\overline{U}$. Therefore, the sequence $\{M_n(U)\}$ converges to

$$\frac{1}{\pi}\sup_{z\in U}\left\{\iint_U \left|S(w, z)\right| dxdy\right\} < \frac{1}{\pi}\sup_{z\in U}\left\{\iint_{\mathbb{C}} \left|S(w, z)\right| dxdy\right\}.$$

We claim that this supremum is finite. Indeed, for fixed $z \in \mathbb{C}$ we have that $S(w, z)$ is in $L^1(\mathbb{C})$, since it has simple poles at 0, 1 and z, and is $O\big(|w|^{-3}\big)$ near ∞. Finiteness follows then from the compactness of $\overline{U}$. With this at hand, we obtain $n_1 \in \mathbb{N}$ such that for all $n \geqslant n_1$ we have

$$\left\| f^{\mu_n(1)} - \mathrm{Id}\right\|_{C^0(U)} \leq M(U)\, \|\mu_n\|_\infty$$

with

$$M(U) = \frac{2}{\pi}\sup_{z\in U}\left\{\iint_U \left|S(w, z)\right| dxdy\right\},$$

where we have used the fact that $x \mapsto x/(1-x^2)$ is tangent to the identity at the origin.

Finally, we restrict both H_n and $G_n = H_n \circ f^{\mu_n(1)}$ to V, and estimate its uniform distance. With this purpose, let $\delta > 0$ be the Euclidean distance between the boundaries ∂V and ∂U (which are disjoint compact sets), that is, $\delta = d\big(\partial V, \partial U\big) = \min\big\{|z - w| : z \in \partial V, w \in \partial U\big\}$. Again by Proposition 11.1, there exists $n_0 \geqslant n_1$ in $\mathbb{N}$ such that for all $n \geqslant n_0$ we have $V \subset f^{\mu_n(1)}(U)$ and moreover

$$f^{\mu_n(1)}(U) \supseteq B(z, \delta/2) \quad \text{for all } z \in V.$$

If we consider the restriction of H_n to V we have

$$\begin{aligned}\|H_n - G_n\|_{C^0(V)} &\leq \left\|H_n'\right\|_{C^0(V)} \left\|f^{\mu_n(1)} - \mathrm{Id}\right\|_{C^0(U)}\\ &\leq \left\|H_n'\right\|_{C^0(V)} M(U)\, \|\mu_n\|_\infty .\end{aligned}$$

Finally, by Cauchy's standard estimates, we deduce for all $z \in V$

$$\begin{aligned}\left|H_n'(z)\right| &= \left|\frac{1}{2\pi i}\int_{\partial B(z,\delta/2)} \frac{H_n(w)}{(w-z)^2}\,dw\right| \leq \frac{2\,\|H_n\|_{C^0(f^{\mu_n(1)}(U))}}{\delta}\\ &= \frac{2\,\|G_n\|_{C^0(U)}}{\delta} \leq \frac{2R}{\delta} \quad \text{for all } n \geqslant n_0.\end{aligned}$$

In other words,

$$\left\|H_n'\right\|_{C^0(V)} \leq \frac{2R}{d\big(\partial V, \partial U\big)} \quad \text{for all } n \geqslant n_0,$$

and then we obtain for all $n \geqslant n_0$ that

$$\frac{\left\|H_n - G_n\right\|_{C^0(V)}}{\|\mu_n\|_\infty} \leq \frac{R}{d\big(\partial V, \partial U\big)}\,\frac{4}{\pi}\,\sup_{z\in U}\left\{\iint_U \left|S(w,z)\right| dxdy\right\}.$$

Therefore, it is enough to consider

$$C(U) = \frac{4}{\pi}\sup_{z\in U}\left\{\iint_U \left|S(w,z)\right| dxdy\right\}.$$

□

Theorem 11.4 will be crucial in Chap. 13, in order to shadow renormalization orbits of C^3 critical circle maps with suitable C^ω critical commuting pairs (see Theorem 13.4). We remark that Theorem 11.4 is applicable to many other situations—see for example the recent paper Clark and Trejo (2020, Section 5.5).

Exercises

Exercise 11.1 We start with an elementary but important fact. Consider an ellipse in the complex plane centered at the origin, whose major axis makes an angle θ with the (positive) real axis. Let L be the length of this major axis, and let ℓ be the length of the minor one. If

$$z \mapsto w_1\, z + w_2\, \overline{z}$$

is an $\mathbb{R}$-linear transformation that maps the given ellipse onto a round circle, consider

$$\mu = \frac{w_2}{w_1},$$

and note that μ is well defined, since the pair w_1, w_2 is unique up to multiplication by a nonzero constant. Show that the *eccentricity* of the ellipse, i.e., the ratio $L/\ell \in [1, +\infty)$, equals

$$\frac{1+|\mu|}{1-|\mu|},$$

and that the angle θ coincides with $\pi/2 + \arg(\mu)/2$ (*Hint:* Take $w_1 = e^{-i\theta}(L^{-1} + \ell^{-1})$ and $w_2 = e^{i\theta}(L^{-1} - \ell^{-1})$, so that $\mu = e^{(2\theta-\pi)i}\,\dfrac{L/\ell - 1}{L/\ell + 1}$).

Let U, V, W be domains in the complex plane. In the next four exercises, the reader may assume that $f : V \mapsto W$ and $g : U \mapsto V$ are K-quasiconformal C^1 diffeomorphisms.

Exercise 11.2 Conclude from Exercise 11.1 that, for every $z \in U$, the differential of g at z maps each ellipse centered at the origin with eccentricity

$$\frac{1+|\mu_g(z)|}{1-|\mu_g(z)|} \le K,$$

and whose major axis makes an angle $\pi/2 + \arg\big(\mu_g(z)\big)/2$ with the (positive) real axis, onto a round circle.[1]

Exercise 11.3 Prove the identity (11.2).

Exercise 11.4 Prove that

$$\begin{aligned}\mu_{f\circ g}(z) &= \frac{\mu_g(z) + \mu_f\big(g(z)\big)\,\overline{\partial g(z)}/\partial g(z)}{1 + \mu_f\big(g(z)\big)\,\partial\overline{g}(z)/\partial g(z)}\\ &= \frac{\mu_g(z) + \mu_f\big(g(z)\big)\,\overline{\partial g(z)}/\partial g(z)}{1 + \overline{\mu_g(z)}\,\mu_f\big(g(z)\big)\,\overline{\partial g(z)}/\partial g(z)}\end{aligned}$$

for every $z \in U$.

[1] Recall that a K-quasiconformal homeomorphism has a derivative almost everywhere with respect to the Lebesgue measure. By Exercise 11.2, such derivative maps an ellipse of eccentricity at most K onto a circle. Hence a quasiconformal map defines a (measurable) field of ellipses with bounded eccentricity, which is mapped into a field of circles by the derivatives. By Theorem 11.1, the converse is also true: any measurable field of ellipses with bounded eccentricity comes from a quasiconformal homeomorphism in this way.

Exercise 11.5 Using the previous exercise, show that if f is holomorphic, then $\mu_{f \circ g} = \mu_g$ (note that this is consistent with the fact that postcomposing g with a conformal map f does not change which ellipse gets mapped to a circle). On the other hand, if g is holomorphic, show that

$$\mu_f \circ g = \left(\frac{g'}{|g'|}\right)^2 \mu_{f \circ g}\,.$$

In particular, $|\mu_f \circ g| = |\mu_{f \circ g}|$, which is consistent with the fact that precomposing f with a conformal map g can change the direction but not the eccentricity of an ellipse that is mapped to a circle.

Exercise 11.6 Using Exercise 11.4, prove that the inverse of a K-quasiconformal diffeomorphism is K-quasiconformal, and that the composition of a K_1-quasiconformal diffeomorphism with a K_2-quasiconformal diffeomorphism is K_1K_2-quasiconformal.

Exercise 11.7 Let $V \subset \mathbb{C}$ be a domain and let μ be a Beltrami coefficient on V, that is, $\mu : V \to \mathbb{D}$ is a measurable function such that $|\mu(z)| \leq (K-1)/(K+1)$ almost everywhere in V, for some $K \geqslant 1$. The *pull-back* of μ under a conformal map $f : U \to V$ is defined as

$$(f^*\mu)(z) = \left(\frac{(f^{-1})'(f(z))}{|(f^{-1})'(f(z))|}\right)^2 \mu(f(z))\,.$$

Naturally, we say that μ is f-*invariant* if $U = V$ and $f^*\mu = \mu$ in V. If this is the case, show that if the K-quasiconformal homeomorphism $h : V \to h(V)$ is given by Theorem 11.1, then the conjugate $h \circ f \circ h^{-1}$ is holomorphic.

The following example is borrowed from the book of Carleson and Gamelin (1993, Ch. VI.4).

Exercise 11.8 Let $A \subset \mathbb{C}$ be an open annulus centered at the origin, let $t \in (0, 1)$ and consider the Beltrami coefficient μ_t on A given by

$$\mu_t(z) = t\left(\frac{z}{|z|}\right)^2.$$

(i) Show that each μ_t is invariant under any rotation of the annulus A.
(ii) Show that, for each $t \in (0, 1)$, the homeomorphism

$$h_t(z) = |z|^{2t/(1-t)} z = z^{1/(1-t)}\,\bar{z}^{t/(1-t)}$$

is a solution of the Beltrami equation $\bar{\partial} h_t = \partial h_t\, \mu_t$ in A.

Exercise 11.9 An *annular Riemann surface* is a Riemann surface S whose fundamental group $\pi_1(S)$ is isomorphic to $\mathbb{Z}$. Using the Uniformization Theorem, show that any annular Riemann surface is conformally equivalent either to $\mathbb{C} \setminus \{0\}$, $\mathbb{D} \setminus \{0\}$ or to an annulus $A_{r,R} = \{z \in \mathbb{C} : r < |z| < R\}$. In the last case, show that the ratio R/r is unique (*Hint:* Let S be an annular Riemann surface which is not biholomorphic to the punctured plane. By the Uniformization Theorem, S is conformally equivalent to a quotient of the upper half-plane $\mathbb{H}$ by a group Γ of Möbius transformations, which acts freely and properly discontinuously on $\mathbb{H}$. Being isomorphic to $\pi_1(S)$, the group Γ must be generated by a *single* transformation $\psi : \mathbb{H} \to \mathbb{H}$, i.e., $\Gamma = \{\psi^n\}_{n\in\mathbb{Z}}$. Now discuss on the number of fixed points of ψ in $\overline{\mathbb{H}}$, noting that ψ has no fixed points in $\mathbb{H}$ (since Γ acts freely on $\mathbb{H}$), and at most two in $\overline{\mathbb{H}}$).

Exercise 11.10 The purpose of this exercise is to find local solutions to the Beltrami equation $\overline{\partial} f = \mu \partial f$ in the special case when μ is a polynomial in z and $\overline{z}$ with complex coefficients, say

$$\mu(z) = \sum_{i,j=0}^{N} a_{ij} z^i \overline{z}^j . \tag{11.3}$$

The idea is to first seek a formal solution written in power-series as follows:

$$f(z) = \sum_{m,n=0}^{\infty} c_{m,n} z^m \overline{z}^n , \tag{11.4}$$

where the coefficients $c_{m,n} \in \mathbb{C}$ are to be determined.

(i) Plugging (11.3) and (11.4) into the Beltrami equation and comparing coefficients, show that for all $\ell, k \geqslant 0$ we have

$$(\ell + 1)c_{k,\ell+1} = \sum_{i,j=0}^{N} (k + 1 - i)a_{ij} c_{k+1-i,\ell-j} .$$

Here and below, we adopt the convention that $c_{m,n} = 0$ whenever m or n is negative.

(ii) If $A = \max |a_{ij}|$, show using (i) that for all $m \geqslant 0$ and $n \geqslant 1$ we have

$$\left|c_{m,n}\right| \leq \frac{A}{n} \sum_{i,j=0}^{N} (m + 1 - i) \left|c_{m+1-i,n-1-j}\right| .$$

(iii) Now suppose the coefficients $c_{m,0}$ are given, and we know that there exists $\lambda > 0$ such that $|c_{m,0}| \leq \lambda^m$ for all $m \geqslant 0$. Using (ii) and induction in n, prove that for all $m, n \geqslant 0$ we have

$$\left|c_{m,n}\right| \leq \binom{m+n}{n} N^{2m} A_n \lambda^{m+n} .$$

(iv) Deduce from (iii) that the series in (11.4) has a positive radius of convergence, and therefore the resulting f solves the Beltrami equation for μ in a neighborhood of the origin.

Chapter 12
Lipschitz Estimates for Renormalization

Our goal in this chapter is to prove a modulus of continuity for the renormalization operator defined in Chap. 10. Our main result is Theorem 12.2, which establishes a Lipschitz estimate for renormalization, when restricted to suitable bounded pieces of topological conjugacy classes of C^3 critical commuting pairs with irrational rotation number and negative Schwarzian derivative. This is a rather technical and difficult chapter, and the reader may skip it on a first reading (just saving the statement of Theorem 12.2 for later use). Our exposition in the whole chapter follows closely the original work by Guarino et al. (2018, Sections 5–10).

12.1 Lipschitz Estimates for Controlled Commuting Pairs

In order to state the main result of the present chapter, we need a couple of definitions. The bounded pieces mentioned in the introduction are defined as follows.

Definition 12.1 Let $K > 1$ and let $\zeta = (\eta, \xi)$ be a normalized C^3 critical commuting pair which is renormalizable with some period $a \in \mathbb{N}$. We say that ζ is *K-controlled* if the following seven conditions are satisfied:

- $1/K \le \xi(0) \le K$;
- $\xi(0) - \eta\big(\xi(0)\big) \ge 1/K$;
- $\eta^{a-1}\big(\xi(0)\big) - \eta^{a}\big(\xi(0)\big) \ge 1/K$;
- $\eta^{a}\big(\xi(0)\big) \ge 1/K$;
- $\eta^{a+1}\big(\xi(0)\big) \le -1/K$;
- $\|\xi\|_{C^3([-1,0])} \le K$ and $\|\eta\|_{C^3([0,\xi(0)])} \le K$;
- $D\eta(x) \ge 1/K$ for all $x \in \big[\eta^{a}(\xi(0)), \xi(0)\big]$.

Of course if ζ is K_0-controlled and $K_1 \ge K_0$, then ζ is also K_1-controlled.

E. de Faria, P. Guarino, *Dynamics of Circle Mappings*, IMPA Monographs,
https://doi.org/10.1007/978-3-031-67495-2_12

Definition 12.2 For $K > 1$ let $\mathscr{K} = \mathscr{K}(K)$ be the space of normalized C^3 critical commuting pairs which are K-controlled. For $K > 1$ and $a \in \mathbb{N}$ let $\mathscr{K}_a(K)$ be the space of normalized C^3 critical commuting pairs which are renormalizable with period a and K-controlled.

From the real bounds (Theorem 6.4) we know that after a *finite* number of renormalizations, every C^4 critical circle map with arbitrary irrational rotation number gives rise to a *controlled* commuting pair. More precisely, we have the following.

Theorem 12.1 *There exists a universal constant $K_0 > 1$ with the following property: for any given C^4 critical circle map f with irrational rotation number there exists $n_0 = n_0(f) \in \mathbb{N}$ such that the critical commuting pair $\mathscr{R}^n(f)$ is K_0-controlled for any $n \geq n_0$.*

The C^4 smoothness hypothesis is needed for the critical commuting pair $\mathscr{R}^n(f)$ to be C^3 bounded for n large enough. As mentioned before, our main result in this chapter is the following.

Theorem 12.2 (Lipschitz Estimate) *For any given $K > 1$, there exist two constants $\varepsilon_0 = \varepsilon_0(K) \in (0, 1)$ and $L = L(K) > 1$ with the following property. Let ζ_0 and ζ_1 be two infinitely renormalizable normalized C^3 critical commuting pairs which are K-controlled, let both ζ_0 and ζ_1 have negative Schwarzian, and let $\rho(\zeta_0) = \rho(\zeta_1) \in [0, 1]\backslash\mathbb{Q}$ and $d_2(\zeta_0, \zeta_1) < \varepsilon_0$. Then we have*

$$d_2\big(\mathscr{R}(\zeta_0), \mathscr{R}(\zeta_1)\big) \leq L\, d_2(\zeta_0, \zeta_1),$$

where d_2 denotes the C^2 distance in the space of C^2 critical commuting pairs.

Let us make a few comments before entering the proof of Theorem 12.2.

Remark 12.1 One might guess that the condition $\lfloor 1/\rho(\zeta_0)\rfloor = \lfloor 1/\rho(\zeta_1)\rfloor$ should be enough in order to compare the commuting pairs $\mathscr{R}(\zeta_0)$ and $\mathscr{R}(\zeta_1)$. Unfortunately, this is not the case since there is no bound for the expansion of the renormalization operator along different topological classes (even sharing the same period of renormalization, see Proposition 12.1 for precise estimates). This is to be expected if we remember that renormalization acts as the Gauss map on the rotation number (as in (10.1)), and that the Gauss map has unbounded derivative on (0, 1).

Remark 12.2 All estimates performed in this chapter rely heavily on Yoccoz's Lemma 7.3, and that is why we require the negative Schwarzian condition in Theorem 12.2. But recall that, given a C^3 multicritical circle map f with a critical point c_i, we know from Proposition 6.2 that the critical commuting pair $\mathscr{R}_i^n(f)$ has negative Schwarzian for sufficiently large n. Therefore, this assumption in Theorem 12.2 is harmless for the applications we have in mind.

Remark 12.3 It is not difficult to prove Theorem 12.2 if one considers an irrational rotation number of bounded type (say, bounded by M) allowing the Lipschitz constant L to depend on M (see Exercise 12.2). The main point in Theorem 12.2

is that the constant L does *not* depend on the number of compositions defining the renormalization operator.

The remainder of this chapter is devoted to the proof of Theorem 12.2. As mentioned in the introduction, this is a rather technical and difficult chapter, and the reader may skip it on a first reading, just saving the statement of Theorem 12.2 for later use.

12.2 Standard Families

Fix $K_0 > 1$ and let $\mathscr{K}$ be the space of normalized C^3 critical commuting pairs which are K_0-controlled (Definition 12.1). We will consider in this section a C^3 critical commuting pair $\zeta = (\eta, \xi)$ with negative Schwarzian that belongs to $\mathscr{K}$, which is renormalizable with period $a \in \mathbb{N}$. For such a pair, we will construct/define its corresponding *standard family* (see Fig. 12.1).

For $\ell \in \{0, \ldots, a\}$, let $x_\ell = \eta^\ell(\xi(0))$. Note that $x_\ell \in I_\eta = [0, \xi(0)]$ for all $\ell \in \{0, \ldots, a\}$. Denote by I_ℓ, $\ell \in \{1, \ldots, a\}$, the fundamental domains of η given by $I_\ell = [\eta^\ell(\xi(0)), \eta^{\ell-1}(\xi(0))]$. By the commuting condition, $I_1 = \xi(I_\xi) = \xi([-1, 0])$.

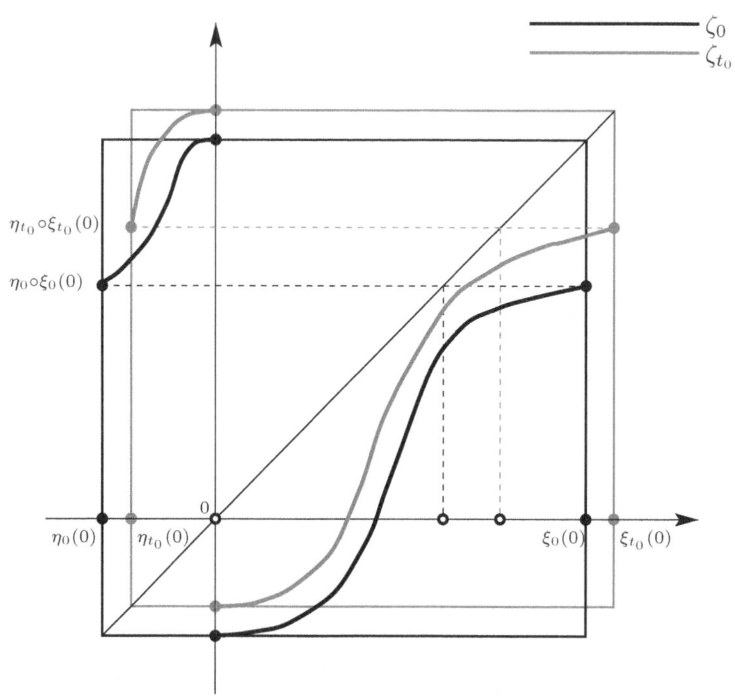

Fig. 12.1 Standard families of critical commuting pairs (in this figure, the period of ζ_0 is equal to 3, while the period of ζ_{t_0} is 6)

12.2.1 Glueing Procedure and Translations

Using the same notation as in the proof of Lemma 10.1, we have the following fact.

Lemma 12.1 *There exists $s_0 = s_0(\mathscr{K}) > 0$ such that, for any $\zeta = (\eta, \xi) \in \mathscr{K}$, both components of $A \setminus \{\eta(0)\}$ and both components of $B \setminus \{\xi(0)\}$ have Euclidean length greater than or equal to s_0.*

Proof There exist positive constants δ and ρ (depending only on K_0) such that both components of $C \backslash \{0\}$ have Euclidean length greater than or equal to δ, $\inf_C\{D\phi\} > \rho$ and $\inf_C\{D\psi\} > \rho$. Then it is enough to take $0 < s_0 < (\delta\rho)^{2d+1}$, where the integer $2d+1$ is the criticality of η and ξ at the origin. □

Still in the notation of the proof of Lemma 10.1, let $M = V_- \cup V_+/\sim$, where $x \sim y$ if $x \in A$, $y \in B$ and $\widehat{\xi}(x) = \widehat{\eta}(y)$. Note that $\eta(0) \sim \xi(0)$ by the commuting condition (2) in Definition 10.1. Let $p : V_- \cup V_+ \to M$ be the canonical projection for the identification $\sim$, and note that M is a compact boundaryless one-dimensional C^3 manifold, since the map $\widehat{\eta}^{-1} \circ \widehat{\xi} : A \to B$ is a C^3 diffeomorphism.

Lemma 12.2 *There exists a C^3 diffeomorphism $\psi : M \to \mathbf{S}^1$ such that defining $P : V_- \cup V_+ \to \mathbf{S}^1$ as $P = \psi \circ p$ we have that for all $x, y \in A \cap I_\xi$, for all $x, y \in B \cap I_\eta$ and for all $x, y \in (I_\xi \cup I_\eta) \backslash (A \cup B)$,*

$$\frac{|x-y|}{K} \leq d\big(P(x), P(y)\big) \leq K|x-y|$$

for some universal constant $K = K(\mathscr{K}) > 1$, where d denotes the Euclidean distance in the unit circle.

From now on let $P : V_- \cup V_+ \to \mathbf{S}^1$ be the C^3 map defined in Lemma 12.2. Given $t \in \mathbb{R}$ we define the *translation* by t on $I_\xi \cup I_\eta$ to be the C^3 map $T : I_\xi \cup I_\eta \times \mathbb{R} \to I_\xi \cup I_\eta$ given by

$$\big(P \circ T_t\big)(x) = e^{2\pi i t} P(x),$$

that is, $T(x,t) = T_t(x) = P^{-1}\big(e^{2\pi i t} P(x)\big)$, whenever is clear which preimage under P we choose for points in $P(A)$. In particular T_0 is the identity on $I_\xi \cup I_\eta$. Note also that

$$\frac{\partial T}{\partial t}(x,t) = \frac{1}{DP\big(T_t(x)\big)} \quad \text{and} \quad \frac{\partial T}{\partial x}(x,t) = \frac{DP(x)}{DP\big(T_t(x)\big)}.$$

From Lemma 12.2, we get that $1/K \leq \frac{\partial T}{\partial t}(x,t) \leq K$ for all $x \in I_\xi \cup I_\eta$.

12.2.2 Standard Families of Commuting Pairs

By Condition (5) in Definition 10.1, the discontinuous piecewise smooth map $\tilde{f}_\zeta : I_\xi \cup I_\eta \to I_\xi \cup I_\eta$ given by

$$\tilde{f}_\zeta(x) = \begin{cases} \xi(x) \text{ for } x \in I_\xi \\ \eta(x) \text{ for } x \in I_\eta \end{cases}$$

projects under p to a C^3 homeomorphism of the quotient manifold M, and then it projects under P to a C^3 critical circle map f_ζ in $\boldsymbol{S}^1$.

By Lemmas 12.1 and 12.2 above, the Euclidean length of both components of $P(A) \setminus \{f_\zeta(P(0))\}$ in $\boldsymbol{S}^1$ is bounded from below by some positive constant l_0, universal in $\mathscr{K}$. For $t \in W = (-l_0, l_0)$ let $f_t : \boldsymbol{S}^1 \to \boldsymbol{S}^1$ be the C^3 critical circle map given by $f_t(z) = e^{2\pi i t} f_\zeta(z)$, and note that $f_0 = f_\zeta$. Since the critical value of f_t (which is $e^{2\pi i t} f_\zeta(P(0))$) belongs to $P(A)$ we can lift each f_t up to a C^3 critical commuting pair $\zeta_t = (\eta_t, \xi_t)$ with

$$\xi_t(x) = (T_t \circ \xi_0)(x) = T(\xi_0(x), t) \quad \text{and} \quad \eta_t(x) = (T_t \circ \eta_0)(x) = T(\eta_0(x), t).$$

Note that

$$\frac{\partial \xi_t}{\partial t}(x) = \frac{1}{DP(\xi_t(x))} \quad \text{and} \quad \frac{\partial \eta_t}{\partial t}(x) = \frac{1}{DP(\eta_t(x))}.$$

Lemma 12.3 *There exists $K = K(\mathscr{K}) > 1$ such that $|t|/K \leq d_2(\zeta_0, \zeta_t) \leq K|t|$ for all $t \in W$.*

Now let $W_a \subset W$ be the set of all $t \in W$ such that ζ_t is renormalizable with period a, that is,

$$W_a = \left\{ t \in W : \left\lfloor \frac{1}{\rho(\zeta_t)} \right\rfloor = \left\lfloor \frac{1}{\rho(\zeta_0)} \right\rfloor = a \right\}.$$

Lemma 12.4 *There exists $a_0 = a_0(\mathscr{K}) \in \mathbb{N}$ such that if $a \geq a_0$ we have that $\overline{W_a} \subset W$. If we denote the boundary points of W_a by $-w_-^a$ and w_+^a, that is, $W_a = [-w_-^a, w_+^a]$, we have that*

$$\eta_{-w_-^a}^{a+1}\left(\xi_{-w_-^a}(0)\right) = 0 \quad \textit{and} \quad \eta_{w_+^a}^{a}\left(\xi_{w_+^a}(0)\right) = 0.$$

Proof By Lemma 12.2, there exists a universal upper bound $K > 0$ for the first derivative of P in $V_- \cup V_+$. By Yoccoz's Lemma 7.3, it is enough to take $a_0 \gtrsim (K/|W|)^{1/2}$ in order to have $|W| \gtrsim K/a_0^2$. The assertion about the boundary of W_a follows by combinatorics. □

Corollary 12.1 *Let $a_0 = a_0(\mathscr{K})$ be given by Lemma 12.4. Let ζ be a normalized C^3 critical commuting pair that belongs to $\mathscr{K}$ which is renormalizable with period $a \geq a_0$. Given $x \in \big[0, \eta^a(\xi(0))\big]$, there exists $t_x \leq 0$ in $W_a(\zeta)$ such that $\eta^a_{t_x}\big(\xi_{t_x}(0)\big) = x$.*

Finally, let $V = [-v_-, v_+] \subset W_a$ defined by

$$\eta^{a+1}_{-v_-}\big(\xi_{-v_-}(0)\big) = -1/K_0^2 \quad \text{and} \quad \eta^a_{v_+}\big(\xi_{v_+}(0)\big) = 1/K_0^2 \,.$$

Lemma 12.5 *For any $t \in V$ and any $k \in \{1, \dots, a-1\}$, the C^3 diffeomorphism $\eta_t^{a-k} : I_k(t) \to I_a(t)$ has universally bounded distortion.*

Recall that $I_\ell(t) = \big[x_\ell(t), x_{\ell-1}(t)\big]$, for all $\ell \in \{1, \dots, a\}$.

Proof Combine Koebe distortion principle (Lemma 5.2) with the K-control. □

Lemma 12.6 *Let $a_0 = a_0(\mathscr{K})$ be given by Lemma 12.4. Let $\zeta_0 = (\eta_0, \xi_0)$ and $\zeta_1 = (\eta_1, \xi_1)$ be two normalized C^3 critical commuting pairs that belong to $\mathscr{K}$ which are renormalizable with the same period $a \geq a_0$. Then there exists $t_0 \in V(\zeta_0) \subset W_a(\zeta_0)$ such that*

$$\eta^a_{t_0}\big(\xi_{t_0}(0)\big) = \eta_1^a\big(\xi_1(0)\big) \quad \textit{and} \quad d_2(\zeta_0, \zeta_{t_0}) \leq K d_2(\zeta_0, \zeta_1) \,,$$

where the constant $K = K(\mathscr{K}) > 1$ is given by Lemma 12.3.

Proof We may suppose that $\eta_0^a\big(\xi_0(0)\big) \geq \eta_1^a\big(\xi_1(0)\big)$, that is, $\eta_1^a\big(\xi_1(0)\big)$ belongs to the interval $\big[1/K_0, \eta_0^a\big(\xi_0(0)\big)\big] \subset [1/K_0, K_0]$. By Corollary 12.1 there exists $t_0 < 0$ in $V(\zeta_0)$ such that $\eta^a_{t_0}\big(\xi_{t_0}(0)\big) = \eta_1^a\big(\xi_1(0)\big)$. Note that $\eta^{a+1}_{t_0}\big(\xi_{t_0}(0)\big) \leq \eta_0^{a+1}\big(\xi_0(0)\big) \leq -1/K_0 < -1/K_0^2$. Now let $K = K(\mathscr{K}) > 1$ be given by Lemma 12.3. We claim that $|t_0| \leq K d_2(\zeta_0, \zeta_1)$. Indeed, if $|t_0| > K d_2(\zeta_0, \zeta_1)$ we would have $\xi_{t_0} < \xi_1$ and $\eta_{t_0} < \eta_1$ in the corresponding intersections of domains, but this implies that $\eta^a_{t_0}\big(\xi_{t_0}(0)\big) < \eta_1^a\big(\xi_1(0)\big)$ which is a contradiction. Then $|t_0| \leq K d_2(\zeta_0, \zeta_1)$ and we are done. □

12.2.3 Renormalization of Standard Families

As before, fix $K_0 > 1$ and let $\mathscr{K}$ be the space of normalized C^3 critical commuting pairs which are K_0-controlled (Definition 12.1). Again, we consider in this section a normalized C^3 critical commuting pair $\zeta = (\eta, \xi)$ in $\mathscr{K}$ with negative Schwarzian, which is renormalizable with some period $a \in \mathbb{N}$. Let $V(\zeta)$ be the parameter interval for the standard family around ζ constructed in Sect. 12.2.2, and consider the one-parameter family of C^3 critical commuting pairs given by $G_t = p\mathscr{R}(\zeta_t)$ for each $t \in V$; that is, G_t is the pre-renormalization of ζ_t (Definition 10.3). See Fig. 12.2.

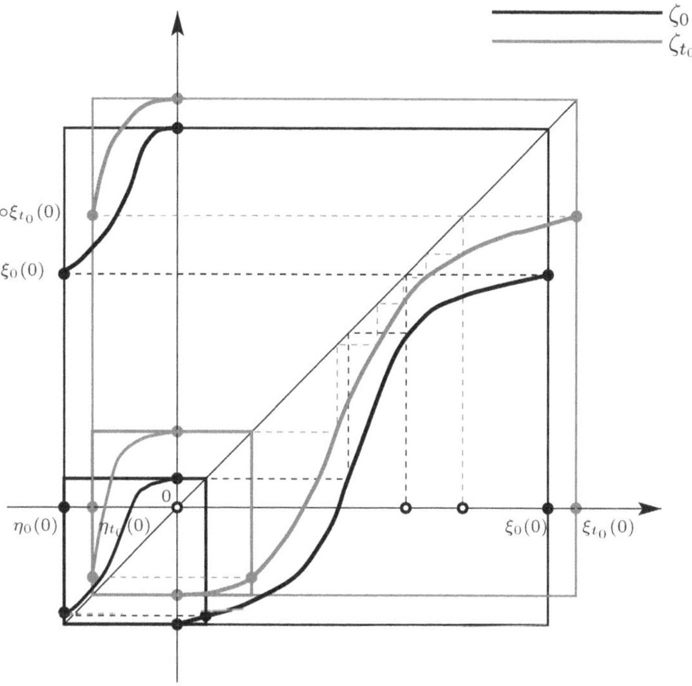

Fig. 12.2 Both critical commuting pairs of Fig. 12.1, and their renormalizations

Proposition 12.1 *There exists $K = K(\mathscr{K}) > 1$ such that for all $t \in V$ and for all x in the domain of G_t we have*

$$\frac{\partial G_t}{\partial t}(x) \asymp a^3 \quad \textit{if } x < 0, \textit{ and } \quad \frac{\partial G_t}{\partial t}(x) \asymp 1 \quad \textit{if } x > 0.$$

Proof We claim first that for $t \in V$ and $x \in I_{\xi_t}$ we have the identity

$$\frac{\partial G_t}{\partial t}(x) = \frac{\partial \xi_t}{\partial t}(x)\, D\eta_t^a\big(\xi_t(x)\big) + \sum_{k=1}^{a} \frac{\partial T}{\partial t}\Big(\eta_0\big(\eta_t^{k-1}\big(\xi_t(x)\big)\big), t\Big) D\eta_t^{a-k}\big(\eta_t^k\big(\xi_t(x)\big)\big). \tag{12.1}$$

Indeed, fix $x \in I_{\xi_t}$ and for each $j \in \{0, 1, \ldots, a\}$ let $y_j(t) = \eta_t^j\big(\xi_t(x)\big)$. Note that $y_0(t) = \xi_t(x)$ and $y_a(t) = G_t(x)$ for $x < 0$. Since $y_{j+1}(t) = \eta_t\big(y_j(t)\big) = T\big(\eta_0\big(y_j(t)\big), t\big)$ for all $j \in \{0, 1, \ldots, a-1\}$ we see that

$$\begin{aligned} y_{j+1}'(t) &= y_j'(t)\frac{\partial T}{\partial x}\big(\eta_0\big(y_j(t)\big), t\big) D\eta_0\big(y_j(t)\big) + \frac{\partial T}{\partial t}\big(\eta_0\big(y_j(t)\big), t\big) \qquad (12.2)\\ &= y_j'(t) D\eta_t\big(y_j(t)\big) + \frac{\partial T}{\partial t}\big(\eta_0\big(y_j(t)\big), t\big), \end{aligned}$$

since from $\eta_t(x) = T\big(\eta_0(x), t\big)$ we get $D\eta_t(x) = \frac{\partial T}{\partial x}\big(\eta_0(x), t\big)D\eta_0(x)$. By induction on (12.2) we obtain that for all $j \in \{1, \dots, a\}$,

$$\begin{aligned} y_j'(t) &= y_0'(t)\prod_{l=0}^{j-1} D\eta_t\big(y_l(t)\big) + \sum_{k=1}^{j-1}\frac{\partial T}{\partial t}\big(\eta_0\big(y_{k-1}(t)\big), t\big)\prod_{l=k}^{j-1} D\eta_t\big(y_l(t)\big) \\ &\quad + \frac{\partial T}{\partial t}\big(\eta_0\big(y_{j-1}(t)\big), t\big) \\ &= y_0'(t)\, D\eta_t^j\big(y_0(t)\big) + \sum_{k=1}^{j-1}\frac{\partial T}{\partial t}\big(\eta_0\big(y_{k-1}(t)\big), t\big) D\eta_t^{j-k}\big(y_k(t)\big) \\ &\quad + \frac{\partial T}{\partial t}\big(\eta_0\big(y_{j-1}(t)\big), t\big) \\ &= y_0'(t)\, D\eta_t^j\big(y_0(t)\big) + \sum_{k=1}^{j}\frac{\partial T}{\partial t}\big(\eta_0\big(y_{k-1}(t)\big), t\big) D\eta_t^{j-k}\big(y_k(t)\big). \end{aligned}$$

In particular,

$$\frac{\partial G_t}{\partial t}(x) = y_a'(t) = y_0'(t)\, D\eta_t^a\big(y_0(t)\big) + \sum_{k=1}^{a}\frac{\partial T}{\partial t}\big(\eta_0\big(y_{k-1}(t)\big), t\big) D\eta_t^{a-k}\big(y_k(t)\big),$$

and then we obtain for all $t \in V$ and all $x \in I_{\xi_t}$ the desired identity (12.1). Now by Lemma 12.2, the K_0-control and Lemma 12.5 we have

$$\begin{aligned} 0 \le \frac{\partial \xi_t}{\partial t}(x) D\eta_t^a\big(\xi_t(x)\big) &= \left(\frac{D\eta_0\big(\eta_t^{a-1}(\xi_t(x))\big) DP\big(\eta_0\big(\eta_t^{a-1}(\xi_t(x))\big)\big)}{DP\big(\xi_t(x)\big) DP\big(\eta_t^a(\xi_t(x))\big)}\right) D\eta_t^{a-1}\big(\xi_t(x)\big) \\ &\le K\, D\eta_0\big(\eta_t^{a-1}(\xi_t(x))\big) D\eta_t^{a-1}\big(\xi_t(x)\big) \le K\frac{|I_a(t)|}{|I_1(t)|} \le K. \end{aligned}$$

On the other hand, for all $k \in \{1, \dots, a\}$, we have

$$\frac{\partial T}{\partial t}\big(\eta_0\big(\eta_t^{k-1}(\xi_t(x))\big), t\big) = \frac{1}{DP\big(\eta_t^k(\xi_t(x))\big)} \in \left[\frac{1}{K}, K\right],$$

again by Lemma 12.2. Therefore, it follows from (12.1) that for any $x < 0$ we have

$$\frac{\partial G_t}{\partial t}(x) \asymp \sum_{k=1}^{a-1} D\eta_t^{a-k}\big(\eta_t^k\big(\xi_t(x)\big)\big), \quad \text{whenever } a > 1.$$

Again by Lemma 12.5 (bounded distortion) and the K_0-control we have that

$$\frac{\partial G_t}{\partial t}(x) \asymp \sum_{k=1}^{a-1} \frac{|I_a(t)|}{|I_k(t)|} \asymp \sum_{k=1}^{a-1} \frac{1}{|I_k(t)|}.$$

Therefore, by Yoccoz's Lemma 7.3 we obtain

$$\frac{\partial G_t}{\partial t}(x) \asymp \sum_{k=1}^{a-1} \min\{k, a-k\}^2 \asymp a^3 \quad \text{for any } x < 0.$$

Finally, recall that for $x \in \big[0, \eta_t^a\big(\xi_t(0)\big)\big]$ we have $G_t(x) = \eta_t(x)$ and then

$$\frac{\partial G_t}{\partial t}(x) = \frac{\partial \eta_t}{\partial t}(x) = \frac{1}{DP\big(\eta_t(x)\big)} \in \left[\frac{1}{K}, K\right]$$

by Lemma 12.2. □

With Proposition 12.1 at hand, we obtain the following.

Corollary 12.2 *There exists $K = K(\mathscr{K}) > 1$ such that for all $t \in V$ and $x, y \in I_{\xi_t}$ we have*

$$\frac{\left|\frac{\partial G_t}{\partial t}(x)\right|}{\left|\frac{\partial G_t}{\partial t}(y)\right|} \leq K.$$

In particular,

$$\frac{\big|G_t(x) - G_0(x)\big|}{\big|G_t(y) - G_0(y)\big|} = \frac{\big|\eta_t^a\big(\xi_t(x)\big) - \eta_0^a\big(\xi_0(x)\big)\big|}{\big|\eta_t^a\big(\xi_t(y)\big) - \eta_0^a\big(\xi_0(y)\big)\big|} \leq K$$

for all $t \in V \setminus \{0\}$ and $x, y \in I_{\xi_t} \cap I_{\xi_0} = \big[\max\{\eta_0(0), \eta_t(0)\}, 0\big]$.

12.3 Orbit Deformations

We begin this section with the following fact.

Lemma 12.7 *Given $K > 1$, there exists $a_0 = a_0(K) \in \mathbb{N}$ with the following property. Let $\zeta = (\eta, \xi)$ be a normalized C^3 critical commuting pair with negative Schwarzian which is K-controlled and renormalizable with some period $a \geq a_0$. Then there exists a unique p in I_η such that $\big|\eta(p) - p\big| \leq \big|\eta(x) - x\big|$ for all $x \in I_\eta$. Moreover, the point p belongs to the interior of I_η, $D\eta(p) = 1$ and $D^2\eta(p) < 0$.*

Proof Since ζ is renormalizable we know that $x > \eta(x)$ for all $x \in I_\eta$. From the continuity of η and the compactness of its domain I_η, we obtain the existence of a point p such that $0 < \left|\eta(p) - p\right| \leq \left|\eta(x) - x\right|$ for all $x \in I_\eta$.

We claim first that if $a_0 > K^2$ and $a \geq a_0$, then p belongs to the interior of I_η. Indeed, note first that the (positive) difference $\mathrm{Id} - \eta$ equals $|I_\xi|$ at the origin, and equals $\left|\xi(I_\xi)\right|$ at the point $\xi(0)$. In both cases it is greater than $1/K$, by the K-control hypothesis. If p is one of the boundary points of I_η, we would have $\left|\eta(x) - x\right| \geq 1/K$ for all $x \in I_\eta$, and since the period of ζ is a, we would have $a/K < |I_\eta|$. On the other hand, again by the K-control hypothesis, we have $a_0 > K^2 > K|I_\eta|$ and then $|I_\eta| < a_0/K$, which gives the desired contradiction.

With the claim at hand, we clearly have $D\eta(p) = 1$ and $D^2\eta(p) \leq 0$. Uniqueness of p follows at once from the Minimum Principle (Lemma 5.1). Now we claim that $D^2\eta(p)$ is strictly negative. Indeed, if $D^2\eta(p) = 0$ we would have $D^3\eta(p) = S\eta(p) < 0$, and then it would exist $\delta_0 > 0$ such that $D^2\eta(x) > 0$ for all $x \in (p - \delta_0, p)$. But then it would exist $0 < \delta_1 \leq \delta_0$ such that $\left|\eta(x) - x\right| < \left|\eta(p) - p\right|$ for all $x \in (p - \delta_1, p)$, which gives the desired contradiction. □

Remark 12.4 We can slightly improve the statement of Lemma 12.7: there exists $K_0 = K_0(K) > 1$ such that $D^2\eta(p) < -1/K_0$. Indeed, the fact that $D^2\eta(p)$ is uniformly bounded away from zero (by a constant depending only on K) follows from (the proof of) Yoccoz's Lemma 7.3.

Throughout this section, fix $K > 1$ and let $\mathscr{K}$ be the space of normalized C^3 critical commuting pairs which are K-controlled (Definition 12.1). Let $\zeta = (\eta, \xi)$ and $\tilde{\zeta} = (\tilde{\eta}, \tilde{\xi})$ be two C^3 critical commuting pairs with negative Schwarzian that belong to $\mathscr{K}$ which are renormalizable with the same period $a \geq a_0$, where $a_0 \in \mathbb{N}$ is given by Lemma 12.7. Denote by $\varepsilon > 0$ the C^2 distance between ζ and $\tilde{\zeta}$, that is, $\varepsilon = d_2(\zeta, \tilde{\zeta})$. We will assume that $\varepsilon < \varepsilon_0$, where $\varepsilon_0 > 0$ will be fixed later in this section (see the proof of Claim 12.1 below, during the proof of Lemma 12.11). Moreover, we will only consider in this section the special situation when

(1) $I_\eta = I_{\tilde{\eta}}$ and $I_\xi = I_{\tilde{\xi}} = [-1, 0]$,
(2) $p = \tilde{p}$, where $D\eta(p) = D\tilde{\eta}(\tilde{p}) = 1$ (see Lemma 12.7).

Let $H : I_\eta \to [-\varepsilon, \varepsilon] \subset \mathbb{R}$ be defined by $H(x) = \eta(x) - \tilde{\eta}(x)$, and let

$$h = H(p).$$

Observe that for every $x \in I_\eta$ we have

$$\left|H(x)\right| \leq |h| + \varepsilon(x - p)^2,$$

and

$$\left|DH(x)\right| \leq \varepsilon|x - p|. \tag{12.3}$$

Indeed, given $x \in I_\eta$ there exists $y \in I_\eta$ such that $DH(x) = D^2H(y)(x-p)$ and then $|DH(x)| = |D^2H(y)||x-p| \le \varepsilon|x-p|$, and there exists also $z \in [p,x] \subset I_\eta$ such that $H(x) = h + DH(z)(x-p)$ and then $|H(x)| \le |h| + |DH(z)||x-p| \le |h| + \varepsilon(x-p)^2$.

As before, we will use the following notation. For $\ell \in \{0,\ldots,a\}$, let $x_\ell = \eta^\ell(\xi(0))$. Note that $x_\ell \in I_\eta = [0,\xi(0)]$ for all $\ell \in \{0,\ldots,a\}$. Define $\tilde{x}_\ell = \tilde{\eta}^\ell(\tilde{\xi}(0))$ similarly. Denote by I_ℓ, $\ell \in \{1,\ldots,a\}$, the fundamental domains of η given by $I_\ell = [\eta^\ell(\xi(0)), \eta^{\ell-1}(\xi(0))]$. By the commuting condition, $I_1 = \xi(I_\xi) = \xi([-1,0])$. Define $\tilde{I}_\ell$ similarly. Let us state some consequences of Yoccoz's Lemma (Lemma 7.3).

Lemma 12.8 *Let ζ be a C^3 K-controlled critical commuting pair which has negative Schwarzian and is renormalizable with some period $a \ge a_0$, where a_0 is given by Lemma 12.7. Let $N \in \{1,\ldots,a\}$ be defined by $p \in I_{N+1}$, that is, $x_{N+1} \le p \le x_N$, where p is given by Lemma 12.7. Then we have*

1. *$N \asymp a$, i.e., there exist two constants $\delta_0 = \delta_0(\mathscr{K})$ and $\delta_1 = \delta_1(\mathscr{K})$ with $0 < \delta_0 \le \delta_1 < 1$ such that $\delta_0\, a \le N \le \delta_1\, a$;*
2. *$|I_\ell| \asymp \dfrac{1}{\ell^2}$ for $1 \le \ell \le N$ and $|I_\ell| \asymp \dfrac{1}{(a-\ell)^2}$ for $N \le \ell \le a-1$;*
3. *$|x_\ell - p| \asymp \dfrac{N-\ell}{\ell N} = \dfrac{1}{\ell} - \dfrac{1}{N} < \dfrac{1}{\ell}$ for all $\ell \le \min\{\lfloor \frac{a}{2} \rfloor, N-1\}$.*

Proof To prove Item (1) we claim that $|I_{N+1}| \asymp p - \eta(p)$. Indeed, note first that $p - \eta(p) \le |I_{N+1}| + |I_{N+2}|$. Being adjacent fundamental domains of η, we know from Yoccoz's Lemma that $|I_{N+1}| \asymp |I_{N+2}|$, and then $p - \eta(p) \le (1+K_0)\,|I_{N+1}|$ for some $K_0 = K_0(K)$. On the other hand, we have

$$
\begin{aligned}
\frac{|I_{N+2}|}{p-\eta(p)} &= \frac{x_{N+1}-x_{N+2}}{p-\eta(p)} = 1 + \frac{1}{p-\eta(p)} \int_{x_{N+1}}^{p} \big(D\eta(t) - 1\big)\,dt \\
&= 1 + \frac{1}{p-\eta(p)} \int_{x_{N+1}}^{p} (p-t)\big(-D^2\eta(p) + O(t-p)\big)\,dt \\
&\le 1 + \int_{x_{N+1}}^{p} \big(-D^2\eta(p) + O(t-p)\big)\,dt \le 1 + K_0(p - x_{N+1}) \\
&\le 1 + K_0\,|I_{N+1}|\,.
\end{aligned}
$$

Using again $|I_{N+1}| \asymp |I_{N+2}|$, we obtain the claim. Combining the comparability $|I_{N+1}| \asymp p - \eta(p)$ with Yoccoz's Lemma 7.3, we deduce that $|I_{N+1}| \asymp 1/a^2$ (recall that the positive number $p - \eta(p)$ is less than or equal to the length of any fundamental domain of η) and then $N \asymp a$, which implies Item (1). Item (2) follows at once from Item (1) and Yoccoz's lemma. To prove Item (3) note first that by definition of N,

$$
\sum_{j=\ell+1}^{N} |I_j| = |x_\ell - x_N| \le |x_\ell - p| \le |x_\ell - x_{N+1}| = \sum_{j=\ell+1}^{N} |I_j| + |I_{N+1}|
$$

for all $\ell \in \{1, \ldots, N-1\}$. By Item (2) we have $\sum_{j=\ell+1}^{N} |I_j| \asymp \sum_{j=\ell+1}^{N} \frac{1}{j^2}$, that is,

$$\frac{1}{K_0} \sum_{j=\ell+1}^{N} \frac{1}{j^2} \le |x_\ell - p| \le K_0 \left[\sum_{j=\ell+1}^{N} \frac{1}{j^2} + \frac{1}{(N+1)^2} \right]$$

for all $\ell \le \min\left\{\lfloor \frac{a}{2} \rfloor, N-1\right\}$, where $K_0(K) > 1$ is given by Lemma 7.3. From the elementary estimates

$$\frac{N-\ell}{(\ell+1)(N+1)} = \frac{1}{\ell+1} - \frac{1}{N+1} = \int_{\ell+1}^{N+1} \frac{dt}{t^2} \le$$
$$\le \sum_{j=\ell+1}^{N} \frac{1}{j^2} \le \int_{\ell}^{N} \frac{dt}{t^2} = \frac{1}{\ell} - \frac{1}{N} = \frac{N-\ell}{\ell N},$$

we obtain

$$\frac{1}{K_0} \frac{1}{(\ell+1)(N+1)} \le \frac{|x_\ell - p|}{N-\ell} \le K_0 \left(\frac{1}{\ell N} + \frac{1}{(N+1)^2} \right)$$

for all $\ell \le \min\left\{\lfloor \frac{a}{2} \rfloor, N-1\right\}$, which implies Item (3). □

A similar application of Lemma 7.3 is given by the following.

Lemma 12.9 *There exists $K_0 = K_0(\mathscr{K}) > 1$ such that for any $\zeta \in \mathscr{K}$ renormalizable with period $a \in \mathbb{N}$, and for any $b < \lfloor \frac{a}{2} \rfloor$ we have*

$$|x_b - x_{a-b}| \le \frac{K_0}{b}.$$

Note that the constant K_0 does *not* depend on the period a.

Proof By Yoccoz's Lemma 7.3,

$$|x_b - x_{a-b}| = \sum_{\ell=b+1}^{a-b} |I_\ell| \le K \left(\sum_{\ell=b+1}^{a-b} \frac{1}{\min\{\ell, a-\ell\}^2} \right).$$

To finish, note that

$$\sum_{\ell=b+1}^{a-b} \frac{1}{\min\{\ell, a-\ell\}^2} \le \frac{2}{b}.$$

Indeed, by symmetry it is enough to prove that

$$\sum_{\ell=b+1}^{\lfloor a/2 \rfloor} \frac{1}{\min\{\ell, a-\ell\}^2} \leq \frac{1}{b},$$

which follows again from elementary calculus. Indeed,

$$\sum_{\ell=b+1}^{\lfloor a/2 \rfloor} \frac{1}{\min\{\ell, a-\ell\}^2} = \sum_{\ell=b+1}^{\lfloor a/2 \rfloor} \frac{1}{\ell^2} \leq \int_b^{\lfloor a/2 \rfloor} \frac{dt}{t^2} \leq \int_b^{+\infty} \frac{dt}{t^2} = \frac{1}{b}.$$

□

Yet another consequence of Yoccoz's lemma is the following.

Lemma 12.10 *There exist $a_0 = a_0(\mathscr{K}) \in \mathbb{N}$ and $b = b(\mathscr{K}) \in \{1, .., a_0\}$ with the following property. Given $\zeta, \tilde{\zeta} \in \mathscr{K}$ renormalizable with period $a \geq a_0$ we have that*

$$\tilde{x}_{\tilde{N}-b} \geq x_{N-1} \quad \textit{and} \quad \tilde{x}_{\tilde{N}+b} \leq x_{N+2}\,.$$

Recall that we are assuming that $p = \tilde{p}$. The number b given by Lemma 12.10 will be used in Lemmas 12.11, 12.13 and 12.14 below.

Proof Consider

$$a_0 \gg \frac{1}{\delta_0} 2 \frac{K^2}{\delta^2} \quad \text{and} \quad 2 \frac{K^2}{\delta^2} \leq b < \delta_0\, a_0\,,$$

where K is given by Lemma 7.3, $\delta = \min\{\delta_0, 1 - \delta_1\}$ and δ_0, δ_1, in turn, are given by Item (1) of Lemma 12.8. We claim that $|\tilde{x}_{\tilde{N}-b} - \tilde{x}_{\tilde{N}}| \geq |x_{N-1} - p|$. Indeed, on one hand,

$$|\tilde{x}_{\tilde{N}-b} - \tilde{x}_{\tilde{N}}| = \sum_{\ell=\tilde{N}-b+1}^{\tilde{N}} |\tilde{I}_\ell| \geq \frac{1}{K} \sum_{\ell=\tilde{N}-b+1}^{\tilde{N}} \frac{1}{\ell^2} \geq \frac{1}{K} \frac{b}{(\tilde{N}-b+1)(\tilde{N}+1)}.$$

On the other hand,

$$\begin{aligned} |x_{N-1} - p| &\leq |I_N| + |I_{N+1}| \\ &\leq K \left(\frac{1}{\min\{N, a-N\}^2} + \frac{1}{\min\{N+1, a-N-1\}^2} \right) \leq \frac{2K}{\delta^2} \frac{1}{a^2}. \end{aligned}$$

Therefore, it is enough to have

$$b \geq \frac{2K^2}{\delta^2} \frac{1}{a^2} (\tilde{N} - b + 1)(\tilde{N} + 1),$$

which follows from our choice of b (recall that $\tilde{N} \leq \delta_1 a < a$ and then $\tilde{N}+1 \leq a$). Then $|\tilde{x}_{\tilde{N}-b} - \tilde{x}_{\tilde{N}}| \geq |x_{N-1} - p|$ as claimed, and this implies at once that $\tilde{x}_{\tilde{N}-b} \geq x_{N-1}$. The other estimate can be proved in the same way. □

The distance between corresponding critical iterates of ζ and $\tilde{\zeta}$ will be denoted by Δx_ℓ, that is,

$$\Delta x_\ell = \tilde{x}_\ell - x_\ell = \tilde{\eta}^\ell\big(\tilde{\xi}(0)\big) - \eta^\ell\big(\xi(0)\big) \quad \text{for all } \ell \in \{0, 1, \ldots, a\}.$$

Lemma 12.11 *There exists $K = K(\mathscr{K}) > 0$ such that for $\ell \leq \min\big\{\lfloor a/2 \rfloor, N - b, \tilde{N} - b\big\}$ we have*

$$|\Delta x_\ell| \leq K\left(|h|\,\ell + \frac{\varepsilon}{\ell}\right),$$

where b is given by Lemma 12.10.

Proof Let $x_0 = \xi(0) = \tilde{\xi}(0)$ be the common critical value of ξ and $\tilde{\xi}$, which is the right boundary point of $I_\eta = I_{\tilde{\eta}}$. Recall that, by definition, $x_\ell = \eta^\ell(x_0)$ and $\tilde{x}_\ell = \tilde{\eta}^\ell(x_0)$ for all $\ell \in \big\{1, \ldots, a\big\}$. We will consider the case $x_{\lfloor a/2 \rfloor} \leq \tilde{x}_{\lfloor a/2 \rfloor}$. Note that for any $\ell \in \big\{1, \ldots, \lfloor a/2 \rfloor\big\}$ and any $k \in \{0, \ldots, \ell - 1\}$ we have, by combinatorics,

$$x_{a-\ell+k+1} \leq x_{\lfloor a/2 \rfloor+1} < x_{\lfloor a/2 \rfloor} \leq \tilde{x}_{\lfloor a/2 \rfloor} \leq \tilde{x}_{k+1} < \tilde{x}_k\,.$$

Therefore $x_{\lfloor a/2 \rfloor+1} < \eta(\tilde{x}_k)$, and then $x_{a-\ell+k+1} < \eta(\tilde{x}_k)$, that is, both points $\eta\big(\tilde{\eta}^k(x_0)\big)$ and $\tilde{\eta}^{k+1}(x_0)$ lie to the right of the point $x_{a-\ell+k+1}$. In particular the iterate $\eta^{\ell-k-1}$ is well defined in the interval with boundary points $\eta\big(\tilde{\eta}^k(x_0)\big)$ and $\tilde{\eta}^{k+1}(x_0)$. This allows us to use a simple telescopic trick and the mean-value theorem, in order to write for any $\ell \in \big\{1, \ldots, \lfloor a/2 \rfloor\big\}$

$$\begin{aligned} |\Delta x_\ell| &= \left|\sum_{k=0}^{\ell-1} \Big(\eta^{\ell-k-1}\big(\eta\big(\tilde{\eta}^k(x_0)\big)\big) - \eta^{\ell-k-1}\big(\tilde{\eta}^{k+1}(x_0)\big)\Big)\right| \\ &\leq \sum_{k=0}^{\ell-1} \big|D\eta^{\ell-k-1}(y_k)\big|\big|H\big(\tilde{\eta}^k(x_0)\big)\big|, \end{aligned} \tag{12.4}$$

where, for each $k \in \{0, \ldots, \ell-1\}$, the point y_k lies between $\eta\big(\tilde{\eta}^k(x_0)\big)$ and $\tilde{\eta}^{k+1}(x_0)$ (the points $y_0, y_1, \ldots, y_{\ell-1}$ depends also on each fixed ℓ, but we will denote them by y_k to simplify the notation). From Sect. 12.3 and Lemma 12.8 we get that

$$\big|H\big(\tilde{\eta}^k(x)\big)\big| \leq |h| + \frac{K\varepsilon}{(k+1)^2}\,. \tag{12.5}$$

□

For each $k \in \{0, \dots, \ell - 1\}$, let us denote $D_k = \left|D\eta^{\ell-k-1}(y_k)\right|$. Our goal is, therefore, to estimate the sum

$$\left|\eta^{\ell}(x) - \tilde{\eta}^{\ell}(x)\right| \le \sum_{k=0}^{\ell-1} D_k \left(|h| + \frac{K\varepsilon}{(k+1)^2}\right). \tag{12.6}$$

For each $k \in \{0, \dots, \ell - 1\}$ let $m = m(k) \in \{1, \dots, a\}$ be such that $y_k \in I_m(\eta)$, where $I_m(\eta) = \left[\eta^m(x), \eta^{m-1}(x)\right]$ as before. Since we are assuming $x_{\lfloor a/2 \rfloor} \le \tilde{x}_{\lfloor a/2 \rfloor}$ we have that $m \le a/2 + 1$. We claim that $m(k) \asymp k$ for all $k \in \{0, \dots, \ell - 1\}$. More precisely, we have the following.

Claim 12.1 There exists $C = C(\mathscr{K}) > 1$ such that $\frac{k}{C} < m < Ck$ for all $k \in \{0, \dots, \ell - 1\}$ and for all $\ell \in \left\{1, \dots, \lfloor a/2 \rfloor\right\}$.

Proof From Lemma 12.8 we know that $|y_k - p| \asymp \frac{1}{m}$, and then it is enough to prove that $|y_k - p| \asymp \frac{1}{k}$. Recall that $d_2(\zeta, \tilde{\zeta}) < \varepsilon_0$, where $\varepsilon_0 > 0$ will be fixed later in the proof. On one hand $|y_k - p| \le \left|\tilde{\eta}^k(x_0) - p\right| \asymp \frac{1}{k}$. On the other hand, since $\ell \le \min\left\{N - b, \tilde{N} - b\right\}$, the point p does not belong to the interval with boundary points $\eta\left(\tilde{\eta}^k(x_0)\right)$ and $\tilde{\eta}^{k+1}(x_0)$, and then

$$\begin{aligned} |y_k - p| &\ge \min\left\{\left|\tilde{\eta}^{k+1}(x_0) - p\right|, \left|\eta\left(\tilde{\eta}^k(x_0)\right) - p\right|\right\} \\ &= \left|\tilde{\eta}^{k+1}(x_0) - p\right| - \left|\eta\left(\tilde{\eta}^k(x_0)\right) - \tilde{\eta}^{k+1}(x_0)\right| \\ &= \left|\tilde{\eta}^{k+1}(x_0) - p\right| - \left|H\left(\tilde{\eta}^k(x_0)\right)\right|. \end{aligned}$$

From (12.5) we get $\left|H\left(\tilde{\eta}^k(x_0)\right)\right| \le K\left(|h| + \frac{\varepsilon}{(k+1)^2}\right) \le \frac{K}{(k+1)^2}$ since $|h| \le K/a^2$ by Yoccoz's lemma (indeed, by Lemma 7.3, the length of the fundamental domain $\left(\eta(p), p\right)$ is bounded by $1/a^2$, up to a multiplicative constant. That is, both $p - \eta(p)$ and $p - \tilde{\eta}(p)$ are bounded by $1/a^2$ up to a multiplicative constant, and then $|h| \le K/a^2$). Therefore

$$|y_k - p| \ge \frac{1}{K}\left(\frac{1}{k+1} - \frac{K^2}{(k+1)^2}\right) = \frac{1}{K}\left(1 - \frac{K^2}{k+1}\right)\frac{k}{k+1}\frac{1}{k} \ge \frac{1}{4k}$$

if $k \ge 2K^2 + 1$ and then $|y_k - p| \asymp \frac{1}{k}$ in this case. We choose $\varepsilon_0 > 0$ in order to have that if $k \le 2K^2 + 1$, then both $\tilde{\eta}^{k+1}(x)$ and $\eta(\tilde{\eta}^k(x))$ belong to the interval $\left[\tilde{\eta}^{k+2}(x), \tilde{\eta}^k(x)\right]$, and again $|y_k - p| \asymp \frac{1}{k}$ as we wanted to prove. □

We have two claims regarding the values of D_k.

Claim 12.2 There exists $K = K(\mathscr{K}) > 0$ such that for all $k \in \{0, \dots, \ell - 1\}$ and $\ell \in \{1, \dots, a/2\}$ we have $D_k \le K$.

Proof By bounded distortion and Yoccoz's lemma we know that

$$\left|D\eta^{\ell-k-1}(y_k)\right| \asymp \frac{\left|I_{m+\ell-k-1}(\eta)\right|}{\left|I_m(\eta)\right|} \asymp m^2\left|I_{m+\ell-k-1}(\eta)\right|,$$

and then it is enough to prove that $\left|I_{m+\ell-k-1}(\eta)\right| \leq \dfrac{K}{m^2}$. To prove this, we have two cases to consider.

- If $\eta^{\ell-k-1}(y_k) \geq p$, then $\left|I_{m+\ell-k-1}(\eta)\right| \asymp \dfrac{1}{(m+\ell-k-1)^2}$ by Yoccoz's lemma. Since $\ell-k-1 \geq 0$, we are done.
- If $\eta^{\ell-k-1}(y_k) < p$, then $\left|I_{m+\ell-k-1}(\eta)\right| \asymp \dfrac{1}{(a-m-\ell+k+1)^2}$, and since $a-m-\ell \geq 0$ we obtain $\left|I_{m+\ell-k-1}(\eta)\right| \leq K/(k+1)^2$. Since $m \asymp k$ by Claim 12.1, we obtain Claim 12.2.

□

Claim 12.3 There exists $K = K(\mathscr{K}) > 0$ such that, if $k < \dfrac{\ell}{4(C-1)}$, then $D_k \leq K\dfrac{k^2}{\ell^2}$.

Proof Write $m = \lfloor \theta k \rfloor$ with $\frac{1}{C} < \theta < C$ (see Claim 12.1). If $m < k$, we have that $\theta < 1$ and $\ell+m-k-1 = \theta\ell+(1-\theta)\ell-(1-\theta)k-1 = \theta\ell+(1-\theta)(\ell-k)-1 \geq \theta\ell-1 \geq \frac{1}{C}\ell$. Since $\ell+m-k-1 \leq \ell \leq \frac{a}{2}$, we have that

$$D_k \leq K\frac{C^2k^2}{\left(\frac{\ell}{C}\right)^2} \leq K\frac{k^2}{\ell^2}\,.$$

On the other hand, if $m > k$ (that is, $\theta > 1$), we have $m+\ell-k-1 \leq \ell+(\theta-1)k-1 \leq \ell+(C-1)k-1 \leq \ell+\frac{1}{4}\ell-1 \leq \frac{3}{2}a$. Then

$$\left|I_{m+\ell-k-1}(\eta)\right| \asymp \frac{1}{(m+\ell-k-1)^2} \leq \frac{1}{(\ell-1)^2}\,,$$

and so we also have $D_k \leq K\frac{k^2}{\ell^2}$ in this case, since $\frac{1}{3}a < j < \frac{2}{3}a$ implies $\frac{1}{a-j} > \frac{1}{2}\frac{1}{j} > \frac{1}{4}\frac{1}{a-j}$. □

With Claims 12.2 and 12.3 at hand, we are ready to estimate sum (12.6).

$$\sum_{k=0}^{\ell-1} D_k\left(|h|+\frac{K\varepsilon}{(k+1)^2}\right) = |h|\left(\sum_{k=0}^{\ell-1} D_k\right) + K\varepsilon\left(\sum_{k=0}^{\left\lfloor\frac{\ell}{4(C-1)}\right\rfloor}\frac{D_k}{(k+1)^2}\right) + K\varepsilon\left(\sum_{k=\left\lfloor\frac{\ell}{4(C-1)}\right\rfloor+1}^{\ell-1}\frac{D_k}{(k+1)^2}\right)$$

$$\leq K|h|\ell + K\frac{\varepsilon}{\ell^2}\left(\sum_{k=0}^{\left\lfloor \frac{\ell}{4(C-1)} \right\rfloor}\left(\frac{k}{k+1}\right)^2\right)$$

$$+ K\varepsilon\left(\sum_{k=\left\lfloor \frac{\ell}{4(C-1)} \right\rfloor+1}^{\ell-1}\frac{1}{(k+1)^2}\right)$$

$$\leq K|h|\ell + K\frac{\varepsilon}{\ell} + K\frac{\varepsilon}{\ell}.$$

For the last inequality, we have used that both sequences

$$\frac{1}{\ell}\left(\sum_{k=0}^{\left\lfloor \frac{\ell}{4(C-1)} \right\rfloor}\left(\frac{k}{k+1}\right)^2\right) \quad \text{and} \quad \ell\left(\sum_{k=\left\lfloor \frac{\ell}{4(C-1)} \right\rfloor+1}^{\ell-1}\frac{1}{(k+1)^2}\right)$$

remain bounded when ℓ goes to infinity, with constants depending only on C. We have proved Lemma 12.11.

Lemma 12.12 *For every $a \geq 1$, there exists $K_a = K_a(a) > 0$ such that*

$$|\Delta x_a| \leq K_a\,\varepsilon.$$

Proof We have

$$|\Delta_{\ell+1}| = |\tilde{\eta}(\tilde{x}_\ell) - \eta(x_\ell)| = |\eta(\tilde{x}_\ell) - \eta(x_\ell) + H(\tilde{x}_\ell)| \leq D|\Delta x_\ell| + \varepsilon\,,$$

where $D = \max\{D\eta\}$. Therefore,

$$|\Delta x_a| \leq \varepsilon \cdot \sum_{k=0}^{a} D^{a-k}\,,$$

and the lemma is proved. □

The following definition is given for general commuting pairs which are contained in the previously discussed set $\mathscr{K}$ of K_0-controlled commuting pairs.

Definition 12.3 Given $L > 1$ we say that the commuting pairs $\zeta_0 = (\xi_0, \eta_0)$ and $\zeta_1 = (\xi_1, \eta_1)$, with $a_{\zeta_0} = a_{\zeta_1} = a$, are L-synchronized if

$$|\Delta x_a| \leq L \cdot d_2(\zeta_0, \zeta_1).$$

By working just as in the proof of Lemma 12.11, but with backwards iterations, we obtain the following.

Lemma 12.13 *Given $L > 0$ there exists $K = K(\mathscr{K}, L) > 0$ such that if $\zeta, \tilde{\zeta} \in \mathscr{K}$ are L-synchronized with $a_\zeta = a_{\tilde{\zeta}} = a$, then we have*

$$|\Delta x_\ell| \leq K \left(|h| \, (a - \ell) + \frac{\varepsilon}{a - \ell} \right)$$

for all $\ell \in \mathbb{N}$ such that $\max \left\{ \lfloor a/2 \rfloor, N + b, \tilde{N} + b \right\} \leq \ell \leq a$.

Proposition 12.2 *For every $L > 0$ there exists $K = K(\mathscr{K}, L) > 1$ such that the following holds. If $\zeta, \tilde{\zeta} \in \mathscr{K}$ are L-synchronized with $a_\zeta = a_{\tilde{\zeta}} = a$, then we have*

$$|h| \leq K \frac{\varepsilon}{a^2} .$$

Proof Let us suppose that $\tilde{\eta}(p) = \eta(p) + h$ with $h > 0$. We want to prove that, under the synchronization assumption, the ratio $C = a^2 h / \varepsilon$ is uniformly bounded in $\mathscr{K}$. As before, let $N \in \{1, \ldots, a\}$ be defined by $p \in [x_{N+1}, x_N]$. By Yoccoz's Lemma, there exists $K_0 = K_0(\mathscr{K}) > 1$ such that $N = \nu a$ with $1/K_0 \leq \nu \leq 1 - \frac{1}{K_0}$. In the same way let $\tilde{N} = \tilde{\nu} a$ defined by $p \in [\tilde{x}_{\tilde{N}+1}, \tilde{x}_{\tilde{N}}]$ with $1/K_0 \leq \tilde{\nu} \leq 1 - \frac{1}{K_0}$.

By Lemma 12.9 there exists $K_1 = K_1(\mathscr{K}) > 1$ such that $(x_j, \tilde{x}_j) \subset \big(p - K_1/M, p\big)$ when $(1 - \frac{1}{K_0})a \leq j \leq a - M$, and $(x_j, \tilde{x}_j) \subset \big(p, p + K_1/M\big)$ when $Mj \leq a/K_0$ for any $M \in \big\{1, \ldots, \lfloor a/K_0 \rfloor\big\}$. Let $K_2 = K_2(\mathscr{K}) > 1$ be the constant given by Lemma 12.11. By Lemma 12.13 we have

$$|\Delta x_{a-M}| \leq K_3 \left(hM + \frac{\varepsilon}{M} \right) \tag{12.7}$$

for some universal constant $K_3(L, \mathscr{K}) > 1$. Let $K = \max\{K_0, K_1, K_2, K_3\}$ and let us suppose that $a > K(4K + 1)$ (otherwise we are done since $|h| \leq \varepsilon$). Fix $M \in \big\{1, \ldots, \lfloor a/2 \rfloor\big\}$ small enough in order to have

$$0 < \theta = \frac{M}{a} < \frac{1}{K(4K+1)} < 1 .$$

Let $T = \big[p - K/M, p + K/M\big]$ and recall that $(x_j, \tilde{x}_j) \subset T$ for all $j \in \{M, \ldots, a - M\}$. The next three claims will show that if C is big enough, in terms of K and $\theta(K)$, the pairs ζ and $\tilde{\zeta}$ cannot be L-synchronized. □

Claim 12.4 If $C \geq 2 \left(\frac{K}{\theta}\right)^2$, then $\tilde{\eta}(x) \geq \eta(x) + \frac{h}{2}$ for all $x \in T$.

Proof As before

$$\tilde{\eta}(x) - \eta(x) \geq h - \varepsilon (x - p)^2 \geq h - \varepsilon \left(\frac{K}{M} \right)^2 = h - \frac{\varepsilon}{a^2} \left(\frac{K}{\theta} \right)^2 \geq h - \frac{h}{2} = \frac{h}{2} .$$

In the last inequality we have used that $\frac{\varepsilon}{a^2} \leq \frac{h}{2}\left(\frac{\theta}{K}\right)^2$ since $\frac{a^2 h}{\varepsilon} \geq 2\left(\frac{K}{\theta}\right)^2$. □

Note that $0 < \theta < \frac{1}{K(4K+1)}$ implies $1 - 2\theta K^2 - \theta K \in (0, 1)$.

Claim 12.5 If $C > \frac{1}{\theta}\left(\frac{2K^2}{1 - 2\theta K^2 - \theta K}\right)$, then there exists $\ell_0 \in \{M, \ldots, a/K\}$ such that $x_{\ell_0} \leq \tilde{x}_{\ell_0}$.

Proof We will prove first that

$$\left(\frac{a}{K} - M\right)\frac{h}{2} \geq K\left(hM + \frac{\varepsilon}{M}\right). \tag{12.8}$$

Indeed, since $1 - 2\theta K^2 - \theta K > \frac{2K^2}{C\theta}$, we have

$$\frac{1 - 2\theta K^2 - \theta K}{2\theta K} > \frac{K}{C\theta^2}$$

and then

$$hM\left(\frac{1 - 2\theta K^2 - \theta K}{2\theta K}\right) > K\frac{\varepsilon}{M}$$

since $\varepsilon/M = hM/C\theta^2$. From

$$\frac{1 - 2\theta K^2 - \theta K}{2\theta K} = \frac{1}{2}\left(\frac{1}{\theta K} - 1 - 2K\right),$$

we obtain

$$\frac{h}{2}\left(\frac{a}{K} - M\right) - KhM > K\frac{\varepsilon}{M},$$

which implies the desired estimate (12.8). Now, estimate (12.8) combined with Lemma 12.11 gives us

$$\left|x_M - \tilde{x}_M\right| \leq \left(\frac{a}{K} - M\right)\frac{h}{2}. \tag{12.9}$$

With estimate (12.9) at hand, we are ready to prove Claim 12.5. Indeed, let $\ell \in \{M, \ldots, a/K\}$ be such that $p \leq \tilde{x}_\ell < x_\ell \leq p + K/M$ (note that, if no such ℓ exists, we are done). From Claim 12.4 we have

$$\begin{aligned}\tilde{x}_{\ell+1} - x_{\ell+1} &= \tilde{\eta}(\tilde{x}_\ell) - \eta(x_\ell)\\ &\geq h/2 + \eta(\tilde{x}_\ell) - \eta(x_\ell)\\ &= h/2 + D\eta(y_\ell)(\tilde{x}_\ell - x_\ell)\\ &= h/2 + \tilde{x}_\ell - x_\ell + D^2\eta(z_\ell)(y_\ell - p)(\tilde{x}_\ell - x_\ell),\end{aligned}$$

where $y_\ell \in [\tilde{x}_\ell, x_\ell]$ and $z_\ell \in [p, y_\ell]$ are given by the mean-value theorem. Since $D^2\eta(z_\ell) < 0$, $y_\ell - p > 0$ and $\tilde{x}_\ell - x_\ell < 0$,

$$\tilde{x}_{\ell+1} - x_{\ell+1} \geq h/2 + \tilde{x}_\ell - x_\ell\,, \quad \text{that is,} \quad \Delta x_{\ell+1} \geq h/2 + \Delta x_\ell\,.$$

Therefore, if the difference $\tilde{x}_{\ell+1} - x_{\ell+1}$ is still negative, it will be at least $h/2$ closer to zero than the previous difference $\tilde{x}_\ell - x_\ell$. What estimate (12.9) tells us is that we have enough time inside the interval $(p, p + K/M)$ in order to interchange the positions of the critical iterates. We have proved Claim 12.5. □

Claim 12.5 implies that $x_\ell \leq \tilde{x}_\ell$ for all $\ell \in \{\ell_0, \ldots, a - M\}$, since $D\eta > 0$ and $h > 0$. Therefore, by Claim 12.4, we have

$$|\Delta x_{a-M}| \geq \frac{h}{2}\left[a - M - \left(1 - \frac{1}{K}\right)a\right]. \tag{12.10}$$

Our third and last claim tells us that (12.10) contradicts the synchronization assumption. Note that $0 < \theta < \frac{1}{K(4K+1)}$ implies $1 - \theta K(4K + 1) \in (0, 1)$.

Claim 12.6

$$\text{If } C \geq \frac{1}{\theta}\left[\frac{4K^2}{1 - \theta K(4K+1)}\right], \text{ then } 2K\left(hM + \frac{\varepsilon}{M}\right) \leq \frac{h}{2}\left[a - M - \left(1 - \frac{1}{K}\right)a\right].$$

Proof Note first that

$$2K\left(hM + \frac{\varepsilon}{M}\right) = \frac{\varepsilon}{a}\left[2K\left(C\theta + \frac{1}{\theta}\right)\right]$$

and

$$\frac{h}{2}\left[a - M - \left(1 - \frac{1}{K}\right)a\right] = \frac{\varepsilon}{a}\left[\frac{C}{2}\left(\frac{1}{K} - \theta\right)\right].$$

A straightforward computation shows that both conditions

$$C \geq \frac{1}{\theta}\left[\frac{4K^2}{1 - \theta K(4K+1)}\right] \quad \text{and} \quad 2K\left(C\theta + \frac{1}{\theta}\right) \leq \frac{C}{2}\left(\frac{1}{K} - \theta\right)$$

are actually equivalent. □

We are ready to finish the proof of Proposition 12.2. Indeed, by combining estimates (12.7) and (12.10) we have

$$\frac{h}{2}\left[a - M - \left(1 - \frac{1}{K}\right)a\right] \leq \left|x_{a-M} - \tilde{x}_{a-M}\right|$$
$$\leq K\left(hM + \frac{\varepsilon}{M}\right) < 2K\left(hM + \frac{\varepsilon}{M}\right)$$

which contradicts Claim 12.6. Therefore

$$C \leq \max\left\{2\left(\frac{K}{\theta}\right)^2, \frac{1}{\theta}\left(\frac{2K^2}{1 - 2\theta K^2 - \theta K}\right), \frac{1}{\theta}\left[\frac{4K^2}{1 - \theta K(4K + 1)}\right]\right\},$$

that is, the ratio $C = a^2h/\varepsilon$ is bounded by a constant only depending on K and L. We have proved Proposition 12.2.

With Proposition 12.2 at hand, we can improve both Lemmas 12.11 and 12.13 under the synchronization assumption.

Lemma 12.14 *Given $L > 0$, there exists $K = K(\mathscr{K}, L) > 0$ such that if $\zeta, \tilde{\zeta} \in \mathscr{K}$ are L-synchronized with $a_\zeta = a_{\tilde{\zeta}} = a$, then we have*

$$|\Delta x_\ell| \leq \frac{K\varepsilon}{\ell} \quad \textit{for all } 1 \leq \ell \leq \min\left\{\lfloor a/2 \rfloor, N - b, \tilde{N} - b\right\}, \textit{ and}$$

$$|\Delta x_\ell| \leq \frac{K\varepsilon}{a - \ell} \quad \textit{for all } a \geq \ell \geq \max\left\{\lfloor a/2 \rfloor, N + b, \tilde{N} + b\right\}.$$

Moreover, we have the following.

Proposition 12.3 *For every $L > 0$, there exists $K = K(\mathscr{K}, L) > 0$ such that the following holds. If ζ and $\tilde{\zeta}$ are L-synchronized then*

$$|\Delta x_\ell| \leq K\varepsilon \cdot \frac{1}{\ell} \quad \textit{for all } \ell \in \{0, 1, \ldots, a/2\}.$$

and

$$|\Delta x_\ell| \leq K\varepsilon \cdot \frac{1}{a - \ell} \quad \textit{for all } \ell \in \{a/2, \ldots, a\}.$$

Proof By Lemma 12.14, we only need to estimate $|\Delta x_\ell|$ for the intermediate iterates $\min\left\{\lfloor a/2 \rfloor, N - b, \tilde{N} - b\right\} < \ell < \left\{\lfloor a/2 \rfloor, N + b, \tilde{N} + b\right\}$. We will prove only the first part of the statement (the other being the same), that is, we will prove that

$$|\Delta x_\ell| \le K\varepsilon \cdot \frac{1}{\ell} \quad \text{for all } \ell \in \big\{\min\{\lfloor a/2 \rfloor, N-b, \tilde{N}-b\}, \ldots, a/2\big\}.$$

We use the same notation as in the proof of Proposition 12.2. By the choice of θ we know that $M \le \min\big\{\lfloor a/2 \rfloor, N-b, \tilde{N}-b\big\}$ and $a - M \ge \max\big\{\lfloor a/2 \rfloor, N+b, \tilde{N}+b\big\}$.

Recall that $H : I_\eta \to [-\varepsilon, \varepsilon] \subset \mathbb{R}$ is defined as $H(x) = \eta(x) - \tilde{\eta}(x)$. By Proposition 12.2 we have that $\big|H(x)\big| \le \varepsilon\big[\frac{K}{a^2} + (x-p)^2\big]$ and then $\big|H(x)\big| \le \frac{K\varepsilon}{a^2}$ whenever $x \in T$, since for $x \in T$ we have that $|x - p| \le \frac{K}{M} \le \frac{K}{a}$. Therefore, by considering $\alpha = 1 + \frac{K}{a}$ and $\beta = \frac{K\varepsilon}{a^2}$, we obtain that $\Delta x_{\ell+1} \le \alpha\, \Delta x_\ell + \beta$, and then

$$\Delta x_{\ell+n} \le \alpha^n \Delta x_\ell + \beta \sum_{j=0}^{n-1} \alpha^j, \quad \text{for all } 1 \le n \le (\delta_1 - \delta_0)a + 2b.$$

Note that $\sum_{j=0}^{n-1} \alpha^j = \frac{\alpha^n - 1}{\alpha - 1} = \frac{a}{K}(\alpha^n - 1)$. Moreover, since $n < a$, we have that $\alpha^n = (\frac{K}{a} + 1)^n \le e^{\frac{Kn}{a}}$ is bounded. Therefore,

$$\Delta x_{\ell+n} \le \alpha^n \Delta x_\ell + \beta \frac{a}{K}(\alpha^n - 1) \le K\frac{\varepsilon}{\ell}\left[\alpha^n + \frac{\ell}{a}(\alpha^n - 1)\right] \le K\frac{\varepsilon}{\ell}\alpha^n \le K\frac{\varepsilon}{\ell}.$$

Finally, from $\ell \ge M = \theta a$ and $n \le (\delta_1 - \delta_0)a + 2b$, we get that $\frac{n}{\ell}$ is bounded and then $\Delta x_{\ell+n} \le K\frac{\varepsilon}{\ell+n}$, as we wanted to prove. □

For $\ell \in \{1, \ldots, a\}$ let

$$\Delta_\ell = |\Delta x_\ell - \Delta x_{\ell-1}|.$$

Proposition 12.4 *For every $L > 0$, there exists $K = K(\mathcal{K}, L) > 0$ such that the following holds. If ζ and $\tilde{\zeta}$ are L-synchronized then*

$$\Delta_\ell \le K\left(\varepsilon \cdot \frac{\log \ell}{\ell^2} + \varepsilon^2 \cdot \frac{1}{\ell}\right) \quad \textit{for all } \ell \le a/2.$$

and

$$\Delta_\ell \le K\left(\varepsilon \cdot \frac{\log(a-\ell)}{(a-\ell)^2} + \varepsilon^2 \cdot \frac{1}{a-\ell}\right) \quad \textit{for all } \ell \ge a/2.$$

Proof The proof of the second part of this proposition can be obtained as the first part by working backward. (See also the proof of Proposition 12.3.) We will only present the proof of the first part. Note that, for $\ell \ge 1$,

$$\begin{aligned}
\Delta_{\ell+1} =& |[\tilde{\eta}(x_\ell + \Delta x_\ell) - \eta(x_\ell)] - [\tilde{\eta}(x_{\ell-1} + \Delta x_{\ell-1}) - \eta(x_{\ell-1})]| \\
=& |[\eta(x_\ell + \Delta x_\ell) - \eta(x_\ell) + \tilde{\eta}(x_\ell + \Delta x_\ell) - \eta(x_\ell + \Delta x_\ell)] - \\
& [\eta(x_{\ell-1} + \Delta x_{\ell-1}) - \eta(x_{\ell-1}) + \tilde{\eta}(x_{\ell-1} + \Delta x_{\ell-1}) - \eta(x_{\ell-1} + \Delta x_{\ell-1})]| \\
=& |[D\eta(\theta_\ell)\Delta x_\ell + H(x_\ell + \Delta x_\ell)] - \\
& [D\eta(\theta_{\ell-1})\Delta x_{\ell-1} + H(x_{\ell-1} + \Delta x_{\ell-1})]| \\
\leq& |D\eta(\theta_\ell)\Delta x_\ell - D\eta(\theta_{\ell-1})\Delta x_{\ell-1}| + |DH(\theta)\tilde{I}_\ell|
\end{aligned}$$

The intermediate point θ is in $\tilde{I}_\ell$. Hence, by using (12.3), the Yoccoz Lemma 7.3, and Lemma 12.8, we have

$$|DH(\theta)\tilde{I}_\ell| \leq K\varepsilon \cdot \frac{1}{\ell^3}. \tag{12.11}$$

The intermediate point θ_ℓ is in $[x_\ell, x_\ell + \Delta x_\ell]$. Similarly, $\theta_{\ell-1} \in [x_{\ell-1}, x_{\ell-1} + \Delta x_{\ell-1}]$. This allows for the following estimate.

$$\begin{aligned}
|D\eta(\theta_\ell)\Delta x_\ell - D\eta(\theta_{\ell-1})\Delta x_{\ell-1}| \leq& \frac{|I_{\ell+1}|}{|I_\ell|}\Delta_\ell + \left|\left(D\eta(\theta_\ell) - \frac{|I_{\ell+1}|}{|I_\ell|}\right)\Delta x_\ell\right| + \\
& \left|\left(D\eta(\theta_{\ell-1}) - \frac{|I_{\ell+1}|}{|I_\ell|}\right)\Delta x_{\ell-1}\right| \\
\leq& \frac{|I_{\ell+1}|}{|I_\ell|}\Delta_\ell + K(|I_\ell| + |\Delta x_\ell|)|\Delta x_\ell| + \\
& K(|I_\ell| + |\Delta x_{\ell-1}|)|\Delta x_{\ell-1}|
\end{aligned}$$

Using the Yoccoz Lemma 7.3 and Proposition 12.3, we obtain

$$|D\eta(\theta_\ell)\Delta x_\ell - D\eta(\theta_{\ell-1})\Delta x_{\ell-1}| \leq \frac{|I_{\ell+1}|}{|I_\ell|}\Delta_\ell + K\left(\varepsilon\frac{1}{\ell^3} + \varepsilon^2\frac{1}{\ell^2}\right). \tag{12.12}$$

Combining (12.11) and (12.12) with the chain of estimates obtained at the beginning of the proof, we deduce that

$$\Delta_{\ell+1} \leq \frac{|I_{\ell+1}|}{|I_\ell|}\Delta_\ell + K\left(\varepsilon\frac{1}{\ell^3} + \varepsilon^2\frac{1}{\ell^2}\right).$$

After iterating this recursive estimate and using the Yoccoz Lemma 7.3, we finally obtain

$$\Delta_\ell \le K \sum_{k=1}^{\ell-1} \left(\varepsilon \frac{1}{k^3} + \varepsilon^2 \frac{1}{k^2} \right) \cdot \frac{|I_\ell|}{|I_{k+1}|}$$

$$\le K \left(\varepsilon \frac{1}{\ell^2} \sum_{k=1}^{\ell-1} \frac{1}{k} + \varepsilon^2 \frac{1}{\ell^4} \sum_{k=1}^{\ell-1} k^2 \right)$$

$$\le K \left(\varepsilon \cdot \frac{\log \ell}{\ell^2} + \frac{\varepsilon^2}{\ell} \right). \qquad \square$$

12.4 Composition

In this section we will discuss composition of multiple diffeomorphisms. Let $I = [a, b]$ be a compact interval in the real line, and let $\mathscr{D} = \mathrm{Diff}_+^2([a, b])$ be the space of orientation preserving C^2 diffeomorphisms of I, endowed with the C^2-metric. Let $X = C^0(I)$ be the space of continuous functions from $[a, b]$ to the real line, and recall that X is a Banach space when endowed with the sup norm. Just as we did in Chaps. 3 and 5, we consider the *nonlinearity* function $\mathscr{N} : \mathscr{D} \to X$, defined as

$$\mathscr{N}\psi = \frac{D^2\psi}{D\psi} = D \log D\psi\,.$$

Note that $\mathscr{N}$ is a homeomorphism, whose inverse is given by

$$(\mathscr{N}^{-1}\phi)(x) = a + \left(\frac{b-a}{\int_a^b \exp\left(\int_a^s \phi(t)dt\right)ds} \right) \int_a^x \exp\left(\int_a^s \phi(t)dt \right) ds\,,$$

for any $x \in [a, b]$ and any $\phi \in X$. To prove that $\mathscr{N}^{-1}\phi \in \mathscr{D}$ note that $D\mathscr{N}^{-1}\phi > 0$, since $\frac{\partial}{\partial x}\left(\int_a^x \exp\left(\int_a^s \phi(t)dt\right)ds\right) = \exp\left(\int_a^x \phi(t)dt\right) > 0$.

More generally, if $f : I \to \mathbb{R}$ is a C^2 map and x is a regular point of f, we define $\mathscr{N} f(x) = D^2 f(x)/Df(x)$. As the reader can easily prove, the chain rule for the nonlinearity is $\mathscr{N}(f \circ g) = \mathscr{N} f \circ g\, Dg + \mathscr{N} g$, while the kernel of $\mathscr{N}$ is the group of affine transformations. In particular, $\mathscr{N}(A \circ f) = \mathscr{N} f$ whenever A is affine. Note also that the nonlinearity goes to infinity around any non-flat critical point. Elementary properties of nonlinearity can be found in the work of Martens (1998). On bounded sets it is bi-Lipschitz. In particular, we have the following.

Lemma 12.15 *Let B be a bounded set in $X = C^0(I)$. There exists $K = K(B) > 0$ such that for any pair ϕ, ψ in B we have*

$$d_2(\mathscr{N}^{-1}\phi, \mathscr{N}^{-1}\psi) \le K d_{C^0}(\phi, \psi).$$

Proof Use the inverse of the nonlinearity to estimate the C^0 distance between $f = \mathcal{N}^{-1}\phi$ and $g = \mathcal{N}^{-1}\psi$, as in Martens (1998, Lem. 10.2, page 579). This gives $d_{C^0}(\mathcal{N}^{-1}\phi, \mathcal{N}^{-1}\psi) \le K d_{C^0}(\phi, \psi)$. Since both $f = \mathcal{N}^{-1}\phi$ and $g = \mathcal{N}^{-1}\psi$ belong to $\mathrm{Diff}^2_+(I)$ there exists $t_0 \in I$ such that $Df(t_0) = Dg(t_0)$, and then $\log Df(t) - \log Dg(t) = \int_{t_0}^{t} (\phi - \psi)(s)\, ds$ for all $t \in I$. Therefore $d_{C^0}(\log Df, \log Dg) \le |I| d_{C^0}(\phi, \psi)$, and since both f and g are C^1-bounded we get $d_{C^0}(Df, Dg) \le K d_{C^0}(\phi, \psi)$. Finally note that for all $t \in I$ we have

$$\left|(D^2 f - D^2 g)(t)\right| \le \left|(\phi - \psi)(t)\right|\left|Df(t)\right| + \left|(Df - Dg)(t)\right|\left|\psi(t)\right|.$$

□

As explained above, the nonlinearity allows us to identify the set $\mathcal{D}$ of diffeomorphisms with the Banach space $X = C^0(I)$. This defines the *nonlinearity norm* on $\mathcal{D}$: $|f| = \|\mathcal{N} f\|_{C^0}$. The following lemma says that composition of multiple diffeomorphisms on C^1-bounded sets is Lipschitz continuous in the nonlinearity norm. This lemma is an adaptation of the *Sandwich Lemma* in Martens (1998, Lem. 10.5, page 581).

Lemma 12.16 *Given $M > 0$, there exists $K = K(M) > 0$ such that for $f_1, \ldots, f_n$, $g_1, \ldots, g_n$ in $\mathrm{Diff}^3_+\left([0,1]\right)$ satisfying*

- $\sum_{j=1}^{n} \|\mathcal{N} f_j\|_{C^0} \le M,$
- $\sum_{j=1}^{n} \|\mathcal{N} g_j\|_{C^0} \le M,$
- $\sum_{j=1}^{n} \|D\mathcal{N} f_j\|_{C^0} \le M,$
- $\sum_{j=1}^{n} \|D\mathcal{N} g_j\|_{C^0} \le M,$

we have

$$\left\|\mathcal{N}\left(\bigcirc_{j=1}^{n} f_j\right) - \mathcal{N}\left(\bigcirc_{j=1}^{n} g_j\right)\right\|_{C^0} \le K \sum_{j=1}^{n} \|\mathcal{N} f_j - \mathcal{N} g_j\|_{C^0}.$$

In particular,

$$d_{C^2}\left(\bigcirc_{j=1}^{n} f_j, \bigcirc_{j=1}^{n} g_j\right) \le K \sum_{j=1}^{n} \left\|\mathcal{N} f_j - \mathcal{N} g_j\right\|_{C^0}.$$

The branches of renormalizations are compositions of a homeomorphism and multiple diffeomorphisms. The composition of multiple diffeomorphisms can be controlled by Lemma 12.16. To control the effect of the first factor we need the following lemma, a basic property of composition.

Lemma 12.17 *For every $L > 0$, there exists $K = K(L) > 0$ such that the following holds. Let $q, \tilde{q} : [-1, 0] \to [0, 1]$ be C^3 homeomorphisms with one critical point, $Dq(0) = D\tilde{q}(0) = 0$. Let $f, \tilde{f} : [0, 1] \to [0, 1]$ be C^3 diffeomorphisms. If $|q|_{C^3}, |\tilde{q}|_{C^3}, |f|_{C^3}, |\tilde{f}|_{C^3} \le L$ then*

$$d_{C^2}(\tilde{f} \circ \tilde{q}, f \circ q) \le K\, d_{C^2}(\tilde{f}, f) + d_{C^2}(\tilde{q}, q).$$

As before, fix $K_0 > 1$ and let $\mathscr{K}$ be the space of normalized C^3 critical commuting pairs which are K_0-controlled. Let $\zeta = (\eta, \xi)$ and $\tilde{\zeta} = (\tilde{\eta}, \tilde{\xi})$ be two C^3 critical commuting pairs with negative Schwarzian that belong to $\mathscr{K}$ which are renormalizable with the same period $a \in \mathbb{N}$. Denote by $\varepsilon > 0$ the C^2 distance between ζ and $\tilde{\zeta}$, that is, $\varepsilon = d_2(\zeta, \tilde{\zeta})$. We may assume in the computations that $\varepsilon \in (0, 1)$. We will only consider the special situation when

(1) $I_\eta = I_{\tilde{\eta}}$ and $I_\xi = I_{\tilde{\xi}}$,
(2) $p = \tilde{p}$ where $D\eta(p) = D\tilde{\eta}(\tilde{p}) = 1$ (see Lemma 12.7).

For each $\ell \in \{1, \ldots, a-1\}$ let $f_\ell \in \mathrm{Diff}^3_+\big([0,1]\big)$ given by $f_\ell = A_{\ell+1}^{-1} \circ \eta \circ A_\ell$, where $A_\ell : [0,1] \to I_\ell$ is the unique orientation preserving affine diffeomorphism

$$A_\ell(x) = |I_\ell| x + x_\ell = \big(\eta^{\ell-1}(\xi(0)) - \eta^\ell(\xi(0))\big)x + \eta^\ell(\xi(0)).$$

Note that $\bigodot_{\ell=1}^{a-1} f_\ell = A_a^{-1} \circ \eta^{a-1} \circ A_1$ in $\mathrm{Diff}^3_+\big([0,1]\big)$.

Lemma 12.18 *There exists $K = K(\mathscr{K}) > 1$ such that for any ζ in $\mathscr{K}$ renormalizable with period $a \in \mathbb{N}$ we have*

$$\sum_{\ell=1}^{a-1} \big|\mathscr{N} f_\ell(x)\big| \le K \quad \textit{and} \quad \sum_{\ell=1}^{a-1} \big|D(\mathscr{N} f_\ell)(x)\big| \le K \quad \textit{for all } x \in [0,1].$$

Proof Note that $\mathscr{N} f_\ell(x) = \mathscr{N}(\eta \circ A_\ell)(x) = \mathscr{N}\eta\big(A_\ell(x)\big)|I_\ell|$ and that $D(\mathscr{N} f_\ell)(x) = D(\mathscr{N}\eta)(A_\ell(x))|I_\ell|^2$ for all $x \in [0,1]$. Since $\zeta \in \mathscr{K}_a$, we know that $\mathscr{N}\eta$ is C^1-bounded in $\big[\eta^a(\xi(0)), \xi(0)\big]$ and then

$$\sum_{\ell=1}^{a-1} \big|\mathscr{N} f_\ell(x)\big| \le K \sum_{\ell=1}^{a-1} |I_\ell| \le K|I_\eta| \quad \text{and}$$

$$\sum_{\ell=1}^{a-1} \big|D(\mathscr{N} f_\ell)(x)\big| \le K \sum_{\ell=1}^{a-1} |I_\ell|^2 \le K|I_\eta| \sum_{\ell=1}^{a-1} |I_\ell| \le K|I_\eta|^2.$$

□

In the same way, let $\tilde{A}_\ell : [0,1] \to \tilde{I}_\ell$ be the unique orientation preserving affine diffeomorphism, and define $g_\ell = \tilde{A}_{\ell+1}^{-1} \circ \tilde{\eta} \circ \tilde{A}_\ell \in \mathrm{Diff}^3_+\big([0,1]\big)$. The first factors of the renormalizations are controlled by

Lemma 12.19 *There exists $K > 0$ such that*

$$\|A_1^{-1} \circ \xi\|_{C^3}, \ \|\tilde{A}_1^{-1} \circ \tilde{\xi}\|_{C^3} \le K$$

and

$$d_{C^2}\big(A_1^{-1}\circ\xi,\tilde{A}_1^{-1}\circ\tilde{\xi}\big)\le K\varepsilon.$$

Proof The four maps $\xi:[-1,0]\to I_1$, $\tilde{\xi}:[-1,0]\to\tilde{I}_1$, $A_1^{-1}:[0,K]\to\mathbb{R}$ and $\tilde{A}_1^{-1}:[0,K]\to\mathbb{R}$ are C^3-bounded by some constant $M>1$ universal on $\mathscr{K}$. Similar to Lemma 12.17, we get

$$\begin{aligned}d_{C^2}\big(A_1^{-1}\circ\xi,\tilde{A}_1^{-1}\circ\tilde{\xi}\big)&\le K\left(\|A_1^{-1}-\tilde{A}_1^{-1}\|_{C^2}+\|\xi-\tilde{\xi}\|_{C^2}\right)\\&\le K\left(\|A_1^{-1}-\tilde{A}_1^{-1}\|_{C^2}+\varepsilon\right).\end{aligned}$$

Observe that

$$\begin{aligned}\big|A_1^{-1}(x)-\tilde{A}_1^{-1}(x)\big|&=\big||I_1|^{-1}(x-x_1)-|\tilde{I}_1|^{-1}(x-\tilde{x}_1)\big|\\&=\frac{\big|(x-x_1)(\tilde{x}_0-\tilde{x}_1)-(x-\tilde{x}_1)(x_0-x_1)\big|}{|I_1||\tilde{I}_1|}\\&=\frac{\big|x(\tilde{x}_0-x_0)+x(x_1-\tilde{x}_1)+(x_0\tilde{x}_1-\tilde{x}_0x_1)\big|}{|I_1||\tilde{I}_1|}\\&\le K\left(\frac{\Delta x_0+\Delta x_1+|x_0\tilde{x}_1-\tilde{x}_0x_1|}{|I_1||\tilde{I}_1|}\right)\\&\le K\left(\frac{\Delta x_0+\Delta x_1+|x_0||\tilde{x}_1-x_1|+|x_1||x_0-\tilde{x}_0|}{|I_1||\tilde{I}_1|}\right)\\&\le K(\Delta x_0+\Delta x_1)/|I_1||\tilde{I}_1|\le K\varepsilon,\end{aligned}$$

where we used Lemma 12.11. On the other hand,

$$\big|(A_1^{-1})'-(\tilde{A}_1^{-1})'\big|=\big(|\tilde{I}_1|-|I_1|\big)/|I_1||\tilde{I}_1|\le(\Delta_0+\Delta_1)/|I_1||\tilde{I}_1|,$$

and we finish in the same way as before. □

Lemma 12.20 *There exists $K>0$ such that for $\ell\le a$, we have*

$$\|\mathscr{N}f_\ell-\mathscr{N}g_\ell\|_{C^0}\le K\big(\varepsilon\,|I_\ell|+\Delta_\ell+|\Delta x_\ell|\,|I_\ell|\big).$$

Proof Note that

$$|\tilde{A}_\ell x-A_\ell x|\le K\big(|\Delta x_\ell|+\Delta_\ell\big).$$

Therefore,

$$\begin{aligned}|\mathcal{N} f_\ell(x) - \mathcal{N} g_\ell(x)| =& |\mathcal{N} f(A_\ell(x))|I_\ell| - \mathcal{N} g(\tilde{A}_\ell(x))|\tilde{I}_\ell||\\ \le& \big|\mathcal{N} f(A_\ell x)|I_\ell| - \mathcal{N} g(A_\ell x)|\tilde{I}_\ell|\big| +\\ & |D\mathcal{N} g(\theta_\ell)| \cdot (|\Delta x_\ell| + \Delta_\ell) \cdot |\tilde{I}_\ell|\\ \le& K\big(\varepsilon|I_\ell| + \Delta_\ell + (|\Delta x_\ell| + \Delta_\ell)(|I_\ell| + \Delta_\ell)\big)\\ \le& K\big(\varepsilon|I_\ell| + \Delta_\ell + |\Delta x_\ell||I_\ell|\big).\end{aligned}$$

□

Lemma 12.21 *For every $L > 0$, there exists $K = K(\mathcal{K}, L) > 0$ such that the following holds. If ζ and $\tilde{\zeta}$ are L-synchronized, then*

$$\sum_{\ell=1}^{a} \|\mathcal{N} f_\ell - \mathcal{N} g_\ell\|_{C^0} \le K\varepsilon.$$

Proof Let $a_\varepsilon = \left\lfloor \frac{1}{\varepsilon} \right\rfloor$. Assume for a moment that $a \ge a_\varepsilon$. Then Lemma 12.9 implies $|x_{a-a_\varepsilon} - x_{a_\varepsilon}|, |\tilde{x}_{a-a_\varepsilon} - \tilde{x}_{a_\varepsilon}| \le K\varepsilon$. Hence,

$$\begin{aligned}\sum_{a_\varepsilon \le \ell \le a-a_\varepsilon} \|\mathcal{N} f_\ell - \mathcal{N} g_\ell\|_{C^0} \le& \sum_{a_\varepsilon \le \ell \le a-a_\varepsilon} \|\mathcal{N} f_\ell\|_{C^0} + \|\mathcal{N} g_\ell\|_{C^0}\\ \le& \sum_{a_\varepsilon \le \ell \le a-a_\varepsilon} \|\mathcal{N} f\|_{C^0} \cdot |I_\ell| + \|\mathcal{N} g\|_{C^0} \cdot |\tilde{I}_\ell|\\ \le& K\big(|x_{a-a_\varepsilon} - x_{a_\varepsilon}| + |\tilde{x}_{a-a_\varepsilon} - \tilde{x}_{a_\varepsilon}|\big)\\ \le& K\varepsilon.\end{aligned} \tag{12.13}$$

These estimates hold trivially when $a < a_\varepsilon$. Note that

$$\begin{aligned}\sum_{\ell=1}^{a} \|\mathcal{N} f_\ell - \mathcal{N} g_\ell\|_{C^0} =& \sum_{\ell=1}^{a_\varepsilon} \|\mathcal{N} f_\ell - \mathcal{N} g_\ell\|_{C^0} + \sum_{\ell=a_0}^{a-a_\varepsilon} \|\mathcal{N} f_\ell - \mathcal{N} g_\ell\|_{C^0}\\ &+ \sum_{\ell=a-a_\varepsilon}^{a} \|\mathcal{N} f_\ell - \mathcal{N} g_\ell\|_{C^0}.\end{aligned}$$

The middle term is estimated by (12.13). The first (and third) term can be estimated by using Lemma 12.20, Yoccoz's Lemma 7.3, Propositions 12.3 and 12.4. Namely,

$$\sum_{\ell=1}^{a_\varepsilon} \|\mathcal{N} f_\ell - \mathcal{N} g_\ell\|_{C^0} \le K \sum_{\ell=1}^{a_\varepsilon} \varepsilon\,|I_\ell| + \Delta_\ell + |\Delta x_\ell|\,|I_\ell|$$

$$
\begin{aligned}
&\le K \sum_{\ell=1}^{a_\varepsilon} \varepsilon \frac{1}{\ell^2} + \varepsilon \frac{\log \ell}{\ell^2} + \varepsilon^2 \frac{1}{\ell} + \varepsilon \frac{1}{\ell^3} \\
&\le K\varepsilon + K \sum_{\ell=1}^{a_\varepsilon} \varepsilon^2 \frac{1}{\ell} \\
&\le K\varepsilon + K\,\varepsilon^2 \log \frac{1}{\varepsilon} \\
&\le K\varepsilon.
\end{aligned}
$$

The lemma follows. □

The following proposition holds for general critical commuting pairs with negative Schwarzian which are contained in the previously discussed set $\mathscr{K}$; that is, the set of normalized C^3 critical commuting pairs which are K-controlled.

Proposition 12.5 *For every $L > 0$ there exists $K = K(\mathscr{K}, L) > 0$ such that the following holds. If ζ_0 and ζ_1 are L-synchronized then*

$$d_2\big(p\mathscr{R}(\zeta_0), p\mathscr{R}(\zeta_1)\big) \le K d_2(\zeta_0, \zeta_1).$$

Proof There exists $K = K(\mathscr{K}) > 0$ such that the following holds. There exists a diffeomorphism $h : \mathrm{Dom}(\zeta_1) \to \mathrm{Dom}(\zeta_0)$ such that $\zeta = \zeta_0$ and $\tilde{\zeta} = h \circ \zeta_1 \circ h^{-1}$ satisfy the normalizations

(1) $I_\eta = I_{\tilde{\eta}}$ and $I_\xi = I_{\tilde{\xi}}$,
(2) $p = \tilde{p}$ where $D\eta(p) = D\tilde{\eta}(\tilde{p}) = 1$,

needed to apply the results from Sects. 12.3 and 12.4. We may construct the conjugation such that

$$d_{C^3}(h, \mathrm{Id}) \le K d_2(\zeta_0, \zeta_1)$$

and $h|\,\mathrm{Dom}\big(p\mathscr{R}(\zeta_1)\big) = \mathrm{Id}$. This last condition implies

$$p\mathscr{R}(\zeta_1) = p\mathscr{R}(\tilde{\zeta}).$$

In particular, it suffices to prove the proposition for the pairs ζ and $\tilde{\zeta}$.

Let $p\mathscr{R}(\zeta) = (\eta', \xi')$ and $p\mathscr{R}(\tilde{\zeta}) = (\tilde{\eta}', \tilde{\xi}')$. Because, $\xi' = \xi$ and $\tilde{\xi}' = \tilde{\xi}$ it suffices to estimate the distance between η' and $\tilde{\eta}'$.

Let $I_{a+1} = [x_{a+1}, x_a]$ and $A : [0, 1] \to I_{a+1}$ be the orientation preserving affine diffeomorphism. Let

$$F = A^{-1} \circ \eta',$$

and similarly define $G = \tilde{A}^{-1} \circ \tilde{\eta}'$. Now apply Lemmas 12.16 and 12.21 to obtain

$$d_{C^2}(F, G) \leq K\varepsilon,$$

where $\varepsilon = d_2(\zeta, \tilde{\zeta})$. A similar argument as the proof of Lemma 12.19 one obtains $d_2(\eta', \tilde{\eta}') \leq K\varepsilon$. This shows that pre-renormalization is Lipschitz among synchronized pairs. □

12.5 Order

Commuting pairs might have different domains. Any natural definition of *order* between such systems has to include this difference of domains also. There are two cases:

$$\text{Case I: } \eta \circ \xi(0) > 0, \qquad \text{Case II: } \eta \circ \xi(0) < 0.$$

Definition 12.4 Let $\zeta_0 = (\xi_0, \eta_0)$ and $\zeta_1 = (\xi_1, \eta_1)$ be two commuting pairs and $t \geq 0$. If

(1) $\zeta_0(x) + t \leq \zeta_1(x)$ for $x \in \mathrm{Dom}(\zeta_0) \cap \mathrm{Dom}(\zeta_1)$ and
(2) $\eta_0(0) \leq \eta_1(0)$ and $\xi_0(0) \leq \xi_1(0)$

we write

$$\zeta_0 \leq_t \zeta_1.$$

Lemma 12.22 *Let $\zeta_0 = (\xi_0, \eta_0)$ and $\zeta_1 = (\xi_1, \eta_1)$ be two commuting pairs such that $\zeta_0 \leq_t \zeta_1$.*

If Case I holds, then:

(1) $a_{\zeta_0} \leq a_{\zeta_1}$,
(2) for $x \in [\eta_1(0), 0]$ and $k = 0, 1, \ldots, a_{\zeta_0}$

$$\eta_0^k \circ \xi_0(x) + t \leq \eta_1^k \circ \xi_1(x).$$

If Case II holds, then:

(1) $a_{\zeta_0} \geq a_{\zeta_1}$,
(2) for $x \in [0, \xi_0(0)]$ and $k = 0, 1, \ldots, a_{\zeta_1}$

$$\xi_0^k \circ \eta_0(x) + t \leq \xi_1^k \circ \eta_1(x).$$

The proof of Lemma 12.22 is different for Case I and Case II. We will only present the proof in Case I.

Proof Let $x \in [0, \xi_0(0)]$. The order condition of Definition 12.4(1) gives the statement of the Lemma for $k = 0$, $\xi_0(x) + t \leq \xi_1(x)$. Property (2) follows inductively. Namely,

$$\begin{aligned}\eta_0^{k+1} \circ \xi_0(x) + t &= \eta_0(\eta_0^k \circ \xi_0(x)) + t \leq \eta_1(\eta_0^k \circ \xi_0(x)) \\ &\leq \eta_1(\eta_1^k \circ \xi_1(x)) \\ &= \eta_1^{k+1} \circ \xi_1(x).\end{aligned}$$

In particular, $\eta_0^{a_{\zeta_0}} \circ \xi_0(x) \leq \eta_1^{a_{\zeta_0}} \circ \xi_1(x)$. This implies, $a_{\zeta_0} \leq a_{\zeta_1}$. □

Pre-renormalization preserves order. Namely, we have the following.

Lemma 12.23 *If $\zeta_0 \leq_t \zeta_1$ and $a_{\zeta_0} = a_{\zeta_1}$, then $p\mathscr{R}(\zeta_0) \leq_t p\mathscr{R}(\zeta_1)$.*

Proof We will only present the proof in Case I. Let $a = a_{\zeta_0} = a_{\zeta_1}$. Observe, $\eta_{p\mathscr{R}(\zeta_0)}(0) = \eta_0(0) \leq \eta_1(0) = \eta_{p\mathscr{R}(\zeta_1)}(0)$. Hence, the left side of the domains of the pre-renormalizations satisfy the order condition of Definition 12.4(2). Consider the right side of the domains of the pre-renormalizations,

$$\xi_{p\mathscr{R}(\zeta_0)}(0) + t = \eta_0^a \circ \xi_0(0) + t \leq \eta_1^a \circ \xi_1(0) = \xi_{p\mathscr{R}(\zeta_1)}(0), \tag{12.14}$$

where we used Lemma 12.22(2). This means that the right side of the domain of the pre-renormalizations also satisfies the order condition of Definition 12.4(2).

According to Lemma 12.22(2) the estimate (12.14) also holds for any $x \in [\eta_1(0), 0]$, instead of $x = 0$. This means that the pre-renormalization also satisfies the order condition of Definition 12.4(1). □

The following proposition will play a key role in the proof of the Synchronization-Lemma (see Sect. 12.6 below).

Proposition 12.6 *If $\zeta_0 \leq_t \zeta_1$ with $t > 0$ then*

$$\rho_{\zeta_0} \neq \rho_{\zeta_1}.$$

Proof Assume $a_{\zeta_0}(n) = a_{\zeta_1}(n)$ for $n \geq 0$. Applying Lemma 12.23,

$$(p\mathscr{R})^n(\zeta_0) \leq_t (p\mathscr{R})^n(\zeta_1).$$

Note, $\eta_{(p\mathscr{R})^n(\zeta_{0,1})}(0) \to 0$. Hence,

$$0 > \eta_{(p\mathscr{R})^n(\zeta_1)}(0) \geq \eta_{(p\mathscr{R})^n(\zeta_0)}(0) + t \geq \frac{1}{2}t > 0$$

for n large enough, a contradiction. □

12.6 Synchronization

In the next statement we refer to the constant $\varepsilon_0 > 0$ obtained in Sect. 12.3 (see in particular the proof of Claim 12.1, during the proof of Lemma 12.11).

Synchronization-Lemma For any given $K_0 > 1$ there exists $L = L(K_0) > 1$ such that the following holds. *Let ζ_0 and ζ_1 be two C^3 critical commuting pairs which are K_0-controlled, both ζ_0 and ζ_1 have negative Schwarzian, $\rho(\zeta_0) = \rho(\zeta_1) \in [0, 1]\backslash\mathbb{Q}$ and $d_2(\zeta_0, \zeta_1) < \varepsilon_0$. Then ζ_0 and ζ_1 are L-synchronized.*

We omit mention of the hypothesis $d_2(\zeta_0, \zeta_1) < \varepsilon_0$ in the proof presented below, but it is needed to allows us to apply the estimates obtained in Sects. 12.3 and 12.5.

Proof of the Synchronization-Lemma We will only present the proof in Case I. Let $a = a_{\zeta_0} = a_{\zeta_1}$. Choose $a_0 \geq 1$ such that Lemma 12.6 applies. The Synchronization Lemma follows from Lemma 12.12 when $a \leq a_0$. We will assume $a \geq a_0$.

We may assume that $x_a^1 \geq x_a^0$. There exists $K = K(K_0) > 0$ such that the following holds: there exists a diffeomorphism $h : \mathrm{Dom}(\zeta_1) \to \mathrm{Dom}(\zeta_0)$ such that $\zeta = \zeta_0$ and $\tilde{\zeta} = h \circ \zeta_1 \circ h^{-1}$ satisfy the normalizations

$$x_1(\zeta) = x_1(\tilde{\zeta}).$$

We may construct the conjugation such that

$$d_{C^3}(h, \mathrm{Id}) \leq K d_2(\zeta_0, \zeta_1)$$

and $h|\,\mathrm{Dom}\left(p\mathscr{R}(\zeta_1)\right) = \mathrm{Id}$. This last condition implies

$$x_a(\zeta_1) = x_a(\tilde{\zeta}).$$

In particular, it suffices to prove synchronization for the pairs ζ and $\tilde{\zeta}$. Let $\varepsilon = d_2(\zeta, \tilde{\zeta}) \leq K d_2(\zeta_0, \zeta_1)$.

Apply Lemma 12.6 to obtain a commuting pair ζ_{t_0} in the standard family of ζ such that

$$\Delta x_a(\zeta_{t_0}, \tilde{\zeta}) = 0.$$

From Lemma 12.6 we get

$$0 \leq t_0 \leq K\varepsilon. \tag{12.15}$$

Note, if $t_0 > 0$ is much larger than $\varepsilon \geq d_{C^0}(\zeta, \tilde{\zeta})$ then $\xi_{t_0}(x) > \tilde{\xi}(x)$. This would imply $x_a(\zeta_{t_0}) > x_a(\tilde{\zeta})$ because $x_1(\zeta) = x_1(\tilde{\zeta})$. Assume that

$$\tilde{x}_a = x_a + L\varepsilon, \tag{12.16}$$

where, just as before, $x_\ell = \eta^\ell(\xi(0))$ and $\tilde{x}_\ell = \tilde{\eta}^\ell(\tilde{\xi}(0))$ for $\ell \in \{0, \ldots, a\}$. Note also that our assumption $x_a^1 \geq x_a^0$, implies that $\tilde{x}_a \geq x_a$.

We have to show that L is uniformly bounded. From (12.16) and Corollary 12.2 we get for every $x \in [\eta_{p\mathscr{R}(\zeta_{t_0})}(0), 0]$

$$\begin{aligned} p\mathscr{R}(\zeta_{t_0})(x) - p\mathscr{R}(\zeta)(x) &\geq \frac{1}{K}\big(p\mathscr{R}(\zeta_{t_0})(0) - p\mathscr{R}(\zeta)(0)\big) \\ &= \frac{1}{K}\big(p\mathscr{R}(\tilde{\zeta})(0) - p\mathscr{R}(\zeta)(0)\big) \\ &= \frac{1}{K}(\tilde{x}_a - x_a) = \frac{1}{K}L\varepsilon. \end{aligned} \tag{12.17}$$

From Proposition 12.5 we get for every $x \in [\eta_{p\mathscr{R}(\zeta_{t_0})}(0), 0]$

$$\begin{aligned} \big|p\mathscr{R}(\zeta_{t_0})(x) - p\mathscr{R}(\tilde{\zeta})(x)\big| &\leq K d_2(\zeta_{t_0}, \tilde{\zeta}) \\ &\leq K d_2(\zeta, \tilde{\zeta}) + K\varepsilon \\ &\leq K\varepsilon, \end{aligned} \tag{12.18}$$

where we also used (12.15). Combine (12.17) and (12.18) to get for every $x \in [\eta_{p\mathscr{R}(\zeta_{t_0})}(0), 0]$

$$p\mathscr{R}(\tilde{\zeta})(x) \geq p\mathscr{R}(\zeta)(x) + \frac{1}{K}L\varepsilon - K\varepsilon. \tag{12.19}$$

As a matter of fact (12.19) holds for $x \in [-1, 0]$. This follows from the following. Let $x \in [-1, \eta_{p\mathscr{R}(\zeta_{t_0})}(0)]$. Observe that, according to (12.15),

$$\big|[-1, \eta_{p\mathscr{R}(\zeta_{t_0})}(0)]\big| = t_0 \leq K\varepsilon.$$

This implies that

$$\begin{aligned} p\mathscr{R}(\tilde{\zeta})(x) &\geq p\mathscr{R}(\zeta)(\eta_{p\mathscr{R}(\zeta_{t_0})}(0)) + \frac{1}{K}L\varepsilon - K\varepsilon - \max\{Dp\mathscr{R}(\tilde{\zeta})\}t_0 \\ &\geq p\mathscr{R}(\zeta)(x) + \frac{1}{K}L\varepsilon - K\varepsilon. \end{aligned}$$

Hence, for $x \in [-1, 0]$ we have

$$p\mathscr{R}(\tilde{\zeta})(x) \geq p\mathscr{R}(\zeta)(x) + \frac{1}{K}L\varepsilon - K\varepsilon. \tag{12.20}$$

So, when $L \geq 2K^2$ then for the relevant $x < 0$,

$$(p\mathscr{R})^2(\tilde{\zeta})(x) > (p\mathscr{R})^2(\zeta)(x). \tag{12.21}$$

The last part of the proof will show that similar estimates hold for relevant positive points. The goal is to prove $\left(p\mathscr{R}\right)^2(\tilde{\zeta}) \geq_t \left(p\mathscr{R}\right)^2(\zeta)$ for some positive t. The branches on the left side of the second pre-renormalizations, according to (12.21), satisfy the order condition of Definition 12.4(1). The right side of the domains of the second pre-renormalizations do satisfy the order condition of Definition 12.4(2). Namely,

$$\text{Dom}\left((p\mathscr{R})^2(\zeta)\right) \cap \{x \geq 0\} = [0, x_a] \subset [0, \tilde{x}_a] = \text{Dom}\left((p\mathscr{R})^2(\tilde{\zeta})\right) \cap \{x \geq 0\}.$$

What remains is to describe the branches on the right and the domains on the left. Let $x \in \text{Dom}\left((p\mathscr{R})^2(\zeta)\right) \cap \{x \geq 0\} = [0, x_a]$ and for $k \geq 1$ define

$$z_k(x) = \left(p\mathscr{R}(\zeta)\right)^k(x),$$

and similarly, $\tilde{z}_k(x) = \left(p\mathscr{R}(\tilde{\zeta})\right)^k(x)$, Observe,

$$|z_1(x) - \tilde{z}_1(x)| = |p\mathscr{R}(\zeta)(x) - p\mathscr{R}(\tilde{\zeta})(x)| = |\eta(x) - \tilde{\eta}(x)| \leq \varepsilon.$$

Hence, applying (12.20),

$$\begin{aligned}
\tilde{z}_2(x) &= p\mathscr{R}(\tilde{\zeta})(\tilde{z}_1) \\
&\geq p\mathscr{R}(\zeta)(\tilde{z}_1) + \frac{1}{K}L\varepsilon - K\varepsilon \\
&\geq z_2(x) - \max\left(Dp\mathscr{R}(\zeta)\right) \cdot |z_1(x) - \tilde{z}_1(x)| + \frac{1}{K}L\varepsilon - K\varepsilon \\
&\geq z_2(x) + \frac{1}{K}L\varepsilon - K\varepsilon > z_2(x),
\end{aligned}$$

when $L \geq 2K^2$. Let $b = a_{p\mathscr{R}(\zeta)} = a_{p\mathscr{R}(\tilde{\zeta})}$. By repeatedly applying (12.20) with $L \geq 2K^2$, we obtain

$$\left(p\mathscr{R}\right)^2(\tilde{\zeta})(x) = \tilde{z}_b(x) > \left(p\mathscr{R}\right)^2(\zeta)(x) = z_b(x). \tag{12.22}$$

In particular,

$$\begin{aligned}
\text{Dom}\left((p\mathscr{R})^2(\tilde{\zeta})\right) \cap \{x \leq 0\} &= [\tilde{z}_b, 0] \subset [z_b, 0] \\
&= \text{Dom}\left((p\mathscr{R})^2(\zeta)\right) \cap \{x \leq 0\}.
\end{aligned} \tag{12.23}$$

The estimates (12.22) and (12.23) finish the proof of

$$\left(p\mathscr{R}\right)^2(\tilde{\zeta}) \geq_t \left(p\mathscr{R}\right)^2(\zeta),$$

for some $t > 0$. However, this contradicts Proposition 12.6 because $\left(p\mathscr{R}\right)^2(\tilde{\zeta})$ and $\left(p\mathscr{R}\right)^2(\zeta)$ have the same rotation number. This contradiction establishes the synchronization with $L \leq 2K^2$. □

12.7 Lipschitz Estimate

In this section we finally prove Theorem 12.2.

Proof of Theorem 12.2 The Synchronization Lemma from Sect. 12.6 tells us that, for $L = L(\mathscr{K})$, the pairs ζ_0 and ζ_1 are L-synchronized. Now the Lipschitz estimate for renormalization of synchronized pairs, Proposition 12.5, implies a Lipschitz estimate for pre-renormalization along topological classes. The fact that the maps are synchronized implies that the domains of the pre renormalizations are also close. This means that the normalizations will not affect the Lipschitz property. □

Exercises

Exercise 12.1 Let $f_1, \ldots, f_n : \mathbb{R} \to \mathbb{R}$ be C^1 maps with C^1 norm bounded by some constant $B > 0$, and let $g_1, \ldots, g_n : \mathbb{R} \to \mathbb{R}$ be C^0 maps. By induction, prove that

$$\left\| f_n \circ \cdots \circ f_1 - g_n \circ \cdots \circ g_1 \right\|_{C^0} \leq \left(\sum_{j=0}^{n-1} B^j \right) \max_{i \in \{1,\ldots,n\}} \left\| f_i - g_i \right\|_{C^0} .$$

Exercise 12.2 Using the previous exercise, prove a C^0 version of Theorem 12.2 for bounded combinatorics (allowing the Lipschitz constant to depend on the bound, see Remark 12.3). Find the minimum set of hypothesis you need to assume (recall Remarks 12.1 and 12.2).

Chapter 13
Exponential Convergence: The Smooth Case

As discussed in Sect. 10.1, the main motivation behind our study of multicritical circle maps is to understand the *smooth rigidity problem*. To be more precise, the goal is to answer Question 10.1: let h be a topological conjugacy between two multicritical circle maps, say f and g, and assume that h identifies each critical point of f with a corresponding critical point of g having the same criticality. Is h a smooth diffeomorphism? As we saw in Theorems 10.1 and 10.2, this problem has essentially been solved in the case of a single critical point, of an odd integer criticality. Moreover, as we saw in Theorem 10.3, rigidity has also been established for bi-critical circle maps of bounded combinatorics (see also the recent preprint Gorbovickis and Yampolsky 2021). As explained in Chap. 10, the rigidity problem reduces to proving geometric contraction of renormalization along multicritical circle maps with the same signature. In the remainder of this book (Chaps. 13 and 14) we will survey the main ideas needed to establish such contraction in the unicritical case. We finish this initial paragraph by pointing out that the analytic tools developed in both Chaps. 11 and 12 are crucial for the methods to be discussed in the present chapter.

The following fundamental theorem was obtained by de Faria and de Melo (2000) for rotation numbers of bounded type, and extended by Khmelev and Yampolsky (2006) to cover all irrational rotation numbers.

Theorem 13.1 *There exists a universal constant λ in $(0, 1)$ with the following property. Given two real-analytic unicritical commuting pairs ζ_1 and ζ_2 with the same irrational rotation number and the same criticality, there exists a constant $C > 0$ such that*

$$d_r\big(\mathscr{R}^n(\zeta_1), \mathscr{R}^n(\zeta_2)\big) \leqslant C\lambda^n$$

for all $n \in \mathbb{N}$ and for any $0 \leqslant r < \infty$.

E. de Faria, P. Guarino, *Dynamics of Circle Mappings*, IMPA Monographs,
https://doi.org/10.1007/978-3-031-67495-2_13

The proof of Theorem 13.1 relies on holomorphic methods, and it will be discussed in the next chapter (Chap. 14). In the present chapter we would like to explain how one can use Theorem 13.1 in order to prove exponential contraction of renormalizations for unicritical circle maps with a *finite* degree of smoothness. More precisely, we will explain the proof of the following two results, which are Guarino and de Melo (2017, Th. C) and Guarino et al. (2018, Th. B) respectively.

Theorem 13.2 *There exists a universal constant* $\lambda \in (0, 1)$ *with the following property. Given two* C^3 *unicritical circle maps* f *and* g *with the same irrational rotation number of bounded type and the same odd integer criticality, there exists* $C > 0$ *such that for all* $n \in \mathbb{N}$ *we have*

$$d_0\big(\mathscr{R}^n(f), \mathscr{R}^n(g)\big) \leqslant C\lambda^n .$$

Theorem 13.3 *There exists a universal constant* $\lambda \in (0, 1)$ *with the following property. Given two* C^4 *unicritical circle maps* f *and* g *with the same irrational rotation number and the same odd integer criticality, there exists* $C > 0$ *such that for all* $n \in \mathbb{N}$ *we have*

$$d_2\big(\mathscr{R}^n(f), \mathscr{R}^n(g)\big) \leqslant C\lambda^n .$$

As established in Chap. 10, such exponential convergence of renormalization orbits implies the desired smooth rigidity: topological conjugacies are actually diffeomorphisms. To be more precise, when combined with Theorems 10.4 and 10.5, Theorems 13.2 and 13.3 imply Theorems 10.2 and 10.1 respectively.

13.1 The Shadowing Property

The link between C^r unicritical circle maps and real-analytic ones is given by the following result.

Theorem 13.4 (Shadowing) *There exists a* C^ω*-compact set* $\mathscr{K}$ *of real-analytic unicritical commuting pairs with the following property. For any* $r \geqslant 3$ *there exists a constant* $\lambda = \lambda(r) \in (0, 1)$ *such that, given a* C^r *unicritical circle map* f *with irrational rotation number and odd integer criticality, there exist* $C > 0$ *and a sequence* $\{f_n\}_{n\in\mathbb{N}}$ *contained in* $\mathscr{K}$ *such that*

$$d_{r-1}\big(\mathscr{R}^n(f), f_n\big) \leqslant C\lambda^n \quad \text{for all } n \in \mathbb{N}, \tag{13.1}$$

and such that the pair f_n *has the same rotation number as the pair* $\mathscr{R}^n(f)$ *for all* $n \in \mathbb{N}$.

The compact set $\mathscr{K}$ and the approximations $\{f_n\}$ given by Theorem 13.4 were constructed in Guarino and de Melo (2017, Sections 6 and 7). However, in that work, the exponential convergence (13.1) was only established for the C^0 metric (see Guarino and de Melo 2017, Th. D). These estimates were later extended to the C^{r-1} metric in Guarino et al. (2018, Th. 11.1). We proceed to survey some of the main tools for this construction.

For simplicity, and without loss of generality, let us assume in this section that the critical point of f is of cubic type. The deformations from smooth to analytic commuting pairs needed in order to prove Theorem 13.4 will be done in the complex plane, with the help of Theorem 11.4. With this as our goal, we will first extend both components of the unicritical commuting pair $\mathscr{R}^n(f)$ to open sets in the complex plane. This is achieved by the following result, which is Guarino and de Melo (2017, Th. 6.1) (given a bounded interval I of the real line we denote its Euclidean length by $|I|$, and for any $\alpha > 0$ we denote by $N_\alpha(I)$ the $\mathbb{R}$-symmetric topological disk

$$N_\alpha(I) = \left\{z \in \mathbb{C} : d(z, I) < \alpha|I|\right\},$$

where d denotes the Euclidean distance in the complex plane).

Theorem 13.5 *There exist universal constants $\lambda \in (0, 1)$, and $\alpha > 0$ and $\beta > 0$ with the following property. Let f be a C^3 unicritical circle map with irrational rotation number and cubic critical point. For all $n \geqslant 1$, denote by (η_n, ξ_n) the components of the critical commuting pair $\mathscr{R}^n(f)$. Then there exist constants $n_0 \in \mathbb{N}$ and $C > 0$ such that for each $n \geqslant n_0$ both ξ_n and η_n extend (after normalization) to $\mathbb{R}$-symmetric orientation-preserving C^3 maps defined in $N_\alpha([-1, 0])$ and $N_\alpha([0, \xi_n(0)])$ respectively, and the following seven properties are satisfied:*

1. *ξ_n and η_n each have a unique critical point at the origin, which is of cubic type;*
2. *The extensions η_n and ξ_n commute in $B(0, \lambda)$, that is, both compositions $\eta_n \circ \xi_n$ and $\xi_n \circ \eta_n$ are well defined in $B(0, \lambda)$, and they coincide;*
3. *$N_\beta(\xi_n([-1, 0])) \subset \xi_n(N_\alpha([-1, 0]))$;*
4. *$N_\beta([-1, (\eta_n \circ \xi_n)(0)]) \subset \eta_n(N_\alpha([0, \xi_n(0)]))$;*
5. *$\eta_n(N_\alpha([0, \xi_n(0)])) \cup \xi_n(N_\alpha([-1, 0])) \subset B(0, \lambda^{-1})$;*
6. *We have*

$$\max_{z \in N_\alpha([-1,0]) \setminus \{0\}} \left\{ \frac{|\bar{\partial}\xi_n(z)|}{|\partial\xi_n(z)|} \right\} \leqslant C\lambda^n;$$

7. *We have*

$$\max_{z \in N_\alpha([0,\xi_n(0)]) \setminus \{0\}} \left\{ \frac{|\bar{\partial}\eta_n(z)|}{|\partial\eta_n(z)|} \right\} \leqslant C\lambda^n.$$

In the language of Chap. 11, the last two items of Theorem 13.5 say that the Beltrami coefficients of the corresponding extensions of ξ_n and η_n are exponentially small in n. An important tool used in Guarino and de Melo (2017, Section 6) in order to prove Theorem 13.5 is the notion of *asymptotically holomorphic* maps, that we review in the next section.

13.1.1 Extended Lifts of Critical Circle Maps

In this section we lift a critical circle map to the real line, and then we extend this lift in a suitable way to a neighborhood of the real line in the complex plane (see Definition 13.2 below).

In order to do this, let $A : \mathbb{C} \setminus \{0\} \to \mathbb{C} \setminus \{0\}$ be the map corresponding to the parameters $a = 0$ and $b = 1$ in the Arnold family defined in Sect. 6.1.2. Recall that the lift of A to the complex plane, under the canonical universal covering map $\pi : \mathbb{C} \to \mathbb{C} \setminus \{0\}$ given by $\pi(z) = e^{2\pi i z}$, is the entire map $\widetilde{A} : \mathbb{C} \to \mathbb{C}$ given by

$$\widetilde{A}(z) = z - \frac{1}{2\pi} \sin(2\pi z).$$

Note that A preserves the unit circle, and its restriction $A : S^1 \to S^1$ is a real-analytic critical circle map. The critical point of A is placed at 1, and is of cubic type (the critical point is also a fixed point for A).

Now let f be a C^3 critical circle map with a single critical point (which is placed at the point 1, and is of cubic type), and let $\widetilde{f}$ be the unique lift of f under π satisfying $D\widetilde{f}(0) = 0$ and $0 < \widetilde{f}(0) < 1$. It is not difficult to prove (see Exercise 13.1) that there exist C^3 orientation preserving circle diffeomorphisms h_1 and h_2, with $h_1(1) = 1$ and $h_2(1) = f(1)$, such that the following diagram commutes.

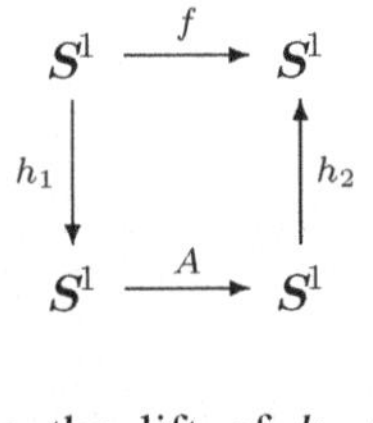

For each $s \in \{1, 2\}$ let $\widetilde{h_i}$ be the lift of h_s to the real line determined by $\widetilde{h}_s(0) \in [0, 1)$. In Proposition 13.1 below we will extend both $\widetilde{h}_1$ and $\widetilde{h}_2$ to complex neighborhoods of the real line, satisfying the following property.

Definition 13.1 Let I be a compact interval in the real line, let U be a neighborhood of I in $\mathbb{C}$ and let $H : U \to \mathbb{C}$ be a C^1 map. We say that H is *asymptotically holomorphic* of order $r \geqslant 1$ in I if for every $z \in I$ we have $\bar{\partial} H(z) = 0$ and moreover

$$\frac{\bar{\partial} H(z)}{(\operatorname{Im} z)^{r-1}} \to 0$$

uniformly as $\operatorname{Im} z$ goes to zero. We say that H is *asymptotically holomorphic* of order r in $\mathbb{R}$ if it is asymptotically holomorphic of order r in compact sets of $\mathbb{R}$.

In the following statement we suppose $r \geqslant 1$, even though we will apply it for $r \geqslant 3$. In the proof we follow the exposition of Graczyk et al. (2005, Lem. 2.1, page 623).

Proposition 13.1 *For each $s = 1, 2$ there exists $H_s : \mathbb{C} \to \mathbb{C}$ of class C^r with the following properties.*

1. *H_s is an extension of $\widetilde{h}_s$: $H_s|_{\mathbb{R}} = \widetilde{h}_s$;*
2. *H_s commutes with unitary horizontal translation: $H_s \circ T = T \circ H_s$;*
3. *H_s is asymptotically holomorphic in $\mathbb{R}$ of order r;*
4. *H_s is $\mathbb{R}$-symmetric: $H_s(\bar{z}) = \overline{H_s(z)}$.*

Moreover there exist $R > 0$ and four domains B_R, U_R, V_R and W_R in $\mathbb{C}$, symmetric about the real line and such that

- $B_R = \{z \in \mathbb{C} : -R < \operatorname{Im} z < R\}$;
- *H_1 is an orientation preserving diffeomorphism between B_R and U_R;*
- $\widetilde{A}(U_R) = V_R$;
- *H_2 is an orientation preserving diffeomorphism between V_R and W_R.*
- *both $\inf_{z \in B_R} |\partial H_1(z)|$ and $\inf_{z \in V_R} |\partial H_2(z)|$ are positive numbers.*

Proof For $z = x + iy \in \mathbb{C}$, with $y \neq 0$, let $P_{x,y}$ be the degree r polynomial map that coincide with $\widetilde{h}_s$ in the $r + 1$ real numbers

$$\left\{x + \frac{j}{r} y\right\}_{j \in \{0,1,\dots,r\}}.$$

Recall that $P_{x,y}$ can be given by the following linear combination (the so-called *Lagrange's form of the interpolating polynomial*):

$$\begin{aligned} P_{x,y}(z) &= \sum_{j=0}^{r} \widetilde{h}_s\big(x + (j/r)y\big) \prod_{\substack{l=0 \\ l \neq j}}^{r} \frac{z - \big(x + (l/r)y\big)}{\big(x + (j/r)y\big) - \big(x + (l/r)y\big)} \\ &= \sum_{j=0}^{r} \widetilde{h}_s\big(x + (j/r)y\big) \prod_{\substack{l=0 \\ l \neq j}}^{r} \frac{z - x - (l/r)y}{\big((j-l)/r\big)y}. \end{aligned}$$

We define $H_s(x + iy) = P_{x,y}(x + iy)$, that is,

$$H_s(x + iy) = P_{x,y}(x + iy) = \sum_{j=0}^{r} \widetilde{h}_s\big(x + (j/r)y\big) \prod_{\substack{l=0 \\ l \neq j}}^{r} \frac{ir - l}{j - l}.$$

After computation we obtain

$$H_s(x+iy) = P_{x,y}(x+iy) = \frac{1}{N}\sum_{j=0}^{r} \frac{(-1)^j}{1+i(j/r)}\binom{r}{j}\widetilde{h}_s\big(x+(j/r)y\big)$$

where

$$N = \sum_{j=0}^{r} \frac{(-1)^j}{1+i(j/r)}\binom{r}{j} \neq 0\,.$$

Note that H_s is as smooth as $\widetilde{h_s}$, and $H_s(x) = \widetilde{h}_s(x)$ for any real number x. As $\widetilde{h}_s$ is a lift, for any $j \in \{0, 1, \ldots, r\}$ we have $\widetilde{h}_s\big(x+1+(j/r)y\big) = \widetilde{h}_s\big(x+(j/r)y\big)+1$, but then $P_{x+1,y}\big(x+1+(j/r)y\big) = P_{x,y}\big(x+(j/r)y\big)+1$ for any $j \in \{0, 1, \ldots, r\}$, and this implies $P_{x+1,y} \circ T = T \circ P_{x,y}$ in the whole complex plane. To prove that H_s is asymptotically holomorphic of order r in $\mathbb{R}$ note that

$$\overline{\partial} H_s(x+iy) = \frac{1}{2N}\sum_{j=0}^{r}(-1)^j\binom{r}{j}\widetilde{h}'_s\big(x+(j/r)y\big)\,,$$

and for each $k \in \{0, \ldots, r\}$,

$$\frac{\partial^k}{\partial y^k}\,\overline{\partial} H_s(x+iy) = \left(\frac{1}{2N}\right)\left(\frac{1}{r^k}\right)\sum_{j=0}^{r}(-1)^j j^k\binom{r}{j}\widetilde{h}_s^{\,(k+1)}\big(x+(j/r)y\big)\,.$$

By using the identity $\sum_{j=0}^{r}(-1)^j j^k\binom{r}{j} = 0$ for each $k \in \{0, \ldots, r-1\}$, we obtain, for every $x \in \mathbb{R}$,

$$\overline{\partial} H_s(x) = \left(\frac{1}{2N}\right)\widetilde{h}'_s(x)\sum_{j=0}^{r}(-1)^j\binom{r}{j} = 0\,,$$

and for each $k \in \{0, \ldots, r-1\}$,

$$\frac{\partial^k}{\partial y^k}\,\overline{\partial} H_s(x) = \left(\frac{1}{2N}\right)\left(\frac{\widetilde{h}_s^{\,(k+1)}(x)}{r^k}\right)\sum_{j=0}^{r}(-1)^j j^k\binom{r}{j} = 0\,.$$

By Taylor's theorem,

$$\lim_{y\to 0}\frac{\overline{\partial} H_s(x+iy)}{y^{r-1}} = 0$$

uniformly on compact subsets of the real line, and hence H_s is asymptotically holomorphic of order r in $\mathbb{R}$. To obtain the symmetry as in the fourth item of the statement, we can take $z \mapsto \big(H_s(z) + \overline{H_s(\bar{z})}\big)/2$, preserving all the other properties. Finally, it is easy to check (see Exercise 13.3) that the Jacobian of H_s at a point x in $\mathbb{R}$ is equal to $|\widetilde{h}_s'(x)|^2 \neq 0$. This gives us a complex neighborhood of the real line where H_s is an orientation preserving diffeomorphism, and the positive constant R. Since we also have $\partial H_s = \widetilde{h}_s'$ on the real line (again, see Exercise 13.3), and each $\widetilde{h}_s$ is the lift of a circle diffeomorphism, we obtain the last item of Proposition 13.1. □

Definition 13.2 The map $F : B_R \to W_R$ defined by $F = H_2 \circ \widetilde{A} \circ H_1$ is called the *extended lift* of the critical circle map f.

$$\begin{array}{ccc} B_R & \xrightarrow{\;F\;} & W_R \\ {\scriptstyle H_1}\big\downarrow & & \big\uparrow{\scriptstyle H_2} \\ U_R & \xrightarrow{\;\widetilde{A}\;} & V_R \end{array}$$

Note the following properties.

- F is C^r in the horizontal band B_R;
- $T \circ F = F \circ T$ in B_R;
- F is $\mathbb{R}$-symmetric, and coincides with $\widetilde{f}$ when restricted to the real line;
- F is asymptotically holomorphic in $\mathbb{R}$ of order r;
- The critical points of F in B_R are the integers (the same as $\widetilde{A}$), and they are of cubic type.

We remark that the extended lift of a real-analytic critical circle map will be C^∞ in the corresponding horizontal strip, but not necessarily holomorphic.

13.1.2 Almost Schwarz Inclusion

To the best of our knowledge, asymptotically holomorphic maps were first used in one-dimensional dynamics by Lyubich in the early nineties (but only published in Lyubich 2019), and later by Graczyk et al. (2005). One of its fundamental properties is Proposition 13.2 below, an *almost Schwarz inclusion*, that we proceed to explain.

Given an open interval $I = (a, b) \subset \mathbb{R}$, consider $\mathbb{C}(I) = \big(\mathbb{C}\backslash\mathbb{R}\big) \cup I$. This domain $\mathbb{C}(I)$ can be naturally endowed with a hyperbolic Riemannian metric. Indeed, by the Riemann mapping theorem we can define on $\mathbb{C}(I)$ a complete and conformal metric of constant curvature equal to -1, just by pulling back the standard Poincaré metric of the unit disk $\mathbb{D}$ by any conformal uniformization. Note that, by symmetry, I is always a hyperbolic geodesic.

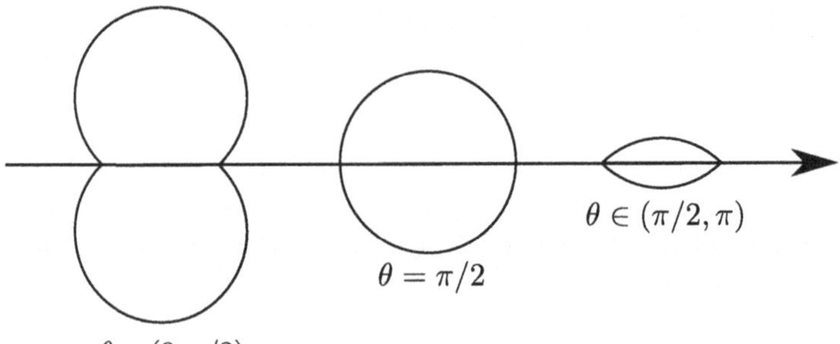

Fig. 13.1 Poincaré disks

For any given $\theta \in (0, \pi)$, let D be the open disk in the plane intersecting the real line along I, and for which the angle from $\mathbb{R}$ to ∂D at the point b (measured anticlockwise) equals θ. Let $D^+ = D \cap \{z : \operatorname{Im} z > 0\}$ and let D^- be the image of D^+ under complex conjugation. Define the *Poincaré disk* of angle θ based on I as $D_\theta(a, b) = D^+ \cup I \cup D^-$, that is, $D_\theta(a, b)$ is the set of points in the complex plane that *view* I under an angle greater or equal than θ. Note that for $\theta = \pi/2$, the Poincaré disk $D_\theta(I)$ is just the Euclidean disk whose diameter is the interval I (see Fig. 13.1).

For each $\theta \in (0, \pi)$ let $\varepsilon(\theta) = \log \tan\big(\pi/2 - \theta/4\big) \in (0, +\infty)$. As it is not difficult to prove (see Exercise 13.6), the Poincaré disk $D_\theta(I)$ coincides with the set of points in $\mathbb{C}(I)$ whose hyperbolic distance to I is less than ε. In particular, we can state the classical Schwarz lemma in the following way: let I and J be two intervals in the real line and let $\phi : \mathbb{C}(I) \to \mathbb{C}(J)$ be a holomorphic map such that $\phi(I) \subset J$. Then for any $\theta \in (0, \pi)$ we have that $\phi\big(D_\theta(I)\big) \subset D_\theta(J)$. The main reason to choose asymptotically holomorphic maps to extend one-dimensional dynamics (recall Proposition 13.1 and Definition 13.2 above) is the following *asymptotic* Schwarz lemma (on its statement, we denote by $\operatorname{diam}\big(D_\theta(a, b)\big)$ the Euclidean diameter of the Poincaré disk $D_\theta(a, b)$).

Proposition 13.2 (Almost Schwarz Inclusion) *Let $h : I \to \mathbb{R}$ be a C^3 diffeomorphism from a compact interval I with non-empty interior into the real line. Let H be any C^3 extension of h to a complex neighborhood of I, which is asymptotically holomorphic of order 3 on I. Then there exist $M > 0$ and $\delta > 0$ such that if $a, b \in I$ are different, $\theta \in (0, \pi)$ and $\operatorname{diam}\big(D_\theta(a, b)\big) < \delta$ then*

$$H\big(D_\theta(a, b)\big) \subseteq D_{\tilde{\theta}}\big(h(a), h(b)\big),$$

where $\tilde{\theta} = \theta - M\,|b - a| \operatorname{diam}\big(D_\theta(a, b)\big)$. Moreover, $\tilde{\theta} > 0$.

A proof of this result can be found in Graczyk et al. (2005, Prop. 2, p. 629). Let us point out that a predecessor of this almost Schwarz inclusion, for real-analytic maps, already appeared in de Faria and de Melo (2000, Lem. 3.3, p. 350), see Lemma 14.6 in Chap. 14.

When combined with Theorem 6.3 (the real bounds), the geometric control given by Proposition 13.2 provides bounds on the quasiconformal distortion of the renormalizations of the previously mentioned extensions (one does not study the dynamics of these extensions, just their geometric behaviour). This control implies Theorem 13.5 (see Guarino and de Melo 2017, Section 6.3, p. 1753 for the computations).

With Theorem 13.5 at hand, the deformations from $\mathscr{R}^n(f)$ to f_n (in order to prove Theorem 13.4) will be done with the help of Theorem 11.4. Our exposition in the remainder of this section (Sect. 13.1) follows closely (Guarino and de Melo 2017, Section 7).

By a *topological disk* we mean an open, connected and simply connected set properly contained in the complex plane. Let $\pi : \mathbb{C} \to \mathbb{C} \setminus \{0\}$ be the holomorphic covering $z \mapsto \exp(2\pi i z)$, and let $T : \mathbb{C} \to \mathbb{C}$ be the horizontal translation $z \mapsto z+1$ (which is a generator of the group of automorphisms of the covering). For any $R > 1$ consider the *band*

$$B_R = \left\{z \in \mathbb{C} : -\log R < 2\pi \operatorname{Im} z < \log R\right\},$$

which is the universal cover of the *round annulus*

$$A_R = \left\{z \in \mathbb{C} : \frac{1}{R} < |z| < R\right\}$$

via the holomorphic covering π. Since B_R is T-invariant, the translation generates the group of automorphisms of the covering. The restriction $\pi : \mathbb{R} \to S^1 = \partial\mathbb{D}$ is also a covering map, the automorphism T preserves the real line, and again generates the group of automorphisms of the covering.

More generally, an *annulus* is an open and connected set A in the complex plane whose fundamental group is isomorphic to $\mathbb{Z}$. By the Uniformization Theorem (recall Exercise 11.9) such an annulus is conformally equivalent either to the punctured disk $\mathbb{D} \setminus \{0\}$, to the punctured plane $\mathbb{C} \setminus \{0\}$, or to some round annulus $A_R = \left\{z \in \mathbb{C} : 1/R < |z| < R\right\}$. In the last case the value of $R > 1$ is unique, and there exists a holomorphic covering map from $\mathbb{D}$ to A whose group of deck transformations is infinite cyclic, and such that any generator is a Möbius transformation that has exactly two fixed points at the boundary of the unit disk.

Since the deck transformations are Möbius transformations, they are isometries of the Poincaré metric on $\mathbb{D}$ and therefore there exists a unique Riemannian metric on A such that the covering map provided by the Uniformization Theorem is a local isometry. This metric is complete, and in particular, any two points can be joined by a minimizing geodesic. There exists a unique simple closed geodesic in A, whose hyperbolic length is equal to $\pi^2/\log R$. The length of this closed geodesic is therefore a conformal invariant.

We denote by Θ the antiholomorphic involution $z \mapsto 1/\bar{z}$ in the punctured plane $\mathbb{C}\setminus\{0\}$, and we say that a map is $\boldsymbol{S}^1$*-symmetric* if it commutes with Θ. An annulus is $\boldsymbol{S}^1$*-symmetric* if it is invariant under Θ (for instance, the round annulus A_R described above is $\boldsymbol{S}^1$-symmetric). In this case, the unit circle is the *core curve* (the unique simple closed geodesic) for the hyperbolic metric in A. In this section we will deal only with $\boldsymbol{S}^1$-symmetric annulus. In particular any time that some annulus A_0 is contained in some other annulus A_1, we have that A_0 separates the boundary components of A_1 (more technically, the inclusion is *essential* in the sense that the fundamental group $\pi_1(A_0)$ injects into $\pi_1(A_1)$).

Besides Theorem 13.5, the main tool for proving Theorem 13.4 is Theorem 11.4. The proof of Theorem 13.4 will be divided in three sections. Throughout the proof, C will denote a positive constant (independent of $n \in \mathbb{N}$) and n_0 will denote a (large enough) positive integer. At first, let $n_0 \in \mathbb{N}$ given by Theorem 13.5. Moreover let us use the following notation: $W_1 = N_\alpha\big([-1, 0]\big)$, $W_2 = W_2(n) = N_\alpha\big([0, \xi_n(0)]\big)$, $W_0 = B(0, \lambda)$ and $\mathscr{V} = B(0, \lambda^{-1})$, where $\alpha > 0$ and $\lambda \in (0, 1)$ are the universal constants given by Theorem 13.5. Recall that $\eta_n(0) = -1$ for all $n \geqslant 1$ after normalization.

13.1.3 A Bidimensional Glueing Procedure

From Theorem 13.5 we have the following.

Lemma 13.1 *There exists an $\mathbb{R}$-symmetric topological disk U with:*

$$-1 \in U \subset W_1 \setminus W_0,$$

such that for all $n \geqslant n_0$ the composition:

$$\eta_n^{-1} \circ \xi_n : U \to \big(\eta_n^{-1} \circ \xi_n\big)(U)$$

is an $\mathbb{R}$-symmetric orientation-preserving C^3 diffeomorphism.

For each $n \geqslant n_0$ denote by A_n the diffeomorphism $\eta_n^{-1} \circ \xi_n$. Note that $\|\mu_{A_n}\|_\infty \leqslant C\lambda^n$ in U for all $n \geqslant n_0$, and that the domains $\big\{A_n(U)\big\}_{n \geqslant n_0}$ are uniformly bounded since they are contained in $\cup_j W_2^j$. Fix $\varepsilon > 0$ and $\delta > 0$ such that the rectangle:

$$V = \big(-1-\varepsilon, -1+\varepsilon\big) \times \big(-i\delta, i\delta\big)$$

is compactly contained in U, and apply Theorem 11.4 to the sequence of $\mathbb{R}$-symmetric orientation-preserving C^3 diffeomorphisms:

$$\{A_n : U \to A_n(U)\}_{n \geqslant n_0}$$

to obtain a sequence of $\mathbb{R}$-symmetric biholomorphisms:

$$\left\{B_n : V \to B_n(V)\right\}_{n \geqslant n_0}$$

such that

$$\left\| A_n - B_n \right\|_{C^0(V)} \leqslant C\lambda^n \quad \text{for all } n \geqslant n_0.$$

By combining Theorem 13.5 with the commuting condition, we obtain the following configuration.

Lemma 13.2 *For each $n \geqslant n_0$ there exist three $\mathbb{R}$-symmetric topological disks $V_i(n)$ for $i \in \{1, 2, 3\}$ with the following five properties:*

- $0 \in V_1(n) \subset W_0$;
- $(\eta_n \circ \xi_n)(0) = (\xi_n \circ \eta_n)(0) = \xi_n(-1) \in V_2(n) \subset W_2$;
- $\xi_n(0) \in V_3(n) \subset W_2$;
- *When restricted to $V_1(n)$, both η_n and ξ_n are orientation-preserving three-fold C^3 branched coverings onto V and $V_3(n)$ respectively, with a unique critical point at the origin;*
- *Both restrictions $\xi_n|_V$ and $\eta_n|_{V_3(n)}$ are orientation-preserving C^3 diffeomorphisms onto $V_2(n)$.*

In particular $\eta_n^{-1} \circ \xi_n$ is an orientation-preserving C^3 diffeomorphism from V onto $V_3(n)$ for all $n \geqslant n_0$.

For each $n \geqslant n_0$ let $U_1(n)$, $U_2(n)$ and $U_3(n)$ be three $\mathbb{R}$-symmetric topological disks such that

- $\overline{U_1(n)}$, $\overline{U_2(n)}$ and $\overline{U_3(n)}$ are pairwise disjoint;
- $V \bigcap U_j(n) = \emptyset$ and $V_i(n) \bigcap U_j(n) = \emptyset$ for $i, j \in \{1, 2, 3\}$;
- $\overline{U_1(n)} \subset W_1$ and $\overline{U_2(n)} \bigcup \overline{U_3(n)} \subset W_2$;

and such that

$$\mathscr{U}_n = \text{interior}\left[V \bigcup \left(\bigcup_{i=1}^{i=3} V_i(n) \right) \bigcup \left(\bigcup_{j=1}^{j=3} \overline{U_j(n)} \right) \right]$$

is an $\mathbb{R}$-symmetric topological disk (see Fig. 13.2). Note that

$$\overline{I_{\xi_n} \cup I_{\eta_n}} \subset \mathscr{U}_n \subset W_1 \cup W_2 \quad \text{for all } n \geqslant n_0,$$

and that $\mathscr{U}_n \setminus \left(\overline{V \cup V_1(n) \cup V_2(n) \cup V_3(n)}\right)$ has three connected components, which are precisely $U_1(n)$, $U_2(n)$ and $U_3(n)$. By Theorem 13.5 we can choose $U_1(n)$, $U_2(n)$ and $U_3(n)$ in order to also have:

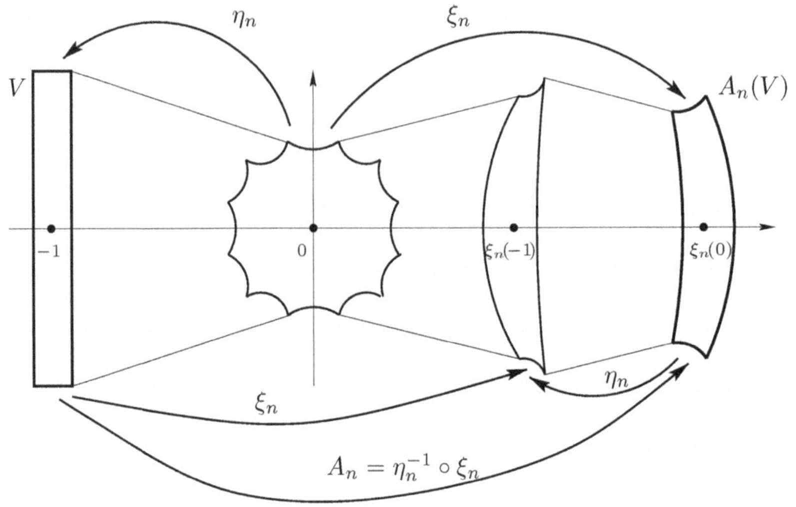

Fig. 13.2 The domain $\mathscr{U}_n$

$$\overline{N_\delta\big([-1,0]\big) \cup N_\delta\big([0,\xi_n(0)]\big)} \subset \mathscr{U}_n \quad \text{for all } n \geqslant n_0,$$

for some universal constant $\delta > 0$, independent of $n \geqslant n_0$. Note also that each $\mathscr{U}_n$ is uniformly bounded since it is contained in $N_\alpha\big([-1,K]\big)$, where $\alpha > 0$ is given by Theorem 13.5, and $K > 1$ is the universal constant given by the real bounds.

For each $n \geqslant n_0$ let $\mathscr{T}_n$ be an $\mathbb{R}$-symmetric topological disk such that:

- V, $V_1(n)$, $V_2(n)$ and $B_n(V)$ are contained in $\mathscr{T}_n$,
- $\mathscr{T}_n \setminus \big(V \cup B_n(V)\big)$ is connected and simply connected,
- The Hausdorff distance between $\overline{\mathscr{T}_n}$ and $\overline{\mathscr{U}_n}$ is less than or equal to

$$\big\| A_n - B_n \big\|_{C^0(V)} \leqslant C\lambda^n,$$

Lemma 13.3 *For each $n \geqslant n_0$ there exists an orientation-preserving $\mathbb{R}$-symmetric C^3 diffeomorphism $\Phi_n : \mathscr{U}_n \to \mathscr{T}_n$ such that*

- $\Phi_n \equiv \mathrm{Id}$ *in the interior of $V \cup \overline{U_1(n)} \cup V_1(n)$, in particular $\Phi_n(0) = 0$.*
- $B_n = \Phi_n \circ \big(\eta_n^{-1} \circ \xi_n\big) \circ \Phi_n^{-1}$ *in V, that is, $\Phi_n \circ A_n = B_n \circ \Phi_n$ in V.*
- $\big\| \Phi_n - \mathrm{Id} \big\|_{C^0(\mathscr{U}_n)} \leqslant C\lambda^n$.
- $\|\mu_{\Phi_n}\|_\infty \leqslant C\lambda^n$ *in $\mathscr{U}_n$.*

Proof For each $n \geqslant n_0$ we have $\|A_n - B_n\|_{C^0(V)} \leqslant C\lambda^n$ and therefore

$$\big\| \mathrm{Id} - \big(B_n \circ A_n^{-1}\big)\big\|_{C^0\left(V_3(n)\right)} \leqslant C\lambda^n.$$

If we define $\Phi_n|_{V_3(n)} = B_n \circ A_n^{-1}$ we also have $\|\mu_{\Phi_n}\|_\infty = \|\mu_{A_n^{-1}}\|_\infty$ in $V_3(n)$, which is equal to $\|\mu_{A_n}\|_\infty$ in V. In particular $\|\mu_{\Phi_n}\|_\infty \leqslant C\lambda^n$ in $V_3(n)$, and then we define Φ_n in the whole $\mathscr{U}_n$ by interpolating $B_n \circ A_n^{-1}$ in $V_3(n)$ with the identity in the interior of $V \cup \overline{U_1(n)} \cup V_1(n)$. □

Consider the seven topological disks:

$$X_1(n) = \text{interior}\left(V \cup \overline{U_1(n)} \cup V_1(n)\right) \subset W_1 \cap \mathscr{U}_n\,,$$

$$X_2(n) = \text{interior}\left(V_1(n) \cup \overline{U_2(n)} \cup V_2(n) \cup \overline{U_3(n)} \cup V_3(n)\right) \subset W_2 \cap \mathscr{U}_n\,,$$

$$\widehat{X}_1(n) = \{z \in X_1(n) : \xi_n(z) \in \mathscr{U}_n\}\,, \qquad \widehat{X}_2(n) = \{z \in X_2(n) : \eta_n(z) \in \mathscr{U}_n\}\,,$$

$$\widehat{\mathscr{T}_n} = \Phi_n\big(\widehat{X}_1(n)\big) \cup \Phi_n\big(\widehat{X}_2(n)\big) \subset \mathscr{T}_n\,,$$

$$Y_1(n) = X_1(n) \cap \Phi_n\big(\widehat{X}_1(n)\big) \quad \text{and} \quad Y_2(n) = X_2(n) \cap \Phi_n\big(\widehat{X}_2(n)\big).$$

Note that V, $V_1(n)$ and $B_n(V)$ are contained in $\widehat{\mathscr{T}_n}$ for all $n \geqslant n_0$. Moreover, we have the following two corollaries of Theorem 13.5.

Lemma 13.4 *There exists $\delta > 0$ such that for all $n \geqslant n_0$ we have:*

$$N_\delta\big([-1, 0]\big) \subset Y_1(n) \quad \textit{and} \quad N_\delta\big([0, \xi_n(0)]\big) \subset Y_2(n)\,.$$

Lemma 13.5 *Both:*

$$\sup_{n \geqslant n_0} \left\{ \sup_{z \in Y_1(n)} \left\{\det\big(D\xi_n(z)\big)\right\} \right\} \quad \textit{and} \quad \sup_{n \geqslant n_0} \left\{ \sup_{z \in Y_2(n)} \left\{\det\big(D\eta_n(z)\big)\right\} \right\}$$

are finite, where $\det(\cdot)$ *denotes the determinant of a square matrix.*

Let

$$\widehat{\xi}_n : \Phi_n\big(\widehat{X}_1(n)\big) \to \big(\Phi_n \circ \xi_n\big)\big(\widehat{X}_1(n)\big) \text{ defined by } \widehat{\xi}_n = \Phi_n \circ \xi_n \circ \Phi_n^{-1},$$

and

$$\widehat{\eta}_n : \Phi_n\big(\widehat{X}_2(n)\big) \to \big(\Phi_n \circ \eta_n\big)\big(\widehat{X}_2(n)\big) \text{ defined by } \widehat{\eta}_n = \Phi_n \circ \eta_n \circ \Phi_n^{-1}.$$

Since each Φ_n is an $\mathbb{R}$-symmetric C^3 diffeomorphism, the pair $\big(\widehat{\eta}_n, \widehat{\xi}_n\big)$ restricts to a critical commuting pair with the same rotation number as (η_n, ξ_n), and the same criticality (that we are assuming to be cubic, in order to simplify). Note also that $\widehat{\eta}_n(0) = -1$ for all $n \geqslant n_0$. Moreover, from Lemma 13.5 and $\big\|\Phi_n - \text{Id}\big\|_{C^0(\mathscr{U}_n)} \leqslant C\lambda^n$ we have

$$\left\|\xi_n - \widehat{\xi}_n\right\|_{C^0\left(Y_1(n)\right)} \leqslant C\lambda^n \quad \text{and} \quad \left\|\eta_n - \widehat{\eta}_n\right\|_{C^0\left(Y_2(n)\right)} \leqslant C\lambda^n \quad \text{for all } n \geqslant n_0.$$

Therefore, it is enough to shadow the sequence $\left(\widehat{\eta}_n, \widehat{\xi}_n\right)$ in the domains $Y_1(n)$ and $Y_2(n)$, instead of (η_n, ξ_n) (the shadowing sequence will be constructed in Sect. 13.1.5 below). The main advantage of working with the sequence $\left(\widehat{\eta}_n, \widehat{\xi}_n\right)$ is precisely the fact that $\widehat{\eta}_n^{-1} \circ \widehat{\xi}_n$ is univalent in V for all $n \geqslant n_0$ (since it coincides with B_n). In particular, we can choose each topological disk $\mathscr{U}_n$ and $\mathscr{T}_n$ defined above with the additional property that, identifying V with $B_n(V)$ via the biholomorphism B_n, we obtain from $\mathscr{T}_n$ an abstract annular Riemann surface $\mathscr{S}_n$ (with the complex structure induced by the quotient).

Let us denote by $p_n : \mathscr{T}_n \to \mathscr{S}_n$ the canonical projection. The projection of the real line, $p_n(\mathbb{R} \cap \mathscr{T}_n)$, is real-analytic diffeomorphic to the unit circle $\boldsymbol{S}^1$. We call it the *equator* of $\mathscr{S}_n$.

Since complex conjugation leaves $\mathscr{T}_n$ invariant and commutes with B_n, it induces an antiholomorphic involution $F_n : \mathscr{S}_n \to \mathscr{S}_n$ acting as the identity on the equator $p_n(\mathbb{R} \cap \mathscr{T}_n)$. Note that F_n has a continuous extension to $\partial \mathscr{S}_n$ that switches the boundary components.

Since $\mathscr{S}_n$ is obviously neither biholomorphic to $\mathbb{D} \setminus \{0\}$ nor to $\mathbb{C} \setminus \{0\}$, we have $\operatorname{mod}(\mathscr{S}_n) < \infty$ for all $n \geqslant n_0$, where $\operatorname{mod}(\cdot)$ denotes the conformal modulus of an annular Riemann surface (recall Sect. 11.1.1 and also Exercise 11.9). For each $n \geqslant n_0$ let R_n in $(1, +\infty)$ be given by

$$R_n = \exp\left(\operatorname{mod}(\mathscr{S}_n)/2\right).$$

In other words, $\mathscr{S}_n$ is conformally equivalent to $A_{R_n} = \left\{z \in \mathbb{C} : R_n^{-1} < |z| < R_n\right\}$. Any biholomorphism between $\mathscr{S}_n$ and A_{R_n} must send the equator $p_n\left(\mathbb{R} \cap \mathscr{T}_n\right)$ onto the unit circle $\boldsymbol{S}^1$ (because the equator is invariant under the antiholomorphic involution F_n, and the unit circle is invariant under the antiholomorphic involution $z \mapsto 1/\bar{z}$ in A_{R_n}). Let $\Psi_n : \mathscr{S}_n \to A_{R_n}$ be the conformal uniformization determined by $\Psi_n\left(p_n(0)\right) = 1$, and let $P_n : \mathscr{T}_n \to A_{R_n}$ be the holomorphic surjective local diffeomorphism

$$P_n = \Psi_n \circ p_n$$

(see Fig. 13.3). Note that $P_n(0) = 1$ and $P_n(\mathscr{T}_n \cap \mathbb{R}) = \boldsymbol{S}^1$ for all $n \geqslant n_0$. Moreover $P_n(z)\overline{P_n(\bar{z})} = 1$ for all $z \in \mathscr{T}_n$ and all $n \geqslant n_0$. From now on we forget about the abstract cylinder $\mathscr{S}_n$.

Lemma 13.6 *There exist two constants $\delta > 0$ and $C > 1$ such that for all $n \geqslant n_0$ and for all $z \in N_\delta\left([-1, \widetilde{\xi}_n(0)]\right)$ we have $z \in \widehat{\mathscr{T}_n} \subset \mathscr{T}_n$ and:*

$$\frac{1}{C} < \left|P_n'(z)\right| < C\,.$$

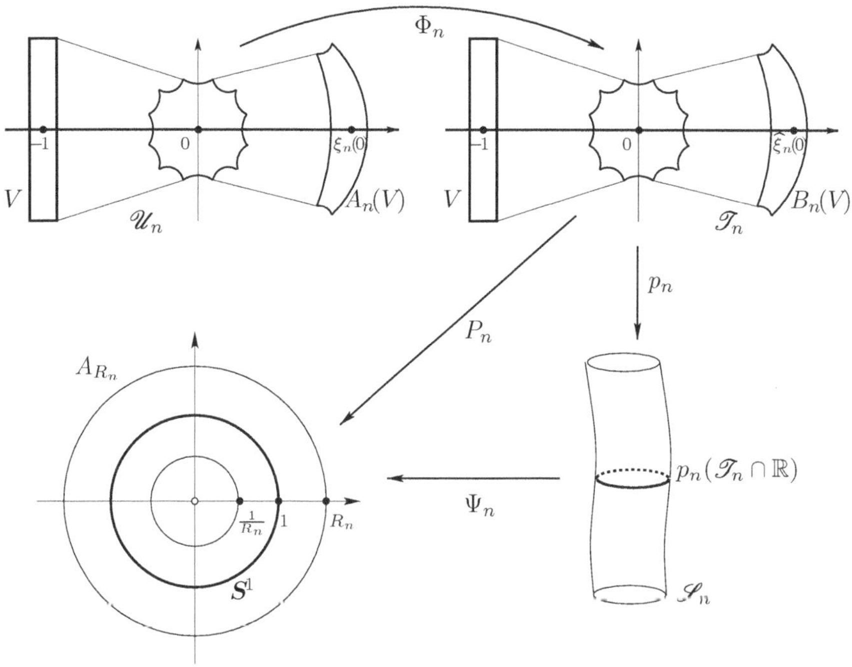

Fig. 13.3 Bidimensional glueing procedure

Proof By the real bounds there exists a universal constant $C_0 > 1$ such that for each $n \geqslant n_0$ there exists $w_n \in \left[-1, \widetilde{\xi}_n(0)\right]$ such that

$$\frac{1}{C_0} < \left|P_n'(w_n)\right| < C_0\,.$$

We need to construct a definite complex domain around $\left[-1, \widetilde{\xi}_n(0)\right]$ where P_n has universally bounded distortion. Again by the real bounds there exist $\delta > 0$ and $l \in \mathbb{N}$ with the following properties. For each $n \geqslant n_0$ there exists $z_1, z_2, \ldots, z_{k_n} \in \left[-1, \widetilde{\xi}_n(0)\right]$ with $k_n < l$ for all $n \geqslant n_0$ such that

- $\left[-1, \widetilde{\xi}_n(0)\right] \subset \bigcup_{i=1}^{k_n} B(z_i, \delta)$.
- $B(z_i, 2\delta) \subset \widehat{\mathscr{T}_n} \subset \mathscr{T}_n$ for all $i \in \{1, \ldots, k_n\}$.
- $P_n|_{B(z_i, 2\delta)}$ is univalent for all $i \in \{1, \ldots, k_n\}$.

By convexity we have for all $n \geqslant n_0$ and for all $i \in \{1, \ldots, k_n\}$ that

$$\sup_{v,w \in B(z_i,\delta)} \left\{\frac{\left|P_n'(v)\right|}{\left|P_n'(w)\right|}\right\} \leqslant \exp\left(\sup_{w \in B(z_i,\delta)} \left\{\frac{\left|P_n''(w)\right|}{\left|P_n'(w)\right|}\right\}\right),$$

and by Koebe distortion theorem (see for instance Carleson and Gamelin 1993, Section I.1, Theorem 1.6) we have

$$\sup_{w\in B(z_i,\delta)}\left\{\frac{|P_n''(w)|}{|P_n'(w)|}\right\}\leqslant\frac{2}{\delta}\quad\text{for all } n\geqslant n_0 \text{ and for all } i\in\{1,\ldots,k_n\}.$$

□

Now we project each commuting pair $(\widetilde{\eta}_n,\widetilde{\xi}_n)$ from $\widehat{\mathscr{T}_n}$ to the round annulus A_{R_n}.

Proposition 13.3 (Glueing Procedure) *The pair*

$$\widehat{\xi}_n:\Phi_n\big(\widehat{X}_1(n)\big)\to\mathscr{T}_n\quad\textit{and}\quad\widehat{\eta}_n:\Phi_n\big(\widehat{X}_2(n)\big)\to\mathscr{T}_n$$

projects under P_n to a well-defined orientation-preserving C^3 map

$$G_n:P_n\big(\widehat{\mathscr{T}_n}\big)\subset A_{R_n}\to A_{R_n}.$$

For each $n\geqslant n_0$, $P_n(\widehat{\mathscr{T}_n})$ is a Θ-invariant annulus with positive and finite modulus. Each G_n is $\mathbf{S}^1$-symmetric and, when restricted to the unit circle, it produces a C^3 critical circle map $g_n:\mathbf{S}^1\to\mathbf{S}^1$ with cubic critical point at $P_n(0)=1$, and with rotation number $\rho(g_n)=\rho\big(\mathscr{R}^n(f)\big)$. In other words, the following diagram commutes.

$$\begin{array}{ccc}
\widehat{\mathscr{T}_n}\subset\mathscr{T}_n & \xrightarrow{(\widehat{\eta}_n,\widehat{\xi}_n)} & \mathscr{T}_n\\
\downarrow{\scriptstyle P_n} & & \downarrow{\scriptstyle P_n}\\
P_n\big(\widehat{\mathscr{T}_n}\big)\subset A_{R_n} & \xrightarrow{G_n} & A_{R_n}
\end{array}$$

Moreover, the unique critical point of G_n in $P_n\big(\widehat{\mathscr{T}_n}\big)$ is the one in the unit circle. Finally, we have

$$\left|\bar{\partial}G_n(z)\right|\leqslant C\lambda^n\,|\partial G_n(z)|\quad\textit{for all } z\in P_n\big(\widehat{\mathscr{T}_n}\big)\setminus\{1\},$$

that is, $\|\mu_{G_n}\|_\infty\leqslant C\lambda^n$ in $P_n\big(\widehat{\mathscr{T}_n}\big)$.

Proof This follows from

- The construction of $\mathscr{U}_n$ and $\mathscr{T}_n$;
- The property $B_n=\Phi_n\circ\big(\eta_n^{-1}\circ\xi_n\big)\circ\Phi_n^{-1}$ in V;
- The commuting condition in $V_1(n)$;
- The symmetry $P_n(z)\overline{P_n(\overline{z})}=1$ for all $z\in\mathscr{T}_n$ and all $n\geqslant n_0$;
- The fact that $P_n:\mathscr{T}_n\to A_{R_n}$ is holomorphic, $P_n(0)=1$ and $P_n(\mathscr{T}_n\cap\mathbb{R})=\mathbf{S}^1$ for all $n\geqslant n_0$.

□

Note that each g_n belongs to the smooth conjugacy class obtained with the glueing procedure described in Sect. 10.2 applied to the C^3 critical commuting pair $(\widehat{\eta}_n, \widehat{\xi}_n)$. In the next section we will construct a sequence of real-analytic critical circle maps, with the desired combinatorics, that extend to holomorphic maps exponentially close to G_n in a definite annulus around the unit circle (see Proposition 13.4 below).

13.1.4 Main Perturbation

The goal of this section is to construct the following sequence of perturbations.

Proposition 13.4 (Main Perturbation) *There exist a constant $r > 1$ and a sequence of holomorphic maps defined in the annulus A_r:*

$$\{H_n : A_r \to \mathbb{C}\}_{n \geqslant n_0}$$

such that for all $n \geqslant n_0$ the following holds.

- $A_r \subset P_n(\widehat{\mathscr{T}_n}) \subset P_n(\mathscr{T}_n) = A_{R_n}$;
- $\left\| H_n - G_n \right\|_{C^0(A_r)} \leqslant C\lambda^n$;
- $H_n(A_r) \subset \big(G_n \circ P_n\big)\big(\widehat{\mathscr{T}_n}\big) \subset P_n(\mathscr{T}_n) = A_{R_n}$;
- *H_n preserves the unit circle and, when restricted to the unit circle, H_n produces a real-analytic critical circle map $h_n : \boldsymbol{S}^1 \to \boldsymbol{S}^1$ such that*
  - *The unique critical point of h_n is at $P_n(0) = 1$, and is of cubic type;*
  - *The critical value of h_n coincide with the one of g_n, that is, $h_n(1) = g_n(1) \in P_n(V \cap \mathbb{R})$;*
  - $\rho(h_n) = \rho(g_n) = \rho\big(\mathscr{R}^n(f)\big) \in \mathbb{R} \setminus \mathbb{Q}$.
- *The unique critical point of H_n in A_r is the one in the unit circle.*

The remainder of this section is devoted to proving Proposition 13.4. We will not perturb the maps G_n directly (basically because they are non invertible). Instead, we will decompose them (see Lemma 13.7 below), and then we will perturb on their *coefficients* (see the definition after the statement of Lemma 13.7). Those perturbations will be done, again, with the help of Theorem 11.4.

As before, let $A : \mathbb{C} \setminus \{0\} \to \mathbb{C} \setminus \{0\}$ be the map corresponding to the parameters $a = 0$ and $b = 1$ in the Arnold family defined in Sect. 6.1.2. Recall that A preserves the unit circle, and its restriction $A : \boldsymbol{S}^1 \to \boldsymbol{S}^1$ is a real-analytic critical circle map. The critical point of A is placed at 1, and is of cubic type (the critical point is also a fixed point for A).

Lemma 13.7 *For each $n \geqslant n_0$ there exist*

- *a real number $S_n > 1$,*

- *an orientation-preserving* C^3 *diffeomorphism* $\psi_n : P_n\big(\widehat{\mathscr{T}_n}\big) \to A_{S_n}$ *which is symmetric about* $\mathbf{S}^1$*, and*
- *a biholomorphism* $\phi_n : A(A_{S_n}) \to \big(G_n \circ P_n\big)(\widehat{\mathscr{T}_n})$*, also symmetric about* $\mathbf{S}^1$*,*

such that $G_n = \phi_n \circ A \circ \psi_n$ *in* $P_n\big(\widehat{\mathscr{T}_n}\big)$*. In other words, the following diagram commutes.*

$$\begin{array}{ccc} P_n\big(\widehat{\mathscr{T}_n}\big) & \xrightarrow{\;G_n\;} & \big(G_n \circ P_n\big)\big(\widehat{\mathscr{T}_n}\big) \\ \Big\downarrow{\scriptstyle \psi_n} & & \Big\uparrow{\scriptstyle \phi_n} \\ A_{S_n} & \xrightarrow{\;A\;} & A(A_{S_n}) \end{array}$$

Proof For each $n \geqslant n_0$ let $S_n > 1$ such that $A(A_{S_n})$ is a Θ-invariant annulus with

$$\operatorname{mod}\big(A(A_{S_n})\big) = \operatorname{mod}\big((G_n \circ P_n)(\widehat{\mathscr{T}_n})\big).$$

In particular there exists a biholomorphism $\phi_n : A(A_{S_n}) \to \big(G_n \circ P_n\big)(\widehat{\mathscr{T}_n})$ that commutes with Θ. Each ϕ_n preserves the unit circle and we can choose it such that $\phi_n(1) = G_n(1)$, that is, ϕ_n takes the critical value of A into the critical value of G_n. Since both G_n and A are three-fold branched coverings around their critical points and local diffeomorphisms away from them, the equation $G_n = \phi_n \circ A \circ \psi_n$ induces an orientation-preserving C^3 diffeomorphism $\psi_n : P_n\big(\widehat{\mathscr{T}_n}\big) \to A_{S_n}$, that commutes with Θ and such that $\psi_n(1) = 1$, that is, ψ_n takes the critical point of G_n into the one of A. The fact that ψ_n is smooth at 1 with non-vanishing derivative follows from the fact that the critical points of G_n and A have the same criticality. □

The diffeomorphisms ψ_n and ϕ_n are called the *coefficients* of G_n in $P_n\big(\widehat{\mathscr{T}_n}\big)$.

As already mentioned, the idea to prove Proposition 13.4 is to perturb each diffeomorphism ψ_n with Theorem 11.4. In order to control the C^0 size of those perturbations, we will need some geometric control. With this as our goal, we state and prove four lemmas before entering into the proof of Proposition 13.4.

Lemma 13.8 *We have*

$$1 < \inf_{n \geqslant n_0} \{R_n\} \quad \text{and} \quad \sup_{n \geqslant n_0} \{R_n\} < +\infty\,.$$

Proof This follows at once from Lemma 13.6. □

Lemma 13.9 *For all* $n \geqslant n_0$ *both* $P_n(\widehat{\mathscr{T}_n})$ *and* $\big(G_n \circ P_n\big)(\widehat{\mathscr{T}_n})$ *are* Θ*-invariant annulus with finite modulus. Moreover there exists a universal constant* $K > 1$ *such that*

$$\frac{1}{K} < \operatorname{mod}\big(P_n(\widehat{\mathscr{T}_n})\big) < K \quad \text{for all } n \geqslant n_0\,.$$

Proof By Lemma 13.8 we know that $R = \sup_{n \geqslant n_0}\{R_n\}$ is finite, and since for all $n \geqslant n_0$ both $P_n(\widehat{\mathscr{T}_n})$ and $(G_n \circ P_n)(\widehat{\mathscr{T}_n})$ are contained in the corresponding A_{R_n}, we obtain at once that both $P_n(\widehat{\mathscr{T}_n})$ and $(G_n \circ P_n)(\widehat{\mathscr{T}_n})$ have finite modulus, and also that $\sup_{n \geqslant n_0}\{\operatorname{mod}(P_n(\widehat{\mathscr{T}_n}))\}$ is finite. Just as in Lemma 13.8, the fact that $\inf_{n \geqslant n_0}\{\operatorname{mod}(P_n(\widehat{\mathscr{T}_n}))\}$ is positive follows from Lemmas 13.4 and 13.6. □

Lemma 13.10 *There exists a constant $r_0 > 1$ such that $\overline{A_{r_0}} \subset P_n(\widehat{\mathscr{T}_n})$ for all $n \geqslant n_0$.*

Proof By the invariance with respect to the antiholomorphic involution $z \mapsto 1/\bar{z}$, the unit circle is the core curve (the unique closed geodesic for the hyperbolic metric) of each annulus $P_n(\widehat{\mathscr{T}_n})$. Since $\inf_{n \geqslant n_0}\{\operatorname{mod}(P_n(\widehat{\mathscr{T}_n}))\}$ is positive, the statement is well-known (see for instance McMullen 1994, Ch. 2, Theorem 2.5). □

Lemma 13.11 *We have*

$$s = \inf_{n \geqslant n_0}\{S_n\} > 1 \quad \textit{and} \quad S = \sup_{n \geqslant n_0}\{S_n\} < +\infty.$$

Proof Since $\mu_{\psi_n} = \mu_{G_n}$ in $P_n(\widehat{\mathscr{T}_n})$, we have $\|\mu_{\psi_n}\|_\infty \leqslant C\lambda^n$ in $P_n(\widehat{\mathscr{T}_n})$ for all $n \geqslant n_0$. By the geometric definition of quasiconformal homeomorphisms (Definition 11.1),

$$\left(\frac{1 - C\lambda^n}{1 + C\lambda^n}\right)\operatorname{mod}\left(P_n(\widehat{\mathscr{T}_n})\right) \leqslant 2\log(S_n) \leqslant \left(\frac{1 + C\lambda^n}{1 - C\lambda^n}\right)\operatorname{mod}\left(P_n(\widehat{\mathscr{T}_n})\right)$$

for all $n \geqslant n_0$, and then we are done by Lemma 13.9. □

With this geometric control at hand, we are ready to prove Proposition 13.4.

Proof of Proposition 13.4 Let $r_0 > 1$ given by Lemma 13.10 (recall that $\overline{A_{r_0}} \subset P_n(\widehat{\mathscr{T}_n})$ for all $n \geqslant n_0$), and fix $r \in (1, (1 + r_0)/2)$. How small $r - 1$ must be will be determined in the course of the argument (see Lemma 13.12 below). For any $r \in (1, (1 + r_0)/2)$ consider $\underline{r} = r_0 - (r - 1) \in ((1 + r_0)/2, r_0)$. The sequence of $\boldsymbol{S}^1$-symmetric C^3 diffeomorphisms

$$\{\psi_n : A_{r_0} \to \psi_n(A_{r_0})\}_{n \geqslant n_0}$$

satisfy the hypothesis of Theorem 11.4 since

- $\mu_{\psi_n} = \mu_{G_n}$ in $P_n(\widehat{\mathscr{T}_n})$ and therefore $\|\mu_{\psi_n}\|_\infty \leqslant C\lambda^n$ for all $n \geqslant n_0$, and
- $\psi_n(A_{r_0}) \subset A_{S_n} \subset A_S$ for all $n \geqslant n_0$ (see Lemma 13.11 above).

Apply Theorem 11.4 to the bounded domain $A_{\underline{r}}$, compactly contained in A_{r_0}, to obtain a sequence of $\boldsymbol{S}^1$-symmetric biholomorphisms

$$\{\widehat{\psi}_n : A_{\underline{r}} \to \widehat{\psi}_n(A_{\underline{r}})\}_{n \geqslant n_0}$$

such that

$$\left\|\widehat{\psi}_n - \psi_n\right\|_{C^0(A_{\underline{r}})} \leqslant C\lambda^n \quad \text{for all } n \geqslant n_0.$$

Fix n_0 big enough to have $\widehat{\psi}_n(A_{\underline{r}}) \subset A_{S_n}$. We may assume that each $\widehat{\psi}_n$ fixes the point 1 (just as ψ_n does) by considering

$$z \mapsto \left(\frac{1}{\widehat{\psi}_n(1)}\right)\widehat{\psi}_n(z).$$

Since $\left|\widehat{\psi}_n(z)\right| \leqslant S$ for all $z \in A_{\underline{r}}$ and for all $n \geqslant n_0$ (where $S \in (1, +\infty)$ is given by Lemma 13.11) and since $\left|\widehat{\psi}_n(1) - 1\right| \leqslant C\lambda^n$ for all $n \geqslant n_0$, we know that this new map (that we will still denote by $\widehat{\psi}_n$ to simplify) satisfy all the properties that we want for $\widehat{\psi}_n$, and also fixes the point $z = 1$.

For each $n \geqslant n_0$ consider the holomorphic map $H_n : A_{\underline{r}} \to \mathbb{C}$ defined by $H_n = \phi_n \circ A \circ \widehat{\psi}_n$. We have

- $H_n(A_{\underline{r}}) \subset \left(G_n \circ P_n\right)\left(\widehat{\mathscr{T}_n}\right) \subset A_{R_n}$.
- H_n is $\boldsymbol{S}^1$-symmetric and therefore it preserves the unit circle.
- When restricted to the unit circle, H_n produces a real-analytic critical circle map $h_n : \boldsymbol{S}^1 \to \boldsymbol{S}^1$.
- The unique critical point of H_n in $A_{\underline{r}}$ is the one in the unit circle, which is at $P_n(0) = 1$, and is of cubic type.
- The critical value of H_n coincide with the one of G_n, that is, $H_n(1) = G_n(1) \in P_n(V \cap \mathbb{R})$.

□

We divide in three lemmas the rest of the proof of Proposition 13.4. We need to prove first that, for a suitable $r > 1$, H_n is C^0 exponentially close to G_n in the annulus A_r (Lemma 13.12 below), and then that we can choose each H_n with the desired combinatorics for its restriction h_n to the unit circle (Lemma 13.13 below). This last perturbation will change the critical value of each H_n (it will not coincide with the one of G_n any more). We will finish the proof of Propositions 13.4 with Lemma 13.14, that allows us to keep the critical point of H_n at the point $P_n(0) = 1$, and to place the critical value of H_n at the point $g_n(1)$ for all $n \geqslant n_0$. This will be important in the following subsection, the last one of this section.

Lemma 13.12 *There exists $r \in \big(1, (1+r_0)/2\big)$ such that in the annulus A_r we have:*

$$\left\|H_n - G_n\right\|_{C^0(A_r)} \leqslant C\lambda^n \quad \textit{for all } n \geqslant n_0.$$

Proof The proof is divided in three claims.
First Claim There exists $\beta > 1$ such that $\overline{A_\beta} \subset A(A_{S_n})$ for all $n \geqslant n_0$.

Indeed, by Lemma 13.11 the round annulus $A_{(1+s)/2}$ is compactly contained in A_{S_n} for all $n \geqslant n_0$, and therefore the annulus $A\big(A_{(1+s)/2}\big)$ is contained in $A(A_{S_n})$

for all $n \geqslant n_0$. Thus we just take $\beta > 1$ such that $\overline{A_\beta} \subset A\big(A_{(1+s)/2}\big)$ and the first claim is proved.

From now on we fix $\alpha \in (1, \beta)$.

Second Claim There exists $r \in \big(1, (1+r_0)/2\big)$ close enough to one in order to simultaneously have $(A \circ \widehat{\psi}_n)(A_r) \subset A_\alpha$ and $(A \circ \psi_n)(A_r) \subset A_\alpha$ for all $n \geqslant n_0$.

Indeed, since $\overline{A_r} \subset A_{\underline{r}}$, $\widehat{\psi}_n$ is holomorphic, and $\widehat{\psi}_n(A_{\underline{r}}) \subset A_{S_n} \subset A_S$ for all $n \geqslant n_0$ (where $S \in (1, +\infty)$ is given by Lemma 13.11), we have by Cauchy's derivative estimate that $\sup_{n \geqslant n_0} \big\{ \big|\widehat{\psi}_n'(z)\big| : z \in A_r \big\}$ is finite. Since each $\widehat{\psi}_n$ preserves the unit circle, and since $\big\| \widehat{\psi}_n - \psi_n \big\|_{C^0(A_{\underline{r}})} \leqslant C\lambda^n$ for all $n \geqslant n_0$, the second claim is proved.

Another way to prove the second claim is by noting that, since $\overline{A_\alpha} \subset A_\beta \subset \overline{A_\beta} \subset A(A_{S_n})$ for all $n \geqslant n_0$, the hyperbolic metric on any annulus $A(A_{S_n})$ and the Euclidean metric are comparable in A_α with universal parameters, that is, there exists a constant $K > 1$ such that

$$\frac{1}{K}\,|z - w| \leqslant d_{A(A_{S_n})}(z, w) \leqslant K|z - w|$$

for all $z, w \in A_\alpha$ and for all $n \geqslant n_0$, where $d_{A(A_{S_n})}$ denote the hyperbolic distance in the annulus $A(A_{S_n})$ (this is well-known, see for instance Carleson and Gamelin 1993, Section I.4, Theorem 4.3). Since each $A \circ \widehat{\psi}_n : A_{\underline{r}} \to A(A_{S_n})$ is holomorphic and preserves the unit circle, we know by the Schwarz lemma that for all $z \in A_{\underline{r}}$ and for all $n \geqslant n_0$ we have:

$$d_{A(A_{S_n})}\left((A \circ \widehat{\psi}_n)(z), \mathbf{S}^1\right) \leqslant d_{A_{\underline{r}}}\left(z, \mathbf{S}^1\right),$$

where $d_{A_{\underline{r}}}$ denote the hyperbolic distance in the annulus $A_{\underline{r}}$. Since all distances $d_{A(A_{S_n})}$ are comparable with the Euclidean distance in A_δ with universal parameters, we have for all $z \in A_{\underline{r}}$ and for all $n \geqslant n_0$ that:

$$d\left((A \circ \widehat{\psi}_n)(z), \mathbf{S}^1\right) \leqslant K d_{A_{\underline{r}}}\left(z, \mathbf{S}^1\right),$$

where d is just the Euclidean distance in the plane. Fix $r \in \big(1, (1+r_0)/2\big)$ close enough to one in order to have that $z \in A_r$ implies $d_{A_{\underline{r}}}\left(z, \mathbf{S}^1\right) < \frac{\alpha - 1}{K\alpha}$ (and therefore $(A \circ \widehat{\psi}_n)(z) \in A_\alpha$ for all $n \geqslant n_0$). Again since $\big\| \widehat{\psi}_n - \psi_n \big\|_{C^0(A_{\underline{r}})} \leqslant C\lambda^n$ for all $n \geqslant n_0$, the second claim is proved.

Third Claim There exists a positive number M such that $\left|\phi_n'(z)\right| < M$ for all $z \in A_\alpha$ and for all $n \geqslant n_0$.

Indeed, recall that $\phi_n\big(A(A_{S_n})\big) = \big(G_n \circ P_n\big)\big(\widehat{\mathscr{T}_n}\big) \subset A_{R_n}$ for all $n \geqslant n_0$. By Lemma 13.8 there exists a (finite) number Δ such that $\phi_n\big(A(A_{S_n})\big) \subset B(0, \Delta)$ for all $n \geqslant n_0$. Since $\overline{A_\alpha} \subset A_\beta \subset \overline{A_\beta} \subset A(A_{S_n})$ for all $n \geqslant n_0$, the third claim follows from Cauchy's derivative estimate.

With the three claims at hand, Lemma 13.12 follows. □

To control the combinatorics after perturbation we use the *monotonicity* of the rotation number.

Lemma 13.13 *Let f be a C^3 critical circle map and let g be a real-analytic critical circle map that extends holomorphically to the annulus*

$$A_R = \left\{ z \in \mathbb{C} : \frac{1}{R} < |z| < R \right\} \quad \textit{for some} \quad R > 1.$$

There exists a real-analytic critical circle map h, with $\rho(h) = \rho(f)$, also extending holomorphically to A_R, where we have

$$\|h - g\|_{C^0(A_R)} \leqslant d_{C^0(S^1)}(f, g).$$

In particular

$$d_{C^r(S^1)}(h, g) \leqslant d_{C^0(S^1)}(f, g) \quad \textit{for any} \quad 0 \leqslant r \leqslant \infty.$$

Proof Let F and G be the corresponding lifts of f and g to the real line satisfying

$$\rho(f) = \lim_{n \to +\infty} \frac{F^n(0)}{n} \quad \text{and} \quad \rho(g) = \lim_{n \to +\infty} \frac{G^n(0)}{n}.$$

Consider the band $B_R = \{z \in \mathbb{C} : -\log R < 2\pi \operatorname{Im} z < \log R\}$, which is the universal cover of the annulus A_R via the holomorphic covering $z \mapsto e^{2\pi i z}$. Let $\delta = \|F - G\|_{C^0(\mathbb{R})}$, and for any t in $[-1, 1]$ let $G_t : B_R \to \mathbb{C}$ defined as $G_t = G + t\delta$. Each G_t preserves the real line, and its restriction is the lift of a real-analytic critical circle map. Moreover, each G_t commutes with unitary horizontal translation in B_R. Note that $\|G_t - G\|_{C^0(B_R)} = |t|\delta \leqslant \|F - G\|_{C^0(\mathbb{R})}$ for any $t \in [-1, 1]$. Moreover for any $x \in \mathbb{R}$ the family $\{G_t(x)\}_{t \in [-1,1]}$ is monotone in t, and we have $G_{-1}(x) \leqslant F(x) \leqslant G_1(x)$. In particular there exists $t_0 \in [-1, 1]$ such that

$$\lim_{n \to +\infty} \frac{G_{t_0}^n(0)}{n} = \rho(F),$$

and we define h as the projection of G_{t_0} to the annulus A_R. □

After the perturbation given by Lemma 13.13 we still have the critical point of h_n placed at 1, but its critical value is no longer placed at $g_n(1)$ (however they are exponentially close). To finish the proof of Proposition 13.4 we need to fix this, without changing the combinatorics of h_n in S^1. Until now each H_n is S^1-symmetric, in the sense that it commutes with $z \mapsto 1/\bar{z}$ in the annulus A_r. We will loose this property in the following perturbation, which turns out to be the last one.

Lemma 13.14 *For each $n \geqslant n_0$ consider the (unique) Möbius transformation M_n which maps the unit disk $\mathbb{D}$ onto itself fixing the basepoint $z = 1$, and which maps*

$H_n(1)$ to $G_n(1)$. Then there exists $\rho \in (1, r)$ such that $\overline{A_\rho} \subset M_n(A_r)$ for all $n \geqslant n_0$. Moreover for each $n \geqslant n_0$ we have:

$$\left\| M_n \circ H_n \circ M_n^{-1} - G_n \right\|_{C^0(A_\rho)} \leqslant C\lambda^n .$$

Note that, when restricted to the unit circle, each M_n gives rise to an orientation-preserving real-analytic diffeomorphism which is, as Lemma 13.14 indicates, C^∞-exponentially close to the identity.

Proof Consider the biholomorphism $\psi : \mathbb{H} \to \mathbb{D}$ given by $\psi(z) = \frac{z-i}{z+i}$, whose inverse $\psi^{-1} : \mathbb{D} \to \mathbb{H}$ is given by $\psi^{-1}(z) = i\left(\frac{1+z}{1-z}\right)$. Note that ψ maps the vertical geodesic of equation $\{z \in \mathbb{H} : \operatorname{Re} z = 0\}$ onto the interval $(-1, 1)$ in $\mathbb{D}$. Since ψ and ψ^{-1} are Möbius transformations, both extend uniquely to corresponding biholomorphisms of the entire Riemann sphere. The extension of ψ is a real-analytic diffeomorphism between the compactification of the real line and the unit circle, which maps the point at infinity to the point $z = 1$. For each $n \geqslant n_0$ consider the real number t_n defined by

$$t_n = \psi^{-1}\big(G_n(1)\big) - \psi^{-1}\big(H_n(1)\big) = 2i\left(\frac{G_n(1) - H_n(1)}{\big(1 - G_n(1)\big)\big(1 - H_n(1)\big)}\right).$$

Each t_n is finite since for all $n \geqslant n_0$ both $G_n(1)$ and $H_n(1)$ are not equal to one. Moreover we claim that:

$$\inf_{n \geqslant n_0} \left\{ \left| G_n(1) - 1 \right| \right\} > 0 \quad \text{and} \quad \inf_{n \geqslant n_0} \left\{ \left| H_n(1) - 1 \right| \right\} > 0 .$$

Indeed, since we have $\left|H_n(1) - G_n(1)\right| \leqslant C\lambda^n$ for all $n \geqslant n_0$, is enough to prove that $\inf_{n \geqslant n_0} \left\{ \left| G_n(1) - 1 \right| \right\} > 0$, and this follows by Lemma 13.6 since $1 = P_n(0)$ and $G_n(1) = P_n(-1)$ for all $n \geqslant n_0$. In particular, again using $\left|H_n(1) - G_n(1)\right| \leqslant C\lambda^n$ for all $n \geqslant n_0$, we see that $|t_n| \leqslant C\lambda^n$ for all $n \geqslant n_0$. From the explicit formula

$$M_n(z) = \frac{(2i - t_n)z + t_n}{(2i + t_n) - t_n z} = \left(\frac{z - \left(\frac{t_n}{t_n - 2i}\right)}{1 - \left(\frac{t_n}{t_n + 2i}\right) z}\right)\left(\frac{2i - t_n}{2i + t_n}\right) \quad \text{for all } n \geqslant n_0,$$

we see that the pole of each M_n is at the point $z_n = 1 + i(2/t_n)$, and since $|t_n| \leqslant C\lambda^n$ for all $n \geqslant n_0$, we can take n_0 so large that $z_n \in \mathbb{C} \setminus \overline{B(0, 2R)}$, where $R = \sup_{n \geqslant n_0}\{R_n\} < +\infty$ is given by Lemma 13.8. A straightforward computation gives

$$\big(M_n - \mathrm{Id}\big)(z) = \frac{t_n(z-1)^2}{(2i + t_n) - t_n z} \quad \text{for all } n \geqslant n_0,$$

and therefore

$$\| M_n - \mathrm{Id} \|_{C^0(A_R)} \leqslant C\lambda^n \quad \text{for all } n \geqslant n_0.$$

In particular, for any fixed $\rho \in (1, r)$ we can choose n_0 so large as to have $\overline{A_\rho} \subset M_n(A_r)$ for all $n \geqslant n_0$. Moreover given any $z \in A_\rho$ we have

$$\begin{aligned}\left(M_n \circ H_n \circ M_n^{-1} - G_n\right)(z) &= (M_n - \mathrm{Id})\left((H_n \circ M_n^{-1})(z)\right) \\ &\quad + (H_n - G_n)(z) + \left(H_n\left(M_n^{-1}(z)\right) - H_n(z)\right).\end{aligned}$$

From this it follows that

$$\begin{aligned}\left\| M_n \circ H_n \circ M_n^{-1} - G_n \right\|_{C^0(A_\rho)} &\leqslant \| M_n - \mathrm{Id} \|_{C^0(H_n(A_r))} \\ &\quad + \| H_n - G_n \|_{C^0(A_\rho)} + \| H_n \|_{C^1(A_r)} \left\| M_n^{-1} - \mathrm{Id} \right\|_{C^0(A_\rho)}.\end{aligned}$$

Since $H_n(A_r) \subset A_R$ and $A_\rho \subset A_r \subset A_R$, each of the three terms

$$\| M_n - \mathrm{Id} \|_{C^0\left(H_n(A_r)\right)}, \quad \| H_n - G_n \|_{C^0(A_\rho)} \quad \text{and} \quad \left\| M_n^{-1} - \mathrm{Id} \right\|_{C^0(A_\rho)}$$

is less than or equal to $C\lambda^n$ for all $n \geqslant n_0$.

Finally, since each H_n is holomorphic and we have $\overline{A_r} \subset A_{\underline{r}}$ and $H_n(A_{\underline{r}}) \subset \left(G_n \circ P_n\right)\left(\widehat{\mathscr{T}_n}\right) \subset A_{R_n} \subset A_R$ for all $n \geqslant n_0$, we obtain from Cauchy's derivative estimate that

$$\sup_{n \geqslant n_0} \left\{ \| H_n \|_{C^1(A_r)} \right\} < \infty,$$

and therefore

$$\left\| M_n \circ H_n \circ M_n^{-1} - G_n \right\|_{C^0(A_\rho)} \leqslant C\lambda^n \quad \text{for all } n \geqslant n_0.$$

□

With Lemma 13.14 at hand we are done since $\left(M_n \circ H_n \circ M_n^{-1}\right)(1) = G_n(1)$. We have finished the proof of Proposition 13.4.

13.1.5 The Shadowing Sequence

We are about to finish Sect. 13.1. Let us recall what we have done: in Sect. 13.1.3 we constructed a suitable sequence $\{G_n\}_{n \geqslant n_0}$ of S^1-symmetric C^3 extensions of C^3 critical circle maps g_n to some annulus $P_n(\widehat{\mathscr{T}_n})$. When lifted with the corresponding projection P_n (also constructed in Sect. 13.1.3), each g_n gives rise to a C^3 critical commuting pair $(\widehat{\eta}_n, \widehat{\xi}_n)$ exponentially close to $\mathscr{R}^n(f)$ and having the same combinatorics at each step (moreover, with complex extensions C^0-exponentially close to the ones of $\mathscr{R}^n(f)$ produced in Theorem 13.5; see Proposition 13.3 above for more properties).

In Sect. 13.1.4 we perturbed each G_n in a definite annulus A_r, in order to obtain a sequence of real-analytic critical circle maps, each of them having the same combinatorics as the corresponding $\mathscr{R}^n(f)$, that extend to holomorphic maps H_n exponentially close to G_n in A_r (see Proposition 13.4 above for more properties). Both the critical point and the critical value of each H_n coincide with the ones of the corresponding G_n. More precisely, the critical point of each H_n is at $P_n(0) = 1 \in P_n(V_1(n)) \cap S^1$, and its critical value is at $H_n(1) = G_n(1) \in P_n(V) \cap S^1 = P_n(B_n(V)) \cap S^1$. Recall also that $H_n(A_r) \subset P_n(\mathscr{T}_n)$ for all $n \geqslant n_0$.

In this section we lift each $H_n : A_r \to A_{R_n}$ via the holomorphic projection $P_n : \mathscr{T}_n \to A_{R_n}$ in the canonical way: let $\alpha > 0$ such that for all $n \geqslant n_0$ we have that

$$\overline{N_\alpha([-1,0]) \cup N_\alpha([0, \widehat{\xi}_n(0)])} \subset \widehat{\mathscr{T}_n}\,,$$

and that $P_n(N_\alpha([-1,0]) \cup N_\alpha([0,\widehat{\xi}_n(0)]))$ is an annulus contained in A_r and containing the unit circle (the existence of such α is guaranteed by Lemmas 13.4 and 13.6). Let us use the more compact notation $Z_1(n) = N_\alpha([-1,0])$ and $Z_2(n) = N_\alpha([0,\widehat{\xi}_n(0)])$. For each $n \geqslant n_0$ let $\widetilde{\eta}_n : Z_2(n) \to \mathscr{T}_n$ be the $\mathbb{R}$-preserving holomorphic map defined by the following two conditions:

$$H_n \circ P_n = P_n \circ \widetilde{\eta}_n \;\text{ in } Z_2(n), \qquad \text{and} \qquad \widetilde{\eta}_n(0) = -1\,.$$

In the same way let $\widetilde{\xi}_n : Z_1(n) \to \mathscr{T}_n$ be the $\mathbb{R}$-preserving holomorphic map defined by the two conditions

$$H_n \circ P_n = P_n \circ \widetilde{\xi}_n \;\text{ in } Z_1(n), \qquad \text{and} \qquad \widetilde{\xi}_n(0) = \widehat{\xi}_n(0)\,.$$

Thus, we have the following commutative diagram:

$$\begin{array}{ccc} Z_1(n) \cup Z_2(n) \subset \mathscr{T}_n & \xrightarrow{(\widetilde{\eta}_n,\, \widetilde{\xi}_n)} & \mathscr{T}_n \\ \Big\downarrow P_n & & \Big\downarrow P_n \\ A_r \subset A_{R_n} & \xrightarrow{\quad H_n \quad} & A_{R_n} \end{array}$$

In the next proposition we summarize the main properties of this lift, which are all straightforward.

Proposition 13.5 (The Shadowing Sequence) *For each $n \geqslant n_0$ the pair $f_n = (\widetilde{\eta}_n, \widetilde{\xi}_n)$ restricts to a real-analytic critical commuting pair with domains $I(\widetilde{\xi}_n) = [\widetilde{\eta}_n(0), 0] = [-1, 0]$ and $I(\widetilde{\eta}_n) = [0, \widetilde{\xi}_n(0)] = [0, \widehat{\xi}_n(0)]$, and such that $\rho(f_n) = \rho(\widehat{\eta}_n, \widehat{\xi}_n) = \rho(\mathscr{R}^n(f)) \in \mathbb{R} \setminus \mathbb{Q}$. Moreover $\widetilde{\xi}_n$ and $\widetilde{\eta}_n$ extend to holomorphic maps in $Z_1(n)$ and $Z_2(n)$ respectively where we have:*

- *$\widetilde{\xi}_n$ has a unique critical point in $Z_1(n)$, which is at the origin and of cubic type;*
- *$\widetilde{\eta}_n$ has a unique critical point in $Z_2(n)$, which is at the origin and of cubic type;*
- $\|\widetilde{\xi}_n - \widehat{\xi}_n\|_{C^0(Z_1(n) \cap \Phi_n(\widehat{X}_1(n)))} \leqslant C\lambda^n$*;*
- $\|\widetilde{\eta}_n - \widehat{\eta}_n\|_{C^0(Z_2(n) \cap \Phi_n(\widehat{X}_2(n)))} \leqslant C\lambda^n$*.*

With Propositions 13.5 at hand, Theorem 13.4 follows directly from the following consequence of Montel's theorem.

Lemma 13.15 *Let α be a constant in $(0, 1)$ and let $\mathscr{V}$ be an $\mathbb{R}$-symmetric bounded topological disk such that $[-1, \alpha^{-1}] \subset \mathscr{V}$. Let W_1 and W_2 be topological disks whose closure is contained in $\mathscr{V}$ and such that $[-1, 0] \subset W_1$ and $[0, \alpha^{-1}] \subset W_2$. Denote by $\mathscr{K}$ the set of all normalized real-analytic critical commuting pairs $\zeta = (\eta, \xi)$ satisfying the following three conditions.*

- $\eta(0) = -1$ *and* $\xi(0) \in [\alpha, \alpha^{-1}]$*;*
- $\alpha\left|\eta([0, \xi(0)])\right| \leqslant \left|\xi([-1, 0])\right| \leqslant \alpha^{-1}\left|\eta([0, \xi(0)])\right|$*;*
- *Both ξ and η extend to holomorphic maps (with a unique cubic critical point at the origin) defined in W_1 and W_2 respectively, where we have*
  1. $N_\alpha(\xi([-1, 0])) \subset \xi(W_1)$*;*
  2. $N_\alpha(\eta([0, \xi(0)])) \subset \eta(W_2)$*;*
  3. $\xi(W_1) \cup \eta(W_2) \subset \mathscr{V}$*.*

Then $\mathscr{K}$ is C^ω-compact.

13.2 Bounding the C^{r-1} Metric

In the previous section we have proved the C^0 version of Theorem 13.4. The details required to bootstrap this estimate to the C^{r-1} metric can be found in Guarino et al. (2018, Section 11). Here we just want to mention that the key point for such bootstrapping argument is the following general fact from complex analysis.

Proposition 13.6 *Let I be a compact interval in the real line with non-empty interior, and let U be an open set in the complex plane containing I. Fix some $M > 0$, and consider the family*

$$\mathscr{F} = \{f : U \to \mathbb{C} \text{ holomorphic: } \|f\|_{C^0(U)} \leqslant M\}.$$

Then for any $k \in \mathbb{N}$ and any $\alpha \in (0, 1)$, there exists $L = L(k, \alpha, M) > 0$ such that

$$\|f\|_{C^k(I)} \leqslant L\left(\|f\|_{C^0(I)}\right)^{\alpha} \quad \text{for all } f \in \mathscr{F},$$

where, as usual,

$$\|f\|_{C^k(I)} = \sup_{\substack{z \in I \\ n \in \{0,1,\ldots,k\}}} \{|f^{(n)}(z)|\}.$$

In the proof of Proposition 13.6 below, we follow the exposition of Lyubich (1999, Lem. 11.5).

Proof Let V be a bounded Jordan domain containing the interval I, and compactly contained in U (as usual, a *Jordan domain* is an open, connected and simply connected set of the complex plane, whose boundary is a Jordan curve).

Consider a continuous function $h : \overline{V} \to [0, 1]$ satisfying

- h is harmonic and positive in the annulus $V \setminus I$,
- $h \equiv 0$ on ∂V and $h \equiv 1$ on I.

The existence of such a function h is a particular case of *Dirichlet's problem*.

To begin the proof, suppose first that $M = 1$, and let $f : U \to \mathbb{C}$ be a holomorphic function such that $|f(z)| \leqslant 1$ for all $z \in U$. Let $\varepsilon = \|f\|_{C^0(I)} \leqslant 1$, and note that

$$\log|f| \leqslant h \log \varepsilon \tag{13.2}$$

on $\partial(V \setminus I) = I \cup \partial V$. Since f is holomorphic, $\log|f|$ is harmonic where $f \neq 0$ and subharmonic in the whole domain V, and since h is harmonic in $V \setminus I$, we get from the maximum principle that inequality (13.2) also holds inside the annulus $V \setminus I$, that is, $|f(z)| \leqslant \varepsilon^{h(z)}$ for all $z \in V$. Given $\alpha \in (0, 1)$, let $W = \{z \in V : h(z) \in (\alpha, 1]\}$, and note that W is a Jordan domain containing I, compactly contained in V, and such that $h(z) = \alpha$ for all $z \in \partial W$. Since $\varepsilon \in [0, 1]$, we have $|f(z)| \leqslant \varepsilon^{h(z)} \leqslant \varepsilon^{\alpha}$ for all $z \in W$, that is,

$$\|f\|_{C^0(W)} \leqslant \left(\|f\|_{C^0(I)}\right)^{\alpha}.$$

The next step is just the standard application of Cauchy's integral formulas. Indeed, let $\delta \in (0, 1)$ be such that $\overline{B(z, \delta)} \subset W$ for all $z \in I$. Then for any $z \in I$ and any $n \in \{0, 1, \ldots, k\}$, we have

$$\begin{aligned}\left|f^{(n)}(z)\right| &= \left|\frac{n!}{2\pi i}\int_{\partial B(z,\delta)} \frac{f(w)}{(w-z)^{n+1}}\,dw\right| = \frac{n!}{2\pi}\left|\int_0^{2\pi} \frac{f(z+\delta e^{i\theta})}{(\delta e^{i\theta})^{n+1}}\, i\delta e^{i\theta}\,d\theta\right| \\ &\leqslant \frac{n!}{2\pi}\frac{1}{\delta^n}\int_0^{2\pi}\left|f(z+\delta e^{i\theta})\right|d\theta \leqslant \frac{n!}{\delta^n}\left(\sup_{w\in\partial B(z,\delta)}\{|f(w)|\}\right).\end{aligned}$$

Defining $L_1 = k!/\delta^k$, we obtain

$$\|f\|_{C^k(I)} \leqslant L_1\,\|f\|_{C^0(W)} \leqslant L_1\left(\|f\|_{C^0(I)}\right)^{\alpha}.$$

Therefore, Proposition 13.6 is true for the case $M = 1$. For the general case, note that for any $f \in \mathscr{F}$ we have

$$\begin{aligned}\|f\|_{C^k(I)} &= M\,\|f/M\|_{C^k(I)} \\ &\leqslant M\,L_1\left(\|f/M\|_{C^0(I)}\right)^{\alpha} = M^{1-\alpha}\,L_1\left(\|f\|_{C^0(I)}\right)^{\alpha},\end{aligned}$$

and therefore it is enough to consider $L = M^{1-\alpha} L_1$. □

13.3 Proof of the Exponential Convergence

In this section we briefly explain how to combine Theorems 13.4,13.1 and 12.2 in order to obtain Theorem 13.3 (the proof of Theorem 13.2, given in Guarino and de Melo 2017, Section 4, is a little bit easier by the bounded combinatorics condition).

Sketch of the proof of Theorem 13.3 Let f and g be two C^4 unicritical circle maps with the same irrational rotation number $\rho(f) = \rho(g) = [a_0, a_1, \ldots]$ and with the same odd integer criticality. By Theorem 13.4 (the shadowing theorem), there exist a C^{ω}-compact set $\mathscr{K}$ of real analytic unicritical commuting pairs, two constants $\lambda_0 \in (0, 1)$ and $C_0 > 1$, and two sequences $\{f_m\}_{m\in\mathbb{N}}$ and $\{g_m\}_{m\in\mathbb{N}}$ contained in $\mathscr{K}$ such that for all $m \in \mathbb{N}$ we have $\rho(f_m) = \rho(g_m) = [a_m, a_{m+1}, \ldots]$ and moreover,

$$d_3\left(\mathscr{R}^m(f), f_m\right) \leqslant C_0\,\lambda_0^m \quad\text{and}\quad d_3\left(\mathscr{R}^m(g), g_m\right) \leqslant C_0\,\lambda_0^m. \tag{13.3}$$

With this, Proposition 6.2 and Theorem 12.1 at hand, it is not difficult to prove that the commuting pairs $\mathscr{R}^{j+m}(f)$, $\mathscr{R}^{j+m}(g)$, $\mathscr{R}^{j}(f_m)$ and $\mathscr{R}^{j}(g_m)$ are K-controlled and have negative Schwarzian for some constant $K > 1$, for m sufficiently large (say, $m > m_0$ for some $m_0 \in \mathbb{N}$) and for all $j \in \mathbb{N}$. Note that at this point we need the C^4 smoothness required in the statement of Theorem 13.3, to be able to obtain C^3-bounds for renormalization (see Guarino et al. 2018, Section 12 for the details).

Let $L = L(K) > 1$ be given by Theorem 12.2. Let $\delta \in (0, 1)$ be sufficiently close to one (to be determined in the course of the argument), and for each $n \in \mathbb{N}$ let $m = m(n) \in \mathbb{N}$ be given by $m = \lfloor \delta n \rfloor$. Combining Theorem 12.2 with (13.3) we obtain for all $m > m_0$ that

$$d_2\big(\mathscr{R}^n(f), \mathscr{R}^{n-m}(f_m)\big) \leqslant L^{n-m}\, d_2\big(\mathscr{R}^m(f), f_m\big) \tag{13.4}$$
$$\leqslant C_0\, L^{n-m}\lambda_0^m \leqslant \left(\frac{LC_0}{\lambda_0}\right)\left(L^{1-\delta}\lambda_0^{\delta}\right)^n$$

for all $n \in \mathbb{N}$ such that $m = \lfloor \delta n \rfloor > m_0$. Let $C_1 = LC_0/\lambda_0$ and $\lambda_1 = L^{1-\delta}\lambda_0^{\delta}$, and note that λ_1 belongs to $(0, 1)$ for δ sufficiently close to one. Replacing f with g, we also get

$$d_2\big(\mathscr{R}^n(g), \mathscr{R}^{n-m}(g_m)\big) \leqslant C_1\, \lambda_1^n \tag{13.5}$$

for all $n \in \mathbb{N}$ such that $m = \lfloor \delta n \rfloor > m_0$.

By Theorem 13.1, there exist constants $C_2 > 1$ and $\lambda_2 \in (0, 1)$ (both uniform in $\mathscr{K}$) such that

$$d_2\big(\mathscr{R}^{n-m}(f_m), \mathscr{R}^{n-m}(g_m)\big) \leqslant C_2\lambda_2^{n-m} \leqslant C_2(\lambda_2^{1-\delta})^n \tag{13.6}$$

for all $n \in \mathbb{N}$ and $m \in \mathbb{N}$. Finally, define $\lambda = \max\{\lambda_1, \lambda_2^{1-\delta}\} \in (0, 1)$ and $C = 2C_1 + C_2 > 1$. Combining (13.4), (13.5) and (13.6) we obtain

$$d_2\big(\mathscr{R}^n(f), \mathscr{R}^n(g)\big) \leqslant C\lambda^n \quad \text{for all } n \in \mathbb{N} \text{ such that } m = \lfloor \delta n \rfloor > m_0.$$

See Guarino et al. (2018, Section 12) for more details. □

13.4 The Attractor of Renormalization

As we have seen in Sect. 10.2 (recall (10.1)), the action of the renormalization operator on the continued fraction expansion of the rotation number is given by a *left shift*, that we denote by σ as customary. More precisely, given a critical commuting pair ζ with rotation number $\rho(\zeta) = [a_0, a_1, a_2, \dots]$ we have that

$$\rho\big(\mathscr{R}(\zeta)\big) = \sigma\big([a_0, a_1, a_2, \dots]\big) = [a_1, a_2 \dots]. \tag{13.7}$$

For real-analytic critical circle maps with a single critical point of some odd integer criticality, Yampolsky (2001, Th. A) was able to establish the existence of a *horseshoe-like* attractor for renormalization. More precisely, he proved the following result.

Theorem 13.6 (Horseshoe-Like Attractor) *There exists a pre-compact $\mathscr{R}$-invariant set Λ, which is homeomorphic to $\mathbb{N}^{\mathbb{Z}}$, consisting of real-analytic unicritical commuting pairs with irrational rotation number, such that the action of $\mathscr{R}|_{\Lambda}$ is topologically conjugate to the two-sided shift σ acting on $\mathbb{N}^{\mathbb{Z}}$ (the action being taken over the continued fraction expansion of the rotation number, as in* (13.7) *above). Moreover, any given real-analytic pair with irrational rotation number converges to the closure of Λ.*

As we have seen along this chapter (see also Chap. 14), such convergence is geometric, and it holds for C^4 pairs, not necessarily real-analytic (and for C^3 pairs with bounded combinatorics as well). For the proof of Theorem 13.6 we refer the reader to the original paper by Yampolsky (2001).

Exercises

Exercise 13.1 Prove the existence of two diffeomorphisms h_1 and h_2, as stated at the beginning of Sect. 13.1.1 (*Hint:* see Guarino and de Melo 2017, Lem. 6.2).

Exercise 13.2 Show that the sum or product of asymptotically holomorphic maps is also asymptotically holomorphic. The inverse of an asymptotically holomorphic diffeomorphism is asymptotically holomorphic. Composition of asymptotically holomorphic maps is asymptotically holomorphic.

Exercise 13.3 Let I be a compact interval in the real line and let $h : I \to \mathbb{R}$ be a C^1 map. Let U be a neighborhood of I in $\mathbb{C}$ and let $H : U \to \mathbb{C}$ be an asymptotically holomorphic extension of h of order 1 (as in Definition 13.1). Show that $\partial H(z) = h'(z)$ for every $z \in I$, and then the Jacobian of H at $z \in I$ equals $|h'(z)|^2$ (recall the identity (11.2)).

Exercise 13.4 In the upper half-plane $\mathbb{H} = \{z \in \mathbb{C} : \operatorname{Im} z > 0\}$ consider the vertical geodesic γ of equation $\{\operatorname{Re} z = 0\}$. Given $\varepsilon > 0$, show that the set of points in $\mathbb{H}$ whose hyperbolic distance to γ is less than ε is given by the cone

$$\left\{ z \in \mathbb{H} : \frac{\operatorname{Re} z}{\operatorname{Im} z} < \tan \alpha \right\},$$

where the Euclidean angle α is related to ε by the formula

$$\varepsilon = \frac{1}{2} \log \left(\frac{1 + \sin \alpha}{1 - \sin \alpha} \right).$$

Exercise 13.5 For $I = (-1, 1)$, show that $\phi : \mathbb{H} \to \mathbb{C}(I)$ given by

$$\phi(z) = \frac{z^2 + 1}{z^2 - 1}$$

is a biholomorphism between the upper half-plane $\mathbb{H}$ and $\mathbb{C}(I)$, that maps the vertical geodesic γ of equation $\{\operatorname{Re} z = 0\}$ onto I (*Hint:* Note that ϕ can be written as the composition

$$E_2 \circ T_{1/2} \circ I \circ T_{-1} \circ Q ,$$

where $Q(z) = z^2$, and the remaining four maps are the Möbius transformations $T_{-1}(z) = z - 1$, $I(z) = 1/z$, $T_{1/2}(z) = z + 1/2$ and $E_2(z) = 2z$).

Exercise 13.6 Let I be a bounded open interval in the real line. For any given $\theta \in (0, \pi)$ consider $\varepsilon(\theta) = \log \tan\big(\pi/2 - \theta/4\big) \in (0, +\infty)$. Show that the Poincaré disk $D_\theta(I)$ coincides with the set of points in $\mathbb{C}(I)$ whose hyperbolic distance to I is less than ε (*Hint:* Note first that it is enough to deal with the case $I = (-1, 1)$. Indeed, given $a < b$, the map $z \mapsto \big((b - a)z + a + b\big)/2$ is an isometry between $\mathbb{C}(-1, 1)$ and $\mathbb{C}(a, b)$, that preserves Euclidean angles. Note that the cone of Exercise 13.4 is mapped under the biholomorphism ϕ of Exercise 13.5 onto a Poincaré disk. Relate the Euclidean angle of the cone with the one of the corresponding Poincaré disk).

Chapter 14
Renormalization: Holomorphic Methods

In this final chapter we will survey some of the complex-analytic ideas that play a decisive role in the theory of (multi)critical circle maps. Since these ideas are quite deep, the narrative to follow is by necessity very sketchy. However, we provide a complete proof of a fundamental theorem in this area: the *complex bounds* (Theorem 14.4).

The use of holomorphic methods in the study of renormalization and rigidity of one-dimensional dynamical systems was started by Sullivan in the mid-eighties (see Sullivan 1992). Since the theory for circle maps follows in parallel with the corresponding theory for unimodal maps, and borrows substantially from it, we need to talk a bit about the latter first.

For the general theory of complex dynamics we refer the reader to the books Carleson and Gamelin (1993), de Faria and de Melo (2008), McMullen (1994), and Milnor (2006).

14.1 Sullivan's Program

We have already mentioned the general ansatz relating renormalization convergence and rigidity. If we are given two topologically conjugate one-dimensional maps f and g which are infinitely renormalizable (say with some restrictions on their combinatorics), and if we know that the C^r distances between their successive renormalizations contract to zero *at an exponential rate*, then the conjugacy between f and g should actually be smooth (for critical circle maps, recall here Theorems 10.4 and 10.5). Hence the goal becomes to establish exponential contraction of renormalizations. The strategy laid down by Sullivan (1992) (and explained in greater detail in de Melo and van Strien 1993, Ch. VI) to achieve this goal can be roughly described as follows.

E. de Faria, P. Guarino, *Dynamics of Circle Mappings*, IMPA Monographs,
https://doi.org/10.1007/978-3-031-67495-2_14

1. First get geometric bounds on the orbits of the critical points of the (real) one-dimensional systems. These so-called *real a priori bounds* should be robust enough that, even if we start with maps which have only a mild, finite degree of smoothness, their successive renormalizations will converge C^0 exponentially fast to the subspace consisting of real-analytic maps.
2. Use such real a priori bounds to show that the topological conjugacy between the two systems has slightly more geometric regularity than being merely continuous: it is actually *quasisymmetric* (at least when restricted to the post-critical sets of both systems).
3. Complexify the given real dynamical systems (when they are real-analytic), in other words, find suitable complex-analytic extensions of these systems.
4. Using the real bounds in (1) and the mild geometric control in (2), get *complex a priori bounds* for the complexified systems. These bounds are usually bounds on the moduli of certain annuli (typically fundamental domains for the complexified systems). Such bounds yield a strong form of compactness.
5. Extend the renormalization operator to the complexified dynamical systems. This operator will, in a suitable domain, be a compact operator due to step (4).
6. Use the bounds and compactness in (3) and a suitable infinite-dimensional version of Schwarz's lemma to establish the desired contraction property of the underlying renormalization operator.

In the context of (real-analytic) unimodal maps of the interval, Sullivan realized that the relevant complex-analytic dynamical systems are *quadratic-like maps* (or more generally polynomial-like maps), and was therefore able to use the theory developed by Douady and Hubbard (1985) for such maps. Recall that a quadratic-like map is a proper, degree two holomorphic branched covering map $F : U \to V$ between two topological disks $U, V \subset \mathbb{C}$ with U compactly contained in V, branched at a unique critical point $c \in U$. The *modulus* of F is by definition the conformal modulus of the annulus $V \setminus U$. The set $\mathcal{K}_F = \bigcap_{n \geqslant 0} F^{-n}(V)$ is called the *filled-in Julia set* of F. It is a totally invariant set under the dynamics, and it is compact due to the fact that F is proper. Every point in $U \setminus \mathcal{K}_F$ has a finite orbit that eventually lands in the outer annulus $V \setminus U$. This annulus therefore works as a *fundamental domain* for the dynamics outside the filled-in Julia set. A central fact about quadratic-like maps is the *straightening theorem* of Douady and Hubbard (1985): every quadratic like map is quasiconformally conjugate to an actual quadratic polynomial map.

A quadratic-like map $F : U \to V$ is said to be *renormalizable* if one can find a sub-disk $D \subset U$ compactly contained in U and containing c and an integer $p \geqslant 2$ such that $F^p|_D : D \to F^p(D) \subset V$ is well-defined, and again a quadratic-like map. This new map, with p smallest possible and suitably rescaled (via a complex affine map), is called the *first renormalization* of F, and denoted $\mathcal{R}F$. The number p is called the *renormalization period* of F, denoted $p(F)$. If all successive renormalizations $\mathcal{R}^2 F = \mathcal{R}(\mathcal{R}F), \ldots, \mathcal{R}^n F = \mathcal{R}(\mathcal{R}^{n-1} F), \ldots$ are well-defined, then we say that F is *infinitely renormalizable*. If in addition all periods $p_n = p(\mathcal{R}^n F)$ form a bounded sequence, we say that F infinitely renormalizable of *bounded type*. The *complex bounds* proved by Sullivan guarantee that if one starts with a real-analytic, infinitely renormalizable quadratic unimodal map f of bounded type on the real line, then after a finite number N of iterations, the renormalized unimodal maps $\mathscr{R}^n f$ will be restrictions of quadratic-like maps F_n with $F_{n+1} = \mathscr{R} F_n$ for all $n \geqslant N$, and moreover the moduli $\mathrm{mod}(\mathscr{R}^n F)$ $(n \geqslant N)$ will be bounded from below. In particular, the sequence $(\mathscr{R}^n F)_{n \geqslant N}$ will be a pre-compact family (in the topology of uniform convergence on compacta), and every limit of such renormalization sequence will be a quadratic-like map. Here and throughout, all holomorphic maps considered commute with complex conjugation, i.e., are symmetric about the real axis.

The crucial feature of quadratic-like maps in this theory, very closely related to the straightening theorem, is that they are amenable to what Sullivan calls a *pull-back* argument. If $F_i : U_i \to V_i$, $i = 0, 1$, are two symmetric, topologically conjugate quadratic-like maps, and if h is a quasisymmetric homeomorphism of the real line which sends the post-critical set of F_0 to the post-critical set of F_1, then F_0 and F_1 are quasiconformally conjugate. More precisely, there exists a quasiconformal homeomorphism $H : V_0 \to V_1$ such that $H \circ F_0 = F_1 \circ H$; in addition, the quasiconformal dilatation of H depends only on the conformal moduli $\mathrm{mod}(V_i \setminus U_i)$ $(i = 0, 1)$ and on the quasisymmetric distortion of h.

The existence of such a conjugacy already allows us to speak of the *quasiconformal* or *Teichmüller* distance between F_0 and F_1, defined as

$$d_T(F_0, F_1) = \inf_{\phi} \log \frac{1 + \|\mu_\phi\|_\infty}{1 - \|\mu_\phi\|_\infty}, \tag{14.1}$$

the infimum being taken over all quasiconformal conjugacies ϕ between F_0 and F_1. This is in fact a pseudo-distance: its value will be zero whenever the two maps are *conformally* conjugate. It turns out that the Julia set of an (symmetric) infinitely renormalizable quadratic-like map carries no *invariant line fields* (equivalently, no non-zero invariant Beltrami differentials). This is another consequence of the straightening theorem. Thus, for every quasiconformal conjugacy ϕ as above we have that μ_ϕ vanishes a.e. on the (filled-in) Julia set of F_0. In particular, when calculating $\|\mu_\phi\|_\infty$ in the right-hand side of (14.1), we only need to look at the values of $\mu_\phi(z)$ for $z \in V_0$.

It is immediate from the definition that the Teichmüller distance is weakly contracted under renormalization: any conjugacy between F_0 and F_1 restricts to a conjugacy between $\mathscr{R}(F_0)$ and $\mathscr{R}(F_1)$.

Now, let H be a quasiconformal conjugacy between F_0 and F_1, say the one constructed via the pull-back argument. Its Beltrami differential $\mu_H = \overline{\partial} H / \partial H$ is invariant under F_0, and therefore it can be used to generate a *path* of (pairwise qc-conjugate) quadratic-like maps joining F_0 to F_1. To see this, define $\mu_t = t\mu_H$ for all $t \in \mathbb{C}$ such that $|t| < \|\mu_H\|_\infty^{-1}$ then integrate each μ_t using the measurable Riemann mapping theorem to get a (normalized) quasiconformal homeomorphism H_t, and then define $F_t = H_t \circ F_0 \circ H_t^{-1}$. Such a path is called a *Beltrami path* joining F_0 to F_1.

As one can see from the definitions given so far, renormalization maps Beltrami paths to Beltrami paths. Some Beltrami paths are more efficient than others, in the sense that they are close to being "geodesics" in the Teichmüller metric. It will usually be the case that a very efficient Beltrami path joining F_0 to F_1 will be mapped to an inefficient Beltrami path joining $\mathscr{R}(F_0)$ to $\mathscr{R}(F_1)$: the image path "coils". It turns out that one can put this coiling property into more quantitative terms, and the result is a form of Schwarz's lemma in infinite dimensions.[1]

There are some difficulties with carrying out the details of this approach. One is the fact that the domain and range of a quadratic-like map vary with the map itself, so it is hard to set up the renormalization procedure as an actual operator on a space of maps defined over a fixed domain. Another difficulty is the fact that, if we are given two quadratic-like maps and they both restrict to the same quadratic unimodal map on the line, then they should be regarded as essentially the same dynamical system; however, their Teichmüller distance, according to the definition given above, will not be zero! Sullivan soon realized that a way to circumvent these difficulties is to take an inverse limit of the dynamics off the filled-in Julia set. To wit, if $F : U \to V$ is the given quadratic-like map, one considers the inverse system

$$\cdots \to F^{-(n+1)}(V \setminus \mathscr{K}_F) \to F^{-n}(V \setminus \mathscr{K}_F) \to \cdots F^{-1}(V \setminus \mathscr{K}_F) \to V \setminus \mathscr{K}_F ,$$

where each map, being a restriction of F, is an unbranched 2-to-1 holomorphic covering. The inverse limit of this system, denoted $\mathscr{L}(F)$, is a *Riemann surface lamination* in a natural way. This object is locally homeomorphic to the product of a disk by a Cantor set, and the chart transitions are holomorphic on the leaves. The construction is canonical in the sense that, if F varies (but stays in the same topological conjugacy class), then topologically $\mathscr{L}(F)$ does not change at all. Only its conformal structure changes. Moreover, a quasiconformal conjugacy between two such maps induces a homeomorphism between the two corresponding laminations which is quasiconformal on each leaf. Hence, one can speak of the (*moduli space* or) *Teichmüller space* of such lamination. It then follows that renormalization induces an operator on such Teichmüller space.

Using these ideas, Sullivan was able to carry out the strategy outlined in steps (1)–(6) above almost completely in the bounded-type case. We say "almost" because in step (6) he was forced to settle for something less than exponential contraction. Sullivan made an ingenious use of the theory of Riemann surface laminations, and used the Teichmüller theory of such objects (which he largely developed on the fly) to prove a (non-uniform) version of Schwarz's lemma in this context, which in turn allowed him to prove renormalization convergence without a rate. The exponential convergence of renormalizations for bounded type infinitely renormalizable maps was finally achieved by McMullen (1994, 1996) by a different route, using his theory of *rigidity of towers*.

[1] However, we warn the reader that the renormalization "operator" is *not* a complex-analytic operator.

Remark 14.1 The theory of Riemann surface laminations is a beautiful subject in its own right. See Ghys (1999) for a nice exposition.

14.2 Holomorphic Commuting Pairs

In his PhD thesis, de Faria (1992) took up the task of carrying out as much as possible of Sullivan's program in the context of critical circle maps with a single critical point of cubic type. Steps (1) and (2) of Sullivan's strategy were already in place due to the works of Herman and Świątek (Theorem 6.3) and Yoccoz (Theorem 7.2 in the unicritical case).

The key to the remaining steps is an analogue of the quadratic-like maps of Douady and Hubbard, a holomorphic dynamical system that somehow extends the real commuting pairs arising as successive renormalizations of a critical circle map. This is the central contribution of de Faria (1992) and of the subsequent paper de Faria (1999). Here are the relevant definitions, taken almost verbatim from de Faria and de Melo (2000, p. 346).

Definition 14.1 By a *bowtie* we mean a 4-tuple $(\mathcal{O}_\xi, \mathcal{O}_\eta, \mathcal{O}_\nu, \mathcal{V})$ of simply-connected domains in the complex plane such that:

(a) Each $\mathcal{O}_\gamma$ is a Jordan domain whose closure is contained in $\mathcal{V}$;
(b) We have $\overline{\mathcal{O}}_\xi \cap \overline{\mathcal{O}}_\eta = \{0\} \subseteq \mathcal{O}_\nu$;
(c) The sets $\mathcal{O}_\xi \setminus \mathcal{O}_\nu$, $\mathcal{O}_\eta \setminus \mathcal{O}_\nu$, $\mathcal{O}_\nu \setminus \mathcal{O}_\xi$ and $\mathcal{O}_\nu \setminus \mathcal{O}_\eta$ are non-empty and connected.

Definition 14.2 Let $(\mathcal{O}_\xi, \mathcal{O}_\eta, \mathcal{O}_\nu, \mathcal{V})$ be a bowtie. A holomorphic commuting pair Γ with domain $\mathcal{U} = \mathcal{O}_\xi \cup \mathcal{O}_\eta \cup \mathcal{O}_\nu$ and co-domain $\mathcal{V}$ is the dynamical system generated by three holomorphic maps $\xi : \mathcal{O}_\xi \to \mathbb{C}$, $\eta : \mathcal{O}_\eta \to \mathbb{C}$ and $\nu : \mathcal{O}_\nu \to \mathbb{C}$ satisfying the following conditions (see Fig. 14.1).

H_1 Both ξ and η are univalent onto $\mathcal{V} \cap \mathbb{C}(\xi(J_\xi))$ and $\mathcal{V} \cap \mathbb{C}(\eta(J_\eta))$ respectively, where $J_\xi = \mathcal{O}_\xi \cap \mathbb{R}$ and $J_\eta = \mathcal{O}_\eta \cap \mathbb{R}$. (Notation: $\mathbb{C}(I) = (\mathbb{C} \setminus \mathbb{R}) \cup I$.)
H_2 The map ν is a three-fold branched cover onto $\mathcal{V} \cap \mathbb{C}(\nu(J_\nu))$, where $J_\nu = \mathcal{O}_\nu \cap \mathbb{R}$, with a unique critical point at 0.
H_3 We have $\mathcal{O}_\xi \ni \eta(0) < 0 < \xi(0) \in \mathcal{O}_\eta$, and the restrictions $\xi|[\eta(0), 0]$ and $\eta|[0, \xi(0)]$ constitute a critical commuting pair.
H_4 Both ξ and η extend holomorphically to a neighborhood of zero, and we have $\xi \circ \eta(z) = \eta \circ \xi(z) = \nu(z)$ for all z in that neighborhood.
H_5 There exists an integer $m \geqslant 1$, called the *height* of Γ, such that $\xi^m(a) = \eta(0)$, where a is the left endpoint of J_ξ; moreover, $\eta(b) = \xi(0)$, where b is the right endpoint of J_η.

The relevant dynamical system here, which we will still denote by Γ, is the pseudo-semigroup generated by the three maps ξ, η, ν. The interval $J = [a, b]$ is called the *long dynamical interval* of Γ, whereas $\Delta = [\eta(0), \xi(0)]$ is the *short*

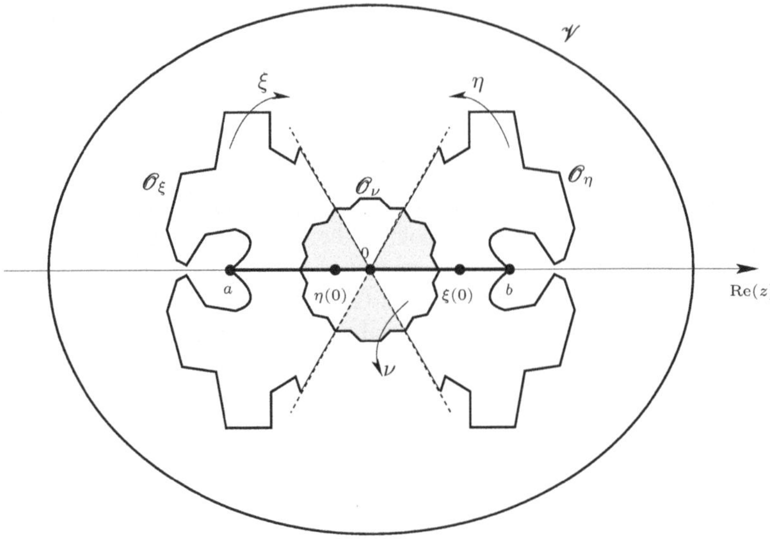

Fig. 14.1 A holomorphic commuting pair

dynamical interval of Γ. They are both forward invariant under the dynamics, as the reader can easily check. The *rotation number* of Γ is by definition the rotation number of the critical commuting pair of Γ obtained by restriction to the real line (condition H_3). We say that the holomorphic commuting pair Γ has *geometric boundaries* if $\partial\mathscr{U}$ and $\partial\mathscr{V}$ are *quasicircles*.[2]

Remark 14.2 Examples of holomorphic commuting pairs with arbitrary rotation number and arbitrary heights can be constructed directly from the Arnold family. This is carefully done in de Faria (1999, §4), and the construction will be reproduced in Sect. 14.4. We should also point out that there is nothing special about cubic critical points. Holomorphic commuting pairs can be defined so as to have a critical point with any odd-power criticality whatever. To see how this is done, the reader should consult the thesis by Vieira (2015) (see also Yampolsky 2017).

It turns out that the holomorphic pair Γ can be renormalized: the first renormalization of the critical commuting pair of Γ extends in a natural way to a holomorphic pair $\mathscr{R}(\Gamma)$ with the same co-domain $\mathscr{V}$. See Prop. 2.3 in de Faria (1999) for the detailed construction of $\mathscr{R}(\Gamma)$. Renormalization is defined in such a way that the restriction of the renormalized holomorphic pair $\mathscr{R}(\Gamma)$ to the real line is the critical commuting pair that represents the renormalization of the critical commuting pair $(\xi|_{[\eta(0),0]}, \eta|_{[0,\xi(0)]})$.

[2] A quasicircle, we recall, is the boundary of a quasidisk, which in turn is the image of a round disk under a quasiconformal homeomorphism of the plane.

14.3 Pull-Back Argument

The first main result in de Faria (1992) (or de Faria 1999) is the following analogue of Sullivan's *pull-back argument*.

Theorem 14.1 (Pull-Back Argument) *Let Γ and Γ' be holomorphic pairs with geometric boundaries and let $h : J \to J'$ be a quasisymmetric conjugacy between the restrictions of Γ and Γ' to their respective long dynamical intervals J and J'. Then there exists a quasiconformal conjugacy $H : \mathcal{V} \to \mathcal{V}'$ between Γ and Γ' which is an extension of h.*

The proof is more involved than that of the original pull-back argument, for the following reason. In the quadratic-like case, we know by the straightening theorem of Douady–Hubbard that every quadratic-like map is quasiconformally conjugate to a quadratic polynomial, and the latter does not have wandering domains (due to Sullivan's no-wandering-domains theorem, see Sullivan 1985). Hence quadratic-like maps do not have wandering domains. By contrast, holomorphic pairs could in principle have wandering domains. To deal with their putative existence, one needs to use a form of quasiconformal surgery (something called the *qc-sewing lemma* of L. Bers, see de Faria 1999, Lem. 3.2). Wandering domains are only ruled out *a posteriori*, combining Theorem 14.1 with the fact that holomorphic pairs constructed from the Arnold family do not carry such domains (see de Faria 1999, Th. 4.2).

14.4 Existence and Limit-Set qc-Rigidity

We have defined holomorphic commuting pairs as complex dynamical systems satisfying certain axioms (see Definition 14.2), but it is not clear at this point in our narrative whether such objects exist. Hence we take the time to construct explicit examples with arbitrary rotation numbers and arbitrary heights. The construction presented below is taken almost verbatim from de Faria (1999, §4). When combined with the pull-back argument, these examples also yield two important properties of holomorphic commuting pairs: a no-wandering-domains theorem for such objects and the absence of invariant line fields in their limit (or Julia) sets.

14.4.1 Construction of Examples

The examples are extracted from our old friend, the *complex Arnold family*. For each $0 \leqslant \theta < 1$, let $E_\theta : \mathbb{C} \to \mathbb{C}$ be the entire mapping given by $E_\theta(z) = z + \theta - \frac{1}{2\pi}\sin(2\pi z)$. Such maps indeed belong to the Arnold family; in fact, we have $E_\theta = F_{\theta,1}$ in the notation introduced in Chap. 6.

Since $E_\theta \circ T = T \circ E_\theta$, where T is the translation $z \mapsto z + 1$, E_θ is the lift to the complex plane of a holomorphic self-mapping of the cylinder, $f_\theta : \mathbb{C}/\mathbb{Z} \cong \mathbb{C}^* \hookleftarrow$. Moreover, the restriction $E_\theta|\mathbb{R}$ maps the real axis onto itself and satisfies $E'_\theta(x) \geqslant 0$ for all $x \in \mathbb{R}$, and equality holds iff $x \in \mathbb{Z}$ (these constitute all the critical points of E_θ). Therefore the restriction $f_\theta|\boldsymbol{S}^1$ is a critical circle homeomorphism with rotation number, say, $\rho(\theta)$. We have already seen in Chap. 4 that $\theta \mapsto \rho(\theta)$ is a continuous, non-decreasing map of $[0, 1)$ onto itself such that the interval $\rho^{-1}(t) \subseteq [0, 1)$ degenerates to a point whenever t is irrational.

With the family $\{E_\theta\}$ at hand we will construct examples of holomorphic commuting pairs with arbitrary rotation number and arbitrary height. More precisely, we shall prove the following theorem.

Theorem 14.2 *For each $n \geqslant 0$ and each θ such that $\rho(\theta)$ has a continued fraction expansion of length at least $n + 1$, the real commuting pair determined by $(f_\theta^{q_n}, f_\theta^{q_{n+1}})$ extends to a holomorphic commuting pair $\Gamma_{n,\theta}$ with geometric boundaries. The family $\{\Gamma_{n,\theta}\}$ runs through all possible pairs of combinatorial invariants at least once, and for each $(m, \rho) \in \mathbb{N} \times [0, 1)$ with $m \geqslant 2$ there exist countably many $(n, \theta) \in \mathbb{N} \times [0, 1)$ such that $\Gamma_{n,\theta}$ has height m and rotation number ρ.*

The main analytic tool to be used in the proof of Theorem 14.2 is the following growth estimate.

Lemma 14.1 *There exist a positive constant C_0 and a positive monotone non-decreasing function $\varphi(s)$ defined for $s \geqslant 0$ such that if $|y| \geqslant \varphi(|x|)$ then $|E_\theta(x + iy)| \geqslant C_0 \exp(\pi|y|)$.*

Proof When $\theta = 0$, a straightforward computation yields

$$\begin{aligned} |E_0(x+iy)|^2 &= \frac{1}{4\pi^2}\cosh^2(2\pi y) + \left[x^2 + y^2 - \frac{1}{4\pi^2}\cos^2(2\pi x)\right] \\ &\quad - \frac{1}{\pi}\left[x \sin(2\pi x)\cosh(2\pi y) + y\cos(2\pi x)\sinh(2\pi y)\right] . \end{aligned}$$

The first expression between brackets is positive as soon as, say, $|y| \geqslant 1$, while the second is dominated by $(|x| + |y|)\cosh(2\pi y)$. Thus, if $|y| \geqslant 1$ we have

$$|E_0(x+iy)|^2 \geqslant \frac{1}{4\pi^2}\left[\cosh(2\pi y) - 4\pi(|x| + |y|)\right]\cosh(2\pi y) . \tag{14.2}$$

Now, let

$$\varepsilon(t) = \frac{1}{4\pi}\cosh(2\pi t) - t - 1 .$$

This is a strictly convex function that reaches a minimum value at a certain $t_0 > 0$ such that $\varepsilon(t_0) < 0$. Hence for each $s \geqslant 0$ there exists a unique $\overline{\varphi}(s) > t_0$ such that $\varepsilon(\overline{\varphi}(s)) = s$. Since $\varepsilon(t)$ is strictly increasing for $t \geqslant t_0$, so is $\overline{\varphi}(s)$ for $s \geqslant 0$, and $t \geqslant \overline{\varphi}(s)$ implies $\varepsilon(t) \geqslant s$. Setting $\varphi(s) = \max\{1, \overline{\varphi}(s)\}$ and observing that the expression between brackets in (14.2) is equal to $4\pi[\varepsilon(|y|) + 1 - |x|]$, we deduce that if $|y| \geqslant \varphi(|x|)$ then

$$|E_0(x + iy)|^2 \geqslant \frac{1}{\pi} \cosh(2\pi|y|) \geqslant \frac{1}{2\pi} \exp(2\pi|y|) . \tag{14.3}$$

On the other hand, when $0 < \theta < 1$ we have $E_\theta(z) = E_0(z) + \theta$, so that $|E_\theta(z)| \geqslant |1 - |E_0(z)|^{-1}|.|E_0(z)|$. Therefore, if $|y| \geqslant \varphi(|x|)$, we have by (14.3)

$$|E_\theta(x + iy)| \geqslant \frac{1}{\sqrt{2\pi}} \left[1 - e^{-\pi}\sqrt{2\pi}\right] \exp(\pi|y|) ,$$

so the desired inequality is proved in all cases if we take $C_0 = \frac{1}{\sqrt{2\pi}} \left[1 - e^{-\pi}\sqrt{2\pi}\right]$.

□

We divide the work required to prove Theorem 14.2 into a few steps. Let us fix θ for the time being and write $\rho(\theta) = [a_0, a_1, \ldots, a_n, \ldots]$. We conform with the notation established in earlier chapters, so that, in its irreducible form, $\frac{p_n}{q_n} - [a_0, a_1, \ldots, a_{n-1}]$ satisfies $p_0 = 0$, $q_0 = 1$; $p_1 = 1$, $q_1 = a_0$ and, for $n \geqslant 1$, $p_{n+1} = a_n p_n + p_{n-1}, q_{n+1} = a_n q_n + q_{n-1}$.

We need a brief geometric description of the map E_θ. The pre-image of the real axis under E_θ consists of $\mathbb{R}$ itself together with the family of analytic curves

$$\mathscr{S}_\pm^{(k)} : x = k \pm \frac{1}{2\pi} \arccos\left[\frac{-2\pi|y|}{\sinh(2\pi y)}\right],$$

where $k \in \mathbb{Z}$, arising as solutions to $\operatorname{Im} E_\theta(x + iy) = 0$. For each $k \in \mathbb{Z}$, the curves $\mathscr{S}_+^{(k)}$ and $\mathscr{S}_-^{(k)}$ meet at the critical point $c_k = k$, and are both asymptotic to the vertical lines $x = k \pm \frac{1}{4}$. Notice that each c_k is a critical point of cubic type. In the upper half-plane $\mathbb{C}^+$, let V_k be the simply-connected region bounded by the arcs $\mathscr{S}_+^{(k-1)} \cap \mathbb{C}^+$ and $\mathscr{S}_-^{(k)} \cap \mathbb{C}^+$ and the interval $[k-1, k] \subseteq \mathbb{R}$. Then $E_\theta|V_k$ is univalent onto $\mathbb{C}^+$; we let $\phi_k : \mathbb{C}^+ \to V_k$ denote the corresponding inverse. Similarly, let $W_k \subseteq \mathbb{C}^+$ be the simply-connected region bounded by $\mathscr{S}_-^{(k)} \cap \overline{\mathbb{C}}^+$ and $\mathscr{S}_+^{(k)} \cap \mathbb{C}^+$, observe that $E_\theta|W_k$ is univalent onto $\mathbb{C}^-$ and let $\psi_k : \mathbb{C}^- \to W_k$ be the corresponding inverse.

Now let $A_n \subseteq \mathbb{C}^+$ be the unique connected component of $(E_\theta^{q_n})^{-1}(\mathbb{C}^+)$ whose closure contains the point $T^{-p_{n+1}} \circ E_\theta^{q_{n+1}}(0) \in \mathbb{R}$. Similarly, let $B_n \subseteq \mathbb{C}^+$ be the unique connected component of $(E_\theta^{q_{n+1}})^{-1}(\mathbb{C}^+)$ such that $T^{-p_n} \circ E_\theta^{q_n}(0) \in \overline{B}_n$. We have either $A_n \subseteq V_0$ and $B_n \subseteq V_1$ or $A_n \subseteq V_1$ and $B_n \subseteq V_0$, depending on whether n is even or odd, respectively (Fig. 14.2 illustrates the even case).

Lemma 14.2 *For each $n \geqslant 0$ there exists a unique q_n-tuple $(k_1, k_2, \ldots, k_{q_n})$ with $0 = k_1 \leqslant k_2 \leqslant \cdots \leqslant k_{q_n} \leqslant p_n + 1$ such that $A_n = \phi_{k_1} \circ \phi_{k_2} \circ \cdots \circ \phi_{k_{q_n}}(\mathbb{C}^+)$. A similar statement holds for B_n.*

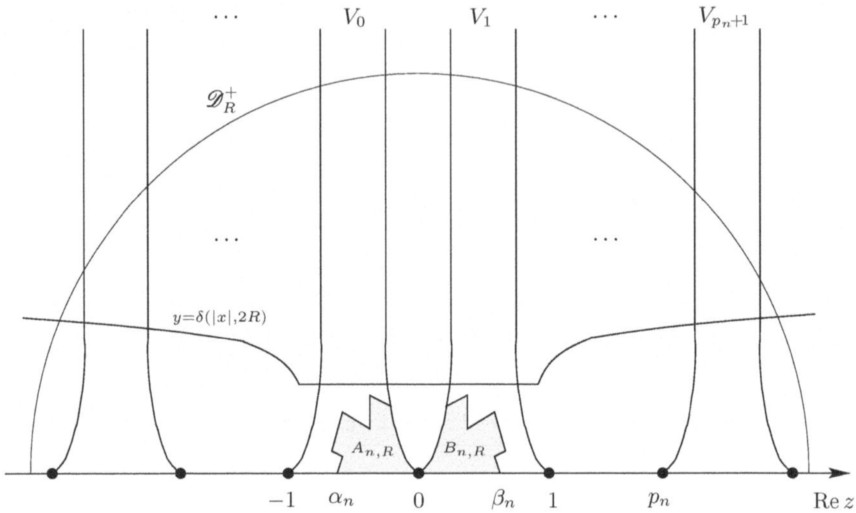

Fig. 14.2 Building holomorphic pairs

Proof This is an easy consequence of the fact that $0 \leqslant E_\theta^j(0) < p_n + 1$ for $j = 0, 1, \ldots, q_n$, for all $n \geqslant 0$, which in turn follows from the very definitions of p_n, q_n. □

Lemma 14.3 *Let f be a circle homeomorphism with $\rho(f) = [a_0, a_1, \ldots, a_n, \ldots]$, let $c \in \mathbf{S}^1$, and for each $n \geqslant 1$ let $J_n \subseteq \mathbf{S}^1$ be the closed interval of endpoints c and $f^{q_{n-1}-q_n}(c)$ containing $f^{q_{n-1}}(c)$. If $j < q_n$ is such that $f^{-j}(c)$ belongs to J_n, then $j \leqslant 0$.*

Proof This reduces to a purely combinatorial statement about rigid rotations, and is left as an exercise for the reader. □

Let us use the notation $\langle \alpha, \beta \rangle$ to represent a closed interval on the line with endpoints α and β, irrespective of order.

Lemma 14.4 *For each $n \geqslant 0$ we have $\overline{A}_n \cap \mathbb{R} = \langle \alpha_n, 0 \rangle$ and $\overline{B}_n \cap \mathbb{R} = \langle 0, \beta_n \rangle$, where $\alpha_0 = -1$, $\beta_0 = \alpha_1$ and for $n \geqslant 1$ the points $\alpha_n, \beta_n \in \mathbb{R}$ are uniquely determined by the requirements: $T^{-p_n} \circ E_\theta^{q_n}(\alpha_n) = T^{-p_{n-1}} \circ E_\theta^{q_{n-1}}(0)$ and $T^{-p_{n+1}} \circ E_\theta^{q_{n+1}}(\beta_n) = T^{-p_n} \circ E_\theta^{q_n}(0)$.*

Proof Consider $f = f_\theta$ and take c to be the critical point of f_θ. Then Lemma 14.3 says that there can be no critical points for $f_\theta^{q_n}$ in the interior of J_n, for by the chain rule these are precisely the pre-images $f_\theta^{-j}(c)$ with $0 \leqslant j < q_n$. The result follows. □

Given $R > 0$, let $\mathscr{D}_R = \{z : |z| < R\}$ and let $A_{n,R}$ be the unique connected component of $(T^{-p_n} \circ E_\theta^{q_n})^{-1}(\mathscr{D}_R^+)$ contained in A_n. Let $B_{n,R}$ be similarly defined. If R is sufficiently large ($R > p_n+1$ is good enough) we see that $\overline{A}_{n,R} \cap \mathbb{R} = \overline{A}_n \cap \mathbb{R}$ and $\overline{B}_{n,R} \cap \mathbb{R} = \overline{B}_n \cap \mathbb{R}$ for $n \geqslant 0$. It is clear that both $A_{n,R}$ and $B_{n,R}$ are Jordan domains, in fact quasidisks, and that they are mapped respectively by $T^{-p_n} \circ E_\theta^{q_n}$ and $T^{-p_{n+1}} \circ E_\theta^{q_{n+1}}$ bijectively onto $\mathscr{D}_R^+$.

Lemma 14.5 *For every sufficiently large R we have $\overline{A}_{n,R} \subseteq \mathscr{D}_R \cap \overline{\mathbb{C}^+}$ and $\overline{B}_{n,R} \subseteq \mathscr{D}_R \cap \overline{\mathbb{C}^+}$.*

Proof For s, R positive numbers, let

$$\delta(s, R) = \varphi(s) + \frac{1}{\pi} \log^+(C_0^{-1} R) ,$$

where φ and C_0 are given by Lemma 14.1. Then $|y| \geqslant \delta(|x|, R)$ implies $|E_\theta(x + iy)| \geqslant R$, which in turn means that $E_\theta(x+iy) \in \mathbb{C} \setminus \mathscr{D}_R$. Therefore, for each $k \in \mathbb{Z}$ we have

$$\phi_k(\overline{\mathscr{D}_R^+}) \subseteq \overline{V}_k \cap \{x + iy : y \leqslant \delta(|x|, R)\} .$$

Let $V_{k,R}$ denote this last intersection. Since $\delta(s, R)$ has logarithmic growth in R, every sufficiently large R satisfies the inequality $R > p_n + 1 + \delta(p_n + 1, 2R)$; for a given R as such, if $0 \leqslant k \leqslant p_n + 1$ and z is any point in $V_{k,2R}$ with $z = x + iy$, then

$$|z| \leqslant |x| + \delta(|x|, 2R) \leqslant p_n + 1 + \delta(p_n + 1, 2R) < R ,$$

and so it follows that $z \in \mathscr{D}_R \cap \overline{\mathbb{C}^+}$. Thus, if $0 \leqslant k \leqslant p_n + 1$ then $\phi_k(\overline{\mathscr{D}_{2R}^+}) \subseteq \mathscr{D}_R \cap \overline{\mathbb{C}^+} \subseteq \overline{\mathscr{D}_{2R}^+}$. Since $T^{p_n}(\overline{\mathscr{D}_R^+}) \subseteq \overline{\mathscr{D}_{2R}^+}$, taking the q_n-tuple $(k_1, k_2, \ldots, k_{q_n})$ as in Lemma 14.2 we deduce that

$$\overline{A}_{n,R} = \phi_{k_1} \circ \phi_{k_2} \circ \cdots \circ \phi_{k_{q_n}}(T^{p_n}\overline{\mathscr{D}_R^+}) \subseteq \phi_{k_1} \circ \phi_{k_2} \circ \cdots \circ \phi_{k_{q_n}}(\overline{\mathscr{D}_{2R}^+}) \subseteq \mathscr{D}_R \cap \overline{\mathbb{C}^+} .$$

This proves the first inclusion; the second is proved in similar fashion. □

Remark 14.3 Observe that if we define $\mathscr{U}_{n,R} = \phi_{k_2} \circ \phi_{k_3} \circ \cdots \circ \phi_{k_{q_n}}(\mathscr{D}_R^+)$ and set $A'_{n,R} = \phi_1(\mathscr{U}_{n,R})$ and $A''_{n,R} = \psi_0\left(\sigma(\mathscr{U}_{n,R})\right)$, where $\sigma : \mathbb{C} \to \mathbb{C}$ is complex conjugation, then the above argument applies *mutatis mutandis* to yield $\overline{A'_{n,R}} \subseteq \mathscr{D}_R \cap \overline{\mathbb{C}^+}$, $\overline{A''_{n,R}} \subseteq \mathscr{D}_R \cap \overline{\mathbb{C}^+}$ as well, for every sufficiently large R and all $n \geqslant 0$.

Proof of Theorem 14.2 Given $n \geqslant 0$, let $R_n > 0$ be large enough for the conclusion of Lemma 14.5 to hold. Let $\xi_n = T^{-p_n} \circ E_\theta^{q_n}$ and $\eta_n = T^{-p_{n+1}} \circ E_\theta^{q_{n+1}}$ and let $\mathscr{O}_{\xi_n}, \mathscr{O}_{\eta_n} \subseteq \mathbb{C}$ be the symmetric Jordan domains (quasidisks) such that $\mathscr{O}_{\xi_n}^+ = A_{n,R_n}$, $\mathscr{O}_{\eta_n}^+ = B_{n,R_n}$. Then ξ_n and η_n commute, and $\overline{\mathscr{O}_{\xi_n}}, \overline{\mathscr{O}_{\eta_n}} \subseteq \mathscr{D}_{R_n}$, by Lemma 14.5. The restrictions $\xi_n|\mathscr{O}_{\xi_n}$ and $\eta_n|\mathscr{O}_{\eta_n}$ are univalent and onto their images, which by Lemma 14.3 are $\mathscr{D}_{R_n} \cap \mathbb{C}(\langle \xi_n(\alpha_n), \xi_n(0)\rangle)$ and $\mathscr{D}_{R_n} \cap \mathbb{C}(\langle \eta_n(0), \eta_n(\beta_n)\rangle)$, respectively. Also, let $\mathscr{O}_{\nu_n} \subseteq \mathbb{C}$ be the connected component of $\xi_n^{-1}(\mathscr{O}_{\eta_n})$ containing the origin and let $\nu_n = \xi_n \circ \eta_n$. Then the restriction $\nu_n|\mathscr{O}_{\nu_n}$ is a holomorphic three-fold branched covering map onto its image, $\nu_n(\mathscr{O}_{\nu_n}) = \mathscr{D}_{R_n} \cap \mathbb{C}(\langle \eta_n(0), \xi_n(0)\rangle)$. Moreover, by the remark following Lemma 14.5, we have

$$\overline{\mathscr{O}_{\nu_n}^{+}} \subseteq \overline{A}_{n,R} \cup \overline{A'_{n,R}} \cup \overline{A''_{n,R}} \subseteq \mathscr{D}_{R_n} \cap \overline{\mathbb{C}^{+}} ,$$

and so $\overline{\mathscr{O}_{\nu_n}} \subseteq \mathscr{D}_{R_n}$. It follows at once that $(\mathscr{O}_{\xi_n}, \mathscr{O}_{\eta_n}, \mathscr{O}_{\nu_n}, \mathscr{D}_{R_n})$ is a *bowtie*.

Now we claim that this bowtie together with the maps ξ_n, η_n, ν_n determine a holomorphic commuting pair $\Gamma_{n,\theta}$ with geometric boundaries, up to orientation, with rotation number $\rho(\Gamma_{n,\theta}) = [a_{n+1} + 1, a_{n+2}, \ldots]$ and height given by $m(\Gamma_{0,\theta}) = a_0$ when $n = 0$, and by $m(\Gamma_{n,\theta}) = a_n + 1$ when $n > 0$. We have indirectly checked all conditions in Definition 14.2, except perhaps condition H_5. We check it for $n > 0$; the case $n = 0$ is just as easy. Using the commutativity of T with E_θ, Lemma 14.2 and the recurrence relations defining p_{n+1} and q_{n+1}, we get

$$\begin{aligned} \xi_n^{a_n+1}(\alpha_n) &= (T^{-p_n} \circ E_\theta^{q_n})^{a_n} (T^{-p_n} \circ E_\theta^{q_n}(\alpha_n)) \\ &= T^{-p_{n+1}} \circ E_\theta^{q_{n+1}}(0) = \eta_n(0) . \end{aligned}$$

Similarly, we have $\eta_n(\beta_n) = \xi_n(0)$. Thus condition H_5 is satisfied too, and $m = a_n + 1$ is the height of $\Gamma_{n,\theta}$. The statement on rotation numbers is clear. □

Remark 14.4 Because holomorphic commuting pairs can be renormalized, once R_0 is chosen so that the above construction works for $n = 0$, we may take $R_n = R_0$ thereafter. If this is done then, for each $n \geqslant 0$, $\Gamma_{n+1,\theta}$ becomes the first renormalization of $\Gamma_{n,\theta}$ up to linear rescaling.

14.4.1.1 Limit Set qc-Rigidity

When combined with the results of the previous section, Theorem 14.2 yields two crucial properties of holomorphic commuting pairs, which we express as follows.

Theorem 14.3 *Let Γ be a holomorphic commuting pair with geometric boundaries and irrational rotation number. Then Γ has no wandering domains and admits no non-trivial, symmetric, invariant Beltrami differentials entirely supported in its limit set.*

This theorem allows holomorphic commuting pairs to be parametrized by conformal structures supported on the outer annulus of a fixed model. The properties of holomorphic commuting pairs stated in this theorem are extracted from corresponding ones found naturally in the family $\{f_\theta\}$ of self-maps of the cylinder $\mathbb{C}^*$ introduced before.

Proof of Theorem 14.3 Combining Theorem 14.1 with Theorem 14.2, we know that Γ is conjugate to $\Gamma_{0,\theta}$ for some θ by a quasiconformal homeomorphism H. Let μ be a Γ-invariant Beltrami differential with support in J_Γ. Then $\mu' = H^*\mu$ is $\Gamma_{0,\theta}$-invariant. Spreading μ' through the entire complex plane via the mappings defining $\Gamma_{0,\theta}$ we get a Beltrami differential ν invariant under both E_θ and $T^{-1} \circ E_\theta^{a_0}$, and therefore invariant under T also. Thus ν projects down to a Beltrami differential on the cylinder which is f_θ-invariant and supported in J_{f_θ}. By Theorem 11.3, this Beltrami differential must vanish almost everywhere, and so $\mu \equiv 0$ a.e. also. A similar argument, which we leave as an exercise, rules out wandering domains. □

14.5 Complex Bounds

Another important fact about holomorphic commuting pairs is that the class of such objects contains all limits of successive renormalizations of a critical circle map (or critical commuting pair). Moreover, we have *complex bounds* for renormalization, in the following sense.

Theorem 14.4 (Complex Bounds) *Let $f : S^1 \to S^1$ be a real-analytic critical circle map with arbitrary irrational rotation number. Then there exists $n_0 = n_0(f)$ such that for all $n \geqslant n_0$ the n-th renormalization of f extends to a holomorphic pair with geometric boundaries whose fundamental annulus has conformal modulus bounded from below by a universal constant.*

This theorem establishes Step (4) of Sullivan's strategy described at the beginning of this chapter. It also provides another proof of existence of holomorphic commuting pairs, independent of the explicit constructions we performed in Sect. 14.4.

Remark 14.5 It is important to observe that, in Theorem 14.4, although the bound on the fundamental annulus of the holomorphic pair corresponding to a deep renormalization of f is bounded from below by a universal constant, the geometric boundaries can become very bad (i.e., they are quasi-circles with qc-distortion constant that can go to infinity with n)—unless the rotation number is of bounded type, in which case the bounds in question depend only on the least upper bound on the coefficients of the continued fraction development of $\rho(f)$. When $\rho(f)$ is a number of *unbounded* type, the limits of renormalization will contain maps with parabolic fixed points, and the proper study of renormalization in these cases requires the notion of *cylinder renormalization* introduced by Yampolsky (see for instance Yampolsky 2002).

We think of the unit circle $S^1 = \mathbb{R}/\mathbb{Z}$ as embedded in the infinite cylinder $\mathbb{C}/\mathbb{Z}$, and we use on latter the conformal metric induced from the standard Euclidean metric $|dz|$ of the complex plane via the exponential map $\exp(z) = e^{2\pi i z}$. Note that $\operatorname{Im} z$ is well-defined for every $z \in \mathbb{C}/\mathbb{Z}$ (it is simply the imaginary part of any one of its pre-images under the exponential).

The main step in the proof of Theorem 14.4 is to establish a geometric estimate showing that, for all sufficiently large n, the appropriate inverse branch of $f^{q_{n+1}}$ maps a sufficiently large disk around the n-th renormalization domain $I_n \cup I_{n+1}$ well within itself. Here, "sufficiently large" means large with respect to the size of $I_n \cup I_{n+1}$. For each $m \geqslant 1$, let $D_m \subset \mathbb{C}/\mathbb{Z}$ denote the disk having as one of its diameters the interval $[f^{q_{m+1}}(c), f^{q_m - q_{m+1}}(c)] \subset S^1$ containing the critical point c. Note[3] that $\operatorname{diam}(D_m)$ is comparable with $|I_m|$: this follows from the real a priori bounds (Theorem 6.3). The geometric estimate is the following (the statement is taken almost verbatim from de Faria and de Melo (2000, Prop. 3.2)).

Proposition 14.1 *There exist universal constants B_1 and B_2 and for each $N \geqslant 1$ there exists $n(N)$ such that for all $n \geqslant n(N)$ the inverse branch $f^{-q_{n+1}+1}$ taking $f^{q_{n+1}}(I_n)$ back to $f(I_n)$ is well-defined over $\Omega_{n,N} = (D_{n-N} \setminus S^1) \cup f^{q_{n+1}}(I_n)$ and it is univalent there, and for all $z \in \Omega_{n,N}$ we have*

$$\frac{\operatorname{dist}\left(f^{-q_{n+1}+1}(z), f(I_n)\right)}{|f(I_n)|} \leqslant B_1 \left(\frac{\operatorname{dist}(z, I_n)}{|I_n|}\right) + B_2 . \tag{14.4}$$

As stated, Theorem 14.4 was proved in de Faria and de Melo (2000, §3). But the story behind it is a bit more involved. The first version of the complex bounds in the present context was proved in de Faria (1992) (also de Faria 1999) under two further assumptions on f, namely

(i) the rotation number of f is of bounded type;
(ii) f is an *Epstein map*.

We say that a real analytic circle map is *Epstein* if its lift to the real line has a holomorphic extension F to a neighborhood of the real axis in the complex plane in such a way that F has inverse branches which are globally defined in the upper (or lower) half-plane. The main examples of Epstein circle maps are the maps in the Arnold family introduced earlier (see Sect. 6.1.2). The proof presented in de Faria (1992, 1999) makes use of the so-called *sector theorem* of Sullivan (see Sullivan 1992; the version used in the circle case is in fact the one proved in de Faria 1998). However, the sector theorem can only be used under the bounded type assumption (i).

That assumption was removed by Yampolsky (1999), using a special case of Proposition 14.1. Assuming that the map f is Epstein, he exploits in full the idea of *Poincaré neighborhood trapping*, already explained in Sect. 13.1.2 and that we briefly recall now. Let $J \subseteq \mathbb{R}$ be a bounded open interval, and write $\mathbb{C}(J) = \mathbb{C} \setminus (\mathbb{R} \setminus J)$. If $\phi : \mathbb{C}(J) \to \mathbb{C}(\phi(J))$ is a real-symmetric holomorphic map, then ϕ maps each Poincaré disk $D_\theta(J) = \{z : \operatorname{angle}(z, J) \geqslant \theta\}$ into a corresponding Poincaré neighborhood $D_\theta(\phi(J))$ with the same angle θ. Here, $0 < \theta < \pi$ and

[3] It is easy to see that $[f^{q_{m+1}}(c), f^{q_m - q_{m+1}}(c)] \supset I_m \cup I_{m+1}$.

angle(z, J) denotes the angle at z under which z views the interval J. This simple but fundamental fact is easily seen to be a consequence of Schwarz's lemma.

The Poincaré neighborhood trapping idea used in Yampolsky's approach works because he is assuming that f is Epstein. But if we abandon the latter hypothesis, then this tool is no longer directly applicable. In order to prove Theorem 14.4, one needs the following "relaxed" version of Poincaré neighborhood trapping. The statement is taken almost verbatim from de Faria and de Melo (2000, Lem. 3.3), but with an important modification introduced by Yampolsky (2019, Lem. 4.4).

Lemma 14.6 *For every small $a > 0$, there exists $\theta(a) > 0$ satisfying $\theta(a) \to 0$ and $a/\theta(a) \to 0$ as $a \to 0$, such that the following holds. Let $F : \mathbb{D} \cap \mathbb{C}([0, a]) \to \mathbb{C}$ be univalent and symmetric about the real axis, and assume $F(0) = 0$, $F(a) = a$. Then for all $\theta \geqslant \theta(a)$ we have $F(D_\theta((0, a))) \subseteq D_{(1-a^{1+\delta})\theta}((0, a))$, where $0 < \delta < 1$ is an absolute constant.*

This lemma is applicable to other situations—see for example Clark et al. (2017). It is a precursor to the more general *almost Schwarz inclusion* lemma for asymptotically holomorphic maps due to Graczyk et al. (2005, Prop. 2), stated in the previous chapter (Proposition 13.2).

We will derive Lemma 14.6 as a consequence of an elegantly simple result due to Gaidashev and Gorbovickis (2021) which is an improvement over Yampolsky's aforementioned version. In what follows, we will employ the following additional notation. For each $\tau > 0$, we will write $\omega(\tau) = 2 \arctan \tau$. We also let $\mathbb{D}(J)$ denote the doubly-slit *disk* obtained by intersecting $\mathbb{C}(J)$ with the open disk of radius 1 centered at the midpoint of J.

At this point it is convenient to restate the Poincaré neighborhood trapping idea as the following *Schwarzian inclusion principle*. We have already seen this principle in Sect. 13.1.2.

Lemma 14.7 *If $\psi : \mathbb{C}(J) \to \mathbb{C}(J)$ is a real-symmetric holomorphic map, then for each θ we have $\psi(D_\theta(J)) \subseteq D_\theta(J)$.*

Now, the Gaidashev–Gorbovickis version of the almost Schwarzian inclusion principle can be stated as follows.

Lemma 14.8 *Let J be a bounded open interval on the real line, let $b = |J|/2$ be such that $b < 1$, and let $\phi : \mathbb{D}(J) \to \mathbb{C}(J)$ be a real-symmetric holomorphic map. Then for each $\tau > 0$ such that $D_{\omega(\tau)}(J) \subset \mathbb{D}(J)$, we have*

$$\phi(D_{\omega(\tau)}(J)) \subseteq D_{\omega(\tau_*)}(J) , \tag{14.5}$$

where

$$\tau_* = \frac{\tau^2 - b^2}{\tau(1 + b^2)} .$$

Proof We first note that every such map ϕ can be factored as $\phi = \psi \circ G$, where G maps $\mathbb{D}(J)$ univalently onto $\mathbb{C}(J)$ and $\psi = \phi \circ G^{-1}$ maps $\mathbb{C}(J)$ into itself. The rough idea of the proof is to choose G suitably so that it maps every set of the form $D_\theta(J)$ contained in $\mathbb{D}(J)$ into a slightly larger set $D_{\theta'}(J)$ (with θ' slightly smaller than θ), which will then be mapped into itself by ψ, by Lemma 14.7.

We may assume, without loss of generality, that J is symmetric about the origin, i.e., $J = (-b, b)$. The map G that does the job is given by

$$G(z) = \frac{(b^2+1)z}{z^2+1} .$$

As the reader may easily check as an exercise, G is indeed real-symmetric and univalent in $\mathbb{D}(J)$, it fixes the (boundary) points b and $-b$, and maps $\mathbb{D}(J)$ onto $\mathbb{C}(J)$.

Now, given $\tau > 0$ such that $D_{\omega(\tau)}(J) \subset \mathbb{D}(J)$, let us look for the smallest Poincaré neighborhood $D_\theta(J)$ that contains $G(D_{\omega(\tau)}(J))$. This is tantamount to finding the smallest angle under which a point on $\partial G(D_{\omega(\tau)}(J))$ views the interval $J = (-b, b)$. Given any $z \in \mathbb{C} \setminus \mathbb{R}$, the angle under which z views the interval $(-b, b)$ is $\arg(R(z))$, where

$$R(z) = \frac{z-b}{z+b} .$$

Now, an easy calculation shows that $R(F(z)) = -R(z)R(b^2 z)$, so that

$$\arg(R(F(z))) = \arg(-R(z)) + \arg(R(b^2 z)) . \tag{14.6}$$

For every $z \in \partial D_{\omega(\tau)}(J)$, we have $\arg(R(z)) = \omega(\tau)$ (which means in particular that $\arg(-R(z))$ is constant). Hence, for such boundary points, the left-hand side of (14.6) will be smallest when $\arg(R(b^2 z)))$ is minimal. This occurs at the points $\pm z_\tau = \pm ib/\tau \in \partial D_{\omega(\tau)}(J)$ (see Fig. 14.3). But

$$G(z_\tau) = \frac{(b^2+1)z_\tau}{z_\tau^2+1} = \frac{ib\tau(b^2+1)}{\tau^2-b^2} \in i\mathbb{R} .$$

Thus, the point $G(z_\tau)$ sits vertically above the origin, on the imaginary axis. Hence the angle θ under which this point views the interval $(-b, b)$ is twice the angle under which it views the interval $(0, b)$, and the latter has tangent equal to

$$\tau_* = \frac{b}{\mathrm{Im}\,(G(z_\tau))} = \frac{\tau^2-b^2}{\tau(1+b^2)} ,$$

Therefore $\theta = 2\arctan\tau_* = \omega(\tau_*)$. This establishes (14.5) and finishes the proof.

□

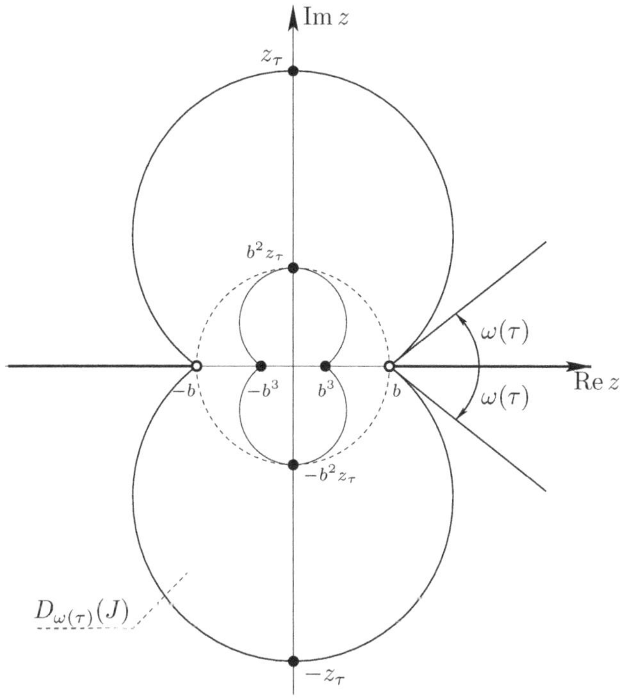

Fig. 14.3 The minimum value of $\arg(R(b^2 z))$ for $z \in \partial D_{\omega(\tau)}(J)$ is attained at the points $\pm z_\tau = \pm ib/\tau$

With this lemma at hand, we are now in a position to give a detailed proof of Lemma 14.6.

Proof of Lemma 14.6 We apply Lemma 14.8 with $J = (0, a)$, so that here $b = a/2$, and with $\phi = F$. Given $\theta > 0$ such that $D_\theta(J) \subset \mathbb{D}(J)$, let $\tau = \tan(\theta/2)$—hence, in the notation introduced above, we have $\theta = \omega(\tau)$. Note that we must have $\tau > a$ for this inclusion to hold. We may assume also that $\tau < 1$. According to Lemma 14.8, we have $F(D_\theta(J)) \subseteq D_{\theta_*}(J)$, where $\theta_* = \omega(\tau_*)$ and

$$\tau_* = \frac{4\tau^2 - a^2}{\tau(4 + a^2)} .$$

Now we have

$$\frac{\tau_*}{\tau} = 1 - \left(\frac{a}{\tau}\right)^2 \frac{\tau^2 + 1}{4 + a^2} > 1 - \frac{1}{2}\left(\frac{a}{\tau}\right)^2 .$$

Thus, if $\tau \geqslant a^{1/3}$ then

$$\frac{\tau_*}{\tau} > 1 - \frac{1}{2}a^{4/3}\,. \tag{14.7}$$

So now we know how to bound τ_* from below in terms of τ, but we need to translate this into a bound for θ_* in terms of θ. At this point, we simply observe that the function $h(x) = \arctan(x)/x$ is monotone decreasing for $x > 0$ (a calculus exercise). Hence we have

$$\frac{\theta_*}{\theta} = \frac{\omega(\tau_*)}{\omega(\tau)} = \frac{\arctan(\tau_*)}{\arctan(\tau)} = \frac{\tau_*}{\tau}\frac{h(\tau_*)}{h(\tau)} > \frac{\tau_*}{\tau}\,.$$

Combining this with (14.7), we deduce that

$$\theta_* > \left(1 - \frac{1}{2}a^{4/3}\right)\theta\,. \tag{14.8}$$

This estimate holds provided $\tau \geqslant a^{1/3}$, that is to say, provided $\theta \geqslant \theta(a)$, where

$$\theta(a) = 2\arctan\sqrt[3]{a}\,.$$

The latter function clearly satisfies $\theta(a) \to 0$ and $a/\theta(a) \to 0$ as $a \to 0$. In summary, we have just established what we wanted, namely, that if $\theta \geqslant \theta(a)$, then

$$F(D_\theta((0,a))) \subset D_{(1-a^{1+\delta})\theta}((0,a))\,,$$

where $\delta = 1/3$. This completes the proof of Lemma 14.6. □

We will also need to know some general facts about complex analytic maps that are very close to maps with a parabolic fixed-point. In other words, complex-analytic versions of the *almost parabolic maps* we encountered before.

Definition 14.3 Let $J \subseteq \mathbb{R}$ be an interval, and let $\theta > 0$. A holomorphic univalent map $\phi : D_\theta(J) \to \mathbb{C}$ is called *almost parabolic* if the following conditions are satisfied.

(a) ϕ is symmetric about the real axis.
(b) $\phi|J$ is monotone without fixed points.
(c) ϕ has positive Schwarzian derivative on J.
(d) $J \cap \phi(J)$ is non-empty.

If Δ_ϕ is the interval $J \setminus \phi(J)$, the largest $a = a(\phi) > 0$ such that $\phi^{a-1}(\Delta_\phi) \subseteq J$ is called the *length* of ϕ.

For our purposes, the most important example of a complex almost parabolic map is the inverse branch of a high first return f^{q_n} of a critical circle map f (or one of its renormalizations), in the situation where a_n is large, that is, the rotation number is "almost rational".

In the notation just introduced, the fundamental inequality of Yoccoz proved in Chap. 7 (see Lemma 7.3) can be restated as follows.

Lemma 14.9 (Yoccoz) *There exists $C_\sigma > 1$ such that for each $\phi \in \mathscr{F}_\sigma$ and for each $0 \leqslant j \leqslant a-1$ we have*

$$\frac{1}{C_\sigma m(j)^2} \leqslant |\phi^j(\Delta_\phi)| \leqslant \frac{C_\sigma}{m(j)^2},$$

where $m(j) = \min\{j+1, a-j\}$.

Given $0 < \sigma < 1$, we denote by $\mathscr{F}_\sigma$ the family of all complex almost parabolic maps ϕ such that $|\Delta_\phi| \geqslant \sigma|J|$ and $|\phi^{a-1}(\Delta_\phi)| \geqslant \sigma|J|$, and also normalized so that

$$[0,1] = \Delta_\phi \cup \phi(\Delta_\phi) \cup \cdots \cup \phi^{a-1}(\Delta_\phi).$$

Every element of the family $\mathscr{F}_\sigma$ whose length is sufficiently large has two fixed points, symmetric about the real axis. These fixed points are necessarily attracting, due to the positive Schwarzian property of the maps in $\mathscr{F}_\sigma$. More precisely, we have the following fact. Let us denote by $\mathbb{H}$ the upper half-plane.

Lemma 14.10 *Given $0 < \sigma < 1$, there exist $C > 0$ and $a_0 > 0$ such that, if $\phi \in \mathscr{F}_\sigma$ has length $a = a(\phi) > a_0$, then there exist two attracting fixed points $z_+ \in \mathbb{H} \cap \mathrm{dom}(\phi)$ and $z_- = \overline{z}_+$ with*

$$\frac{1}{Ca} \leqslant \mathrm{Im} z_+ \leqslant \frac{C}{a}.$$

Moreover, if $|z - z_+| \leqslant C/a$ then $|z - \phi(z)| \leqslant C/a^2$.

Proof The proof follows from Yoccoz's Lemma, the saddle-node bifurcation and a normality argument. □

The family $\mathscr{F}_\sigma$ is normal in the sense of Montel, and every limit map *not in* $\mathscr{F}_\sigma$ is a map with a parabolic (indifferent) fixed point on the real axis, whose multiplier is necessarily equal to one.

Lemma 14.11 *Given a compact set $W \subset \mathbb{H}$ and an open set $D \supset [0,1]$ in the plane, there exist $N_* > 0$, $\theta_* > 0$ and $a_* > 0$ with the following property. For each $\phi \in \mathscr{F}_\sigma$ such that $a(\phi) \geqslant a_*$ and $\theta(\phi) < \theta_*$, the domain of ϕ contains W, and for each $z \in W$ there exists $n < N_*$ such that $\phi^n(z) \in D$.*

Proof If the statement is false, we find sequences $N_k \to \infty$, $a_k \to \infty$ and $\theta_k \to 0$, maps $\phi_k \in \mathscr{F}_\sigma$ with $\theta(\phi_k) = \theta_k$ and $a(\phi_k) = a_k$ (whose domains contain W), and points $z_k \in W$ such that $\phi_k^n(z_k)$, whenever defined, does not belong to D for all $n \leqslant N_k$. Since $\mathscr{F}_\sigma$ is normal and W is compact, we may assume that the sequence ϕ_k converges uniformly on compact subsets of $\mathbb{H}$ to a map $\phi : \mathbb{H} \to \mathbb{H}$ and that $z_k \to z \in W$. Applying Lemma 2.4 to each ϕ_k, we deduce that ϕ has a fixed-point

$x_0 \in [0, 1]$. By the Denjoy–Wolff theorem, $\phi^n(z) \to x_0$ as $n \to \infty$. Hence there exists N such that $\phi^N(z) \in D$. But then $\phi_k^N(z_k) \in D$ also, for all sufficiently large k, a contradiction. □

In what follows, we will fix $f : S^1 \to S^1$ as in the statement of Theorem 14.4. We will assume wherever necessary that f is normalized so that its critical point is $c = 1 \in S^1$. Since f is real-analytic, it extends to a holomorphic map $f : A_R \to \mathbb{C}/\mathbb{Z}$, where A_R is the annulus $\{z \in \mathbb{C}/\mathbb{Z} : |\mathrm{Im}\, z| < R\}$. Making R smaller if necessary, we may assume that f has no critical points outside S^1. Using again Koebe's distortion theorem, it is easy to see that there exists $R_0 > 0$ such that, if $z \in S^1$ and $f(z)$ is at a distance $> R_0$ from the critical value of f, then the inverse branch f^{-1} which maps $f(z)$ back to z is well-defined and univalent on the disk $D(f(z), R_0)$.

On an intuitive level, the key to the proof of Theorem 14.4 is to show that for all sufficiently large n the n-th renormalization of f satisfies an inequality of the form $|\mathscr{R}^n(f)(z)| \geqslant C|z|^3$ on a neighborhood of the origin, where C is a universal constant. Thus, the relevant inverse branches of $\mathscr{R}^n(f)$ behave as cube roots, mapping a large disk about the origin well within itself, giving rise to a holomorphic pair. The proof depends on Proposition 14.1 stated above.

For our purposes, the main consequence of Lemma 14.6 is the following.

Lemma 14.12 *For each $n \geqslant 1$ there exist $K_n \geqslant 1$ and $\theta_n > 0$, with $K_n \to 1$ and $\theta_n \to 0$ as $n \to \infty$, such that for all $\theta \geqslant \theta_n$ and all $1 \leqslant j \leqslant q_{n+1}$ the inverse branch f^{-j+1} mapping $f^j(I_n)$ back to $f(I_n)$ is well-defined over $D_\theta(f^j(I_n))$ and maps this neighborhood univalently into $D_{\theta/K_n}(f(I_n))$.*

Proof Let $d_n = \max_{1\leqslant j\leqslant q_{n+1}} |f^j(I_n)|$; from the real bounds, these numbers go to zero exponentially with n. Take $\delta > 0$ as in Lemma 14.6, and let K_n be given by

$$K_n^{-1} = \prod_{j=1}^{q_{n+1}} \left(1 - |f^j(I_n)|^{1+\delta}\right) .$$

Then define $\theta_n = K_n\theta(d_n)$, where $\theta(\cdot)$ is the function in Lemma 14.6. Note that

$$\log K_n \leqslant C \sum_{j=1}^{q_{n+1}} |f^j(I_n)|^{1+\delta} \leqslant C d_n^\delta .$$

Therefore $K_n \to 1$ and $\theta_n \to 0$ as required. Also, $d_n/\theta_n \to 0$.

Now fix j as in the statement and suppose $\theta \geqslant \theta_n$. Define inductively $\vartheta_0 = \theta$ and $\vartheta_{i+1} = (1 - |f^{j-i}(I_n)|^{1+\delta})\vartheta_i$ for $i = 0, 1, \ldots, j-2$, and note that $\vartheta_{j-1} \geqslant \theta/K_n$. Moreover,

$$\mathrm{diam}\left(D_{\vartheta_i}\left(f^{j-i}(I_n)\right)\right) \leqslant 2\,\frac{|f^{j-i}(I_n)|}{\sin\vartheta_i} \leqslant \frac{C' d_n}{\theta(d_n)} \ll R_0 .$$

Therefore f^{-1} is well-defined and univalent over $D_{\vartheta_i}(f^{j-i}(I_n))$, and by Lemma 14.6 we have the inclusion $f^{-1}(D_{\vartheta_i}(f^{j-i}(I_n))) \subseteq D_{\vartheta_{i+1}}(f^{j-i-1}(I_n))$. This completes the proof. □

Remark 14.6 The same result holds if we replace I_n by any interval $J \supseteq I_n$ such that the map $f^{q_{n+1}-1} : f(J) \to f^{q_{n+1}}(J)$ is a diffeomorphism.

We will need four lemmas concerning the sequence $\{D_m\}$ introduced earlier. The first is an easy consequence of Lemma 14.12 and the above remark.

Lemma 14.13 *There exists $m_0 \geqslant 1$ such that for all $m \geqslant m_0$ the inverse branch f^{-q_m+1} taking $f^{q_m}(I_m)$ back to $f(I_m)$ is well-defined and univalent in D_m, and*

$$\frac{\operatorname{diam}(f^{-q_m+1}(D_m))}{|I_m|} \leqslant C\,\frac{\operatorname{diam}(f(D_m))}{|f(I_m)|}\,.$$

The second is the analogue of Yampolsky (1999, Lem. 4.1).

Lemma 14.14 *There exist $\varepsilon_1 > 0$ and $m_1 \geqslant m_0$ such that for all $m \geqslant m_1$ and each $w \in f^{-q_{m+1}}(D_m) \setminus D_m$ we have (a)* $\operatorname{dist}(w, I_m) \leqslant C|I_m|$ *and (b) for each $x \in I_m$, $\varepsilon_1 < |\arg(w - x)| < \pi - \varepsilon_1$.*

Proof The same proof given in Yampolsky (1999) applies here. Invariance of Poincaré neighborhoods is replaced by *quasiinvariance*, using Lemma 14.12. □

The third is the analogue of Yampolsky (1999, Lem. 4.4). It provides us with the tools we need for the inductive step in the proof of Proposition 14.1. The situation is depicted in Fig. 14.4.

Lemma 14.15 *There exist $\varepsilon_2 > 0$ and $m_2 \geqslant m_0$ such that the following holds for all $m \geqslant m_2$. Let $\zeta \in D_m \setminus D_{m+1}$ be a point not on the circle, and let $\zeta' = f^{-q_m}(\zeta)$ and $\zeta'' = f^{-q_{m+2}}(\zeta')$. Then we have either $\zeta'' \in D_{m+1}$, or else* $\operatorname{dist}(\zeta'', I_{m+1}) \leqslant C|I_m|$ *and $\varepsilon_2 < \arg(\zeta'' - x) < \pi - \varepsilon_2$ for all $x \in I_m \cup I_{m+1}$.*

Proof Once again, the proof given by Yampolsky (1999, Lem. 4.4) can be repeated here, *mutatis mutandis.* □

Notation Given a point $\zeta \in \mathbb{C}$ and an interval $J = (a, b) \subseteq \mathbb{R}$, we denote by $\operatorname{angle}(\zeta, J)$ the smallest of the angles $\pi - \arg(\zeta - a)$ and $\arg(\zeta - b)$.

The fourth is a consequence of de Faria and de Melo (2000, Lem. 2.5).

Lemma 14.16 *There exist universal constants $N_* > 0$ and $a_* > 0$ and some $m_3 > 0$ with the following property. For all $m \geqslant m_3$ such that $a_{m+1} > a_*$ and each $w \in V_m = f^{-q_{m+1}}(D_m) \setminus D_m$, there exists $1 < i < N_*$ such that the iterate $(f^{-q_{m+1}})^i(w)$ is well-defined and belongs to D_m.*

Proof Lift $f^{-q_{m+1}}$ to the real line and normalize it so that $I_m \setminus I_{m+2}$ becomes the interval $[0, 1]$ to get an almost parabolic map ϕ_m. Note that ϕ_m belongs to the normal family $\mathscr{F}_\sigma$ introduced earlier, for some σ depending only on the real bounds. Let

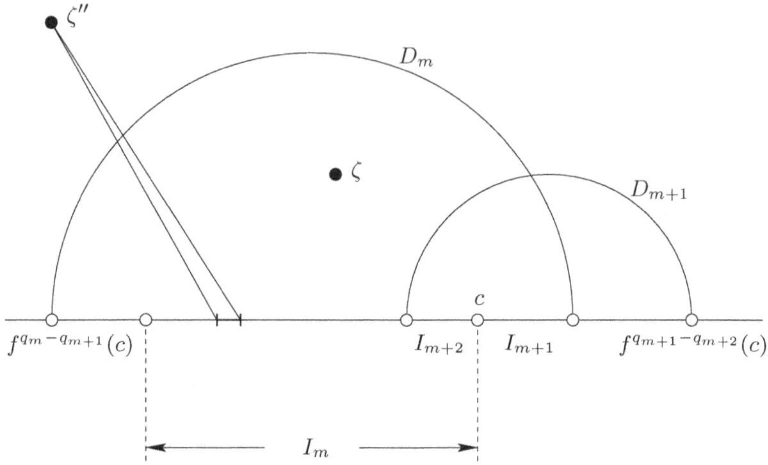

Fig. 14.4 Poincaré-neighborhood trapping in action

W_m be the image of V_m under such normalization. It is an easy matter to check that $f^{-q_{m+1}}(D_m) \cap S^1 \subseteq D_m \cap S^1$, so that V_m does not intersect S^1, and that $W_m^+ = W_m \cap \mathbb{H}$ is compactly contained in $\mathbb{H}$. Therefore, by Lemma 14.14, there exists a fixed compact set $W \subset \mathbb{H}$ such that $W_m^+ \subseteq W$ for all sufficiently large m. Similarly, the normalized copies of D_m contain a fixed open set $D \supseteq [0, 1]$ for all sufficiently large m. Hence we can take N_* and a_* as given by Lemma 14.11. □

Proof of Proposition 14.1 We will start with a point z in the disk D_{n-N}. For the argument to work, n will have to be sufficiently large. We start taking $n > N + \max\{m_1, m_2, m_3\}$, where m_1, m_2 and m_3 are given respectively by Lemmas 14.14 to 14.16. Let us denote by J_{-i} the interval $f^{q_{n+1}-i}(I_n)$. Also, given z, let $z_{-i} = f^{-i}(z)$ be the corresponding pre-images of z.

The proof runs by finite induction in the range $n - N \leqslant m \leqslant n$. Let m be the largest with the property that $z \in D_m$, and keep in mind that $\operatorname{dist}(z, I_n) \asymp |I_m|$. Consider those moments $i_1 < i_2 < \cdots < i_\ell$ in the backward orbit $\{J_{-i}\}$ before the first return to I_{m+1} such that $J_{-i_k} \subseteq I_m$. Then, there are two possibilities.

The first possibility is that $z_{-i_\ell} \notin D_m$. In this case there exists a smallest $k \leqslant \ell$ such that $z_{-i_s} \notin D_m$ for $s = k, k+1, \ldots, \ell$. We claim that $|J_{-i_k}| \asymp |I_n|$. This is clear from the real bounds if $\ell = a_{m+1} \leqslant a_*$, where a_* is given by Lemma 14.16. If on the other hand $\ell > a_*$, then again by Lemma 14.16 we must have $\ell - k < N_*$, and the claim follows from Exercise 6.3 (or the original result in de Faria and de Melo 2000, Lem. 2.2). Therefore, by Lemma 14.14,

$$\frac{\operatorname{dist}(z_{-i_k}, J_{-i_k})}{|J_{-i_k}|} \leqslant C \frac{|I_m|}{|J_{-i_k}|} \leqslant C' \frac{\operatorname{dist}(z, I_n)}{|I_n|}. \tag{14.9}$$

Moreover, angle $(z_{-i_k}, J_{-i_k}) \geqslant \varepsilon_1$, so there exists $\theta = \theta(\varepsilon_1, N)$ such that $z_{-i_k} \in D_\theta(J_{-i_k})$. Now, if n is sufficiently large, $\theta_n < \theta$ and we can use Lemma 14.12 to get that $z_{-q_{n+1}+1} \in D_{\theta/K_n}(f(I_n))$. This gives us

$$\frac{\operatorname{dist}(z_{-q_{n+1}+1}, f(I_n))}{|f(I_n)|} \leqslant C'' K_n \frac{\operatorname{dist}(z_{-i_k}, J_{-i_k})}{|J_{-i_k}|}, \tag{14.10}$$

and this together with (14.9) yields the proposition in this case.

The second possibility is that $\zeta = z_{-i_\ell} \in D_m$, and we can assume that $\zeta \notin D_{m+1}$ (otherwise the induction step is complete). In this case, consider $\zeta' = f^{-q_m}(\zeta)$ and $\zeta'' = f^{-q_{m+2}}(\zeta')$ and the corresponding interval $J'' = f^{-q_m - q_{m+2}}(J_{-i_\ell})$, and apply Lemma 14.15. Then either $\zeta'' \in D_{m+1}$, in which case the induction step is complete, or else dist $(\zeta'', I_{m+1}) \leqslant C|I_m|$ and angle $(\zeta'', J'') \geqslant \varepsilon_1$, in which case we can apply the same argument leading to (14.9) and (14.10).

If the backward orbit survives all the steps of the induction, this means that in the end $z_{-q_{n+1}+q_{n-1}} \in D_{n-1}$. By Lemma 14.13, the image of D_{n-1} under $f^{-q_{n-1}+1}$ has diameter comparable to $|f(I_n)|$, so the first member of (14.10) is simply bounded by an absolute constant. So in any case we have (14.4). □

Proof of Theorem 14.4 First we remark that, since f is a cubic critical circle map, there exists a neighborhood Ω of the critical point of f such that the restriction $f : \Omega \to f(\Omega)$ is of the form $f = \psi \circ Q \circ \phi$ where ψ and ϕ are univalent maps with universally bounded distortion, with $\phi(0) = 0$, and Q is the map $z \mapsto z^3$.

Let B_1, B_2 be the constants of Proposition 14.1 and let us fix a large integer N. How large N must be will be determined in the course of the argument. By Proposition 14.1, if $n \geqslant n(N)$ then inequality (14.4) holds for all $z \in \Omega_{n,N}$. Making n larger still if necessary, we have $f^{-q_{n+1}+1}(D_{n-N}) \subseteq f(\Omega)$. By the above remark, the branch of f^{-1} mapping $f(I_n)$ back to I_n is the composition of three maps: a univalent map fixing zero, a cube root, and another univalent map with bounded distortion. Using this fact and inequality (14.4), we have

$$\frac{\operatorname{diam}(f^{-q_{n+1}}(D_{n-N}))}{|I_n|} \leqslant C \sqrt[3]{B_1 \frac{\operatorname{diam}(D_{n-N})}{|I_n|} + B_2}, \tag{14.11}$$

for some universal constant $C > 0$.

Now, as we know from the real bounds (Theorem 6.3), there exist universal constants $K_2 > K_1 > 1$ such that, for all sufficiently large n,

$$K_1^N \leqslant \frac{\operatorname{diam}(D_{n-N})}{|I_n|} \leqslant K_2^N. \tag{14.12}$$

If $N \geqslant 1$ is the smallest integer greater than $3 \log(2C\sqrt[3]{B_1 + B_2})/2 \log K_1$, we can check from (14.11) and the first inequality in (14.12) that

$$\frac{\operatorname{diam}(f^{-q_{n+1}}(D_{n-N}))}{|I_n|} < \frac{1}{2}\frac{\operatorname{diam}(D_{n-N})}{|I_n|}.$$

Note that N depends only on B_1, B_2, K_1, C, and is therefore universal. From these facts, we see at once that for all sufficiently large n the topological disk $f^{-q_{n+1}}(D_{n-N})$ is compactly contained in D_{n-N}, and moreover

$$\operatorname{mod}\,(D_{n-N} \setminus f^{-q_{n+1}}(D_{n-N})) \geqslant \mu\,, \tag{14.13}$$

where $\mu > 0$ is a universal constant. With these basic geometric bounds at hand, we can easily construct the holomorphic pair to which $\mathscr{R}^n(f)$ extends, in the following way. For more details on this construction, see de Faria (1999, Section 4).

First, let $\hat{f}$ be the standard lift of f to the real line. For each n, let $\lambda_n = T^{-p_n} \circ \hat{f}^{q_n}(0)$ and denote by Λ_n the linear map $x \mapsto \lambda_n x$. Take the topological disks $\widetilde{\mathscr{V}}_n = D_{n-N}$ and $\widetilde{\mathscr{O}}_{\eta_n} = f^{-q_{n+1}}(D_{n-N})$ in the cylinder and consider their lifted and normalized copies in $\mathbb{C}$, namely $\mathscr{V}_n = \Lambda_n^{-1}(\exp^{-1}(\widetilde{\mathscr{V}}_n))$ and $\mathscr{O}_{\eta_n} = \Lambda_n^{-1}(\exp^{-1}(\widetilde{\mathscr{O}}_{\eta_n}))$ (here, $\exp^{-1}$ denotes the inverse branch of the exponential that maps the critical point $c = 1 \in S^1$ of f to the origin). Then consider the map

$$\eta_n = \Lambda_n^{-1} \circ T^{-p_{n+1}} \circ \hat{f}^{q_{n+1}} \circ \Lambda_n \,:\, \mathscr{O}_{\eta_n} \to \mathbb{C}\,.$$

The geometric estimates proved above show that $\mathscr{O}_{\eta_n}$ is compactly contained in $\mathscr{V}_n$, while $\eta_n(\mathscr{O}_{\eta_n}) \subseteq \mathscr{V}_n$ holds by construction. We define the domains $\mathscr{O}_{\xi_n}$, $\mathscr{O}_{\nu_n}$ and the maps $\xi_n : \mathscr{O}_{\xi_n} \to \mathbb{C}$, $\nu_n : \mathscr{O}_{\nu_n} \to \mathbb{C}$ in a similar way. We obtain in this fashion a holomorphic pair Γ_n whose underlying critical commuting pair is precisely the n-th renormalization $\mathscr{R}^n(f)$. In addition, (14.13) shows that the conformal modulus of Γ_n is bounded from below by μ. Finally, it is straightforward to check that all the above topological disks have piecewise analytic boundaries, consisting of finitely many analytic arcs meeting at definite angles, so Γ_n has geometric boundaries. This completes the proof of Theorem 14.4. □

14.6 McMullen's Dynamic Inflexibility Theorem

Let f and g be two real-analytic critical circle maps and let h be a quasisymmetric conjugacy between f and g, mapping the critical point c_f of f to the critical point c_g of g. Suppose h is $C^{1+\epsilon}$ at the critical point c_f, for some $\epsilon > 0$. Then, it is not difficult to prove (using the real bounds) that the C^0 distance between $\mathscr{R}^n f$ and $\mathscr{R}^n g$ converges to zero exponentially fast as $n \to \infty$ (and this, as we have already seen, at least in the bounded type case leads to C^1 rigidity). Now, one way to guarantee that h is $C^{1+\epsilon}$ at the critical point is if we know that h extends to a quasiconformal homeomorphism H (conjugating, say, the holomorphic extensions of f and g on a small neighborhood of their critical points) which happens to be $C^{1+\alpha}$-*conformal* at the critical point c_f, in the following sense.

Definition 14.4 We say that a map $\phi : \widehat{\mathbb{C}} \to \widehat{\mathbb{C}}$ is $C^{1+\alpha}$-*conformal* at $p \in \widehat{\mathbb{C}}$ (for some $\alpha > 0$) if the complex derivative $\phi'(p)$ exists and we have

$$\phi(z) = \phi(p) + \phi'(p)(z - p) + O(|z - p|^{1+\alpha})$$

for all z near p.

McMullen (1996) developed a powerful theory that yields in particular a criterion for a conjugacy between two holomorphic dynamical systems to be $C^{1+\alpha}$-conformal at a point. His definition of holomorphic dynamical system is very broad, encompassing rational or transcendental maps, Kleinian groups, etc, as well as all possible geometric limits of such systems.

In order to state McMullen's criterion, we need some preparatory definitions. Our exposition here is borrowed from de Faria and de Melo (2000, §7).

Let us denote by $\mathscr{V}(\widehat{\mathbb{C}} \times \widehat{\mathbb{C}})$ the set of all analytic hypersurfaces of $\widehat{\mathbb{C}} \times \widehat{\mathbb{C}}$. We topologize $\mathscr{V}(\widehat{\mathbb{C}} \times \widehat{\mathbb{C}})$ as follows. If $F \subseteq \widehat{\mathbb{C}} \times \widehat{\mathbb{C}}$ is a hypersurface, its boundary $\partial F = \overline{F} \setminus F$ is closed in $\widehat{\mathbb{C}} \times \widehat{\mathbb{C}}$. Hence, given $F \in \mathscr{V}(\widehat{\mathbb{C}} \times \widehat{\mathbb{C}})$ and a sequence $F_i \in \mathscr{V}(\widehat{\mathbb{C}} \times \widehat{\mathbb{C}})$, declare $F_i \to F$ if

(a) $\partial F_i \to \partial F$ in the Hausdorff metric on closed subsets of $\widehat{\mathbb{C}} \times \widehat{\mathbb{C}}$;
(b) For each open set $U \subseteq \widehat{\mathbb{C}} \times \widehat{\mathbb{C}}$ there exist $f, f_i : U \to \mathbb{C}$ such that $U \cap F = f^{-1}(0)$, $U \cap F_i = f_i^{-1}(0)$, each f, f_i vanishes to order one on F, F_i respectively, and the sequence f_i converges uniformly to f on compact subsets of U.

Define a set to be closed in $\mathscr{V}(\widehat{\mathbb{C}} \times \widehat{\mathbb{C}})$ if it contains the limits of all its convergent sequences. As McMullen shows in McMullen (1996, Ch. 9), the space $\mathscr{V}(\widehat{\mathbb{C}} \times \widehat{\mathbb{C}})$ with this topology is separable and metrizable.

Definition 14.5 A *holomorphic dynamical system* is a subset $\mathscr{F} \subseteq \mathscr{V}(\widehat{\mathbb{C}} \times \widehat{\mathbb{C}})$. The elements of $\mathscr{F}$ are its *holomorphic relations*.

One is primarily interested in *closed* holomorphic dynamical systems, in other words, those which are closed subsets of $\mathscr{V}(\widehat{\mathbb{C}} \times \widehat{\mathbb{C}})$. The *geometric topology* on the space of all closed holomorphic dynamical systems is by definition the Hausdorff topology on the space of closed subsets of $\mathscr{V}(\widehat{\mathbb{C}} \times \widehat{\mathbb{C}})$. As proved in McMullen (1996, Ch. 9), the geometric topology is typically non-Hausdorff (hence non-metrizable), but it is always sequentially compact.

We also need the following notions introduced by McMullen.

1. *Deep point* Given a compact set $\Lambda \subseteq \mathbb{C}$ and a positive number δ, we say that a point $p \in \Lambda$ is a δ-*deep point* of Λ if for every $r > 0$ the largest disk contained in $D(p, r)$ which does not intersect Λ has radius $\leqslant r^{1+\delta}$.

2. *Saturation* Given a holomorphic dynamical system $\mathscr{F}$, we define its saturation $\mathscr{F}^{\mathrm{sat}}$ to be the closure in $\mathscr{V}(\widehat{\mathbb{C}}\times\widehat{\mathbb{C}})$ of the set whose elements are the intersections $F \cap U$, where $F \in \mathscr{F}$ and $U \subseteq \widehat{\mathbb{C}} \times \widehat{\mathbb{C}}$ is open.
3. *Nonlinearity* A holomorphic dynamical system $\mathscr{F} \subseteq \mathscr{V}(\widehat{\mathbb{C}} \times \widehat{\mathbb{C}})$ is said to be non-linear if it does not leave invariant a parabolic line field in $\widehat{\mathbb{C}}$.
4. *Twisting* A (closed) holomorphic dynamical system $\mathscr{F} \subseteq \mathscr{V}(\widehat{\mathbb{C}} \times \widehat{\mathbb{C}})$ is said to be twisting if every holomorphic dynamical system quasiconformally conjugate to $\mathscr{F}$ is non-linear.
5. *Uniform twisting* A family $\{\mathscr{F}_\alpha\}$ of holomorphic dynamical systems is said to be uniformly twisting if every geometric limit of the family of saturations $\{\mathscr{F}_\alpha^{\mathrm{sat}}\}$ is a twisting dynamical system.
6. *The family* $(\mathscr{F}, \Lambda)$ Given $\mathscr{F} \subseteq \mathscr{V}(\widehat{\mathbb{C}} \times \widehat{\mathbb{C}})$ and a compact set Λ in the Riemann sphere, we define a family $(\mathscr{F}, \Lambda)$ of holomorphic dynamical systems in the following way. For each baseframe ω in the convex-hull $\mathrm{ch}(\Lambda)$ of Λ in hyperbolic 3-space, let T_ω be the fractional linear transformation that sends ω onto the standard baseframe ω_0 at $(0, 1) \in \mathbb{C} \times \mathbb{R}_+ \equiv \mathbb{H}^3$. Define $(\mathscr{F}, \omega)$ to be the dynamical system $T_\omega^*(\mathscr{F})$, the pull-back of $\mathscr{F}$ by T_ω. Then let $(\mathscr{F}, \Lambda)$ be the family of all $(\mathscr{F}, \omega)$ as ω ranges through the baseframes in $\mathrm{ch}(\Lambda)$.

Now we have everything we need to state McMullen's dynamic inflexibility theorem. The proof is given in McMullen (1996, p. 166).

Theorem 14.5 (Dynamic Inflexibility) *Let $\mathscr{F} \subseteq \mathscr{V}(\widehat{\mathbb{C}} \times \widehat{\mathbb{C}})$ be a holomorphic dynamical system and let $\Lambda \subseteq \widehat{\mathbb{C}}$ be a compact set. If $(\mathscr{F}, \Lambda)$ is uniformly twisting and $\phi : \widehat{\mathbb{C}} \to \widehat{\mathbb{C}}$ is a K-quasiconformal conjugacy between $\mathscr{F}$ and another holomorphic dynamical system $\mathscr{F}'$, then for each δ-deep point $p \in \Lambda$ the map ϕ is $C^{1+\alpha}$ conformal at p, for some $\alpha > 0$.* □

14.7 Proof of Exponential Convergence

With McMullen's dynamic inflexibility theorem at hand, we prove Theorem 13.1. By the complex bounds, every sufficiently high renormalization of a real-analytic critical commuting pair extends to a holomorphic commuting pair with good geometric control. Moreover, a quasisymmetric conjugacy between two such renormalized critical commuting pairs (mapping critical point to critical point) extends to a quasiconformal conjugacy between the corresponding renormalized holomorphic commuting pairs, by the pull-back argument. All one has to do, then, is to prove two things: (a) that the critical point of a holomorphic commuting pair is δ-deep for some $\delta > 0$; and (b) that the full holomorphic dynamical system generated by a holomorphic commuting pair is uniformly twisting in its limit set. The precise statements—modulo the notion of *good geometric control*, which we do not define here—are as follows.

Theorem 14.6 (Deep Critical Point) *Let Γ be a holomorphic pair with arbitrary rotation number and limit set $\mathcal{K}_\Gamma$. Then there exists $\delta > 0$ such that the critical point of Γ is a δ-deep point of $\mathcal{K}_\Gamma$.*

Theorem 14.7 (Small Limit Sets Everywhere) *Let Γ be a holomorphic pair with good geometric control and irrational rotation number of bounded type, and let $\mathcal{K}_\Gamma$ be its limit set. Then for each $z_0 \in \mathcal{K}_\Gamma$ and each $r > 0$ there exists a pointed domain (U, y) with $|z_0 - y| \asymp r$ and $\mathrm{diam}(U) \asymp r$, and there exist some iterate of Γ mapping (U, y) onto a pointed domain $(V, 0)$ univalently with bounded distortion. In particular, U contains a conformal copy of some renormalization of Γ whose limit set has size commensurable with r.*

These results are exact analogues of results obtained by McMullen in the context of (bounded-type, infinitely renormalizable) quadratic-like maps. Used in combination with Theorem 14.5, they yield the exponential convergence of renormalizations of Theorem 13.1 in the bounded type case. Theorem 14.6 was proved in de Faria and de Melo (2000) as stated here, without any assumption on the rotation number (other than being irrational). In that same paper, Theorem 14.7 is stated and proved under the assumption that the rotation number is an irrational of bounded combinatorial type. This assumption was removed by Khmelev and Yampolsky (2006). When the sequence of partial quotients of the continued-fraction development of the rotation number is *unbounded*, renormalization orbits may accumulate on commuting pairs having a fixed point (being in particular non-renormalizable). Such fixed point is necessarily *parabolic* (with multiplier one), since the limiting pair is accumulated by pairs with no fixed points. Roughly speaking, the idea developed by Khmelev and Yampolsky was to apply the theory of *parabolic bifurcations* (see Douady 1994; Shishikura 1998, 2000 and references therein) to holomorphic commuting pairs, in order to understand the geometry of the domain of definition of pairs with arbitrarily small rotation number. With this at hand, the authors were able in the end to adapt, to the unbounded type case, the proof of Theorem 13.1 for the bounded type case explained above, see Khmelev and Yampolsky (2006, secs. 6 and 7).

14.8 Hyperbolicity of Renormalization

In the previous section we have finally established Theorem 13.1, which assures exponential convergence of renormalization of *real-analytic* critical commuting pairs with the same irrational rotation number and the same odd type at the critical point. As explained in Chap. 13, this dynamical picture can be promoted to critical commuting pairs with a *finite degree of smoothness*, as in Theorems 13.2 and 13.3. These two results can be regarded as the state of the art concerning exponential convergence of renormalization of critical circle maps (with a single critical point). As explained in Chap. 10, they imply the rigidity results Theorems 10.1 and 10.2.

At this point, one would like to discuss the *hyperbolicity* of renormalization (in the sense of Smale, i.e., uniform contraction/expansion on the tangent bundle). To give a meaning to this problem, one first needs to endow the phase-space of the renormalization operator with a smooth structure (a Banach manifold structure) on which $\mathscr{R}$ is (Fréchet) differentiable. As it turns out, this is a difficult problem that obstructs the hyperbolicity discussion directly in the space of critical commuting pairs. To overcome this problem, at least for real-analytic pairs, a crucial idea in this area was developed by Yampolsky (2002, 2003). Roughly speaking, Yampolsky's idea was to replace the renormalization operator $\mathscr{R}$, acting on the space of commuting pairs, with an analytic operator, the *cylinder* renormalization operator, defined on a complex-analytic Banach manifold. This operator was constructed in Yampolsky (2002, Section 7), while hyperbolicity of periodic orbits and the construction of the corresponding stable manifolds were given in Yampolsky (2002, secs. 8 and 9). Finally, hyperbolicity of the whole horseshoe-like attractor for the cylinder renormalization operator was obtained in Yampolsky (2003) and re-obtained in Khmelev and Yampolsky (2006, Section 8).

It would take us too far afield to discuss the aforementioned hyperbolicity in any real depth. Therefore, in the discussion to follow, we will merely describe the two main renormalization schemes used by Yampolsky: *cylindrical renormalization* (roughly described above) and *parabolic renormalization*. The latter is required to treat the unbounded type case.

14.8.1 Cylindrical and Parabolic Renormalizations

We forewarn the reader that what follows is a simplified description of the tools introduced by Yampolsky in the study of renormalization of critical circle maps covering the *unbounded type* case. In this short subsection we can hardly do justice to the wealth of ideas involved, and no details are given. Indeed, in order to make the discussion complete, we would need to go way beyond the scope of this book. For the most part, we conform with the notation used by Yampolsky in his papers.

As we have seen earlier in this book, the process of renormalizing a circle map around a given point p requires us to cut the circle at two consecutive closest returns of the orbit of p to p and then consider the first return map to the resulting interval. If we try to glue the interval in question to get a new smooth circle and a new circle map, we find that there is no canonical way of identifying such smooth boundaryless one-dimensional manifold with the standard affine unit circle $\boldsymbol{S}^1 = \mathbb{R}/\mathbb{Z}$. Hence, by using this approach, we are not able to define the renormalization operator on a space of circle maps, but have instead to deal with commuting pairs.

A different, clever procedure was introduced by Yampolsky (2002) to circumvent this difficulty in the analytic case. This procedure is called *cylinder renormalization*, and the rough idea is as follows. An analytic critical circle map f has a holomorphic extension to a neighborhood of $\boldsymbol{S}^1$ inside the cylinder $\mathbb{C}/\mathbb{Z}$. A sufficiently deep renormalization (without rescaling) of f around its critical point c is given by a pair of the form $(f^{q_n}, f^{q_{n+1}})$ which extends to a holomorphic commuting pair (η, ξ) in a small neighborhood of c. If the partial quotient a_{n+1} of the continued fraction expansion of $\rho(f)$ is very large, then η is an almost parabolic map, and as such it has two repelling fixed points, symmetric about the unit circle. Joining then by a simple smooth arc ℓ, we look at the "crescent" region bounded by the closed curve $\ell \cup \eta(\ell)$ and consider the first return map to this region. Glueing the two arcs by η, we get a new cylinder C_f bi-holomorphically equivalent to $\mathbb{C}/\mathbb{Z}$, and through this equivalence the first return map to C_f becomes a new analytic map defined on a new neighborhood of S^1 inside $\mathbb{C}/\mathbb{Z}$. What makes this procedure work is the fact that the conformal identification of C_f with the standard cylinder $\mathbb{C}/\mathbb{Z} \simeq \mathbb{C}^*$ is unique up to post-composition with a linear map of the form $z \mapsto \lambda z$ (a conformal automorphism of $\mathbb{C}^*$); it is therefore unique once we normalize one point—say, by sending c to $1 \in S^1$). *In this way, we get a renormalization operator acting directly on cylinder maps (rather than on holomorphic commuting pairs).*

However, even with the cylinder renormalization operator in place, we still have problems when trying to use, say, compactness arguments. When the rotation number is of unbounded type, holomorphic pairs arising as limits along a subsequence of renormalizations will develop parabolic fixed points, and these limits are not renormalizable in the sense we defined. To circumvent this difficulty, Yampolsky borrowed the idea of *parabolic renormalization* from the works of Douady (1987) and Shishikura (1998, 2000), adapting it to the context of critical circle maps. Parabolic renormalization can be viewed as a limiting case of cylindrical renormalization. Conversely, and perhaps more surprisingly, cylindrical renormalization can be thought of as a natural unfolding of parabolic renormalization as one goes through a parabolic bifurcation (say in a one-parameter family of maps). Due to this, it is more convenient to first describe parabolic renormalization, and this is what we will do next.

14.8.1.1 Parabolic Renormalization: Fatou Coordinates

In order to define parabolic renormalization, we first need to introduce a few facts concerning the local structure of a holomorphic map in a neighborhood of a parabolic fixed point. Rather than doing this in full generality, we only examine the case of specific interest to us.

Let $\phi : W \to \mathbb{C}$ be a holomorphic univalent map defined on a topological disk symmetric about the real axis, and suppose ϕ is a Epstein map, so that ϕ maps $W^+ = W \cap \mathbb{C}^+$ (respectively $W^- = W \cap \mathbb{C}^-$) onto $\mathbb{C}^+$ (respectively $\mathbb{C}^-$). We assume that ϕ has a parabolic fixed point $p \in \mathbb{R}$, so that locally around p the map ϕ can be written as $\phi(z) = z + a(z-p) + O((z-p)^2)$. Then one can show (see Yampolsky 2001, p. 554) the following:

(i) There exist symmetric topological disks $U^A, U^R \subset \mathbb{C}$ with $\overline{U^A} \cap \overline{U^R} = \{p\}$ such that $U^A \cup U^R$ is a punctured neighborhood of p and we have

$$\phi(\overline{U^A}) \subset U^A \cup \{p\}\,, \quad \phi^{-1}(\overline{U^R}) \subset U^R \cup \{p\}\,,$$

as well as

$$\bigcap_{j=0}^{\infty} \phi^j(\overline{U^A}) = \{p\} = \bigcap_{j=0}^{\infty} \phi^{-j}(\overline{U^R}) .$$

(ii) There exist holomorphic univalent maps $\Phi^A : U^A \to \mathbb{C}$ and $\Phi^R : U^R \to \mathbb{C}$ such that

$$\Phi^A(\phi(z)) = \Phi^A(z) + 1 , \quad \text{for all } z \in U^A ,$$

as well as

$$\Phi^R(\phi(z)) = \Phi^R(z) - 1 , \quad \text{for all } z \in U^R .$$

(iii) The quotient Riemann surfaces $C^A = U^A/\phi$ and $C^R = U^R/\phi$ are both conformally equivalent to $\mathbb{C}/\mathbb{Z}$.

The region U^A is called an *attracting petal* for ϕ, and the region U^R is called a *repelling petal* for ϕ. The maps Φ^A and Φ^R conjugate the dynamics of $\phi|_{U^A}$ and $\phi|_{U^R}$, respectively, to translations, and are unique up to post-composition with translations. They are called *Fatou coordinates* for ϕ. The cylinders C^A and C^R are called *Écalle–Voronin cylinders* of ϕ.

Now, let $\pi_A : U^A \to C^A$ and $\pi_R : U^R \to C^A$ be the natural projections. Note that the cylinders C^A and C^R both have natural equators, namely the quotients $E^A = (U^A \cap \mathbb{R})/\phi$ and $E^R = (U^R \cap \mathbb{R})/\phi$. They also each have a north pole and a south pole. Fix a base point $z^A \in E^A$ and identify C^A with $\mathbb{C}/\mathbb{Z}$ via the unique bi-holomorphic map $C^A \to \mathbb{C}/\mathbb{Z}$ that sends E^A onto $\mathbb{R}/\mathbb{Z}$ and maps z^A to 0 (and north pole to north pole and south pole to south pole). Likewise, fix a base point $z^R \in E^R$ and identify C^R with $\mathbb{C}/\mathbb{Z}$ via the unique bi-holomorphic map $C^R \to \mathbb{C}/\mathbb{Z}$ that sends E^R onto $\mathbb{R}/\mathbb{Z}$ and maps z^R to 0 (and again preserves north and south poles). By a *transfer isomorphism* we mean a bi-holomorphic map $\tau : C^A \to C^R$ which sends E^A onto E^R, preserving the natural orientation of these circles provided by the identifications just described. Thus τ corresponds to a unique rotation of the cylinder $\mathbb{C}/\mathbb{Z}$ by an angle $\theta \in \mathbb{R}/\mathbb{Z}$. Accordingly, we write $\tau = \tau_\theta$.

We are now ready to define the parabolic renormalization of a holomorphic pair $\zeta = (\eta, \xi)$ in which the map $\phi = \eta$ has a parabolic fixed point. The procedure will produce a one-parameter family of holomorphic pairs, one for each choice of $\theta \in \mathbb{R}/\mathbb{Z}$. Let us fix inverse branches π_A^{-1} and π_R^{-1} of both projections. Let $N \geqslant 1$ be such that $\eta^N \circ \eta(0) \in \pi_A^{-1}(C^A)$, and let $M \geqslant 1$ be the smallest integer with the property that

$$\eta^M \circ \pi_R^{-1} \circ \tau_\theta \circ \pi_A \circ \eta^N \circ \xi(0) \in [\eta(0), 0] \subseteq \mathbb{R} .$$

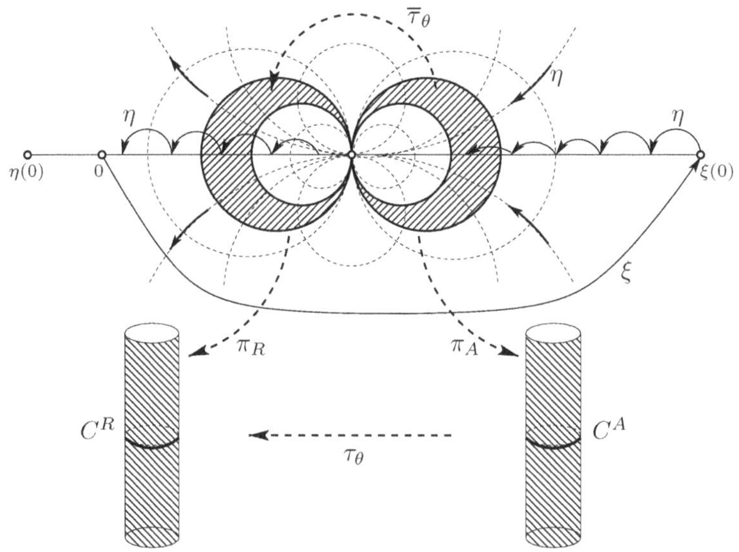

Fig. 14.5 Parabolic renormalization. Here, $\overline{\tau}_\theta = \pi_R^{-1} \circ \tau_\theta \circ \pi_A$

Thus, we have the situation depicted in Fig. 14.5. Therefore consider the composition

$$\gamma = \eta^M \circ \pi_R^{-1} \circ \tau_\theta \circ \pi_A \circ \eta^N \circ \xi .$$

It is possible to prove that γ has a well-defined extension to a neighborhood of the interval $[\eta(0), 0]$ which is independent of the choices of inverse branches π_A^{-1}, π_R^{-1}. The *parabolic renormalization of* $\zeta = (\eta, \xi)$ *corresponding to* θ is defined to be the normalized commuting pair

$$\mathscr{P}_\theta\zeta = \left(\widetilde{\gamma|_{[\eta(0),0]}}, \widetilde{\eta|_{[0,\gamma(0)]}}\right) .$$

At this point we may ask: What is the connection between parabolic renormalization as just described and the notion of renormalization of holomorphic commuting pairs previously defined? One answer is provided by Proposition 14.2 below.

14.8.1.2 Douady Coordinates

Let us now fix a map $\eta_0 \in \mathscr{E}$ in the Epstein class. We consider here small perturbations[4] of η_0 in $\mathscr{E}$. We assume to start with that η_0 is a parabolic map, i.e., η_0

[4] The topology on $\mathscr{E}$ is taken to be the *Carathéodory topology*. See for instance McMullen (1996, p. 75).

has a unique parabolic fixed point $p \in \mathbb{R}$ with multiplier equal to 1. Then for every sufficiently small perturbation η of η_0 in $\mathscr{E}$ the parabolic fixed point p of η_0 splits into a pair of repelling fixed points for η, say $p_\eta^+ \in \mathbb{C}^+$ and $p_\eta^- \in \mathbb{C}^-$, symmetric about the real axis ($p_\eta^- = \overline{p_\eta^+}$). Let λ_η^+ and λ_η^- be the multipliers of p_η^+ and p_η^-, respectively. Then we have $\lambda_\eta^\pm \to 1$ as $\eta \to \eta_0$. Some simple considerations involving the notion of *holomorphic index* of a fixed point (see Exercise 14.5) imply that

$$\frac{1}{1-\lambda_\eta^+} + \frac{1}{1-\lambda_\eta^-} \to 0\,, \quad \text{as } \eta \to \eta_0\,. \tag{14.14}$$

Taken together, these facts imply that $\arg(1-\lambda_\eta^\pm) \to 0$ as $\eta \to \eta_0$. One can show that there exists a neighborhood $\mathscr{U}(\eta_0) \subset \mathscr{E}$ with the following property. For every $\eta \in \mathscr{U}(\eta_0)$ with $|\arg(1-\lambda_\eta^\pm)| \leqslant \pi/4$ there exist topological disks U_η^A and U_η^R such that $U_\eta^A \cup U_\eta^R$ is a neighborhood of p, and univalent maps $\Phi_\eta^A : U_\eta^A \to \mathbb{C}$ and $\Phi_\eta^R : U_\eta^R \to \mathbb{C}$ (unique up to post-composition with translations) which conjugate the dynamics of η with translations by -1 and $+1$ respectively, that is to say

$$\Phi_\eta^A(\eta(z)) = \Phi_\eta^A(z) + 1 \ \text{ and } \ \Phi_\eta^R(\eta(z)) = \Phi_\eta^R(z) - 1\,.$$

As before, the quotient Riemann surfaces $C_\eta^A = U_\eta^A/\eta$ and $C_\eta^R = U_\eta^R/\eta$ are both bi-infinite cylinders, i.e. conformally equivalent to the standard cylinder $\mathbb{C}/\mathbb{Z}$. The maps Φ_η^A and Φ_η^R are called the *Douady coordinates* for η.

It is possible to prove that, as $\eta \to \eta_0$ (in the Carathéodory topology), one has $\Phi_\eta^A \to \Phi^A$ and $\Phi_\eta^R \to \Phi^R$ uniformly on compact subsets of U^A and U^R, respectively—in other words, the Douady coordinates of η converge to the Fatou coordinates of η_0. This form of continuity implies the following result.

Proposition 14.2 *Let $\zeta_k = (\eta_k, \xi_k)$, $k \geqslant 1$, be a sequence of renormalizable commuting pairs in the Epstein class, and let $\zeta = (\eta, \xi)$ be a parabolic commuting pair such that $\zeta_k \to \zeta$ as $k \to \infty$ (in particular, the rotation numbers $\rho(\zeta_k)$ converge to 0). Suppose also that the renormalizations $\mathscr{R}\zeta_k$ converge to another commuting pair $\overline{\zeta}$. Then there exists $\theta \in \mathbb{R}/\mathbb{Z}$ such that $\mathscr{P}_\theta \zeta = \overline{\zeta}$.*

This proposition establishes the desired connection between the standard notion of renormalization and the notion of parabolic renormalization. Putting the two renormalization schemes together is a sort of "compactification" of the renormalization operator, and with this at hand one can adapt McMullen's theory to the context of critical circle maps through the use of towers of holomorphic commuting pairs.

14.8.1.3 Cylindrical Renormalization

We close this chapter with some brief words about the cylinder renormalization operator introduced by Yampolsky (2002). As we saw in Sect. 10.2, a commuting pair represents a whole conjugacy class of critical circle maps, and therefore the renormalization operator acts on such classes. The concept introduced by Yampolsky has two main advantages. First, the cylinder renormalization operator acts directly on maps, rather than on their conjugacy classes. Second, it extends to an analytic operator in the Banach manifold $\boldsymbol{B}_V$ of analytic critical circle maps defined in some equatorial neighborhood V of the circle $\mathbb{R}/\mathbb{Z}$ inside the cylinder $\mathbb{C}/\mathbb{Z}$.

Let us be a bit more formal. By a *cylinder map* we mean a map $f : V \to \mathbb{C}/\mathbb{Z}$, where $V \subseteq \mathbb{C}/\mathbb{Z}$ is an equatorial neighborhood, which is holomorphic, has a unique (cubic) critical point at $0 \in \mathbb{R}/\mathbb{Z}$, and is such that the restriction $f|_{\mathbb{R}/\mathbb{Z}}$ maps the equator $\mathbb{R}/\mathbb{Z}$ homeomorphically onto its image $f(\mathbb{R}/\mathbb{Z})$. The space of all cylinder maps with domain V is denoted $\boldsymbol{B}_V$. With the help of the implicit function theorem, this can be shown to be a complex Banach manifold—whose real slice $\boldsymbol{B}_V^{\mathbb{R}}$, consisting of those $f \in \boldsymbol{B}_V$ which preserve the equator, is precisely the space of real-analytic critical circle maps having a complex analytic extension to V. See Yampolsky (2002, pp. 23–24) for details.

We say that a cylinder map $f \in \boldsymbol{B}_V$ is *cylinder renormalizable* if there exist $q > 1$ and an equatorial neighborhood $W \subset \mathbb{C}/\mathbb{Z}$ if the following holds:

(1) There exist two f-periodic points $p_1, p_2 \in V$ of period q and a simple arc ℓ joining them such that $f^q(\ell)$ is also a simple arc and $f^q(\ell) \cap \ell = \{p_1, p_2\}$.
(2) The union $\ell \cup f^q(\ell)$ bounds a simply connected region $C_f \subset V$ on which f^q is univalent—called a *fundamental crescent for* f—and the inverse $f^{-q}|_{f^q(C_f)}$ extends to C_f as a univalent map.
(3) The quotient of $\mathscr{C}_f(p_1, p_2) = \overline{C_f \cup f^q(C_f)} \setminus \{p_1, p_2\}$ by the action of f^q is a Riemann surface conformally equivalent to the bi-infinite cylinder $\mathbb{C}/\mathbb{Z}$, i.e., there exists a bi-holomorphism $\psi : \mathscr{C}_f(p_1, p_2) \to \mathbb{C}/\mathbb{Z}$.
(4) If $z \in \overline{C_f} \setminus \{p_1, p_2\}$ and its positive orbit $\{f^j(z) : j \geqslant 1\}$ intersects $\overline{C_f}$, let $n(z) = \min\{j \geqslant 1 : f^j(z) \in \overline{C_f}\}$, and let $V_f = \{z \in \overline{C_f} : n(z) < \infty\}$. Then set $R_{C_f} : V_f \to \overline{C_f}$ by $R_{C_f}(z) = f^{n(z)}(z)$ (note that this is a first return map).
(5) Let $\tilde{V}_f = V_f / f^q \subset \mathscr{C}_f(p_1, p_2)$ and $\tilde{R}_{C_f} = R_{C_f} / f^q$ (the quotient space and quotient map, respectively).
(6) Putting $W_f = \psi(\tilde{V}_f) \subset \mathbb{C}/\mathbb{Z}$ and letting $\hat{f} : W_f \to \mathbb{C}/\mathbb{Z}$ be defined so that the diagram

$$\begin{array}{ccc} \tilde{V}_f & \xrightarrow{\tilde{R}_{C_f}} & \mathscr{C}_f(p_1, p_2) \\ \Big\downarrow{\scriptstyle\psi} & & \Big\downarrow{\scriptstyle\psi} \\ W_f & \xrightarrow[\hat{f}]{} & \mathbb{C}/\mathbb{Z} \end{array}$$

commutes, then $W \subseteq W_f$ and $\hat{f}$ is a cylinder map belonging to $\boldsymbol{B}_W$.

The map $\hat{f}|_W \in \boldsymbol{B}_W$ thus defined is called a *cylinder renormalization of f of period q*. What is the relationship between the notion of cylinder renormalization and the standard notion of renormalization previously introduced in this book? An answer to this question is provided by the following result. Let us agree to call a cylinder map an *analytic critical circle map* if it preserves the equator $\mathbb{R}/\mathbb{Z}$ and its restriction to the equator is a critical circle map.

Proposition 14.3 *Let $f \in \boldsymbol{B}_V$ be an analytic critical circle map, and suppose that its rotation number $\rho(f)$ is irrational. If f is cylinder renormalizable with period $q = q_n$ (where q_n is a return time for f), then the corresponding cylinder renormalization $\hat{f}$ is also a critical circle map, with rotation number $G^n(\rho(f))$ (where G is the Gauss map). Moreover, its renormalization $\mathscr{R}\hat{f}$ is analytically conjugate to $\mathscr{R}^{n+1}f$.*

Note that a given cylinder map $f \in \boldsymbol{B}_V$ can be cylinder renormalized in several different ways. However, if we fix the period q and the equatorial neighborhood W, then every $g \in \boldsymbol{B}_V$ sufficiently close to f will also be cylinder renormalizable with period q, and $\hat{g} \in \boldsymbol{B}_W$. In fact, using the theory of holomorphic motions, Yampolsky shows that $\hat{g}$ depends holomorphically on g. The cylinder renormalization procedure just described can also be defined for holomorphic commuting pairs $\zeta = (\eta, \xi)$ in the Epstein class, and in this context it can be made more canonical because the natural periodic points to use are the (symmetric) fixed points $p_\eta^\pm$ of the maps η. This is the basis for obtaining a well-defined, Fréchet differentiable cylinder renormalization operator $\mathscr{R}_{\text{cyl}}$ whose hyperbolicity properties can then be studied.

There would be much more to be said, but we choose to stop here.

14.8.1.4 Final Remarks

There is a large literature on parabolic bifurcations, on both their local and global aspects, and on parabolic renormalization. The reader interested in delving deeper into this subject may start with Milnor (2006) for the basic local theory and then consult Shishikura (2000), as well as the book by Lanford and Yampolsky (2014) for global aspects. The theory of parabolic renormalization was used by Shishikura in his celebrated proof that the boundary of the Mandelbrot set has Hausdorff dimension equal to 2—see Shishikura (1998)—and more recently by Buff and Chéritat in their construction of Julia sets of positive measure—see Buff and Cheritat (2012).

Exercises

Exercise 14.1 Prove Lemma 14.3.

Exercise 14.2 Complete the proof of Theorem 14.3 by showing that holomorphic commuting pairs have no wandering domains (see Sect. 11.2).

Exercise 14.3 Prove Lemma 14.13.

Exercise 14.4 Prove that the family $\mathscr{F}_\sigma$ of complex almost parabolic maps introduced in Sect. 14.5 is a normal family in the sense of Montel.

Exercise 14.5 Let $f : U \to \mathbb{C}$ be a holomorphic map and suppose $z_0 \in U$ is an isolated fixed point of f. We define the *holomorphic index* of f at z_0 to be the integral

$$i(f; z_0) = \frac{1}{2\pi i} \int_C \frac{dz}{z - f(z)} ,$$

where C is any simple closed curve containing z_0 in its interior, and no other fixed point of f, oriented in the counterclockwise direction.

(i) Show that $i(f; z_0)$ varies continuously with f (in an appropriate sense).
(ii) Let $\lambda = f'(z_0)$ be the multiplier of the fixed point z_0. Show that (a) if $\lambda \neq 1$, then $i(f; z_0) = (1 - \lambda)^{-1}$; and (b) if $\lambda = 1$, then $i(f; z_0) = 0$.
(iii) Let D be a closed topological disk contained in U, and suppose f has exactly N fixed points inside D, say $z_0, z_1, \ldots, z_{N-1}$. Show that

$$\frac{1}{2\pi i} \int_{\partial D} \frac{dz}{z - f(z)} = \sum_{j=0}^{N-1} i(f; z_j) .$$

(iv) Use items (i),(ii) and (iii) to prove (14.14).

Epilogue

We end this book with some remarks, conjectures and open questions on multicritical circle maps.

1. Recall that in Chap. 6 we introduced the notion of *signature* of a multicritical circle map (see Definition 6.2). We may re-state Question 10.1 as follows. Let $f, g : S^1 \to S^1$ be two C^3 multicritical circle maps with the same signature, and let $h : S^1 \to S^1$ be a conjugacy between f and g such that h maps each critical point of f to a corresponding critical point of g. Is h a C^1 diffeomorphism? Are there conditions on the rotation number that make h better than C^1? To the best of our knowledge, no rigidity results are available for maps with $N \geq 3$ critical points.[1] As mentioned in Sect. 6.1.1 (see Remark 6.1), a construction similar to the one developed by Zakeri, in order to prove Theorem 6.1, could be useful as a starting point.
2. What about (multi)critical circle maps with *non-integer* criticalities? Not even the existence of periodic orbits (for renormalization) in the unicritical case has been established yet in full generality (for real-analytic unicritical circle maps with bounded combinatorics, Gorbovickis and Yampolsky (2020) were able to establish both existence and hyperbolicity of the horseshoe-like attractor (recall Sect. 13.4), but only allowing criticalities close enough to an odd integer). For unimodal maps, this problem has been completely solved by Martens (1998) (see also Gorbovickis and Yampolsky 2018), but it is unclear to us whether his methods can be adapted to the circle case.
3. As already mentioned, the full Lebesgue measure set $\mathbb{A} \subset (0, 1)$ of rotation numbers considered in Definition 10.4, for which $C^{1+\alpha}$ rigidity holds, was originally defined in de Faria and de Melo (1999, §4.4). A natural question is: Are these conditions optimal? In other words, is $\mathbb{A}$ the largest set of rotation numbers for which statement (3) in Theorem 10.1 is true?
4. Another way to approach the previous question is to look at the complement of the set $\mathbb{A}$. In Sect. 10.5, a saddle-node surgery technique was used to build C^∞

[1] A recent preprint by Gorbovickis and Yampolsky (2021) contains a proof of the $C^{1+\alpha}$ rigidity problem for real-analytic maps with bounded combinatorics.

E. de Faria, P. Guarino, *Dynamics of Circle Mappings*, IMPA Monographs,
https://doi.org/10.1007/978-3-031-67495-2

counterexamples to $C^{1+\alpha}$ rigidity for each rotation number $\rho = [a_0, a_1, a_2, \ldots]$ satisfying $a_n \geq 2$ for all n and

$$\limsup_{n\to\infty} \frac{1}{n} \log a_n = \infty .$$

Can such counterexamples be built for every rotation number not in $\mathbb{A}$? An analogous question can be asked in the analytic category. In Avila (2013) (recall Theorem 10.6), using parabolic surgery, Avila constructed real-analytic counterexamples to $C^{1+\alpha}$ rigidity for each rotation number in another set of rotation numbers (still properly contained in the complement of $\mathbb{A}$). What is the optimal class of rotation numbers in this case? Is it still the whole complement of $\mathbb{A}$?

5. The problem of global hyperbolicity of the renormalization operator for C^r *unimodal maps* was solved in de Faria et al. (2006), through a combination of the deep holomorphic results obtained by Lyubich (1999) (later improved by Avila and Lyubich 2011) with certain techniques of non-linear functional analysis borrowed from the work of Davie (1996). Can these ideas be adapted to the study of the renormalization of C^r (multi)critical circle maps? There are several difficulties to overcome here, such as to provide a suitable definition of a manifold structure directly in the space of real analytic critical commuting pairs. If this space can be endowed with a Banach manifold structure under which the renormalization operator is hyperbolic, then it is not too difficult to push such hyperbolicity to the space of C^r critical commuting pairs (see Voutaz 2006). Recall, however, that the space of *cylinder maps* defined by Yampolsky (see Sect. 14.8) does have a complex Banach manifold structure, and that, in somewhat imprecise terms, the cylinder renormalization operator does act as a holomorphic map in this space. So perhaps an alternative approach to C^r hyperbolicity exists which avoids commuting pairs altogether.
6. Is it possible to prove Theorems 13.2 and 13.3 without appeal to holomorphic methods? Although quite powerful, the use of holomorphic methods limits the discussion to maps all of whose critical points have *integer criticalities*. See problem (2) above.
7. Is Theorem 10.1 still true if the maps f and g are only C^3?
8. Rigidity in the space of C^2 maps is most likely false. Can one construct explicit examples? What about rigidity in the space of $C^{2+\alpha}$ maps?
9. Let $f : S^1 \to S^1$ be a C^∞ multicritical circle map, all of whose critical points are non-flat. As we know from Sect. 2.3, f is uniquely ergodic. Is f distributionally uniquely ergodic? More precisely, is it true that, for each $k \in \mathbb{N}$, the linear space $\mathscr{D}'_k(f)$ is one-dimensional? As mentioned in Sect. 8.4, this is true for $k = 1$ (see Theorem 8.9). Recall also that every C^∞ diffeomorphism of the circle with irrational rotation number is distributionally uniquely ergodic (Theorem 3.13).
10. Finally, one topic that we did not touch at all in this book is what is commonly referred to by physicists as *mode locking universality*. In a typical (monotone) one-parameter family of (uni)critical circle maps, such as the Arnold family,

the set of parameters for which the rotation number is irrational constitutes a Cantor set (see Fig. 6.2), called the *mode-locking Cantor set*. It is conjectured that the Hausdorff dimension of this Cantor set is a *universal number*—which has been numerically computed to be approximately 0.870..., see Cvitanović et al. (1990). It has been shown by Graczyk and Świątek (1996) that this dimension indeed lies strictly between zero and one. In particular, the Cantor set in question has zero Lebesgue measure. This is in sharp contrast with what happens in typical one-parameter families of circle *diffeomorphisms*: in such cases, Herman (1988) had already shown in the seventies that the corresponding Cantor set has positive measure. Universality of the Hausdorff dimension in the critical case would follow from a careful study of the holonomy of the lamination determined by the stable manifolds of the renormalization operator (acting on a suitable space of critical commuting pairs), presumably in a similar manner as in the corresponding study of holonomy carried out for C^r unimodal maps in de Faria et al. (2006). For more on the empirical study of the scaling geometry of the mode-locking Cantor set, see the work by Cvitanović et al. (1985).

Appendix A
Ergodic Theory of Continued Fractions

In this appendix we briefly discuss the relationship between continued fraction expansions and the ergodic theory of the Gauss map. The reader can find much more about this beautiful subject in the books (Billingsley 1965; Cornfeld et al. 1982; Iosifescu and Kraaikamp 2002). Here we content ourselves to providing a proof of the fact that the set $\mathbb{A} \subset (0, 1)$ given by Definition 10.4 has full Lebesgue measure (see Corollary A.1 and Lemma A.3 below).

A.1 Expansions as Itineraries

For any real number x denote by $\lfloor x \rfloor$ the *integer part* of x, that is, the greatest integer less than or equal to x. Also, denote by $\{x\}$ the *fractional part* of x : $\{x\} = x - \lfloor x \rfloor \in [0, 1)$. Recall from Chap. 1 that the *Gauss map* $G : [0, 1] \to [0, 1]$ is given by

$$G(\rho) = \left\{ \frac{1}{\rho} \right\} \text{ for } \rho \neq 0\,, \text{ and } G(0) = 0.$$

Note that both $\mathbb{Q} \cap [0, 1]$ and $[0, 1] \setminus \mathbb{Q}$ are G-invariant. Under the action of G, all rational numbers in $[0, 1]$ eventually land on the fixed point at the origin (see Exercise A.6). On the other hand, the positive orbit of any irrational number remains in the open set $\bigcup_{k \geq 1} M_k$, where $M_k = \left(\frac{1}{k+1}, \frac{1}{k} \right)$. In this appendix, we briefly discuss the following *dynamical* definition of continued fraction expansions.

Definition A.1 The *continued fraction expansion* of an irrational number in $[0, 1]$ is the sequence given by its itinerary under G, according to the partition $\bigcup_{k \geq 1} M_k$.

More precisely, we identify each irrational number ρ in $[0, 1]$ with the sequence $[a_0, a_1, \ldots]$ defined by $G^n(\rho) \in M_{a_n}$ for all $n \in \mathbb{N}$. In other words, for any $\rho \in$

E. de Faria, P. Guarino, *Dynamics of Circle Mappings*, IMPA Monographs,
https://doi.org/10.1007/978-3-031-67495-2

$(0, 1) \setminus \mathbb{Q}$ and any $n \in \mathbb{N}$ we have that $G^n(\rho) \in \left(\frac{1}{k+1}, \frac{1}{k}\right)$ if, and only if, $a_n = k$. It is easy to see that this definition coincides with the one used along the book. Indeed, if

$$\rho = \cfrac{1}{a_0 + \cfrac{1}{a_1 + \cfrac{1}{\ddots}}}$$

belongs to $\big(1/(k+1), 1/k\big)$, then $a_0 = \left\lfloor \frac{1}{\rho} \right\rfloor = k$ and $G(\rho) = \cfrac{1}{a_1 + \cfrac{1}{a_2 + \cfrac{1}{\ddots}}}$, since

$$1/\rho = a_0 + \cfrac{1}{a_1 + \cfrac{1}{a_2 + \cfrac{1}{\ddots}}}.$$

In particular, the Gauss map acts as a *left shift* on the continued fraction expansion of ρ. Indeed, since for each $k \geq 1$ the restriction $G|_{M_k}$ is an expanding diffeomorphism onto $(0, 1)$, it can be proved (see Exercise A.7) that the map h from $[0, 1] \setminus \mathbb{Q}$ to $\mathbb{N}^{\mathbb{N}}$ identifying each irrational number to its itinerary under G is a homeomorphism (endowing $\mathbb{N}^{\mathbb{N}}$ with the product topology). Therefore, the action of G on $[0, 1] \setminus \mathbb{Q}$ is topologically conjugate to the left shift $\sigma : \mathbb{N}^{\mathbb{N}} \to \mathbb{N}^{\mathbb{N}}$ mapping $[a_0, a_1, \ldots]$ to $[a_1, a_2, \ldots]$:

$$\begin{array}{ccc} [0,1] \setminus \mathbb{Q} & \xrightarrow{\;G\;} & [0,1] \setminus \mathbb{Q} \\ \big\downarrow h & & \big\downarrow h \\ \mathbb{N}^{\mathbb{N}} & \xrightarrow{\;\sigma\;} & \mathbb{N}^{\mathbb{N}} \end{array}$$

A.2 The Gauss Measure and Almost Surely Properties

As we have seen in the fourth part of the present monograph, irrational numbers of *bounded type* play a major role in the rigidity theory of multicritical circle maps. Let us recall here their definition.

Definition A.2 An irrational number $\rho \in [0, 1]$ is of *bounded type* if there exists a constant $K > 0$ such that $a_n < K$ for all $n \in \mathbb{N}$.

The set of numbers of bounded type is dense in $(0, 1)$. Indeed, as it is not difficult to prove (see Exercise A.3), the set of periodic orbits of σ is dense in $\mathbb{N}^{\mathbb{N}}$, which implies that irrational numbers with periodic continued fraction expansion[1] are dense in $(0, 1)$. On the other hand, from the measure-theoretical viewpoint, we have the following result.

Lemma A.1 *The set of numbers of bounded type has zero Lebesgue measure.*

As it is well known (see for instance Mañé 1987), the map G admits an invariant ergodic Borel measure ν (called the *Gauss measure*) given by

$$\nu(A) = \frac{1}{\log 2} \int_A \frac{d\rho}{1+\rho} \quad \text{for any Borel set } A \subset [0, 1].$$

Note that $\nu(0, \rho) = \log(1 + \rho)/\log 2$ for any $\rho \in [0, 1]$, so the coefficient $1/\log 2$ turns ν into a probability measure on the unit interval. If we denote by m the Lebesgue measure on $[0, 1]$, we immediately have

$$\frac{1}{2\log 2} m(A) \leq \nu(A) \leq \frac{1}{\log 2} m(A)$$

for any Borel set $A \subset [0, 1]$. In particular, the Gauss measure ν is equivalent to the Lebesgue measure on $[0, 1]$ (i.e., they share the same null sets).

Proof of Lemma A.1 Consider the increasing sequence $\{K_m\}_{m\in\mathbb{N}}$ of subsets of $[0, 1]$ defined by

$$K_m = \big\{\rho \in [0, 1] \setminus \mathbb{Q} \ : \ \rho = [a_0, a_1, \dots] \quad \text{with } a_n < m \text{ for all } n \in \mathbb{N}\big\}.$$

Note that each K_m is a Cantor set, homeomorphic to $\{1, 2, \dots, m-1\}^{\mathbb{N}}$. The union $\bigcup_{m\in\mathbb{N}} K_m$ coincides with the set of numbers of bounded type. Since the Gauss measure is equivalent to Lebesgue, it is enough to prove that $\nu\big(K_m\big) = 0$ for each $m \in \mathbb{N}$. But this follows at once from the ergodicity of ν under G, since each K_m is a G-invariant set contained in the interval $(1/m, 1)$. □

[1] It is worth mentioning here that the set of irrational numbers with periodic (or *eventually* periodic) continued fraction expansion coincides with the set of *quadratic algebraic* numbers, i.e., the set of roots of all quadratic polynomials with integer coefficients (a proof of this result, due to Lagrange, can be found in Khinchin (1997, Section 10), see also Exercise A.5). Algebraic numbers of higher degree are also interesting from the continued fraction expansion viewpoint, see Khinchin (1997, Section 9) and also Appendix A.3 below.

For the classical proof of Lemma A.1, with no dynamical arguments, we refer the reader to Khinchin (1997, Section 13). A much more precise statement can be obtained from Birkhoff's Ergodic Theorem.

Proposition A.1 *For Lebesgue almost every* ρ *in* $[0, 1]$ *we have that every integer* $k \geq 1$ *must appear infinitely many times in the continued fraction expansion of* $\rho = \big[a_0, a_1, \dots\big]$. *Moreover, if we define*

$$\tau_n(\rho, k) = \frac{1}{n}\,\#\big\{0 \leq j < n : a_j = k\big\},$$

we have that $\big\{\tau_n(\rho, k)\big\}_{n\in\mathbb{N}}$ *converges to the positive value*

$$\frac{1}{\log 2}\,\log\left(1 + \frac{1}{k(k+2)}\right),$$

that only depends on k.

Proof Since

$$\tau_n(\rho, k) = \frac{1}{n}\,\#\big\{0 \leq j < n : G^j(\rho) \in M_k\big\},$$

we deduce from Birkhoff's Ergodic Theorem that

$$\lim_{n\to+\infty} \tau_n(\rho, k) = \nu(M_k) = \frac{1}{\log 2}\,\log\left(1 + \frac{1}{k(k+2)}\right)$$

for ν almost every ρ (and then the same holds for Lebesgue almost every ρ). □

Since the asymptotic frequency given by Proposition A.1 is strictly decreasing in k, one should expect that typical numbers, even having unbounded partial quotients, have slow growth. This is explicitly formulated in the following lemma.

Lemma A.2 *Let* $\{b_n\}_{n\in\mathbb{N}}$ *be any given increasing sequence of positive numbers such that* $\sum_{n\in\mathbb{N}} 1/b_n < \infty$. *For Lebesgue almost every* $\rho = \big[a_0, a_1, \dots\big]$ *in* $[0, 1]$ *we have* $a_n < b_n$ *for all* n *large enough.*

Proof For each $n \in \mathbb{N}$ consider the sets $U_n = \big\{\rho : a_n > b_n\big\}$ and $V_n = \big\{\rho : a_0 > b_n\big\}$. We want to prove that

$$\nu\left(\limsup_{n\to+\infty} U_n\right) = \nu\left(\bigcap_{k\in\mathbb{N}}\bigcup_{n\geq k} U_n\right) = 0. \tag{A.1}$$

Since $G^{-n}(V_n) = U_n$ and $V_n \subset \big(0, 1/b_n\big)$, we have that

$$\nu(U_n) \leq \nu\big(0, 1/b_n\big) = \frac{1}{\log 2} \log\left(1 + \frac{1}{b_n}\right) \leq \frac{1}{\log 2}\frac{1}{b_n}$$

for n large enough. In particular, $\sum_{n\in\mathbb{N}} \nu(U_n) < \infty$ and then (A.1) follows from the Borel–Cantelli Lemma. □

Corollary A.1 *For Lebesgue almost every ρ in* [0, 1] *we have*

$$\lim_{n\to\infty} \frac{1}{n} \log a_n = 0 \quad \textit{and} \quad \limsup_{n\to\infty} \frac{1}{n}\sum_{j=1}^{n} \log a_j < \infty.$$

Lemma A.3 *For Lebesgue almost every ρ in* [0, 1] *we have*

$$\frac{1}{n}\sum_{j=k+1}^{k+n} \log a_j \leq C\big(\rho\big)\left(1 + \log\frac{k}{n}\right) \quad \textit{for all} \quad 0 < n \leq k, \tag{A.2}$$

where $C\big(\rho\big) > 0$ depends on ρ.

As mentioned in Chap. 10 (see Definition 10.4), the set of numbers satisfying both Corollary A.1 and Lemma A.3 simultaneously was first considered in de Faria and de Melo (1999, Section 4.4). Corollary A.1 follows straightforward from Lemma A.2 by taking, say, $b_n = n^{1+\varepsilon}$ for any $\varepsilon > 0$. We proceed now to prove Lemma A.3, following de Faria and de Melo (1999, App. C).

We remark that all probabilistic estimates below will be done for the Gauss measure ν. Note first that the probability p_k that the n-th partial quotient $a_n(\rho)$ be equal to a given integer $k \geq 1$ is

$$p_k = \nu\big(G^{-n}(M_k)\big) = \nu(M_k) = \frac{1}{\log 2}\log\left(1 + \frac{1}{k(k+2)}\right) < \frac{2}{k^2}. \tag{A.3}$$

From this, we see that the probability that $a_n(\rho)$ be at least k is smaller than $4/k$.

We shall now prove Lemma A.3, which establishes Condition (3) in Definition 10.4 for almost all numbers $\rho \in [0, 1]$ with $\omega(t) = C(\rho)(1 - \log t)$, where $C(\rho) > 0$.

Proof Let $E \subset (0, 1)$ be the full Lebesgue measure set of irrational numbers satisfying Lemma A.2 with $b_n = n^2$. In attempting to prove the inequality (A.2) for a given $\rho \in E$, we may assume that k is so large that $a_j(\rho) < j^2$ for all $j \geq k$. The remaining cases, corresponding to the remaining finitely many pairs (n, k), are taken care of by a suitable choice of the constant $C(\rho)$.

Given (n, k), there are two possibilities to consider. The first possibility is that $n^2 < k$. In this case we simply observe that

$$\frac{1}{n}\sum_{j=k+1}^{k+n} \log a_j(\rho) \leq 2\log(k+n) < 5\log\frac{k}{n},$$

for all sufficiently large k.

The second possibility is that $n \le k \le n^2$. Here, we shall prove that *with probability one* the left-hand side of (A.2) is bounded by 10. For this purpose, let us consider the following pathologies.

(a) For a given $m \ge 1$, there are more than $2np_m$ partial quotients $a_{k+i}(\rho)$ with $1 \le i \le n$ such that $a_{k+i}(\rho) = m$ (where p_m is as defined in (A.3)). By an elementary combinatorial argument, we see that this occurs with probability at most

$$\sum_{j=\lceil 2np_m \rceil}^{n} \binom{n}{j} p_m^j (1-p_m)^{n-j} < \left(\frac{e}{4}\right)^{np_m}.$$

The probability that this happens for *some* m in the range $1 \le m \le n^{1/3}$ is therefore smaller than

$$n^{1/3}\left(\frac{e}{4}\right)^{2n^{1/3}} < \frac{1}{n^4}$$

if n is sufficiently large.

(b) There are more than $n^{2/3}$ partial quotients $a_{k+i}(\rho)$ with $1 \le i \le n$ such that $a_{k+i}(\rho) > n^{1/3}$. By a similar reasoning to the one used in (a), we see that this occurs with probability smaller than

$$\left(\frac{e}{4}\right)^{n^{1/3}} < \frac{1}{n^4}$$

if n is sufficiently large.

Therefore, fixing n sufficiently large, the probability that there exists k in the range $n \le k \le n^2$ such that one of the above pathologies occurs for (n,k) is certainly less than $n^2 \times (2/n^4) = 2/n^2$. Since the series $\sum 2/n^2$ converges, again by Borel–Cantelli we deduce that with probability one there are no pathologies for (n,k) if k (and hence n) is sufficiently large.

Now, if there are no pathologies for (n,k), and noting that for $1 \le i \le n$ we have

$$a_{k+i}(\rho) < (k+i)^2 \le (n^2+n)^2 \le 4n^4$$

if k is sufficiently large, we deduce that

$$\begin{aligned}\frac{1}{n}\sum_{j=k+1}^{k+n} \log a_j(\rho) &\le \frac{1}{n}\sum_{m=1}^{\lfloor n^{1/3}\rfloor} (2np_m)\log m + \frac{n^{2/3}}{n}\log(4n^4) \\ &< \sum_{m=1}^{\lfloor n^{1/3}\rfloor} \frac{4\log m}{m^2} + \frac{1}{n^{1/3}}(2\log 2 + 4\log n),\end{aligned}$$

which is less than 10 if n is sufficiently large. This completes the proof. □

We have proved that the set $\mathbb{A}$ from Definition 10.4 has full Lebesgue measure in $(0, 1)$.

A.3 Diophantine Approximations Revisited

We finish this appendix by recalling the definition and basic properties of Diophantine and Liouville numbers.

Definition A.3 An irrational number in $[0, 1]$ is said to be *Diophantine* if there exist constants $C > 0$ and $\delta \geq 0$ such that:

$$\left|\rho - \frac{p}{q}\right| \geq \frac{C}{q^{2+\delta}},$$

for any natural numbers p and $q \neq 0$. Irrational numbers which are not Diophantine are called *Liouville* numbers.

A famous theorem, due to Liouville, asserts that any given *algebraic* number of degree $n \geq 2$ (i.e., the root of a degree n polynomial with integer coefficients) is Diophantine, with exponent $\delta = n - 2$ (see Khinchin 1997, Section 9). From a dynamical viewpoint, Diophantine numbers will reappear in Appendix B, where we will prove that sufficiently smooth circle diffeomorphisms with Diophantine rotation number are smoothly linearizable (more precisely, we will prove Theorem 4.11: if f belongs to $\text{Diff}^{2+\alpha}(S^1)$ and has rotation number ρ which is Diophantine of order δ, then any topological conjugacy between f and the rigid rotation of angle ρ is a $C^{1+\alpha-\delta}$ diffeomorphism, provided $0 \leq \delta < \alpha < 1$).

As we saw in Chap. 4 (see Exercise 4.11), an irrational number is of bounded type if it satisfies Definition A.3 for $\delta = 0$, that is, ρ in $[0, 1]$ is of bounded type if there exists $C > 0$ such that

$$\left|\rho - \frac{p}{q}\right| \geq \frac{C}{q^2},$$

for any natural numbers p and $q \neq 0$. Recall from Lemma A.1 that, despite being uncountable and dense in the unit interval, the set of numbers of bounded type has zero Lebesgue measure. The following lemma says that for any small $\delta > 0$ in Definition A.3 we capture almost every number.

Lemma A.4 *For any given $\delta > 0$, the set*

$$D_\delta = \left\{\rho \in [0, 1] : \exists\ C > 0 \textit{ such that } \left|\rho - \frac{p}{q}\right| \geq \frac{C}{q^{2+\delta}} \quad \forall\ p, q \in \mathbb{N}\right\}$$

has full Lebesgue measure in $[0, 1]$.

In particular, the set of Liouville numbers has zero Lebesgue measure (in fact, its Hausdorff dimension is also equal to zero, see Milnor 2006, Lem. C.7).

Proof Fix some decreasing sequence $\{\varepsilon_n\}_{n\in\mathbb{N}} \subset (0, 1/2)$ such that $\varepsilon_n \to 0$ as n goes to infinity. For each $n \in \mathbb{N}$ and each rational number $p/q \in [0, 1]$ let $I_n(p/q)$ be the open interval centered at p/q with radius $\varepsilon_n/q^{2+\delta}$, and let U_n be the union of these intervals over all rational numbers, i.e.,

$$\begin{aligned} U_n &= \bigcup_{p/q\in[0,1]\cap\mathbb{Q}} I_n(p/q) \\ &= \left\{\rho \in [0, 1] : \exists\ p, q \in \mathbb{N} \text{ such that } \left|\rho - \frac{p}{q}\right| < \frac{\varepsilon_n}{q^{2+\delta}}\right\}. \end{aligned}$$

Note that each U_n is open and dense in the unit interval, and that the sequence $\{U_n\}_{n\in\mathbb{N}}$ is a nested sequence. Since

$$\bigcap_{n\in\mathbb{N}} U_n = [0, 1] \setminus D_\delta\,,$$

it is enough to prove that $\lim_n m(U_n) = 0$. With this purpose, fix some $n \in \mathbb{N}$ and consider, for any positive integer q, the set

$$U_n(q) = \left\{\rho \in [0, 1] : \exists\ p \in \{0, 1, \ldots, q-1, q\} \text{ such that } \left|\rho - \frac{p}{q}\right| < \frac{\varepsilon_n}{q^{2+\delta}}\right\}.$$

From

$$U_n = \bigcup_{q\in\mathbb{N}\setminus\{0\}} U_n(q) \quad \text{and} \quad m\big(U_n(q)\big) = \frac{2\,\varepsilon_n}{q^{1+\delta}}\,,$$

we obtain

$$m(U_n) \leq 2\,\varepsilon_n \sum_{q\geq 1} \frac{1}{q^{1+\delta}}\,,$$

which converges to zero as n goes to infinity (recall that δ is assumed to be strictly positive!). □

Incidentally we have proved that $[0, 1]\setminus D_\delta$ is a *residual* set, in the sense of Baire, since each U_n is open and dense. With minor adaptations, this implies the following fact.

Lemma A.5 *The set of Liouville numbers is residual in* $[0, 1]$, *in the sense of Baire.*

Exercises

Exercise A.1 Show that, when restricted to each interval $M_k = \big(1/(k+1), 1/k\big)$, the Gauss map has a unique fixed point, given by $\rho_k = \left(\sqrt{k^2+4} - k\right)/2$. In particular, if $\varphi = \left(1+\sqrt{5}\right)/2$ denotes the famous *golden ratio*, we have $\rho_1 = \varphi - 1$.

Exercise A.2 For each $k \in \mathbb{N}$, show that the continued fraction expansion of ρ_k equals $[k, k, k, \ldots]$ (note that, for even integer k, this expansion has been obtained in Exercise 1.4).

Exercise A.3 Show that the set of periodic points of the Gauss map is dense in $[0, 1]$.

Exercise A.4 Show that the Gauss map is transitive (in fact, topologically mixing) in the unit interval.

Exercise A.5 Show that any periodic point of the Gauss map is the root of a quadratic polynomial with integer coefficients.

Exercise A.6 Show that $\rho \in (0, 1)$ is a rational number if, and only if, there exists $n \geq 1$ such that $G^n(\rho) = 0$.

Exercise A.7 Prove that the map h from $[0, 1] \setminus \mathbb{Q}$ to $\mathbb{N}^{\mathbb{N}}$ identifying each irrational number to its itinerary under the Gauss map is a homeomorphism.

Exercise A.8 Show that the Gauss measure ν is *invariant* under the Gauss map.

Exercise A.9 Following Chap. 8, the *Lyapunov exponent* of the Gauss map at a given point $\rho \in (0, 1) \setminus \mathbb{Q}$ is defined as

$$\chi_G(\rho) = \lim_{n\to+\infty} \frac{1}{n} \log \left|DG^n(\rho)\right|, \tag{A.4}$$

whenever the limit exists. For any such ρ, let $f : \boldsymbol{S}^1 \to \boldsymbol{S}^1$ be a circle homeomorphism with rotation number ρ and unique invariant measure μ. Using identity (6.45) from Exercise 6.2, show that for any $x \in \boldsymbol{S}^1$ and any $n \in \mathbb{N}$ we have

$$\lim_{n\to+\infty} \frac{\log \mu(I_n)}{n} = -\frac{\chi_G(\rho)}{2},$$

where I_n is the interval with endpoints x and $f^{q_n}(x)$ containing $f^{q_{n+2}}(x)$.

Exercise A.10 Show that $\chi_G(\rho_k) = \log\left|DG(\rho_k)\right| = -2\,\log(\rho_k)$ for all $k \in \mathbb{N}$. In particular, $\chi_G(\rho_1) = 2\log\varphi$, since $\rho_1 = 1/\varphi$.

Exercise A.11 Let $\rho \in (0, 1)$ be such that the limit in (A.4) exists. Show that $\chi_G(\rho) \geq \chi_G(\rho_1)$, with equality if, and only if, $\rho = \rho_1$.

Exercise A.12 Using Birkhoff's Ergodic Theorem, show that the limit in (A.4) exists for Lebesgue almost every $\rho \in (0, 1)$, and equals

$$\int_{[0,1]} \log |DG|\, d\nu = \frac{\pi^2}{6 \log 2} .$$

(*Hint:* Use integration by parts to deduce that

$$\int_{[0,1]} \log |DG|\, d\nu = \frac{2}{\log 2} \int_0^1 \frac{\log(1+\rho)}{\rho}\, d\rho .$$

Solve this integral using Taylor series, and recall that $\sum_{n=1}^{+\infty} (-1)^{n-1} \frac{1}{n^2} = \pi^2/12$).

Exercise A.13 Let $f : S^1 \to S^1$ be a homeomorphism with irrational rotation number $\rho \in (0, 1)$ and unique invariant measure μ. Conclude from Exercise A.9 and A.12 that for Lebesgue almost every ρ we have

$$\lim_{n\to+\infty} \frac{\log \mu(I_n)}{n} = \frac{-\pi^2}{12 \log 2} .$$

In other words, the decay of $\mu(I_n)$ is comparable to $\exp\left(\frac{-\pi^2}{12 \log 2} n\right)$ for almost every rotation number ρ.

Exercise A.14 Following the same ideas, one proves (see for instance Iosifescu and Kraaikamp 2002, Th. 4.1.26) that for Lebesgue almost every ρ we have

$$\lim_{n\to+\infty} \frac{\log q_n(\rho)}{n} = \frac{\pi^2}{12 \log 2} .$$

Conclude from this and Exercise A.13 that for Lebesgue almost every ρ we have

$$\lim_{n\to+\infty} \frac{1}{n} \log \left| \rho - \frac{p_n(\rho)}{q_n(\rho)} \right| = \frac{-\pi^2}{6 \log 2} .$$

Exercise A.15 Prove Lemma A.5.

Appendix B
Cohomological Equations and Smooth Conjugacies

In this appendix we provide a proof of Theorem 4.11, following the original work of Khanin and Teplinsky (2009). With this purpose, fix $\alpha \in (0, 1)$ and let $f \in \text{Diff}^{2+\alpha}(S^1)$ whose irrational rotation number $\rho \in (0, 1)$ is Diophantine of order $\delta \geq 0$, i.e., there exists a constant $C > 0$ such that

$$\left|\rho - \frac{p}{q}\right| \geq \frac{C}{q^{2+\delta}}$$

for any rational number p/q. We assume, moreover, that $\delta < \alpha$.

Besides Chap. 4, we have seen the Diophantine condition several times in the present book, see Chaps. 3, 7, and 8. An equivalent definition can be given as follows: ρ is Diophantine of order δ if there exists a constant $M > 0$ such that $q_{n+1} \leq M\, q_n^{1+\delta}$ for all $n \in \mathbb{N}$ (Exercise 4.10), where $\{q_n\}$ is the sequence of return times of ρ. In particular, note that ρ is of bounded type if, and only if, it is Diophantine of order 0 (Exercise 4.11). Recall, finally, that we have proved in Appendix A that Diophantine numbers in (0, 1) form a meager set (in the sense of Baire) with full Lebesgue measure (see Lemmas A.4 and A.5).

B.1 The Cohomological Equation

As usual, we denote by $C^0(S^1)$ the space of continuous real functions of the circle, endowed with the uniform convergence topology. In this space, we consider the following *cohomological* equation:

$$\log Df = \psi - \psi \circ f\,. \tag{B.1}$$

E. de Faria, P. Guarino, *Dynamics of Circle Mappings*, IMPA Monographs,
https://doi.org/10.1007/978-3-031-67495-2

Note that if there exists a C^1 diffeomorphism h conjugating f with the rigid rotation R_ρ (i.e., $h \circ f = R_\rho \circ h$), then $\psi = \log Dh$ belongs to $C^0(S^1)$ and is a solution of the cohomological equation (B.1).

Conversely, let us assume for a moment that there is a solution $\psi \in C^0(S^1)$ of (B.1). Just as we did in Sect. 4.1, during the proof of Theorem 4.3, we consider $\phi : S^1 \to \mathbb{R}$ given by $\phi(x) = c \cdot e^{\psi(x)}$, where the positive constant c is chosen so that $\int_{S^1} \phi(x)\,dx = 1$ (in other words, $c = 1/\int_{S^1} e^{\psi(x)}\,dx$). From the cohomological equation (B.1), we immediately obtain

$$\phi(x) = \phi\big(f(x)\big)\,Df(x) \tag{B.2}$$

for all $x \in S^1$. With this at hand, we fix some point $w_0 \in S^1$ and consider $h : S^1 \to S^1$ defined by $Dh = \phi$ and $h(w_0) = 1$. More precisely,

$$h(z) = \int_{w_0}^{z} \phi(x)\,dx \quad (\mathrm{mod} 1)\,.$$

Since ϕ is continuous, h is of class C^1. Moreover, from our choice of the constant c above, $\int_{S^1} Dh(x)\,dx = 1$ which implies that h is a homeomorphism, i.e., it has topological degree one. Since Dh is strictly positive, the inverse h^{-1} is also C^1. Therefore, $h \in \mathrm{Diff}^1(S^1)$. Finally, using again that $Dh = \phi$, the functional equation (B.2) implies that $Dh = D(h \circ f)$ in the whole circle. Thus, any given lifts of h and $h \circ f$ to the real line differ by a constant. In other words, the C^1 diffeomorphism h conjugates f with a rigid rotation (note, in particular, that the rotation number ρ equals $\int_{w_0}^{f(w_0)} \phi(x)\,dx$).

Summarizing, we have proved that if there exists a solution of the cohomological equation (B.1) in $C^0(S^1)$, then f is C^1 conjugate to the rigid rotation R_ρ. Our main task, therefore, is to solve equation (B.1). In the setting of Sect. 4.1, this was done by means of the Gottschalk–Hedlund theorem. In this appendix we will follow a different path, after Khanin and Teplinsky (2009).

B.2 Solving the Cohomological Equation

Let us fix some point $w_0 \in S^1$; recall that its orbit $\mathcal{O}_f(w_0) = \{w_i = f^i(w_0) : i \in \mathbb{Z}\}$ is dense in S^1. We define a function $\psi : \mathcal{O}_f(w_0) \to \mathbb{R}$ by the initial condition $\psi(w_0) = 0$ and the recursive formula

$$\psi(w_{i+1}) = \psi(w_i) - \log Df(w_i) \quad \text{for all } i \in \mathbb{Z}. \tag{B.3}$$

Our main task to prove Theorem 4.11 is to establish the following fact.

Proposition B.1 *The function ψ defined above is continuous.*

Since $\mathcal{O}_f(w_0)$ is dense in S^1, Proposition B.1 allows us to extend ψ continuously to the whole circle, obtaining in this way a solution of the cohomological equation (B.1) in $C^0(S^1)$. As it turns out, the estimates we will obtain on our way to prove Proposition B.1 allow us to also prove that the function ϕ defined above is Hölder continuous with (positive) exponent $\alpha - \delta$ (see Appendix B.6.1), thus proving Theorem 4.11.

Remark B.1 In the fourth and final part of this book, we have presented renormalization methods developed to prove that certain topological conjugacies are in fact smooth diffeomorphisms. In order to keep this appendix independent of those tools, we will not mention any renormalization operator here. However, the reader will not fail to notice the overlap with some of the ideas presented in Chap. 10. Indeed, after establishing some distortion estimates in Appendix B.3, we will prove in Appendix B.4 that for any given $f \in \mathrm{Diff}^{2+\alpha}(S^1)$ with irrational rotation number ρ, the sequences $\{f^{q_n}\}_{n\in\mathbb{N}}$ and $\{R_\rho^{q_n}\}_{n\in\mathbb{N}}$ converge together exponentially fast in $\mathrm{Diff}^1(S^1)$, endowed with the standard C^1 metric, where $\{q_n\}_{n\in\mathbb{N}}$ is the sequence of closest return times of ρ (both sequences converging, as expected, to the identity map).

An important remark here is that this exponential convergence holds for *any* irrational rotation number. By assuming that ρ is Diophantine of order δ, with $0 \leq \delta < \alpha$, we will be able to prove in Appendix B.5 and Appendix B.6 that such exponential convergence implies rigidity: any topological conjugacy between f and R_ρ is a $C^{1+\alpha-\delta}$ diffeomorphism. Recall from Chap. 4 that this is certainly *not* the case without a Diophantine condition (see Sect. 4.3).

Finally, let us point out that in Appendix B.6.1 we will establish the following nice estimate. For each $n \in \mathbb{N}$ let $\Lambda_n = |q_n\,\rho - p_n|$, and recall from Chap. 1 that $\{\Lambda_n\}$ decays to zero exponentially fast (as fast as q_{n+1}^{-1}, see Theorem 1.2). As it turns out, if ρ is Diophantine of order δ, there exists $K > 1$ such that $\Lambda_n^{1+\delta} \leq K\,\Lambda_{n+1}$ for all $n \in \mathbb{N}$ (see Remark B.4 in Appendix B.5 below). With this at hand, we will prove in Appendix B.6.1 that

$$\left\| f^{q_n} - R_\rho^{q_n} \right\|_{C^1} = O\big(\Lambda_n^{\frac{\alpha}{1+\delta}}\big) \quad \text{for all } n \in \mathbb{N}.$$

Note that the obtained bound $\Lambda_n^{\frac{\alpha}{1+\delta}}$ depends only on ρ and α; it does not depend on any other data coming from f.

B.3 Distortion Estimates

This section is devoted to establish some fundamental distortion estimates that will be needed in the proof of Theorem 4.11. It is divided in two parts: in Appendix B.3.1 we deal with diffeomorphisms in the $C^{1+\mathrm{BV}}$ class, extending some results obtained in Chap. 3, while in Appendix B.3.2 we deal with $C^{2+\alpha}$ diffeomorphisms, focusing on cross-ratio distortion estimates (in the same spirit as we did in the third part of this book).

B.3.1 Distortion Estimates in C^{1+BV}

Just as in Chap. 3, we consider in this subsection a C^1 diffeomorphism $f : S^1 \to S^1$ with irrational rotation number such that $\log Df$ is a function of bounded variation, say $V = \mathrm{Var}(\log Df)$ (we write $f \in C^{1+\mathrm{BV}}$). For any given $x \in S^1$, let $\mathscr{P}_n(x)$ be the sequence of dynamical partitions constructed in Chap. 6 (see Sect. 6.3.1). Recall that $I_n = I_n(x)$ denotes the interval with endpoints x and $f^{q_n}(x)$ containing $f^{q_{n+2}}(x)$.

Lemma B.1 *For any given $f \in C^{1+BV}$ there exist $\lambda = \lambda(f) \in (0,1)$ and $K = K(f) > 1$ such that $\big|I_{n+m}(x)\big| \le K\,\lambda^m\,\big|I_n(x)\big|$ for any $x \in S^1$ and any $n, m \in \mathbb{N}$.*

Proof As we have seen in Chap. 3, during the proof of Denjoy's Theorem 3.4, $\|\log Df^{q_n}\|_{C^0} \le V$ for any $n \in \mathbb{N}$. Let $\lambda = \lambda(V) \in (0,1)$ and $K = K(V) > 1$ be given by

$$\lambda = \sqrt{\frac{e^V}{1+e^V}} \quad \text{and} \quad K = \sqrt{e^V(1+e^V)}\,.$$

Now fix $x \in S^1$ and $n \in \mathbb{N}$. By combinatorics, $I_n \subsetneq (I_n \setminus I_{n+2}) \cup f^{q_{n+1}}(I_n \setminus I_{n+2})$. Since $\big|f^{q_{n+1}}(I_n \setminus I_{n+2})\big| \le e^V\,|I_n \setminus I_{n+2}|$, we have that $|I_n \setminus I_{n+2}| > |I_n|/(1+e^V)$. Therefore,

$$|I_{n+2}| = |I_n| - |I_n \setminus I_{n+2}| < \left(1 - \frac{1}{1+e^V}\right)|I_n| = \lambda^2\,|I_n|\,.$$

With this at hand, an inductive argument proves Lemma B.1 in the case that m is an even integer. Now, using again $\|\log Df^{q_n}\| \le V$, we have that $e^{-V}\,|I_{n+1}| \le \big|f^{q_n}(I_{n+1})\big| < |I_n|$, since $f^{q_n}(I_{n+1}) \subsetneq I_n$ by combinatorics. From our previous estimates, we know that $|I_{n+m+1}| < \lambda^m\,|I_{n+1}|$ for any even integer m, and then $|I_{n+m+1}| < \lambda^m\,e^V\,|I_n| = K\,\lambda^{m+1}\,|I_n|$, since $K\,\lambda = e^V$. This finishes the proof.

□

Definition B.1 For each $n \in \mathbb{N}$ let

$$\ell_n = \ell_n(f) = \|f^{q_n} - \mathrm{Id}\,\|_{C^0} = \max_{x \in S^1} \big|I_n(x)\big|,$$

and let $\Lambda_n = \ell_n(R_\rho) = |q_n\,\rho - p_n|\,.$

Note that, by minimality, $\ell_n \to 0$. As it easily follows from our previous lemma, the convergence happens at an exponential rate.

Corollary B.1 *For any given* $f \in C^{1+BV}$ *there exist* $\lambda = \lambda(f) \in (0,1)$ *and* $K = K(f) > 1$ *such that* $\ell_{n+m} \leq K\,\lambda^m\,\ell_n$ *for any* $n, m \in \mathbb{N}$.

Proof Let $x \in S^1$ be such that $|I_{n+m}(x)| = \ell_{n+m}$. By definition, $\ell_n \geq |I_n(x)|$ and then Corollary B.1 follows at once from Lemma B.1. □

Lemma B.2 *We have* $\ell_n \geq \Lambda_n$ *for any* $n \in \mathbb{N}$.

Proof On the 2-dimensional torus $\mathbb{T}^2 = S^1 \times S^1$ we consider the product measure $m \times \mu$, where μ denotes the unique f-invariant Borel probability measure. We also fix some $n \in \mathbb{N}$ and consider $\Omega_n = \{(x, y) \in \mathbb{T}^2 : y \in I_n(x)\}$. By Fubini's Theorem,

$$(m \times \mu)(\Omega_n) = \int_{S^1} \mu\big(I_n(x)\big)\, dm(x) = \int_{S^1} m\big(I_n(f^{-q_n}(y))\big)\, d\mu(y).$$

Since $\mu\big(I_n(x)\big) = \Lambda_n$ for all $x \in S^1$, we get

$$\int_{S^1} m\big(I_n(f^{-q_n}(y))\big)\, d\mu(y) = \Lambda_n.$$

Since, by definition, $m\big(I_n(f^{-q_n}(y))\big) \leq \ell_n$ for all $y \in S^1$, we obtain the desired estimate $\ell_n \geq \Lambda_n$. □

Lemma B.3 *For any given* $f \in C^{1+BV}$ *there exists* $K = K(f) > 1$ *such that for any* $x \in S^1$, $n, m \in \mathbb{N}$ *and* $j \in \mathbb{N}$ *with* $0 \leq j \leq q_{n+1}$ *we have*

$$\frac{1}{K}\,\frac{|I_{n+m}(x)|}{|I_n(x)|} \leq \frac{|I_{n+m}(f^j(x))|}{|I_n(f^j(x))|} \leq K\,\frac{|I_{n+m}(x)|}{|I_n(x)|}.$$

Proof This result is a straightforward consequence of Koebe's distortion principle (Lemma 5.2). Indeed, fix some $n \in \mathbb{N}$ and consider the three intervals $L_n = I_n\big(f^{-q_n}(x)\big) \setminus I_{n+m}(x)$, $M_n = I_n(x) \cup I_{n+m}(x)$ and $R_n = I_n\big(f^{q_n}(x)\big)$. Note that if m is an even integer, these intervals are just consecutive fundamental domains for f^{q_n}. In any case, we also consider

$$T_n = L_n \cup M_n \cup R_n = \big[f^{-q_n}(x), f^{2q_n}(x)\big].$$

By combinatorics, the trajectory $\{T_n, f(T_n), \ldots, f^{q_{n+1}-1}(T_n)\}$ has multiplicity of intersection equal to 3. Moreover, for any $0 \leq j \leq q_{n+1}$, the space of $f^j(M_n)$ inside $f^j(T_n)$ is bounded from below, as it follows from $\|\log Df^{q_n}\| \leq V$ and the fact that $|I_{n+m}(x)| \leq K\lambda^m |I_n(x)|$ (recall Lemma B.1). With this at hand, Lemma 5.2 immediately implies Lemma B.3. □

Lemma B.4 *For any given* $f \in C^{1+BV}$ *there exists* $K = K(f) > 1$ *such that* $|I_{n+m}(w_0)|/|I_n(w_0)| \leq K\,\ell_{n+m}/\ell_n$ *for all* $n, m \in \mathbb{N}$.

Proof Fix $n \in \mathbb{N}$ and let $x \in S^1$ be such that $|I_n(x)| = \ell_n$. Let $i \in \{0, 1, \ldots, q_{n+1} + q_n - 1\}$ be such that

$$I_n(x) \subset I_n(w_i) \cup I_n(w_{i+q_n}).$$

Say that $|I_n(w_i)| \geq \ell_n/2$. Then

$$\frac{|I_{n+m}(w_i)|}{|I_n(w_i)|} \leq \frac{2\,\ell_{n+m}}{\ell_n}.$$

By our previous result (Lemma B.3), the same estimate holds for $i = 0$, up to a multiplicative constant depending only on f. Indeed, note that Lemma B.3 is applied at most three times, since $q_{n+1} + 2q_n < 3q_{n+1}$. Finally, note that if $|I_n(w_i)| < \ell_n/2$, we must have $|I_n(w_{i+q_n})| \geq \ell_n/2$, but then $|I_n(w_i)| \geq e^{-V}|I_n(w_{i+q_n})| \geq \ell_n/2e^V$, where $V = \mathrm{Var}(\log Df)$, and the proof goes in the same way. □

B.3.2 Distortion Estimates in $C^{2+\alpha}$

From now on, we assume that $f \in \mathrm{Diff}^{2+\alpha}(S^1)$ for some $\alpha \in (0, 1)$. Following Khanin and Teplinsky (2009), the cross-ratio distortion of f with respect to four pairwise distinct points x_1, x_2, x_3, x_4 will be written in this appendix as

$$\mathrm{CrD}(f;\ x_1, x_2, x_3, x_4) = \frac{\dfrac{f(x_1) - f(x_2)}{x_1 - x_2}\ \dfrac{f(x_3) - f(x_4)}{x_3 - x_4}}{\dfrac{f(x_2) - f(x_3)}{x_2 - x_3}\ \dfrac{f(x_4) - f(x_1)}{x_4 - x_1}},$$

independently of the order of these points. Note that in the particular case $x_2 < x_1 < x_3 < x_4$, this definition coincides with the one used along the book (namely, the b-cross-ratio), whereas the configuration $x_1 < x_4 < x_3 < x_2$ provides the a-cross-ratio (recall Chap. 5). Note also that the previous definition can be extended to the case of points *not necessarily distinct*, by simply replacing a ratio like $\big(f(x_i) - f(x_i)\big)/(x_i - x_i)$ by $Df(x_i)$.

Remark B.2 In order to simplify some of the formulas below, we shall use the notation $\Delta f[x, y] = (f(x) - f(y))/(x - y)$ wherever convenient.

Lemma B.5 *For any given $f \in \mathrm{Diff}^{2+\alpha}(S^1)$ there exists a constant $K = K(f) > 1$ with the following property. Given $n \in \mathbb{N}$ and four points x_1, x_2, x_3, x_4 in $I_n(w_0)$ we have*

$$\big|\log \mathrm{CrD}(f^{q_{n+1}};\ x_1, x_2, x_3, x_4)\big| \leq K\,\ell_n^{\alpha}.$$

In the same way,

$$\left|\log \mathrm{CrD}(f^{q_n};\, x_1, x_2, x_3, x_4)\right| \leq K\, \ell_{n+1}^{\alpha}\,.$$

for any given x_1, x_2, x_3, x_4 in $I_{n+1}(w_0)$.

Remark B.3 As it turns out, the second estimate in Lemma B.5 above also holds for points x_1, x_2, x_3, x_4 in $I_{n-1}(w_0)$. See Khanin and Teplinsky (2009, Lem. 6).

The proof of Lemma B.5, to be given below, will be a combination of the chain rule for the cross-ratio distortion and the following estimate on the distortion of a single iterate.

Lemma B.6 *For any given $f \in \mathrm{Diff}^{2+\alpha}(S^1)$ there exists a constant $K = K(f) > 1$ such that given an interval $I \subset S^1$ and four points $x_1, x_2, x_3, x_4 \in I$ we have*

$$|\log \mathrm{CrD}(f;\, x_1, x_2, x_3, x_4)| \leq K\, |I|^{1+\alpha}. \tag{B.4}$$

Proof Fix some point $a \in I$, let p be the second-order Taylor polynomial of f around a and let r be its remainder term. In other words, $f(x) = p(x) + r(x)$ for all $x \in I$, where $p(x) = f(a) + f'(a)\,(x-a) + \frac{1}{2}\, f''(a)\,(x-a)^2$. For each $i \in \{1, \ldots, 4\}$ let $d_i = x_i - a$, and consider

$$\kappa_{i,j} = \frac{d_i + d_j}{2}\,\mathcal{N} f(a) + \frac{1}{f'(a)}\,\frac{r(x_i) - r(x_j)}{x_i - x_j}\,,$$

for any $i \neq j$ in $\{1, \ldots, 4\}$, where $\mathcal{N} f$ denotes the non-linearity of f (recall Chaps. 3 and 5). A straightforward computation gives

$$\Delta f[x_i, x_j] = \frac{f(x_i) - f(x_j)}{x_i - x_j} = f'(a)\,\big(1 + \kappa_{i,j}\big),$$

for any $i \neq j$ in $\{1, \ldots, 4\}$. We claim that there exists $K = K(f) > 1$ such that $|\kappa_{i,j}| \leq K\,|I|$. Indeed, on one hand, using that r'' is α-Hölder continuous on I and that $r'(a) = r''(a) = 0$, it is not difficult to deduce that

$$\big|\Delta r[x_i, x_j]\big| = \left|\frac{r(x_i) - r(x_j)}{x_i - x_j}\right| \leq K\, |I|^{1+\alpha}\,. \tag{B.5}$$

On the other hand, since f belongs to $\mathrm{Diff}^{2+\alpha}(S^1)$, both $\mathcal{N} f(a)$ and $1/f'(a)$ are bounded, and since we obviously have $|d_i| \leq |I|$, we deduce that $|\kappa_{i,j}| \leq K\,|I|$ for any $i \neq j$ in $\{1, \ldots, 4\}$, as claimed. Now we write

$$\begin{aligned}&\left|\log \mathrm{CrD}(f;\, x_1, x_2, x_3, x_4)\right| = \\ &= |\log \Delta f[x_1, x_2] - \log \Delta f[x_2, x_3] + \log \Delta f[x_3, x_4] - \log \Delta f[x_4, x_1]|\end{aligned}$$

$$= \left| \log \frac{\Delta f[x_1, x_2]}{f'(a)} - \log \frac{\Delta f[x_2, x_3]}{f'(a)} + \log \frac{\Delta f[x_3, x_4]}{f'(a)} - \log \frac{\Delta f[x_4, x_1]}{f'(a)} \right|$$
$$= \left| \log(1+\kappa_{1,2}) - \log(1+\kappa_{2,3}) + \log(1+\kappa_{3,4}) - \log(1+\kappa_{4,1}) \right| .$$

Using that $|\kappa_{i,j}| \leq K\,|I|$ and that $\kappa \mapsto \log(1+\kappa)$ is tangent to the identity at the origin, we deduce that

$$\begin{aligned} &\left| \log \mathrm{CrD}(f;\; x_1, x_2, x_3, x_4) \right| \\ &\quad = \left| \log(1+\kappa_{1,2}) - \log(1+\kappa_{2,3}) + \log(1+\kappa_{3,4}) - \log(1+\kappa_{4,1}) \right| \\ &\quad \leq \left| \kappa_{1,2} - \kappa_{2,3} + \kappa_{3,4} - \kappa_{4,1} \right| + K\,|I|^2 \\ &\quad = \frac{1}{f'(a)} \left| \Delta r[x_1, x_2] - \Delta r[x_2, x_3] + \Delta r[x_3, x_4] - \Delta r[x_4, x_1] \right| + K\,|I|^2 . \end{aligned}$$

Combined with (B.5), this finishes the proof of Lemma B.6. □

With estimate (B.4) at hand, we are ready to prove Lemma B.5.

Proof of Lemma B.5 We will prove the first estimate in the statement, the proof of the second being exactly the same. By Lemma B.6 and the chain rule for the cross-ratio distortion (recall Lemma 5.3),

$$\begin{aligned} &\left| \log \mathrm{CrD}(f^{q_{n+1}};\; x_1, x_2, x_3, x_4) \right| \\ &\qquad \leq \sum_{j=0}^{q_{n+1}-1} \left| \log \mathrm{CrD}(f;\; f^j(x_1), f^j(x_2), f^j(x_3), f^j(x_4)) \right| \\ &\qquad \leq K \sum_{j=0}^{q_{n+1}-1} \left| I_n(w_j) \right|^{1+\alpha} \\ &\qquad \leq K\,\ell_n^{\alpha} \sum_{j=0}^{q_{n+1}-1} \left| I_n(w_j) \right| \leq K\,\ell_n^{\alpha} , \end{aligned}$$

since the intervals $\left\{ I_n(w_j) \right\}_{j=0}^{q_{n+1}-1}$ are pairwise disjoint (recall Sect. 6.3.1). □

To establish further estimates, we define for each $n \in \mathbb{N}$ and $x \in \boldsymbol{S}^1$,

$$M_n(x) = \frac{w_{q_{n+1}} - f^{q_{n+1}}(x)}{w_0 - x} \Bigg/ \frac{f^{q_{n+1}}(x) - w_{q_{n+1}+q_n}}{x - w_{q_n}}$$

and

$$K_n(x) = \frac{w_{q_n} - f^{q_n}(x)}{w_0 - x} \Big/ \frac{f^{q_n}(x) - w_{q_{n+1}+q_n}}{x - w_{q_{n+1}}}.$$

Note that

$$\frac{M_n(x)}{M_n(y)} = \mathrm{CrD}(f^{q_{n+1}};\ w_0, x, w_{q_n}, y) \tag{B.6}$$

and

$$\frac{K_n(x)}{K_n(y)} = \mathrm{CrD}(f^{q_n};\ w_0, x, w_{q_{n+1}}, y)$$

for any x, y in S^1. Thus, by Lemma B.5 we have

$$\frac{M_n(x)}{M_n(y)} = 1 + O(\ell_n^{\alpha}) \tag{B.7}$$

for any x, y in $I_n(w_0)$. Analogously,

$$\frac{K_n(x)}{K_n(y)} = 1 + O(\ell_{n+1}^{\alpha})$$

for any x, y in $I_{n+1}(w_0)$.

Lemma B.7 *We have $M_n(x) = 1 + O(\ell_n^{\alpha})$ for all $x \in I_n(w_0)$, and $K_n(x) = 1 + O(\ell_{n+1}^{\alpha})$ for all $x \in I_{n+1}(w_0)$.*

Proof Note first that from $e^{-V} \le |Df^{q_n}(x)| \le e^{V}$, we deduce at once $e^{-2V} \le |M_n(x)| \le e^{2V}$ (and a similar bound for K_n). Now for each $n \in \mathbb{N}$ let

$$m_n = \sqrt{M_n(w_0)\, M_n(w_{q_n})}\,,$$

and note that

$$m_n = \sqrt{K_n(w_0)\, K_n(w_{q_{n+1}})} = \sqrt{\frac{Df^{q_n}(w_0)}{Df^{q_n}(w_{q_{n+1}})}} = \sqrt{\frac{Df^{q_{n+1}}(w_0)}{Df^{q_{n+1}}(w_{q_n})}}\,,$$

since we obviously have $Df^{q_n}(w_{q_{n+1}})\, Df^{q_{n+1}}(w_0) = Df^{q_{n+1}}(w_{q_n})\, Df^{q_n}(w_0)$ by the chain rule. We remark that

$$M_n(x) = m_n + O(\ell_n^{\alpha}) \quad \text{for all } x \in I_n(w_0)\,, \tag{B.8}$$

and

$$K_n(x) = m_n + O(\ell_{n+1}^{\alpha}) \quad \text{for all } x \in I_{n+1}(w_0)\,. \tag{B.9}$$

For instance, (B.8) follows at once from (B.6), (B.7) and the following identity:

$$M_n(x) = m_n + \frac{M_n^2(x)}{M_n(x)+m_n}\left(1 - \frac{M_n(w_0)}{M_n(x)}\,\frac{M_n(w_{q_n})}{M_n(x)}\right).$$

Now we claim that

$$m_n - 1 = O(\ell_{n+1}^{\alpha})\,.$$

Indeed, by combining Lemma B.4, (B.8) and (B.9) with the following formula (which is left to the reader as an exercise)

$$K_{n+1}(w_{q_n}) - 1 = \frac{\left|I_{n+2}(w_0)\right|}{\left|I_n(w_0)\right|}\left(M_n(w_{q_{n+2}}) - 1\right),$$

we deduce

$$m_{n+1} - 1 = \frac{\left|I_{n+2}(w_0)\right|}{\left|I_n(w_0)\right|}(m_n - 1) + O(\ell_{n+2}^{\alpha})\,, \tag{B.10}$$

where we have used estimate $K_{n+1}(x) = m_{n+1} + O(\ell_{n+2}^{\alpha})$ for $x = w_{q_n}$ (recall Remark B.3). By induction on (B.10), we obtain

$$m_n - 1 = \frac{\left|I_{n+1}(w_0)\right|\left|I_n(w_0)\right|}{\left|I_1(w_0)\right|\left|I_0(w_0)\right|}(m_0 - 1) + \sum_{j=1}^{n}\frac{\left|I_{n+1}(w_0)\right|\left|I_n(w_0)\right|}{\left|I_{j+1}(w_0)\right|\left|I_j(w_0)\right|}\,O(\ell_{j+1}^{\alpha})$$

for all $n \geq 2$. Using again Lemma B.4 and also Corollary B.1, we finally obtain

$$|m_n - 1| \leq K\left(\ell_{n+1}\,\ell_n + \ell_{n+1}^{\alpha}\sum_{j=0}^{n-1}(\lambda^{2-\alpha})^j\right),$$

which implies the claim, and finishes the proof of Lemma B.7. □

B.4 C^1 Exponential Convergence

As before, let $f \in \mathrm{Diff}^{2+\alpha}(S^1)$ with irrational rotation number. As we know from Chap. 3, f is topologically conjugate to a rigid rotation (recall Denjoy's Theorem 3.4). This implies, in particular, that the sequence of circle diffeomorphisms $\{f^{q_n}\}_{n\in\mathbb{N}}$ converges uniformly (i.e., in the C^0 metric) to the identity map. In this section we will prove that such converge actually holds in the C^1 metric, at an exponential rate. With this purpose, for each $n \in \mathbb{N}$ consider

$$\varepsilon_n = \ell_{n-1}^{\alpha} + \frac{\ell_n}{\ell_{n-1}}\,\ell_{n-2}^{\alpha} + \frac{\ell_n}{\ell_{n-2}}\,\ell_{n-3}^{\alpha} + \cdots + \frac{\ell_n}{\ell_1}\,\ell_0^{\alpha} + \frac{\ell_n}{\ell_0}\,,$$

where the sequence $\{\ell_n\}$ comes from Definition B.1. In particular,

$$\varepsilon_{n+1} = \ell_n^{\alpha} + \frac{\ell_{n+1}}{\ell_n}\,\varepsilon_n\,. \tag{B.11}$$

Proposition B.2 *There exists a constant $K = K(f) > 0$ such that*

$$1 - K\,\varepsilon_n \leq \left|Df^{q_n}(x)\right| \leq 1 + K\,\varepsilon_n$$

for all $n \in \mathbb{N}$ and $x \in \mathbf{S}^1$.

In other words, $\log Df^{q_n}(x) = O(\varepsilon_n)$ for all $x \in \mathbf{S}^1$ and $n \in \mathbb{N}$. An immediate consequence of Proposition B.2 is that

$$\left\| f^{q_n} - \mathrm{Id} \right\|_{C^1} = O(\varepsilon_n).$$

Moreover, note that by Corollary B.1 we have $\varepsilon_n = O(\lambda^{\alpha(n-1)})$, and then the sequence $\{\varepsilon_n\}$ goes to zero exponentially fast (see also Lemma B.9 below).

Proof of Proposition B.2 Note first the identities

$$\frac{Df^{q_{n+1}}(w_0)}{M_n(w_0)} = \frac{\left|I_n(w_{q_{n+1}})\right|}{\left|I_n(w_0)\right|} \quad \text{and} \quad \frac{Df^{q_n}(w_0)}{K_n(w_0)} = \frac{\left|I_{n+1}(w_{q_n})\right|}{\left|I_{n+1}(w_0)\right|}\,.$$

Since we obviously have $\left|I_n(w_{q_{n+1}})\right| + \left|I_{n+1}(w_{q_n})\right| = \left|I_n(w_0)\right| + \left|I_{n+1}(w_0)\right|$, we obtain

$$\frac{Df^{q_{n+1}}(w_0)}{M_n(w_0)} - 1 = \frac{\left|I_{n+1}(w_0)\right|}{\left|I_n(w_0)\right|}\left(1 - \frac{Df^{q_n}(w_0)}{K_n(w_0)}\right).$$

In other words,

$$Df^{q_{n+1}}(w_0) - M_n(w_0) = \frac{\left|I_{n+1}(w_0)\right|}{\left|I_n(w_0)\right|}\,\frac{M_n(w_0)}{K_n(w_0)}\big(K_n(w_0) - Df^{q_n}(w_0)\big).$$

By Lemma B.7,

$$Df^{q_{n+1}}(w_0) - 1 = \frac{\left|I_{n+1}(w_0)\right|}{\left|I_n(w_0)\right|}\big(1 - Df^{q_n}(w_0)\big) + O(\ell_n^{\alpha}).$$

By induction,

$$Df^{q_n}(w_0) - 1 =$$

$$(-1)^n \frac{|I_n(w_0)|}{|I_0(w_0)|} \big(Df(w_0) - 1\big) + \sum_{j=0}^{n-1} (-1)^{n+j-1} \frac{|I_n(w_0)|}{|I_{j+1}(w_0)|} \, O(\ell_j^\alpha)\,.$$

When combined with Lemma B.4, our last estimate implies Proposition B.2 for $x = w_0$. Since w_0 is an arbitrary point in S^1, this finishes the proof of Proposition B.2. □

B.4.1 Further Distortion Estimates

Fix some large $n \in \mathbb{N}$, and for each $m \in \mathbb{N}$ consider the set

$$\mathscr{C}(n+m, n) = \big\{ i \in \{0, 1, \ldots, q_{n+m+1} - 1\} : \; I_{n+m}(w_i) \subset I_n(w_0) \big\}$$

and the number $c(n+m, n) = \#\mathscr{C}(n+m, n)$. Note that $c(n, n) = 1$, $c(n+1, n) = a_{n+1}$ and that

$$c(n+m+1, n) = a_{n+m+1}\, c(n+m, n) + c(n+m-1, n)$$

for all $m \geq 2$. We also remark that, by induction on m, it is not difficult to prove that

$$\Lambda_n = c(n+m, n)\, \Lambda_{n+m} + c(n+m-1, n)\, \Lambda_{n+m+1}\,. \tag{B.12}$$

Corollary B.2 *Let $K = K(f) > 0$ be given by Proposition B.2. Then we have*

$$\frac{\ell_n}{\ell_{n+m}} \geq c(n+m, n) \left(1 - K \sum_{k=n}^{n+m-1} a_{k+1}\, \varepsilon_{k+1} \right)$$

for all $n, m \in \mathbb{N}$.

Proof Fix $n, m \in \mathbb{N}$ and let $i, j \in \mathscr{C}(n+m, n)$ with, say, $i < j$. By Lemma 10.4 we know that for each $k \in \{n, \ldots, n+m-1\}$ there exists an integer $b_{k+1} \in \{0, \ldots, a_{k+1}\}$ such that

$$j = i + \sum_{k=n}^{n+m-1} b_{k+1}\, q_{k+1}\,.$$

By Proposition B.2 we have

$$\begin{aligned}\left|I_{n+m}(w_j)\right| &\geq \left|I_{n+m}(w_i)\right| \left(1 - K \sum_{k=n}^{n+m-1} b_{k+1}\,\varepsilon_{k+1}\right)\\ &\geq \left|I_{n+m}(w_i)\right| \left(1 - K \sum_{k=n}^{n+m-1} a_{k+1}\,\varepsilon_{k+1}\right).\end{aligned}$$

Now, from the definition of the set $\mathscr{C}(n+m,n)$ we have the obvious estimate

$$\left|I_n(w_0)\right| \geq \sum_{i\in\mathscr{C}(n+m,n)} \left|I_{n+m}(w_i)\right|.$$

Thus, fixed $i \in \mathscr{C}(n+m,n)$ we have

$$\left|I_n(w_0)\right| \geq c(n+m,n)\left|I_{n+m}(w_i)\right| \left(1 - K \sum_{k=n}^{n+m-1} a_{k+1}\,\varepsilon_{k+1}\right).$$

Now let $w_0 \in \boldsymbol{S}^1$ be such that $\ell_{n+m} = \left|I_{n+m}(w_i)\right|$. Then

$$\ell_n \geq \left|I_n(w_0)\right| \geq c(n+m,n)\,\ell_{n+m} \left(1 - K \sum_{k=n}^{n+m-1} a_{k+1}\,\varepsilon_{k+1}\right).$$

This finishes the proof of Corollary B.2. □

B.5 The Diophantine Condition

All previous estimates in this appendix hold for any irrational rotation number. We will now use the Diophantine condition, in order to establish the following bound.

Proposition B.3 *Let $\lambda \in (0,1)$ be given by Lemma B.1. For any given $\lambda_1 \in (\lambda^{\alpha-\delta},1)$ we have*

$$a_n\,\varepsilon_n = O(\lambda_1^n),$$

where $\rho = [a_0, a_1, \dots]$ is the rotation number of f.

Remark B.4 Before entering the proof of Proposition B.3, consider the sequence $\{\Lambda_n\}$ given by Definition B.1. We claim that if the rotation number ρ is Diophantine of order $\delta \geq 0$, then there exists a constant $K = K(\rho) > 0$ such that

$$\Lambda_n^{1+\delta} \le K\,\Lambda_{n+1} \tag{B.13}$$

for all $n \in \mathbb{N}$. Indeed, recall first that from estimate (1.16) in Theorem 1.2 we have

$$\frac{1}{2\,q_n\,q_{n+1}} < \left|\rho - \frac{p_n}{q_n}\right| < \frac{1}{q_n\,q_{n+1}}$$

for all $n \in \mathbb{N}$. In other words,

$$\frac{1}{2\,q_n} < \Lambda_{n-1} < \frac{1}{q_n}\,.$$

Now, if ρ is Diophantine of order δ, there exists a constant $M = M(\rho) > 0$ such that $q_{n+1} \le M\,q_n^{1+\delta}$ for all $n \in \mathbb{N}$ (recall Exercise 4.10), which implies

$$\Lambda_{n-1}^{1+\delta} < \frac{1}{q_n^{1+\delta}} \le \frac{M}{q_{n+1}} < 2M\,\Lambda_n\,.$$

Thus, considering $K = 2M$ we deduce (B.13). This is how the Diophantine condition will be used in this appendix.

B.5.1 Proof of Proposition B.3

A natural number n is a *divergent level* for f if

$$a_{n+1}\,\varepsilon_{n+1} > \lambda_1^{n+1}\,.$$

We denote by $\{n_i\}_{i\in\mathbb{N}}$ the sequence of *all* divergent levels of f. To prove Proposition B.3 we assume, by contradiction, that there exist *infinitely many* divergent levels.

Lemma B.8 *For all $i \in \mathbb{N}$ we have*

$$\frac{\ell_{n_i}}{\ell_{n_i+1}} > \frac{\lambda_1^{n_i+1}}{2\,\varepsilon_{n_i+1}}\,.$$

Proof Fix some $i \in \mathbb{N}$. Since n_i is a divergent level for f, we can choose an integer $b_{n_i+1} \in \{0, 1, \ldots, a_{n_i+1}\}$ so that $b_{n_i+1}\,\varepsilon_{n_i+1} > \lambda_1^{n_i+1}$ but $b_{n_i+1}\,\varepsilon_{n_i+1} \le 1/(2K)$, where $K > 0$ is given by Proposition B.2 (recall here that both ε_n and λ_1^n go to zero as n goes to infinity). By Proposition B.2,

$$
\begin{aligned}
\left|I_{n_i}(w_0)\right| &> \sum_{j=0}^{b_{n_i+1}-1} \left|I_{n_i+1}(w_{jq_{n_i+1}+q_{n_i}})\right| \\
&\geq \left|I_{n_i+1}(w_{q_{n_i}})\right| \sum_{j=0}^{b_{n_i+1}-1} \left(1 - j\,K\,\varepsilon_{n_i+1}\right) \\
&> \left|I_{n_i+1}(w_{q_{n_i}})\right| b_{n_i+1} \left(1 - K\,\varepsilon_{n_i+1}\,b_{n_i+1}\right),
\end{aligned}
$$

where we have used the elementary estimate $\sum_{j=0}^{n-1} j < n^2$. Since $\varepsilon_{n_i+1}\,b_{n_i+1} \leq 1/(2K)$, we deduce that

$$
\left|I_{n_i}(w_0)\right| > \left|I_{n_i+1}(w_{q_{n_i}})\right| b_{n_i+1}/2\,.
$$

Finally, given $i \in \mathbb{N}$ we choose w_0 so that $\ell_{n_i+1} = \left|I_{n_i+1}(w_{q_{n_i}})\right|$, to obtain

$$
\ell_{n_i} \geq \left|I_{n_i}(w_0)\right| > \ell_{n_i+1}\,b_{n_i+1}/2\,.
$$

Thus

$$
\frac{\ell_{n_i}}{\ell_{n_i+1}} > b_{n_i+1}/2 > \frac{\lambda_1^{n_i+1}}{2\,\varepsilon_{n_i+1}}\,,
$$

since $b_{n_i+1}\,\varepsilon_{n_i+1} > \lambda_1^{n_i+1}$. This finishes the proof of Lemma B.8. □

We claim now that $\varepsilon_{n_i+1} < \ell_{n_i}^{\alpha}$ for sufficiently large i. Indeed, from (B.11) we have

$$
\ell_{n_i}^{\alpha} = \varepsilon_{n_i+1} - \frac{\ell_{n_i+1}}{\ell_{n_i}}\,\varepsilon_{n_i}\,,
$$

and then

$$
\ell_{n_i}^{\alpha} > \varepsilon_{n_i+1}\left(1 - \frac{2\,\varepsilon_{n_i}}{\lambda_1^{n_i+1}}\right)
$$

since $\dfrac{\ell_{n_i+1}}{\ell_{n_i}} < \dfrac{2\,\varepsilon_{n_i+1}}{\lambda_1^{n_i+1}}$ by Lemma B.8. Now, since $\varepsilon_{n_i} = O(\lambda^{\alpha(n_i-1)})$ (recall Appendix B.4) and since $\lambda^{\alpha} < \lambda_1$ by definition, the ratio $\varepsilon_{n_i}/\lambda_1^{n_i}$ decays exponentially fast. This implies that $\varepsilon_{n_i+1} < \ell_{n_i}^{\alpha}$ for sufficiently large i, as claimed.

With the claim at hand, Lemma B.8 implies that

$$\frac{\ell_{n_i}}{\ell_{n_i+1}} > \frac{\lambda_1^{n_i+1}}{2\,\varepsilon_{n_i+1}} > \frac{\lambda_1^{n_i+1}}{2\,\ell_{n_i}^{\alpha}}$$

for sufficiently large i. In other words,

$$\ell_{n_i+1} = O\big(\ell_{n_i}^{1+\alpha}/\lambda_1^{n_i}\big). \tag{B.14}$$

On the other hand, by Corollary B.2 we have

$$\begin{aligned}\frac{\ell_{n_i+1}}{\ell_{n_{i+1}}} &\geq c(n_{i+1}, n_i+1)\left(1 - K\sum_{k=n_i+1}^{n_{i+1}-1} a_{k+1}\,\varepsilon_{k+1}\right)\\ &\geq c(n_{i+1}, n_i+1)\left(1 - K\sum_{k=n_i+1}^{n_{i+1}-1} \lambda_1^{k+1}\right),\end{aligned}$$

where we have used that $a_{k+1}\,\varepsilon_{k+1} \leq \lambda_1^{k+1}$ for all $k \in \{n_i+1, \ldots, n_{i+1}-1\}$. In particular, we have

$$\ell_{n_{i+1}} = O\left(\frac{\ell_{n_i+1}}{c(n_{i+1}, n_i+1)}\right). \tag{B.15}$$

Now, recall that $\lambda_1 \in (0,1)$ was defined by

$$0 < \frac{\log\lambda_1}{\log\lambda} < \alpha - \delta.$$

We then fix some $0 < \tau < \alpha - \delta - \log\lambda_1/\log\lambda$, and note that $\lambda^{\alpha-\delta-\tau} < \lambda_1$. By Corollary B.1 we have

$$\ell_{n_i}^{1+\alpha} = \ell_{n_i}^{1+\delta+\tau}\,\ell_{n_i}^{\alpha-\delta-\tau} = O\big(\ell_{n_i}^{1+\delta+\tau}\,\lambda^{(\alpha-\delta-\tau)n_i}\big)$$

and then, by (B.14), we obtain

$$\ell_{n_i+1} = O\big(\ell_{n_i}^{1+\delta+\tau}\,\lambda^{(\alpha-\delta-\tau)n_i}\,\lambda_1^{-n_i}\big). \tag{B.16}$$

But from (B.15) we have

$$\ell_{n_i} = O\left(\frac{\ell_{n_{i-1}+1}}{c(n_i, n_{i-1}+1)}\right),$$

and since $\lambda^{(\alpha-\delta-\tau)n_i}\,\lambda_1^{-n_i}$ goes to zero as i goes to infinity (recall that $\lambda^{\alpha-\delta-\tau} < \lambda_1$), we deduce from (B.16) that

$$\ell_{n_i+1} \leq \left(\frac{\ell_{n_{i-1}+1}}{c(n_i, n_{i-1}+1)} \right)^{1+\delta+\tau} \tag{B.17}$$

for sufficiently large i. On the other hand, from (B.12) we have $\Lambda_n \leq 2\,c(n+m,n)\,\Lambda_{n+m}$ and then $\Lambda_{n_{i-1}+1} \leq 2\,c(n_i, n_{i-1}+1)\,\Lambda_{n_i}$, while from (B.13) we have $\Lambda_{n_i}^{1+\delta+\tau/2} \leq K\,\Lambda_{n_i+1}\,\Lambda_{n_i}^{\tau/2}$. Therefore,

$$\Lambda_{n_i+1} \geq \left(\frac{\Lambda_{n_{i-1}+1}}{c(n_i, n_{i-1}+1)} \right)^{1+\delta+\tau/2}$$

for sufficiently large i, since Λ_{n_i} goes to zero as i goes to infinity. Combining this estimate with (B.17) we obtain

$$\frac{\log \ell_{n_i+1}}{\log \Lambda_{n_i+1}} \geq \frac{1+\delta+\tau}{1+\delta+\tau/2} \times \frac{\log \ell_{n_{i-1}+1} - \log c(n_i, n_{i-1}+1)}{\log \Lambda_{n_{i-1}+1} - \log c(n_i, n_{i-1}+1)}$$

for large enough i. Note that $K = (1+\delta+\tau)/(1+\delta+\tau/2) > 1$. Moreover, from $\log \Lambda_{n_{i-1}+1} \leq \log \ell_{n_{i-1}+1} < 0 < \log c(n_i, n_{i-1}+1)$ (recall Lemma B.2) we have

$$\frac{\log \ell_{n_i+1}}{\log \Lambda_{n_i+1}} \geq K\,\frac{\log \ell_{n_{i-1}+1}}{\log \Lambda_{n_{i-1}+1}}$$

for sufficiently large i. Thus, the sequence of positive numbers $\{\delta_n\}_{n\in\mathbb{N}}$ given by

$$\delta_n = \frac{\log \ell_n}{\log \Lambda_n}$$

is unbounded. However, by Lemma B.2, we know that $\delta_n \in (0, 1]$ for all $n \in \mathbb{N}$. This contradiction shows that there exist at most finitely many divergent levels. In other words, $a_n\,\varepsilon_n \leq \lambda_1^n$ for sufficiently large n, which finishes the proof of Proposition B.3.

B.6 Proof of Theorem 4.11

With Propositions B.2 and B.3 at hand, we are ready to establish the continuity of the function ψ defined in Appendix B.2.

Proof of Proposition B.1 Fix $i \in \mathbb{Z}$ and some large $n \in \mathbb{N}$, and let $j > i$ be such that $w_j \in I_n(w_i)$. Recall from (B.3) that the function $\psi : \mathcal{O}_f(w_0) \to \mathbb{R}$ is defined by the initial condition $\psi(w_0) = 0$ and the recursive formula

$$\psi(w_{i+1}) = \psi(w_i) - \log Df(w_i) \quad \text{for all } i \in \mathbb{Z}.$$

In particular,

$$\psi(w_i) - \psi(w_j) = \log Df^{j-i}(w_i).$$

Now let $p \in \mathbb{N}$ be such that $w_j \in \mathscr{P}_{n+p}(w_i)$. By Lemma 10.4, for each $k \in \{n, \ldots, n+p-1\}$ there exists an integer $b_{k+1} \in \{0, \ldots, a_{k+1}\}$ such that

$$j = i + q_n + \sum_{k=n}^{n+p-1} b_{k+1}\, q_{k+1}\,.$$

Therefore,

$$\psi(w_i) - \psi(w_j) = \log Df^{q_n + \sum_{k=n}^{n+p-1} b_{k+1}q_{k+1}}(w_i)\,.$$

Using Propositions B.2 and B.3 we obtain

$$\begin{aligned}
\big|\psi(w_i) - \psi(w_j)\big| &\leq \big\| \log Df^{q_n} \big\|_{C^0} + \sum_{k=n}^{n+p-1} \big\| \log Df^{b_{k+1}q_{k+1}} \big\|_{C^0} \\
&= O\left(\varepsilon_n + \sum_{k=n}^{n+p-1} b_{k+1}\, \varepsilon_{k+1} \right) = O\left(\varepsilon_n + \sum_{k=n}^{+\infty} a_{k+1}\, \varepsilon_{k+1} \right) \\
&= O\left(\varepsilon_n + \sum_{k=n}^{+\infty} \lambda_1^{k+1} \right).
\end{aligned}$$

Since this last term goes to zero as n goes to infinity, we deduce that ψ is continuous at w_i, and since i was an arbitrary integer, this finishes the proof of Proposition B.1. □

As explained in Appendix B.1, Proposition B.1 implies that $\psi : \mathscr{O}_f(w_0) \to \mathbb{R}$ can be continuously extended to the whole circle, and that $\phi : \mathbf{S}^1 \to \mathbb{R}$ given by $\phi(x) = e^{\psi(x)} / \int e^{\psi(y)} dy$ is the derivative of a C^1 diffeomorphism that conjugates f with the corresponding rotation. Therefore, we have proved the following result.

Corollary B.3 *Let $\alpha \in (0,1)$ and let $f \in \mathrm{Diff}^{2+\alpha}(\mathbf{S}^1)$ such that its rotation number ρ is Diophantine of order $\delta < \alpha$. Then any topological conjugacy between f and the rigid rotation of angle ρ is a C^1 diffeomorphism.*

B.6.1 Hölder Continuity of the Invariant Density

As before, we denote by μ the unique Borel probability measure on S^1 which is invariant under f, and by m the (normalized) Lebesgue measure on S^1. Recall from Sect. 2.3 that $\mu(A) = m\big(h(A)\big)$ for any Borel set $A \subset S^1$ and any topological conjugacy h between f and the rigid rotation R_ρ. A straightforward consequence of Corollary B.3 is that

$$\mu(A) = \int_A \phi \, dm.$$

The function ϕ is the *density function* of the measure μ. Note that, since ϕ is continuous and strictly positive in S^1, we have $\mu(A) \asymp m(A)$. In other words, the μ-measure of any given interval is comparable to its Euclidean length.[1] Therefore, for smooth diffeomorphisms with Diophantine rotation number, the qualitative notion of minimality can be strengthened to a quantitative one. Indeed, from Birkhoff Ergodic Theorem we know that given any point $x \in S^1$ and any interval $A \subset S^1$ we have

$$\lim_{n\to+\infty} \frac{1}{n} \#\big\{j : 0 \leq j < n \text{ and } f^j(x) \in A\big\} = \mu(A) \asymp m(A).$$

In other words, the *asymptotic frequency* with which any given point visits an open interval is comparable to the Euclidean length of the given interval. These are some of the statistical aspects of the smooth rigidity being discussed here.

In this final subsection we establish Hölder continuity of the invariant density. More precisely, we will prove the following fact.

Proposition B.4 *The function ϕ is Hölder continuous in the whole circle, with Hölder exponent $\alpha - \delta$.*

Before entering the proof of Proposition B.4, we establish the following estimate, announced at the introduction of this appendix.

Lemma B.9 $\varepsilon_n = O(\Lambda_n^{\frac{\alpha}{1+\delta}})$.

Proof Note first that by combining Corollary B.3 with Lemma B.2 and the Mean Value Theorem we obtain that for all $n \in \mathbb{N}$

$$0 < \min_{x\in S^1} \big\{\phi(x)\big\} \leq \frac{\Lambda_n}{\ell_n} \leq 1. \tag{B.18}$$

[1] Recall that this is *not* the case either for diffeomorphisms with Liouvillean rotation number (even real-analytic ones, see Sect. 4.3) or for multicritical circle maps (of any combinatorics and any degree of smoothness, see Sect. 8.2).

In other words, $\ell_n \asymp \Lambda_n$. Since, by definition,

$$\varepsilon_n = \frac{\ell_n}{\ell_0} + \sum_{j=0}^{n-1} \frac{\ell_n}{\ell_{j+1}}\, \ell_j^{\alpha},$$

estimate (B.18) implies that

$$\varepsilon_n = O\left(\Lambda_n + \sum_{j=0}^{n-1} \frac{\Lambda_n}{\Lambda_{j+1}}\, \Lambda_j^{\alpha}\right).$$

But by (B.13) $\Lambda_j^{\alpha} = O\left(\Lambda_{j+1}^{\frac{\alpha}{1+\delta}}\right)$, and then

$$\begin{aligned}\varepsilon_n &= O\left(\Lambda_n + \sum_{j=0}^{n-1} \Lambda_n\, \Lambda_{j+1}^{\frac{\alpha}{1+\delta}-1}\right) = O\left(\Lambda_n + \Lambda_n^{\frac{\alpha}{1+\delta}} \sum_{j=0}^{n-1} \left(\frac{\Lambda_n}{\Lambda_{j+1}}\right)^{1-\frac{\alpha}{1+\delta}}\right)\\ &= O\left(\Lambda_n + \Lambda_n^{\frac{\alpha}{1+\delta}} \sum_{j=0}^{n-1} (\lambda^{1-\frac{\alpha}{1+\delta}})^{n-j-1}\right)\end{aligned}$$

by Corollary B.1. Since $\alpha/(1+\delta) < 1$, the sum $\sum_{j=0}^{n-1} (\lambda^{1-\frac{\alpha}{1+\delta}})^{n-j-1}$ is bounded, which implies Lemma B.9. □

Remark B.5 With Lemma B.9 at hand, we can improve some of our previous estimates. For instance, by combining Lemma B.9 with Proposition B.2 we immediately obtain

$$\| f^{q_n} - \mathrm{Id} \|_{C^1} = O\left(\Lambda_n^{\frac{\alpha}{1+\delta}}\right) = O\left(|q_n \rho - p_n|^{\frac{\alpha}{1+\delta}}\right),$$

which improves our estimates from Appendix B.4. Moreover, since $\left|R_\rho^{q_n} - \mathrm{Id}\right|$ is constant and equal to Λ_n, we deduce that $\| f^{q_n} - R_\rho^{q_n} \|_{C^1} = O\big(\Lambda_n^{\frac{\alpha}{1+\delta}}\big)$ for all $n \in \mathbb{N}$, as announced in Remark B.1.

On the other hand, combining Lemma B.9 with (B.13) (i.e., the Diophantine condition) we obtain

$$a_n\, \varepsilon_n = O\left(\frac{\Lambda_{n-1}}{\Lambda_n}\, \Lambda_n^{\frac{\alpha}{1+\delta}}\right) = O\left(\Lambda_n^{\frac{1}{1+\delta}+\frac{\alpha}{1+\delta}-1}\right) = O\left(\Lambda_n^{\frac{\alpha-\delta}{1+\delta}}\right) = O\left(\Lambda_{n-1}^{\alpha-\delta}\right),$$

thus improving Proposition B.3.

To prove Proposition B.4 consider two points w and w_0 in S^1, and recall that the orbit $\mathscr{O}_f(w_0) = \{w_i = f^i(w_0) : i \in \mathbb{Z}\}$ is dense in S^1. Let $n \in \mathbb{N}$ be such that

$$w \in \overline{I_{n-1}(w_{-q_{n-1}}) \setminus I_n(w_0)}\,.$$

In other words, $w_{q_n} \leq w \leq w_{-q_{n-1}}$. Moreover, let $j \in \{1, \ldots, a_n\}$ be such that $w_{jq_n} \leq w \leq w_{(j+1)q_n}$.

Lemma B.10

$$\big|\phi(w_0) - \phi(w)\big| = O\big((j\Lambda_n)^{\alpha-\delta}\big). \tag{B.19}$$

We remark that (B.19) implies at once Proposition B.4. Indeed, note first that

$$\big|h(w_0) - h(w)\big| \geq \big|h(w_0) - h(w_{jq_n})\big| = \big|R_\rho^{jq_n}\big(h(w_0)\big) - h(w_0)\big| = j\Lambda_n\,.$$

Therefore

$$\big|\phi(w_0) - \phi(w)\big| = O\big((j\Lambda_n)^{\alpha-\delta}\big) = O\big(\big|h(w_0) - h(w)\big|^{\alpha-\delta}\big),$$

and since, by Corollary B.3, h is certainly bi-Lipschitz, we have

$$\big|\phi(w_0) - \phi(w)\big| = O\big(\big|w_0 - w\big|^{\alpha-\delta}\big).$$

Thus, (B.19) implies Proposition B.4 as claimed. We finish Appendix B by proving Lemma B.10.

Proof of Lemma B.10 Continuity of ϕ, combined with the identity $\phi(w_0) - \phi(w) = \phi(w)\big(e^{\psi(w_0)-\psi(w)} - 1\big)$, easily implies that it is enough to prove

$$\big|\psi(w_0) - \psi(w)\big| = O\big((j\Lambda_n)^{\alpha-\delta}\big).$$

With this purpose, we first claim that

$$\big|\psi(w_0) - \psi(w)\big| = O\left(j\varepsilon_n + \sum_{k=n}^{+\infty} a_{k+1}\,\varepsilon_{k+1}\right). \tag{B.20}$$

Indeed, let $N \in \mathbb{N}$ be large enough so that $w_N = f^N(w_0)$ is arbitrarily close to w. Let $p \in \mathbb{N}$ be such that $w_N \in \mathscr{P}_{n+p}(w_0)$. Just as in the proof of Corollary B.2 and Proposition B.1, for each $k \in \{n, \ldots, n+p-1\}$ let $b_{k+1} \in \{0, \ldots, a_{k+1}\}$ be such that

$$N = jq_n + \sum_{k=n}^{n+p-1} b_{k+1}\,q_{k+1}\,.$$

By definition of ψ we have

$$\psi(w_0) - \psi(w_N) = \log Df^{jq_n + \sum_{k=n}^{n+p-1} b_{k+1}q_{k+1}}(w_0)\,,$$

and then, just as in the proof of Proposition B.1, we obtain

$$\big|\psi(w_0) - \psi(w_N)\big| = O\left(j\varepsilon_n + \sum_{k=n}^{+\infty} a_{k+1}\,\varepsilon_{k+1} \right).$$

The obtained bound is finite by Proposition B.3, and does not depend on N. Thus, by continuity of ψ, the same bound applies to $\big|\psi(w_0) - \psi(w)\big|$, which establishes (B.20). It remains to prove that $j\varepsilon_n + \sum_{k\geq n} a_{k+1}\,\varepsilon_{k+1} = O\big((j\Lambda_n)^{\alpha-\delta}\big)$.

On one hand, by Lemma B.9, we can write

$$j\,\varepsilon_n = O\left(j\,\Lambda_n^{\frac{\alpha}{1+\delta}} \right) = O\left((j\,\Lambda_n)^{\alpha-\delta} \left(j\,\Lambda_n^{\frac{\delta}{1+\delta}} \right)^{1-(\alpha-\delta)} \right).$$

But $j\,\Lambda_n^{\frac{\delta}{1+\delta}}$ is bounded. Indeed, note first that

$$j\,\Lambda_n^{\frac{\delta}{1+\delta}} \leqslant a_n\,\Lambda_n^{\frac{\delta}{1+\delta}} < \frac{\Lambda_{n-1}}{\Lambda_n}\,\Lambda_n^{\frac{\delta}{1+\delta}}\,,$$

since $a_n = \Lambda_{n-1}/\Lambda_n - \Lambda_{n+1}/\Lambda_n$. From (B.13) we have $\Lambda_{n-1} = O\left(\Lambda_n^{\frac{1}{1+\delta}}\right)$, and then $\Lambda_{n-1}/\Lambda_n = O\left(\Lambda_n^{\frac{-\delta}{1+\delta}}\right)$. Thus, $j\,\Lambda_n^{\frac{\delta}{1+\delta}}$ is bounded and then

$$j\,\varepsilon_n = O\big((j\,\Lambda_n)^{\alpha-\delta}\big). \tag{B.21}$$

On the other hand,

$$\sum_{k=n}^{+\infty} a_{k+1}\,\varepsilon_{k+1} = O\left(\sum_{k=n}^{+\infty} \frac{\Lambda_k}{\Lambda_{k+1}}\,\Lambda_{k+1}^{\frac{\alpha}{1+\delta}} \right) = O\left(\sum_{k=n}^{+\infty} \Lambda_k^{\alpha-\delta}\,\Lambda_k^{1-(\alpha-\delta)}\,\Lambda_{k+1}^{\frac{\alpha}{1+\delta}-1} \right).$$

Using again (B.13) we obtain $\Lambda_k^{1-(\alpha-\delta)} = O\left(\Lambda_{k+1}^{\frac{1-(\alpha-\delta)}{1+\delta}}\right) = O\left(\Lambda_{k+1}^{1-\frac{\alpha}{1+\delta}}\right)$, and then

$$\sum_{k=n}^{+\infty} a_{k+1}\,\varepsilon_{k+1} = O\left(\sum_{k=n}^{+\infty} \Lambda_k^{\alpha-\delta} \right) = O\left(\Lambda_n^{\alpha-\delta}\right).$$

Combining this last estimate with (B.20) and (B.21) we obtain (B.19). This finishes the proof of Lemma B.10. □

As explained above, Lemma B.10 implies Proposition B.4. Finally, by combining Corollary B.3 with Proposition B.4, we deduce Theorem 4.11 and finish Appendix B.

Appendix C
A Skew Product Over the Gauss Map

In our study of orbit flexibility in Chap. 9, we considered a certain skew product $T : M \to M$, where $M = ([0, 1] \setminus \mathbb{Q}) \times [-1, 1]$. Here, we enlarge it to get a self-map of the rectangle $R = [0, 1] \times [-1, 1]$.

Recall the formula defining T over M, namely

$$T(\rho, \alpha) = (G(\alpha), T_\rho(\alpha)) , \tag{C.1}$$

where G is the Gauss map, and for each (irrational) ρ the fiber map $T_\rho : [-1, 1] \to [-1, 1]$ is given by

$$T_\rho(\alpha) = \begin{cases} -\alpha & \text{for } \alpha \in \big[-1, 0\big] \\ -\dfrac{\alpha}{\rho\, G(\rho)} & \text{for } \alpha \in \big(0, \rho\, G(\rho)\big] \\ \left\{\dfrac{1-\alpha}{\rho}\right\} & \text{for } \alpha \in \big(\rho\, G(\rho), 1\big], \end{cases} \tag{C.2}$$

Thus, to extend T to a self-map of the rectangle R, it suffices to define the fiber maps $T_\rho : [-1, 1] \to [-1, 1]$ also for rational values of ρ. When $\rho \in [0, 1] \cap \mathbb{Q}$ is not of the form $\rho = \frac{1}{n}$, we define T_ρ using the same formulas in (C.2). We also define $T_0 \equiv 0$, and for each $n \in \mathbb{N}$, $T_{1/n} : [-1, 1] \to [-1, 1]$ by $T_{1/n}(\alpha) = -\alpha$ if $\alpha \in [-1, 0]$ and $T_{1/n}(\alpha) = \{n(1-\alpha)\}$ if $\alpha \in (0, 1]$. Once this is done, we can define the extended skew product, which we still denote by T, by the same formula (C.1).

Remark C.1 Although the fiber maps $T_\rho : [-1, 1] \to [-1, 1]$ are not (piecewise) expanding, it is important to observe that the composition of any two of them (with $\rho \neq 0$) *is* expanding. This fact will be very useful in our study of the skew product T.

E. de Faria, P. Guarino, *Dynamics of Circle Mappings*, IMPA Monographs,
https://doi.org/10.1007/978-3-031-67495-2

Our main purpose in this appendix is to examine T from the point of view of ergodic theory. More precisely, our goal is to prove the following result.

Theorem C.1 *The skew product $T : R \to R$ admits a unique invariant Borel probability measure which is absolutely continuous with respect to Lebesgue measure. This invariant measure is ergodic under T, and its support coincides with R.*

C.1 An Absolutely Continuous Invariant Measure

For one-dimensional maps, there is a well-known result called the *Folklore Theorem* (see Mañé 1987, Ch. III, Thm. 1.2 or de Melo and van Strien 1993, Ch. V, Thm. 2.2), that in essence asserts that a piecewise smooth expanding map always admits an absolutely continuous invariant probability measure. In sharp contrast with this fact, a piecewise smooth two-dimensional expanding map may not admit an absolutely continuous invariant measure. For such a measure to exist, additional hypotheses are necessary (see for instance Buzzi 2000; Tsujii 2001 and references therein).

Fortunately, in our case the map T is rather special. The fact that T is a skew product over an expanding map (the Gauss map), combined with the fact that it is a Markov map (see below) which is even affine on the fibers, allows us to reduce the problem to an essentially one-dimensional situation. Indeed, we start this appendix with the following useful property of the family of fiber maps defined above.

Lemma C.1 *Given any sequence $\{\theta_n\}_{n\in\mathbb{N}} \subset [0, 1] \setminus \mathbb{Q}$ consider the sequence of compositions $\left\{\Psi_{\theta_0\cdots\theta_{n-1}}\right\}_{n\geq 1}$ in $[-1, 1]$ given by*

$$\Psi_{\theta_0\cdots\theta_{n-1}} = T_{\theta_0} \circ T_{\theta_1} \circ \cdots \circ T_{\theta_{n-1}} \quad \textit{for all } n \geq 1.$$

Then for any given Borel set $B \subset [-1, 1]$, the sequence $\left\{\lambda\left(\Psi^{-1}_{\theta_0\cdots\theta_{n-1}}(B)\right)\right\}_{n\in\mathbb{N}}$ is convergent, where λ denotes the Lebesgue measure on $[-1, 1]$. Moreover,

$$\theta_0\, G(\theta_0)\, \lambda(B) \leq \lim_{n\to+\infty} \left\{\lambda\left(\Psi^{-1}_{\theta_0\cdots\theta_{n-1}}(B)\right)\right\} \leq \left(2 - \theta_0\, G(\theta_0)\right) \lambda(B)\,.$$

Proof From $\{\theta_n\}$ we build the sequence $\{\tau_n\}_{n\in\mathbb{N}} \subset [0, 1]$ given by

$$\tau_0 = 1,\ \ \tau_1 = 0 \quad \text{and} \quad \tau_{n+2} = \theta_n\, G(\theta_n)\, \tau_n + \left(1 - \theta_n\, G(\theta_n)\right) \tau_{n+1}\,, \qquad \forall\, n \in \mathbb{N}.$$

In other words, $\tau_2 = \theta_0\, G(\theta_0)$ and

$$\tau_n = \theta_0\, G(\theta_0) + \sum_{j=1}^{n-2} (-1)^j \prod_{i=0}^{j} \theta_i\, G(\theta_i) \quad \text{for all } n \geq 3.$$

The sequence $\{\tau_n\}$ clearly converges to some number τ_∞, which satisfies[1]

$$0 < \frac{\theta_0\, G(\theta_0)}{2} < \tau_\infty < \theta_0\, G(\theta_0) < \frac{1}{2}\,.$$

Given a Borel set $B \subset [-1, 1]$ and $n \in \mathbb{N}$ let ℓ_n and r_n in $[0, 1]$ be given by

$$\ell_n = \lambda\big(\Psi^{-1}_{\theta_0\cdots\theta_{n-1}}(B) \cap [-1, 0]\big) \quad \text{and} \quad r_n = \lambda\big(\Psi^{-1}_{\theta_0\cdots\theta_{n-1}}(B) \cap [0, 1]\big).$$

By definition of each T_θ, the following relations hold for all $n \in \mathbb{N}$:

$$\begin{cases} \ell_{n+1} = r_n \\ r_{n+1} = \theta_n\, G(\theta_n)\, \ell_n + \left\lfloor \dfrac{1}{\theta_n} \right\rfloor \theta_n\, r_n = \theta_n\, G(\theta_n)\, \ell_n + \big(1 - \theta_n\, G(\theta_n)\big)\, r_n \end{cases}$$

With this at hand, we easily obtain by induction that for all $n \in \mathbb{N}$ we have

$$\begin{cases} \ell_n = \tau_n\, \ell_0 + (1 - \tau_n)\, r_0 \\ r_n = \tau_{n+1}\, \ell_0 + (1 - \tau_{n+1})\, r_0 \end{cases}$$

In particular, the Lebesgue measure of $\Psi^{-1}_{\theta_0\cdots\theta_{n-1}}(B)$ in $[-1, 1]$ is given by

$$\lambda\big(\Psi^{-1}_{\theta_0\cdots\theta_{n-1}}(B)\big) = (\tau_n + \tau_{n+1})\, \ell_0 + \big(2 - (\tau_n + \tau_{n+1})\big)\, r_0,$$

which converges to $2\big(\tau_\infty\, \ell_0 + (1 - \tau_\infty)\, r_0\big)$ as n goes to infinity. This proves Lemma C.1. □

With Lemma C.1 at hand we have the following result.

Lemma C.2 *The skew product T preserves a probability measure μ_T on the rectangle R which is absolutely continuous (with respect to Lebesgue).*

Proof We only sketch the arguments, as they are quite standard. As before, denote by ν and λ the Gauss measure on $[0, 1]$ and the Lebesgue measure on $[-1, 1]$ respectively. Denote by μ the absolutely continuous (with respect to Lebesgue) Borel measure on the rectangle R given by $\mu = \nu \times \lambda$. In other words, given a Borel set $A \subset R$ we have

$$\mu(A) = \int_{\pi_1(A)} \lambda_\rho(A)\, d\nu(\rho)\,,$$

[1] Remember here that $\theta\, G(\theta) \in (0, 1/2)$ for any $\theta \in [0, 1] \setminus \mathbb{Q}$ (if $\theta < 1/2$ this is obvious since $0 < G(\theta) < 1$; if $\theta > 1/2$, then $\theta\, G(\theta) = 1 - \theta$).

where $\pi_1 : R \to [0, 1]$ is the projection on the first coordinate given by $\pi_1(\rho, \alpha) = \rho$, and where λ_ρ is the Lebesgue measure on the vertical fiber given by ρ, i.e., $\lambda_\rho(A) = \lambda\big(A \cap (\{\rho\} \times [-1, 1])\big)$ for any $\rho \in [0, 1]$.

Given $n \in \mathbb{N}$ and open intervals $I \subset [0, 1]$ and $J \subset [-1, 1]$ we label each point θ_{n-1} of $G^{-n}(I)$ with the n-tuple $\{\theta_0, \dots, \theta_{n-1}\}$ given by $G(\theta_0) \in I$ and $G(\theta_i) = \theta_{i-1}$ for all $i \in \{1, \dots, n-1\}$. With this notation we can write

$$T^{-n}(I \times J) = \bigcup_{\substack{\{\theta_0,\dots,\theta_{n-1}\} \\ G(\theta_0)\in I\,,\, G(\theta_i)=\theta_{i-1}}} \big\{\theta_{n-1}\big\} \times \Psi^{-1}_{\theta_0\dots\theta_{n-1}}(J)\,.$$

From Lemma C.1 we know that

$$\lambda\big(\Psi^{-1}_{\theta_0\cdots\theta_{n-1}}(J)\big) \leq 2\,\lambda(J)$$

holds for any n-tuple, and then

$$\begin{aligned} \mu\big(T^{-n}(I \times J)\big) &= \int_{G^{-n}(I)} \lambda_\rho\big(T^{-n}(I \times J)\big)\, d\nu(\rho) \leq 2\,\lambda(J) \int_{G^{-n}(I)} d\nu(\rho) \\ &= 2\,\lambda(J)\,\nu\big(G^{-n}(I)\big) = 2\,\lambda(J)\,\nu(I) = 2\,\mu(I \times J). \end{aligned}$$

With this at hand we deduce that

$$\big(T_*^n \mu\big)(A) \leq 2\,\mu(A) \quad \text{for any Borel set } A \subset R \text{ and any } n \in \mathbb{N}. \tag{C.3}$$

Finally, consider the sequence of Borel measures on the rectangle R given by

$$\mu_n = \frac{1}{n} \sum_{j=0}^{n-1} T_*^j \mu\,.$$

Since T is a local diffeomorphism around Lebesgue almost every point in R, we deduce that the push-forward under T of any absolutely continuous measure is also absolutely continuous and that, when restricted to absolutely continuous measures, the operator T_* acts continuously in the weak* topology. Let ω be any weak* accumulation point of $\{\mu_n\}$ (recall that $\mu_n(R) = 2$ for all n). By (C.3), $\omega(A) \leq 2\,\mu(A)$ for any Borel set $A \subset R$. Therefore, ω is absolutely continuous with respect to μ, and then it is also absolutely continuous with respect to Lebesgue. In particular, the measure ω is a continuity point of T_*, which implies that it is T-invariant in the usual way. We conclude the proof of Lemma C.2 by taking the *probability* measure $\mu_T = \frac{1}{2}\,\omega$. □

C.2 Markov Property

In order to prove Theorem C.1, it remains to prove that the absolutely continuous invariant probability measure μ_T given by Lemma C.2 is unique, supported on the whole rectangle R and ergodic under T (see Corollary C.1 below).

C.2.1 A Countable Markov Partition

The skew product T admits a countable Markov partition that we presently describe. The basic (open) Markov atoms of the partition are of three different types (see Fig. C.1):

1. The trapezoids $V_{k,\ell}$, with $k \in \mathbb{N}$ and $0 \leq \ell \leq k-1$, given by

$$V_{k,\ell} = \left\{ (\rho, \alpha) \in R : \frac{1}{k+1} < \rho < \frac{1}{k} \, , \; 1-(\ell+1)\rho < \alpha < 1 - \ell\rho \right\} ;$$

2. The triangles

$$U_k = \left\{ (\rho, \alpha) \in R : \frac{1}{k+1} < \rho < \frac{1}{k} \, , \; 0 < \alpha < 1 - k\rho \right\} \quad (k \in \mathbb{N}) ;$$

3. The rectangles

$$R_k = \left\{ (\rho, \alpha) \in R : \frac{1}{k+1} < \rho < \frac{1}{k} \, , \; -1 < \alpha < 0 \right\} \quad (k \in \mathbb{N}) .$$

The map T is one-to-one in each of these Markov atoms, mapping them diffeomorphically onto either $R^+ = (0,1) \times (0,1)$ or $R^- = (0,1) \times (-1,0)$. More precisely, we have $T(U_k) = R^-$, $T(R_k) = R^+$ and $T(V_{k,\ell}) = R^+$, for all k and all ℓ. The collection $\mathscr{P}$ of all such atoms is our *Markov partition* for T.

C.2.1.1 Markov Tiles

Let us write

$$\mathscr{P} = \{W_1, W_2, \ldots, W_m, \ldots\}$$

for an enumeration of the elements of the Markov partition $\mathscr{P}$. For each m, let $\tau_m : R^{\pm} \to W_m$ be the inverse branch of T that takes $T(W_m) = R^{\pm}$ back onto W_m. Then τ_m is a smooth diffeomorphism and we have $\tau_m \circ T = \mathrm{id}_{W_m}$ and $T \circ \tau_m = \mathrm{id}_{R^{\pm}}$.

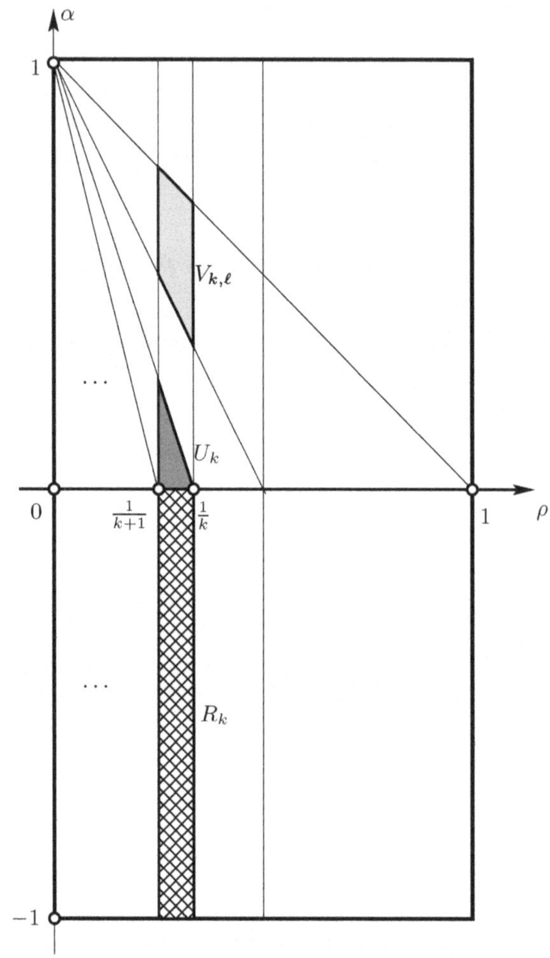

Fig. C.1 The Markov partition for T has three different types of atoms

An n-tuple $(m_1, m_2, \ldots, m_n) \in \mathbb{N}^n$ is said to be *admissible* if the composition $\tau_{m_1} \circ \tau_{m_2} \circ \cdots \tau_{m_n}$ is well-defined (as a map of $R^{\pm}$ into R). For each admissible n-tuple $(m_1, m_2, \ldots, m_n) \in \mathbb{N}^n$, we consider the region (polygon)

$$W_{m_1,m_2,\ldots,m_n} = \tau_{m_1} \circ \tau_{m_2} \circ \cdots \circ \tau_{m_n}(R^{\pm}) .$$

Such region is called a *Markov n-tile*. Note that $T(W_{m_1,m_2,\ldots,m_n}) = W_{m_2,\ldots,m_n}$, so each Markov n-tile is mapped onto a Markov $(n-1)$-tile if $n \geq 2$, or onto $R^{\pm}$ if $n = 1$.

Lemma C.3 *There exist constants $C > 0$ and $0 < \lambda < 1$ such that, for every Markov n-tile $W_{m_1,m_2,\ldots,m_n}$, we have*

$$\mathrm{diam}(W_{m_1,m_2,\ldots,m_n}) < C\lambda^n .$$

Proof This follows at once from the easily verifiable fact that the map $T^2 = T \circ T$ is expanding. □

We denote by $\mathscr{W}$ the collection of all Markov tiles, and for each n we denote by $\mathscr{W}^{(n)}$ the collection of all Markov n-tiles, so that $\mathscr{W} = \bigcup_{n\in\mathbb{N}} \mathscr{W}^{(n)}$. The following easily proven facts are worth keeping in mind here:

MT1. For each n the elements of $\mathscr{W}^{(n)}$ are pairwise disjoint open subsets of R;
MT2. For each n the complement of $\bigcup_{W\in\mathscr{W}^{(n)}} W$ in R is a Lebesgue null-set;
MT3. The union $\bigcup_{W\in\mathscr{W}} \partial W$ is a Lebesgue null-set;
MT4. For each open subset $A \subseteq R$, there exists a collection $\mathscr{C}_A \subseteq \mathscr{W}$ of pairwise disjoint Markov tiles such that $A \setminus \bigcup_{W\in\mathscr{C}_A} W$ has zero Lebesgue measure.

Note that Lemma 9.2 follows at once from the fact that any given open set in R contains the closure of an n-tile (and then it eventually covers the whole rectangle under iteration of T).

C.3 Ergodicity

We are now ready to show that the skew product T is ergodic with respect to the invariant measure μ_T.

C.3.1 Bounding Jacobian Distortion

One path towards proving that T is ergodic is to show that the Jacobians of all inverse branches of iterates of T have uniformly bounded distortion. This follows from Proposition C.1 below. In the proof, we will need the following simple lemma.

Lemma C.4 *Let $k_j > 0$, $b_j \geq 0$ $(j \geq 0)$ be two sequences of real numbers, and assume that $B = \sum_{j=0}^{\infty} \sqrt{b_j} < \infty$. Then for each $n \in \mathbb{N}$ we have*

$$\sum_{j=0}^{n} k_j \min\left\{b_{n-j}\,,\; k_j^{-2}\right\} \leq B\,. \tag{C.4}$$

Proof For each $1 \leq j \leq n$, there are only two possibilities:

(i) $k_j^{-2} < b_{n-j}$: In this case we have

$$k_j \min\{b_{n-j}, k_j^{-2}\} = k_j^{-1} < \sqrt{b_{n-j}}\,.$$

(ii) $k_j^{-2} \geq b_{n-j}$: In this case we have

$$k_j \min\{b_{n-j}, k_j^{-2}\} = k_j b_{n-j} \leq (b_{n-j}^{-1})^{\frac{1}{2}} b_{n-j} = \sqrt{b_{n-j}}\ .$$

From (i) and (ii) it follows that the sum in the left-hand side of (C.4) is bounded by $\sum_{j=0}^{n} \sqrt{b_{n-j}} \leq B$. □

Proposition C.1 *There exists a constant $K > 1$ for which the following holds for all $n \in \mathbb{N}$. If (ρ_0, α_0) and (ρ_0^*, α_0^*) are any two points in the same Markov n-tile, then*

$$\frac{1}{K} \leq \left| \frac{\det DT^n(\rho_0, \alpha_0)}{\det DT^n(\rho_0^*, \alpha_0^*)} \right| \leq K\ . \tag{C.5}$$

Proof First, some preliminary considerations. For definiteness, let $W_{m_1,m_2,\ldots,m_n}$ be the Markov n-tile containing the two points (ρ_0, α_0) and (ρ_0^*, α_0^*). Let us write, for $j = 1, 2, \ldots$, $(\rho_j, \alpha_j) = T^j(\rho_0, \alpha_0)$ and $(\rho_j^*, \alpha_j^*) = T^j(\rho_0^*, \alpha_0^*)$. From the definition of our skew product, we see that

$$\begin{cases} \rho_j = G^j(\rho_0) \\ \alpha_j = T_{\rho_{j-1}} \circ T_{\rho_{j-2}} \circ \cdots \circ T_{\rho_0}(\alpha_0) \end{cases} \tag{C.6}$$

and similar formulas hold for ρ_j^*, α_j^*. Note that, for each $0 \leq j \leq n$, we have $(\rho_j, \alpha_j), (\rho_j^*, \alpha_j^*) \in W_{m_{j+1},\ldots,m_n}$. Hence, by Lemma C.3, for each such j we have

$$|\rho_j - \rho_j^*| \leq \operatorname{diam}(W_{m_{j+1},\ldots,m_n}) < C\lambda^{n-j}$$

Next, for each $0 \leq j \leq n$, let k_j be the unique natural number such that $\frac{1}{k_j+1} < \rho_j, \rho_j^* < \frac{1}{k_j}$, so that $|\rho_j - \rho_j^*| < \frac{1}{k_j^2}$. Combining these two estimates, we can write

$$|\rho_j - \rho_j^*| < \min\left\{C\lambda^{n-j}\ ,\ k_j^{-2}\right\}\ . \tag{C.7}$$

We are now ready to estimate the ratio of determinant Jacobians in (C.5). Using (C.6) and the chain rule, we see that

$$DT^n(\rho_0, \alpha_0) = \begin{bmatrix} \prod_{j=0}^{n-1} G'(\rho_j) & 0 \\ * & \prod_{j=0}^{n-1} T'_{\rho_j}(\alpha_j) \end{bmatrix},$$

and similarly for $DT^n(\rho_0^*, \alpha_0^*)$. Hence the ratio of determinant Jacobians at both points equals

$$\frac{\det DT^n(\rho_0, \alpha_0)}{\det DT^n(\rho_0^*, \alpha_0^*)} = \prod_{j=0}^{n-1} \frac{G'(\rho_j)}{G'(\rho_j^*)} \prod_{j=0}^{n-1} \frac{T'_{\rho_j}(\alpha_j)}{T'_{\rho_j^*}(\alpha_j^*)} \tag{C.8}$$

We proceed to estimate both products in the right-hand side of (C.8).

(i) Since $G'(\xi) = -1/\xi^2$ wherever G is differentiable, each term in the first product is positive, equal to $(\rho_j^*/\rho_j)^2$, and thus we have

$$\left| \log \prod_{j=0}^{n-1} \frac{G'(\rho_j)}{G'(\rho_j^*)} \right| \leq 2 \sum_{j=0}^{n-1} \left| \log \rho_j - \log \rho_j^* \right|$$

The mean value inequality tells us that $\left| \log \rho_j - \log \rho_j^* \right| \leq (k_j + 1)|\rho_j - \rho_j^*|$, and therefore, by (C.7), we have

$$\left| \log \prod_{j=0}^{n-1} \frac{G'(\rho_j)}{G'(\rho_j^*)} \right| \leq 4 \sum_{j=0}^{n-1} k_j \min \left\{ C\lambda^{n-j} , k_j^{-2} \right\} . \tag{C.9}$$

(ii) From the formulas defining the fiber maps T_ρ (see Sect. 9.3.1), we deduce that there are only three possibilities:

$$\frac{T'_{\rho_j}(\alpha_j)}{T'_{\rho_j^*}(\alpha_j^*)} = \begin{cases} 1 , & \text{if } -1 < \alpha_j, \alpha_j^* < 0 \\[2ex] \dfrac{\rho_j^* \rho_{j+1}^*}{\rho_j \rho_{j+1}} , & \text{if } 0 < \alpha_j < 1 - k_j \rho_j \text{ and } 0 < \alpha_j^* < 1 - k_j \rho_j^* \\[2ex] \dfrac{\rho_j^*}{\rho_j} , & \text{if } 1 - k_j \rho_j < \alpha_j < 1 \text{ and } 1 - k_j \rho_j^* < \alpha_j^* < 1 \end{cases}$$

Whichever case occurs, we always have

$$\left| \log \frac{T'_{\rho_j}(\alpha_j)}{T'_{\rho_j^*}(\alpha_j^*)} \right| \leq \left| \log \rho_j - \log \rho_j^* \right| + \left| \log \rho_{j+1} - \log \rho_{j+1}^* \right| .$$

This yields

$$\left| \log \prod_{j=0}^{n-1} \frac{T'_{\rho_j}(\alpha_j)}{T'_{\rho_j^*}(\alpha_j^*)} \right| \leq 2 \sum_{j=0}^{n} \left| \log \rho_j - \log \rho_j^* \right| ,$$

Therefore, using the mean value inequality and (C.7) just as in (i), we deduce that

$$\left|\log \prod_{j=0}^{n-1} \frac{T'_{\rho_j}(\alpha_j)}{T'_{\rho_j^*}(\alpha_j^*)}\right| \leq 4 \sum_{j=0}^{n} k_j \min\left\{C\lambda^{n-j}, k_j^{-2}\right\} . \tag{C.10}$$

Combining the estimates (C.9) and (C.10), we arrive at

$$\left|\log \left(\prod_{j=0}^{n-1} \frac{G'(\rho_j)}{G'(\rho_j^*)} \prod_{j=0}^{n-1} \frac{T'_{\rho_j}(\alpha_j)}{T'_{\rho_j^*}(\alpha_j^*)}\right)\right| \leq 8 \sum_{j=0}^{n} k_j \min\left\{C\lambda^{n-j}, k_j^{-2}\right\} . \tag{C.11}$$

Applying Lemma C.4 with $b_j = C\lambda^j$, we deduce that the sum on the right-hand side of (C.11) is bounded by $B = \sqrt{C}/\left(1 - \sqrt{\lambda}\right)$. Thus, exponentiating both sides of this last inequality, one finally arrives at (C.5), with $K = e^{8B}$. This completes the proof of Proposition C.1. □

C.3.2 A Lebesgue Density Argument

In what follows, we denote by meas(A) the Lebesgue measure of a measurable set $A \subseteq R$.

Lemma C.5 *Let $A \subseteq R^{\pm}$ be a set with positive Lebesgue measure. Then there exists a constant $0 < c_A < 1$ such that, for every Markov n-tile W with $T^n(W) = R^{\pm}$, we have*

$$\frac{\text{meas}(W \cap T^{-n}(A))}{\text{meas}(W)} \geq c_A . \tag{C.12}$$

Proof Since T^n maps W diffeomorphically onto $R^{\pm}$, the change-of-variables formula tells us that

$$\text{meas}(A) = \iint_{W \cap T^{-n}(A)} \left|\det DT^n(\rho, \alpha)\right| \, d\rho d\alpha ,$$

as well as

$$1 = \text{meas}(R^{\pm}) = \iint_{W} \left|\det DT^n(\rho, \alpha)\right| \, d\rho d\alpha .$$

Applying the mean-value theorem for double integrals to both integrals above and using Proposition C.1, we deduce (C.12), with a constant c_A that depends only on meas(A) (and the constant K in (C.5)). □

Lemma C.6 *If $B \subseteq R$ is a set with positive Lebesgue measure, then*

$$\text{meas}\left(R \setminus \bigcup_{n\geq 0} T^{-n}(B)\right) = 0\,.$$

Proof Replacing B by $T^{-1}(B)$ if necessary, we may assume that $B^+ = B \cap R^+$ and $B^- = B \cap R^-$ both have positive measure. Let $\epsilon = \frac{1}{2}\min\{c_{B^+}, c_{B^-}\}$, where $c_{B^\pm}$ are the constants obtained applying Lemma C.5 to $A = B^\pm$.

We argue by contradiction. Suppose $E = R \setminus \bigcup_{n\geq 0} T^{-n}(B)$ is such that meas$(E) > 0$. Let $z \in E$ be a Lebesgue density point of E, and choose $\delta > 0$ so small that the disk $D = D(z, \delta) \subset R$ satisfies

$$\frac{\text{meas}(D \cap E)}{\text{meas}(D)} \geq 1 - \epsilon\,. \tag{C.13}$$

By fact (MT4) stated right after Lemma C.3, there exists a collection $\mathscr{C}$ of pairwise disjoint Markov tiles such that $D = D^* \cup \bigcup_{W\in\mathscr{C}} W$, where D^* has zero Lebesgue measure. For each $W \in \mathscr{C}$, there exists a positive integer m_K such that $T^{m_K}(W) = R^\pm \supseteq B^\pm$. Thus, by Lemma C.5, we have

$$\begin{aligned}\text{meas}\left(W \cap \bigcup_{n\geq 0} T^{-n}(B)\right) &\geq \text{meas}\left(W \cap T^{-m_K}\left(B^\pm\right)\right)\\ &\geq c_{B^\pm}\text{meas}(W) \geq 2\epsilon\ \text{meas}(W)\,.\end{aligned}$$

Since this is true for every Markov tile in $\mathscr{C}$, we deduce that

$$\text{meas}\left(D \cap \bigcup_{n\geq 0} T^{-n}(B)\right) \geq 2\epsilon\text{meas}(D)\,,$$

that is to say,

$$\frac{\text{meas}(D \cap (R \setminus E))}{\text{meas}(D)} \geq 2\epsilon\,. \tag{C.14}$$

But (C.13) and (C.14) are clearly incompatible. This contradiction shows that meas$(E) = 0$, and the lemma is proved. □

Corollary C.1 *Let $A \subset R$ be a Borel set which is T-invariant, i.e., $T^{-1}(A) = A$. If A has positive Lebesgue measure, then it has full Lebesgue measure in the whole rectangle R.*

Proof The invariance $T^{-1}(A) = A$ implies $T^{-n}(A) = A$ for all $n \geq 0$. Since $\text{meas}(A) > 0$, we obtain from Lemma C.6 that

$$\text{meas}(A) = \text{meas}\left(\bigcup_{n\geq 0} T^{-n}(A)\right) = \text{meas}(R)\,.$$

□

C.3.3 End of Proof

With this at hand we can finish the proof of Theorem C.1: Corollary C.1 implies at once that any absolutely continuous probability measure which is invariant under T, is also ergodic under T. Therefore, the measure μ_T given by Lemma C.2 is ergodic. Moreover, since the support of μ_T is itself a T-invariant subset of R with positive Lebesgue measure (because it has full μ_T-measure), Corollary C.1 implies that it must coincide with the whole rectangle R (since it is compact and it has full measure). In particular, μ_T is the unique absolutely continuous probability measure invariant under T, and this concludes the proof of Theorem C.1. We finish this appendix by proving Proposition 9.4.

Proof of Proposition 9.4 Let $B_1, B_2, \ldots, B_j, \ldots$ be a basis for the topology of $R^+ \cup R^-$. For each $j \geq 1$, let $B_j^\infty = \bigcup_{n\geq 0} T^{-n}(B_j)$. Note that each $B_j^\infty \subset R^+ \cup R^-$ is open, and by Lemma C.6 it has full Lebesgue measure in R (in particular, it is also dense in R). Therefore $\mathscr{G}_0 = \bigcap_{j\geq 1} B_j^\infty$ also has full Lebesgue measure in R. Moreover, $\mathscr{G}_0$ is a dense G_δ, hence residual, subset of $R^+ \cup R^-$. Finally, if z is any point in $\mathscr{G}_0$, then its positive orbit $\{T^n(z) :\ n \geq 0\}$ visits every basic set B_j, and therefore is dense in R. □

References

Ahlfors, L. V. [2006]. *Lectures on Quasiconformal Mappings*. 2nd ed. Vol. 38. University Lecture Series. American Mathematical Society.

Arnold, V. I. [1961]. "Small denominators I: Mappings of the circle onto itself". In: *Izv. Akad. Nauk. Math. Series* 25, pp. 213–284.

Arnold, V. I. and A. Avez [1968]. *Ergodic Problems of Classical Mechanics*. The Mathematical Physics Monograph Series. W. A. Benjamin.

Astala, K., T. Iwaniec, and G. Martin [2009]. *Elliptic Partial Differential Equations and Quasiconformal Mappings in the Plane*. Vol. 48. Princeton Mathematical Series. Princeton University Press.

Athanassopoulos, K. [2015]. "Denjoy C^1 diffeomorphisms of the circle and Mc-Duff's question". In: *Expo. Math.* 33, pp. 48–66.

Avila, A. [2013]. "On rigidity of critical circle maps". In: *Bull. Braz. Math. Soc.* 44, pp. 611–619. URL: https://doi.org/10.1007/s00574-013-0027-5.

Avila, A., D. Cheraghi, and A. Eliad [2022]. "Analytic maps of parabolic and elliptic type with trivial centralisers". In: *Annales de l'Institut Henri Poincaré C* 39.4, pp. 885–903. DOI: 10.4171/aihpc/22.

Avila, A. and A. Kocsard [2011]. "Cohomological equations and invariant distributions for minimal circle diffeomorphisms". In: *Duke Math. J.* 158, pp. 501-536. https://doi.org/10.1215/00127094-1345662.

Avila, A. and M. Lyubich [2011]. "The full renormalization horseshoe for unimodal maps of higher degree: exponential contraction along hybrid classes". In: *Publ. Math. IHES* 114, pp. 171–223. https://doi.org/10.1007/s10240-011-0034-2.

Billingsley, P. [1965]. *Ergodic theory and information*. John Wiley & Sons.

Billingsley, P. [1986]. *Probability and Measure*. New York: John Wiley & Sons.

Bohl, P. [1916]. "Über die Hinsichtlich der Unabhängigen und Abhängigen Variabeln Periodische Differentialgleichung Erster Ordnung". In: *Acta Math.* 40, pp. 321–336. https://doi.org/10.1007/BF02418549.

Boshernitzan, M. [1993]. "Dense orbits of rationals". In: *Proc. Amer. Math. Soc.* 117, pp. 1201–1203. https://doi.org/10.1090/S0002-9939-1993-1134622-6.

Boyland, P. [1986]. "Bifurcations of circle maps: Arnol'd tongues, bistability and rotation intervals". In: *Comm. Math. Phys.* 106, pp. 353–381. https://doi.org/10.1007/BF01207252.

Bruin, H., J. Rivera-Letelier, W. Shen, and S. van Strien [2008]. "Large derivatives, backward contraction and invariant densities for interval maps". In: *Inventiones mathematicae* 172.3, pp. 509–533. https://doi.org/10.1007/s00222-007-0108-4.

E. de Faria, P. Guarino, *Dynamics of Circle Mappings*, IMPA Monographs,
https://doi.org/10.1007/978-3-031-67495-2

Buff, Xavier and Arnaud Cheritat [2012]. "Quadratic Julia sets with positive area". In: *Ann. of Math. (2)* 176.2, pp. 673–746. https://doi.org/10.4007/annals.2012.176.2.1.

Buzzi, J. [2000]. "Absolutely continuous invariant probability measures for arbitrary expanding piecewise $\mathbb{R}$-analytic mappings of the plane". In: *Ergod. Th. & Dynam. Sys.* 20, pp. 697–708. https://doi.org/10.1017/S0143385700000377.

Carleson, L. [1967]. "On mappings conformal at the boundary". In: *J. Anal. Math.* 19, pp. 1–13. https://doi.org/10.1007/BF02788706.

Carleson, L. and T. W. Gamelin [1993]. *Complex dynamics*. Universitext. Springer-Verlag. https://doi.org/10.1007/978-1-4612-4364-9.

Chenciner, A., J.-M. Gambaudo, and C. Tresser [1984]. "Une remarque sur la structure des endomorphismes de degré 1 du cercle". In: *C. R. Acad. Sci. Paris* 299, pp. 145–148.

Clark, T., E. de Faria, and S. van Strien [2022]. "Asymptotically holomorphic methods for infinitely renormalizable C^r unimodal maps". In: *Ergodic Theory and Dynamical Systems*, pp. 1–49. https://doi.org/10.1017/etds.2022.72.

Clark, T., S. van Strien, and S. Trejo [2017]. "Complex bounds for real maps". In: *Comm. Math. Phys.* 355, pp. 1001–1119. https://doi.org/10.1007/s00220-017-2958-y.

Clark, T. and S. Trejo [2020]. "The boundary of chaos for interval mappings". In: *Proc. London Math. Soc.* 121, pp. 1427–1467. https://doi.org/10.1112/plms.12372.

Cornfeld, I. P., S. V. Fomin, and Ya. G. Sinaĭ [1982]. *Ergodic theory*. Vol. 245. Grundlehren der Mathematischen Wissenschaften. Springer-Verlag, New York.

Coullet, P. and C. Tresser [1978]. "Itérations d'endomorphismes et groupe de renormalisation". In: *J. Phys. Colloque C* 539, pp. C5–25.

Crovisier, S., P. Guarino, and L. Palmisano [2019]. "Ergodic properties of bimodal circle maps". In: *Ergod. Th. & Dynam. Sys.* 39, pp. 1462–1500. https://doi.org/10.1017/etds.2017.80.

Cvitanović, P., G. H. Gunaratne, and M. J. Vinson [1990]. "On the mode-locking universality for critical circle maps". In: *Nonlinearity* 3, pp. 873–885. https://doi.org/10.1088/0951-7715/3/3/015.

Cvitanović, P., B. Shraiman, and B. Söderberg [1985]. "Scaling laws for mode locking in circle maps". In: *Physica Scrypta* 32, pp. 263–270. https://doi.org/10.1088/0031-8949/32/4/003.

Davie, A. [1996]. "Period doubling for $C^{2+\epsilon}$ mappings". In: *Comm. Math. Phys.* 176, pp. 261–272. https://doi.org/10.1007/BF02099549.

Denjoy, A. [1932]. "Sur les courbes définies par les équations différentielles à la surface du tore". In: *J. Math. Pure et Appl.* 11, pp. 333–375.

Dixon, T. W., T. Gherghetta, and B. G. Kenny [1996]. "Universality in the quasiperiodic route to chaos". In: *Chaos* 6, pp. 32–42. https://doi.org/10.1063/1.166155.

Douady, A. [1987]. "Disques de Siegel et anneaux de Herman". In: *Sém. Bourbaki 1986/87, Astérisque, 1986/87*, pp. 151–172.

Douady, A. [1994]. "Does a Julia set depend continuously on the polynomial?" In: *Complex dynamical systems: The Mathematics Behind the Mandelbrot and Julia Sets*. Ed. by Robert L. Devaney. Vol. 49. Proc. Sympos. Appl. Math. Providence, RI: American Mathematical Scociety, pp. 91–138. https://doi.org/10.1090/psapm/049.

Douady, A. and J. Hubbard [1985]. "On the dynamics of polynomial-like mappings". In: *Ann. Sci. Ec. Norm. Sup.* 18, pp. 287–343. https://doi.org/10.24033/asens.1491.

Douady, R. and J.-C. Yoccoz [1999]. "Nombre de rotation des difféomorphismes du cercle et mesures automorphes (Rotation number of diffeomorphisms of the circle and automorphic measures)". In: *Regul. Chaotic Dyn.* 4.4, pp. 3–24. https://doi.org/10.1070/rd1999v004n04ABEH000129.

Eliasson, H., B. Fayad, and R. Krikorian [2018]. "J.-C. Yoccoz and the theory of circle diffeomorphisms". English. In: *Jean-Christophe Yoccoz. Conference on dynamical systems in memory of Jean-Christophe Yoccoz, Collège de France, Paris, France, May 29–31, 2017*. Paris: Société Mathématique de France (SMF), pp. 55–66.

Epstein, A., L. Keen, and C. Tresser [1995]. "The set of maps $F_{a,b} : x \mapsto x+a+\frac{b}{2\pi}\sin(2\pi x)$ with any given rotation interval is contractible". In: *Comm. Math. Phys.* 173, pp. 313–333. https://doi.org/10.1007/BF02101236.

Estevez, G. and E. de Faria [2018]. "Real bounds and quasisymmetric rigidity of multicritical circle maps". In: *Trans. Amer. Math. Soc.* 370, pp. 5583–5616. https://doi.org/10.1090/tran/7177.

Estevez, G., E. de Faria, and P. Guarino [2018]. "Beau bounds for multicritical circle maps". In: *Indag. Math.* 29, pp. 842–859. https://doi.org/10.1016/j.indag.2017.12.007.

Estevez, G. and P. Guarino [2023]. "Renormalization of bicritical circle maps". In: *Arnold Math. J.* 9.1, pp. 69–104. https://doi.org/10.1007/s40598-022-00199-x.

Estevez, G., D. Smania, and M. Yampolsky [2022]. "Renormalization of analytic multicritical circle maps with bounded type rotation numbers". In: *Bulletin of the Brazilian Mathematical Society, New Series* 53.3, pp. 1053–1071. https://doi.org/10.1007/s00574-022-00295-8.

Farb, B. and D. Margalit [2012]. *A Primer on Mapping Class Groups*. Vol. 49. Princeton Mathematical Series. Princeton University Press, Princeton, NJ, pp. xiv+472.

de Faria, E. [1992]. "Proof of universality for critical circle mappings". PhD thesis. CUNY. URL: https://www.proquest.com/docview/303994168.

de Faria, E. [1996]. "Quasi-symmetric distortion and rigidity of expanding endomorphisms of S^1". In: *Proc. Amer. Math. Soc.* 124, pp. 1949–1957. https://doi.org/10.1090/S0002-9939-96-03218-2.

de Faria, E. [1998]. "On conformal distortion and Sullivan's sector theorem". In: *Proc. Amer. Math. Soc.* 126, pp. 67–74. https://doi.org/10.1090/S0002-9939-98-04487-6.

de Faria, E. [1999]. "Asymptotic rigidity of scaling ratios for critical circle mappings". In: *Ergod. Th. & Dynam. Sys.* 19, pp. 995–1035. https://doi.org/10.1017/S0143385799133959.

de Faria, E. [n.d.]. "Quasi-symmetry and absolute continuity: A note on a theorem of Carleson". unpublished manuscript.

de Faria, E. and P. Guarino [2016]. "Real bounds and Lyapunov exponents". In: *Disc. and Cont. Dyn. Sys. A* 36, pp. 1957–1982. https://doi.org/10.3934/dcds.2016.36.1957.

de Faria, E. and P. Guarino [2021]. "There are no σ-finite absolutely continuous invariant measures for multicritical circle maps". In: *Nonlinearity* 34, pp. 6727–6749. https://doi.org/10.1088/1361-6544/ac1a02.

de Faria, E. and P. Guarino [2022a]. "Dynamics of multicritical circle maps". In: *São Paulo J. Math. Sci.* 16.1, pp. 340–395. https://doi.org/10.1007/s40863-021-00236-1.

de Faria, E. and P. Guarino [2022b]. "Quasisymmetric orbit-flexibility of multicritical circle maps". In: *Ergodic Theory Dynam. Systems* 42.11, pp. 3271–3310. https://doi.org/10.1017/etds.2021.104.

de Faria, E., P. Guarino, and B. Nussenzveig [2024]. "Automorphic measures and invariant distributions for circle dynamics". In: *Mathematische Zeitschrift* 306.2, pp. 1–34. https://doi.org/10.1007/s00209-023-03427-y.

de Faria, E. and W. de Melo [1999]. "Rigidity of critical circle mappings I". In: *J. Eur. Math. Soc.* 1, pp. 339–392. https://doi.org/10.1007/s100970050011.

de Faria, E. and W. de Melo [2000]. "Rigidity of critical circle mappings II". In: *J. Amer. Math. Soc.* 13, pp. 343–370. https://doi.org/10.1090/S0894-0347-99-00324-0.

de Faria, E. and W. de Melo [2008]. *Mathematical tools for one-dimensional dynamics*. Cambridge University Press.

de Faria, E., W. de Melo, and A. Pinto [2006]. "Global hyperbolicity of renormalization for C^r unimodal mappings". In: *Annals of Math.* 164, pp. 731–824. https://doi.org/10.4007/annals.2006.164.731.

de Faria, E. and S. van Strien [2023]. "Dennis Sullivan's Work on Dynamics". In: *The Abel Prize 2018–2022*. The Abel Prize. Springer Cham.

de Faria, E. and C. Tresser [2014]. "On Sloane's persistence problem". In: *Exp. Math.* 23, pp. 363–382. https://doi.org/10.1080/10586458.2014.910849.

Feigenbaum, M. [1978]. "Qualitative universality for a class of nonlinear transformations". In: *J. Statist. Phys.* 19, pp. 25–52. https://doi.org/10.1007/BF01020332.

Feigenbaum, M., L. Kadanoff, and S. Shenker [1982]. "Quasiperiodicity in dissipative systems: A renormalization group analysis". In: *Physica* 5D, pp. 370–386.

Gaidashev, D. and I. Gorbovickis [2021]. "Complex *a priori* bounds for Lorenz maps". In: *Nonlinearity* 34.3, pp. 1263–1287. https://doi.org/10.1088/1361-6544/abbe64.

Gardiner, F. [1987]. *Teichmüller theory and quadratic differentials*. Pure and Applied Mathematics (New York). A Wiley-Interscience Publication. John Wiley & Sons, Inc., New York, pp. xviii+236.

Gardiner, F. and N. Lakic [2000]. *Quasiconformal Teichmuller theory*. Vol. 76. Mathematical Surveys and Monographs. American Mathematical Society, Providence, RI, pp. xx+372. https://doi.org/10.1090/surv/076.

Ghys, E. [1999]. "Laminations par surfaces de Riemann". In: *Dynamique et géométrie complexes (Lyon, 1997)*. Vol. 8. Panor. Synthèses. Soc. Math. France, pp. 49–95.

Goncharuk, N. and M. Yampolsky [2023]. *Renormalization of circle maps and smoothness of Arnold tongues*. arXiv: 2307.15195 [math.DS].

Good, I. J. [1941]. "The fractional dimensional theory of continued fractions". In: 37.3, pp. 199–228. https://doi.org/10.1017/S030500410002171X.

Gorbovickis, I. and M. Yampolsky [2018]. "Renormalization for Unimodal Maps with Non-integer Exponents". In: *Arnold Math J.* 4, pp. 179–191. https://doi.org/10.1007/s40598-018-0089-y.

Gorbovickis, I. and M. Yampolsky [2020]. "Rigidity, universality, and hyperbolicity of renormalization for critical circle maps with non-integer exponents". In: *Ergod. Th. & Dynam. Sys.* 40, pp. 1282–1334. https://doi.org/10.1017/etds.2018.82.

Gorbovickis, I. and M. Yampolsky [2021]. Rigidity of analytic and smooth bi-cubic multicritical circle maps with bounded type rotation numbers. arXiv: 2112.05845 [math.DS].

Gottschalk, W. and G. Hedlund [1955]. *Topological Dynamics*. Vol. 36. AMS Colloquium Publications. Providence, RI: American Mathematical Society.

Graczyk, J., D. Sands, and G. Świątek [2005]. "Decay of geometry for unimodal maps: negative Schwarzian case". In: *Annals of Math.* 161, pp. 613–677. https://doi.org/10.4007/annals.2005.161.613.

Graczyk, J. and G. Świątek [1993]. "Singular measures in circle dynamics". In: *Comm. Math. Phys.* 157, pp. 213–230. https://doi.org/10.1007/BF02099758.

Graczyk, J. and G. Świątek. [1996]. "Critical circle maps near bifurcation". In: *Comm. Math. Phys.* 176, pp. 227–260. https://doi.org/10.1007/BF02099548.

Guarino, P. [2012]. "Rigidity conjecture for C^3 critical circle maps". PhD thesis. IMPA.

Guarino, P., M. Martens, and W. de Melo [2018]. "Rigidity of critical circle maps". In: *Duke Math. J.* 167, pp. 2125–2188. https://doi.org/10.1215/00127094-2018-0017.

Guarino, P. and W. de Melo [2017]. "Rigidity of smooth critical circle maps". In: *J. Eur. Math. Soc.* 19, pp. 1729–1783. https://doi.org/10.4171/JEMS/704.

Guillotin-Plantard, N. and R. Schott [2006]. *Dynamic Random Walks: Theory and Applications*. Amsterdam: Elsevier.

Hall, G. R. [1981]. "A C^∞ Denjoy counterexample". In: *Ergod. Th. & Dynam. Sys.* 1, pp. 261–272. https://doi.org/10.1017/S0143385700001243.

Halmos, P. [1956]. *Lectures on Ergodic Theory*. Chelsea Publishing Company.

Herman, M. [1978]. "Résultats récents sur la conjugaison différentiable". In: *Proceedings of the International Congress of Mathematicians (Helsinki)*, pp. 811–820.

Herman, M. [1979]. "Sur la conjugaison différentiable des difféomorphismes du cercle à des rotations". In: *Publ. Math. IHES* 49, pp. 5–234. https://doi.org/10.1007/BF02684798.

Herman, M. [1988]. "Conjugaison quasi-simétrique des homéomorphismes du cercle à des rotations". Original by Michel Herman here https://www.math.kyotou.ac.jp/~mitsu/Herman/qsconj1/index.html and translation to English https://www.math.univ-toulouse.fr/~cheritat/Herman/Herman_man_1_eng.pdf. URL: https://www.math.univtoulouse.fr/~cheritat/Herman/Herman_man_1.pdf.

Hu, J. and D. Sullivan [1997]. "Topological conjugacy of circle diffeomorphisms". In: *Ergod. Th. & Dynam. Sys.* 17, pp. 173–186. https://doi.org/10.1017/S0143385797061002.

Hubbard, J. [2006]. *Teichmüller Theory and Applications to Geometry, Topology and Dynamics, Volume 1: Teichmüller Theory*. Matrix Editions.

Iosifescu, M. and C. Kraaikamp [2002]. *Metrical theory of continued fractions*. Vol. 547. Mathematics and its Applications. Kluwer Academic Publishers, Dordrecht, pp. xx+383.

Ji, H. [2022]. "Note on Lyapunov Exponent for Critical Circle Maps". In: *Bulletin of the Brazilian Mathematical Society, New Series* 53, pp. 1305–1315. https://doi.org/10.1007/s00574-022-00305-9.

Kadanoff, L. and S. Shenker [1982]. "Critical behavior of a KAM surface: I. Empirical results". In: *J. Statist. Phys.* 27, pp. 631–656. https://doi.org/10.1007/BF01013439.

Katok, A. and E. A. Robinson Jr. [2001]. "Cocycles, cohomology and combinatorial constructions in ergodic theory". In: *Smooth ergodic theory and its applications. Proceedings of the AMS summer research institute, Seattle, WA, USA, July 26–August 13, 1999.* Providence, RI: American Mathematical Society (AMS), pp. 107–173.

Katznelson, Y. [1977]. "Sigma-finite invariant measures for smooth mappings of the circle". In: *J. d'Analyse Math.* 31, pp. 1–18. https://doi.org/10.1007/BF02813295.

Katznelson, Y. and D. Ornstein [1989a]. "The absolute continuity of the conjugation of certain diffeomorphisms of the circle". In: *Ergod. Th. & Dynam. Sys.* 9, pp. 681–690. https://doi.org/10.1017/S0143385700005289.

Katznelson, Y. and D. Ornstein [1989b]. "The differentiability of the conjugation of certain diffeomorphisms of the circle". In: *Ergod. Th. & Dynam. Sys.* 9, pp. 643–680. https://doi.org/10.1017/S0143385700005277.

Keen, L. [1988]. "Dynamics of holomorphic self-maps of $\mathbb{C}^*$". In: *Holomorphic Functions and Moduli I*. Ed. by D. Drasin et al. Vol. 10. MSRI Publications. Springer, pp. 9–30.

Khanin, K. [1991]. "Universal estimates for critical circle mappings". In: *Chaos* 1, pp. 181–186. https://doi.org/10.1063/1.165826.

Khanin, K. and Y. Sinai [1987]. "A new proof of M. Herman's theorem". In: *Commun. Math. Phys.* 112, pp. 89–101. https://doi.org/10.1007/BF01217681.

Khanin, K. and A. Teplinsky [2007]. "Robust rigidity for circle diffeomorphisms with singularities". In: *Invent. Math.* 169, pp. 193–218. https://doi.org/10.1007/s00222-007-0047-0.

Khanin, K. and A. Teplinsky [2009]. "Herman's theory revisited". In: *Invent. Math.* 178, pp. 333–344. https://doi.org/10.1007/s00222-009-0200-z.

Khinchin, A. Ya. [1997]. *Continued fractions.* (reprint of the 1964 edition). Dover Publications, Inc.

Khmelev, D. and M. Yampolsky [2006]. "The rigidity problem for analytic critical circle maps". In: *Mosc. Math. J.* 6, pp. 317–351. https://doi.org/10.17323/1609-4514-2006-6-2-317-351.

Kolmogorov, A. N. [1954]. "On conservation of conditionally periodic motions under small perturbations of the hamiltonian". In: *Dokl. Akad. Nauk, SSSR* 98, pp. 527–530.

Kra, B. and J. Schmeling [2002]. "Diophantine classes, dimension and Denjoy maps". In: *Acta Arithmetica* 105, pp. 323–340. https://doi.org/10.4064/aa105-4-2.

Lanford, O. [1987]. "Renormalization group methods for critical circle mappings with general rotation number". In: *VIIIth Int. Congress on Mathematical Physics (Marseille, 1986).* World Scientific, Singapore, pp. 532–536.

Lanford, O. [1988]. "Renormalization group methods for circle mappings". In: *Nonlinear Evolution and Chaotic Phenomena.* Vol. 176. NATO Adv. Sci. Inst. Series (Ser. B: Phys.) Springer, pp. 25–36.

Lanford, O. and M. Yampolsky [2014]. *Fixed point of the parabolic renormalization operator.* SpringerBriefs in Mathematics. Springer, Cham, pp. viii+111. https://doi.org/10.1007/978-3-319-11707-2.

Lang, S. [1995]. *Introduction to Diophantine Approximations.* (new expanded edition). Springer-Verlag.

Lehto, O. and K. I. Virtanen [1973]. *Quasiconformal Mappings in the Plane.* Springer-Verlag.

Liousse, I. [2004]. "PL homeomorphisms of the circle that are piecewise C^1 conjugate to irrational rotations". In: *Bull. Braz. Math. Soc* 35, pp. 269–280. https://doi.org/10.1007/s00574-004-0014-y.

Lyubich, M. [1999]. "Feigenbaum–Coullet–Tresser universality and Milnor's hairiness conjecture". In: *Annals of Math.* 149, pp. 319–420. https://doi.org/10.2307/120968.

Lyubich, M. [2019]. "Teichmüller space of Fibonacci maps". In: *New Trends in One-Dimensional Dynamics*. Ed. by P. Guarino and M. Pacifico. Vol. 285. Springer Proceedings in Mathematics & Statistics. Springer, pp. 221–237.

MacKay, R. S. [1983]. "A renormalisation approach to invariant circles in area-preserving maps". In: *Physica* 7D, pp. 283–300.

MacKay, R. S. [1993]. *Renormalisation in Area-Preserving Maps*. World-Scientific.

Mañé, R. [1987]. *Ergodic theory and differentiable dynamics*. Springer-Verlag, Berlin, pp. xii+317.

Markeley, M. [1970]. "Homeomorphisms of the circle without periodic points". In: *Proc. London Math. Soc.* 20, pp. 688–698.

Martens, M. [1998]. "The periodic points of renormalization". In: *Annals of Math.* 147, pp. 543–584. https://doi.org/10.2307/120959.

Martens, M., L. Palmisano, and B. Winckler [2018]. "The Rigidity Conjecture". In: *Indag. Math.* 29, pp. 825–830. https://doi.org/10.1016/j.indag.2017.08.001.

McDuff, D. [1981]. "C^1-minimal subsets of the circle". In: *Annales de l'Institut Fourier* 31, pp. 177–193. https://doi.org/10.5802/aif.822.

McMullen, C. T. [1994]. *Complex Dynamics and Renormalization*. Vol. 135. Annals of Math. Studies. Princeton University Press.

McMullen, C. T. [1996]. *Renormalization and 3-manifolds which fiber over the circle*. Vol. 142. Annals of Math. Studies. Princeton University Press.

de Melo, W. and C. Pugh [1994]. "The C^1 Brunovsky Hypothesis". In: *Journal of Differential Equations* 113.2, pp. 300–337.

de Melo, W. and S. van Strien [1993]. *One-dimensional dynamics*. Springer-Verlag.

Milnor, J. [2006]. *Dynamics in One Complex Variable*. Vol. 160. Annals of Math. Studies. Princeton University Press.

Misiurewicz, M. [1986]. "Rotation intervals for a class of maps of the real line into itself". In: *Ergod. Th. & Dynam. Sys* 6, pp. 117–132. https://doi.org/10.1017/S0143385700003321.

Moser, J. [1966]. "A rapidly convergent interation method, II". In: *Ann. Scuola Norm. Sup. di Pisa* 20, pp. 499–535.

Navas, A. and M. Triestino [2013]. "On the invariant distributions of C^2 circle diffeomorphisms of irrational rotation number". In: *Math. Z.* 274.1-2, pp. 315–321. https://doi.org/10.1007/s00209-012-1071-3.

Norton, A. [1999]. "Denjoy's theorem with exponents". In: *Proc. Amer. Math. Soc.* 127, pp. 3111–3118.

Norton, A. [2002]. "Denjoy minimal sets are far from affine". In: *Ergodic Theory Dynam. Systems* 22, pp. 1803–1812.

Ostlund, S. D. Rand, J. Sethna and E. Siggia [1983]. "Universal properties of the transition from quasiperiodicity to chaos in dissipative systems". In: *Physica* 8D, pp. 303–342.

Patterson, S. J. [1976]. "The limit set of a Fuchsian group". In: *Acta Math.* 136.3-4, pp. 241–273. https://doi.org/10.1007/BF02392046.

Petersen, C. [2000]. "The Herman–Świątek theorems with applications". In: *The Mandelbrot Set, Theme and Variations*. Ed. by T. Lei. Vol. 274. London Mathematical Society Lecture Note Series. Cambridge University Press, pp. 211–225.

Petersen, C. and S. Zakeri [2004]. "On the Julia set of a typical quadratic polynomial with a Siegel disk". In: *Ann. of Math.* 159, pp. 1–52. https://doi.org/10.4007/annals.2004.159.1.

Portela, A. [2007]. "New examples of Cantor sets in S^1 that are not C^1-minimal". In: *Bull. Braz. Math. Soc.* 38, pp. 623–633.

Portela, A. [2009]. "Regular interval Cantor sets of S1 and minimality". In: *Bull. Braz. Math. Soc.* 40, pp. 53–75.

Przytycki, F. [1993]. "Lyapunov characteristic exponents are nonnegative". In: *Proc. Amer. Math. Soc.* 119, pp. 309–317. https://doi.org/10.1090/S0002-9939-1993-1186141-9.

Rand, D. [1987]. "Universality and renormalisation in dynamical systems". In: *New Directions in Dynamical Systems*. Ed. by T. Bedford and J. W. Swift. Cambridge University Press.

Rand, D. [1988]. "Global phase space universality, smooth conjugacies and renormalisation: I. The $C^{1+\alpha}$ case". In: *Nonlinearity 1*, pp. 181–202. https://doi.org/10.1088/0951-7715/1/1/007.

Rand, D. [1992]. "Existence, non-existence and universal breakdown of dissipative golden invariant tori: I. Golden critical circle maps". In: *Nonlinearity* 5, pp. 639–662. https://doi.org/10.1088/0951-7715/5/3/002.

Shenker, S. [1982]. "Scaling behaviour in a map of a circle onto itself: empirical results". In: *Physica* 5D, pp. 405–411.

Shishikura, M. [1998]. "The Hausdorff dimension of the boundary of the Mandelbrot set and Julia sets". In: *Annals of Math.* 147, pp. 225–267. https://doi.org/10.2307/121009.

Shishikura, M. [2000]. "'Bifurcation of parabolic fixed points". In: *The Mandelbrot Set, Theme and Variations*. Ed. by T. Lei. Vol. 274. London Mathematical Society Lecture Note Series. Cambridge University Press, pp. 325–364.

Siegel, C. L. [1942]. "Iteration of analytic functions". In: *Ann. of Math.* 43, pp. 607–612. https://doi.org/10.2307/1968952.

Singer, D. [1978]. "Stable orbits and bifurcations of maps of the interval". In: *SIAM J. Appl. Math.* 35, pp. 260–267. https://doi.org/10.1137/0135020.

Stark, J. [1988]. "Smooth conjugacy and renormalisation for diffeomorphisms of the circle". In: *Nonlinearity* 1, pp. 541–575. https://doi.org/10.1088/0951-7715/1/4/004.

Sullivan, D. [1979]. "The density at infinity of a discrete group of hyperbolic motions". In: *Inst. Hautes Etudes Sci. Publ. Math.* 50, pp. 171–202. URL: http://www.numdam.org/item?id=PMIHES_1979__50__171_0.

Sullivan, D. [1983]. "Conformal dynamical systems". In: *Geometric dynamics (Rio de Janeiro, 1981)*. Vol. 1007. Lecture Notes in Math. Springer, Berlin, pp. 725–752. https://doi.org/10.1007/BFb0061443.

Sullivan, D. [1985]. "Quasiconformal homeomorphisms and Dynamics I. Fatou–Julia problem on wandering domains". In: *Ann. of Math.* 122, pp. 401–418. https://doi.org/10.2307/1971308.

Sullivan, D. [1986]. "Quasiconformal homeomorphisms in dynamics, topology, and geometry". In: *Proceedings of the International Congress of Mathematicians (Berkeley)*, pp. 1216–1228.

Sullivan, D. [1992]. "Bounds, quadratic differentials, and renormalization conjectures". In: *AMS Centennial Publications, Vol. II*, pp. 417–466.

Świątek, G. [1988]. "Rational rotation numbers for maps of the circle". In: *Comm. Math. Phys.* 119, pp. 109–128. https://doi.org/10.1007/BF01218263.

Świątek, G. [1992]. "One-dimensional maps and Poincaré metric". In: *Nonlinearity* 5, pp. 81–108. https://doi.org/10.1088/0951-7715/5/1/003.

Trujillo, F. [2020]. "Hausdorff dimension of invariant measures of multicritical circle maps". In: *Ann. Henri Poincaré* 21, pp. 2861–2875. https://doi.org/10.1007/s00023-020-00940-2.

Tsujii, M. [2001]. "Absolutely continuous invariant measures for expanding piecewise linear maps". In: *Invent. Math.* 143, pp. 349–373. https://doi.org/10.1007/PL00005797.

Vieira, A. M. S. [2015]. "Pares holomorfos e a família de Arnol'd generalizada". PhD thesis. IME-USP.

Voutaz, E. [2006]. "Hyperbolicity of the renormalization operator for critical C^r circle mappings". In: *Ergod. Th. & Dynam. Sys.* 26, pp. 585–618. https://doi.org/10.1017/S0143385705000453.

Walters, Peter [1982]. *An introduction to ergodic theory*. Vol. 79. Graduate Texts in Mathematics. Springer-Verlag, New York-Berlin, pp. ix+250.

Yampolsky, M. [1999]. "Complex bounds for renormalization of critical circle maps". In: *Ergod. Th. & Dynam. Sys.* 19, pp. 227–257. https://doi.org/10.1017/S0143385799120947.

Yampolsky, M. [2001]. "The attractor of renormalization and rigidity of towers of critical circle maps". In: *Comm. Math. Phys.* 218, pp. 537–568. https://doi.org/10.1007/PL00005561.

Yampolsky, M. [2002]. "Hyperbolicity of renormalization of critical circle maps". In: *Publ. Math. IHES* 96, pp. 1–41. https://doi.org/10.1007/s10240-003-0007-1.

Yampolsky, M. [2003]. "Renormalization horseshoe for critical circle maps". In: *Comm. Math. Phys.* 240, pp. 75–96. https://doi.org/10.1007/s00220-003-0891-8.

Yampolsky, M. [2017]. "Renormalization of unicritical analytic circle maps". In: *C. R. Math. Rep. Acad. Sci. Canada* 39, pp. 77–89.

Yampolsky, M. [2019]. "Renormalization of bi-cubic circle maps". In: *C. R. Math. Rep. Acad. Sci. Canada* 41, pp. 57–83.

Yoccoz, J.-C. [1984a]. “Conjugaison différentiable des difféomorphismes du cercle dont le nombre de rotation vérifie une condition diophantienne”. In: *Ann. Scient. Éc. Norm. Sup.* 17, pp. 333–359. https://doi.org/10.24033/asens.1475.

Yoccoz, J.-C. [1984b]. “Il n’y a pas de contre-exemple de Denjoy analytique”. In: *C. R. Acad. Sc. Paris* 298, pp. 141–144.

Yoccoz, J.-C. [1995a]. “Centralisateurs et conjugaison différentiable des difféomorphismes du cercle”. In: *Astérisque* 231, pp. 89–242.

Yoccoz, J.-C. [1995b]. “Théorème de Siegel, nombres de Bruno et polynômes quadratiques”. In: *Astérisque* 231, pp. 3–88.

Yoccoz, J.-C. [2002]. “Analytic linearization of circle diffeomorphisms”. In: *Dynamical systems and small divisors*. Springer, pp. 125–173.

Yoccoz, J.-C. [2006]. “Continued fraction algorithms for interval exchange maps: an introduction”. In: *Frontiers in Number Theory, Physics and Geometry, vol. 1: On random matrices, zeta functions and dynamical systems*. Springer-Verlag.

Zakeri, S. [1999]. “Dynamics of cubic Siegel polynomials”. In: *Comm. Math. Phys.* 206, pp. 185–233. https://doi.org/10.1007/s002200050702.

Index

A
Admissible pair, 242
Almost parabolic map, 166, 260
  balanced decomposition, 170
Almost Schwarz inclusion, 336
Ancestor, 229, 230
Annular Riemann surface, 291
Arnold's counterexamples, 96
Arnold tongue, 91
Asymptotically holomorphic map, 332, 333
Atom
  critical, 176
  level, 267
  long, 137
  regular, 180
  saddle-node, 180
  short, 137
Automorphic measure, 64, 217
  ergodicity, 71
  existence, 70
  uniqueness, 71

B
Beau bounds, 150
Beltrami
  equation, 282
  path, 280, 363
Bi-critical circle map, 252
Bi-Schwarzian, 112
  chain rule, 112, 124
Blaschke product, 126, 165
Bounded geometry, 228
Bowtie, 365
Bridge, 178
  balanced decomposition, 180
  primary, 180
  saddle-node, 180
Brjuno
  condition, 97, 98, 101
  number, 98

C
Carathéodory topology, 391
C^1 bounds, 144
Centralizer, 227
Circle homeomorphism
  lift, 25
  quasisymmetrically orbit-rigid, 224
  (p, q)-stable, 93
Closest returns, 6
Coboundary, 66
Cocycle identity, 75, 191
Cohomological equation, 66
Cohomology group, 67
Collet–Eckmann condition, 209
Commuting pair
  critical, 254
    K-controlled, 293
  holomorphic, 365
  infinitely renormalizable, 258
Comparability, 143
Complex bounds, 361, 373
Conformal measure, 64
Conformal modulus, 280

E. de Faria, P. Guarino, *Dynamics of Circle Mappings*, IMPA Monographs,
https://doi.org/10.1007/978-3-031-67495-2

Continued fractions, 7
  convergents, 10
  partial quotients, 10
Coordinates
  Fatou, 390
Critical
  spots, 178, 195
  times, 178
Critical point
  non-flat, 114
  order, 114
  type, 114
Cross-ratio distortion, 111
  chain rule, 111
  weakly bounded, 158
Cross-ratios, 105
  inequality, 116, 131, 186, 195
Cylinder
  Écalle–Voronin, 390
  map, 393
  renormalization, 388

D
Deep point, 385
Denjoy
  Cantor set, 59
    Hausdorff dimension, 60
  example, 49, 55, 72, 219
  sequence, 55
Denjoy–Koksma inequality, 46, 68, 245
Devil staircase, 89
Diophantine
  class, 60
  number, 77, 101
  order, 98, 101
  type, 78
Distribution
  finite-order, 65
  invariant, 65, 66, 218
  Schwartz, 65
Douady coordinates, 391
Dynamical partition, 136
Dynamical system
  holomorphic, 385

E
Écalle–Voronin cylinder, 390
Ellipse field
  measurable, 289
Epstein
  class, 391
  map, 374
Equidistribution, 15
  modulo one, 14
  Weyl's criterion, 15
  Weyl's theorem, 14
Ergodicity
  distributional unique, 67
  w.r.t. Lebesgue measure, 61
Euler–Lambert identity, 236
Exponential covering map, 3
Extended lift, 335

F
Faa-di-Bruno's formula, 270
Family
  Arnold, 90, 128
  one-parameter, 89
  standard, 89
Fatou coordinates, 390
Fiber map, 232
Fine grid, 159
  isomorphism, 161
Fourier
  coefficients, 81
  series, 80

G
Gauss map, 11, 258
Geometrically equivalent orbits, 223
Glueing, 255
  bidimensional, 344

H
Hausdorff dimension, 60
Herman's invariants, 74
Holomorphic index, 392, 395
Holomorphic map
  entire, 101
Homeomorphism
  circle, 25
    lift, 25
    measurably non-singular, 36
    minimal, 37
    rotation number, 26
    (p, q)-stable, 93
  expansive, 40
  K-quasiconformal, 281
  K-quasisymmetric, 158
  quasiconformal, 158, 280
  quasisymmetric, 158
  quasisymmetric distortion, 158

I

Inequality
  Denjoy–Koksma, 46, 68
  Yoccoz, 196
Interval exchange map, 257
Invariant measure
  σ-finite, 191, 195
  Hausdorff dimension, 187, 219
  purely singular, 187
  singular, 93
Irrational number
  bounded type, 101, 164
  Brjuno, 98
  Diophantine, 77, 78, 98
Isothermal coordinates, 282

K

KAM theory, 77
Katznelson criterion, 191
Koebe's distortion principle, 111, 181, 197, 269
Kolmogorov's strategy, 78

L

Laurent
  coefficients, 100
  series, 100
Lemma
  cancellation, 118
  C^m-approximation, 270
  C^∞ realization, 226
  holomorphic approximation, 285
  Kakutani–Rokhlin, 155
  naive distortion, 43
  Weyl's, 281
Liouville number, 101
Lyapunov exponent, 63, 69, 208, 209

M

Map
  almost parabolic, 166
  holomorphic, 83
    entire, 101
  piecewise affine, 38
  quasiconformal, 279, 289
  smooth, 111
Markov map, 232
Measure
  automorphic, 64, 217
  conformal, 64
Mode-locking
  interval, 90
  phenomenon, 90
Multicritical circle map, 126
  beau bounds, 151
  Lyapunov exponent, 209
  quasisymmetric rigidity, 173
  real a priori bounds, 138
  signature, 126
Multiplicity of intersection, 117

N

Nonlinearity, 43, 386
  chain rule, 44
  Koebe's principle, 108

P

Parabolic bifurcation, 387
Parabolic map, 392
Parabolic renormalization, 388, 389
Partition
  auxiliary, 178
  dynamical, 178
Phase-locking
  interval, 90, 94
  phenomenon, 90
Pigeonhole principle, 4
Plateau, 35, 45
Poincaré
  density, 106
  disk, 336
  length, 106
  metric, 335
  neighborhood trapping, 375
Polar ratio, 147
Pull-back argument, 367

Q

Quadratic-like map, 362
  renormalizable, 363
Quasicircle, 165
Quasiconformal homeomorphism, 280, 281
  analytic definition, 281
  geometric definition, 280
Quasiconformal surgery, 165
Quasisymmetric
  homeomorphism, 158
  orbit-flexibility, 224
  orbit-rigidity, 224
Quasisymmetry, 157
  criterion for, 160

R

Real bounds, 135, 138, 362
Renormalization, 252
  ancestor, 230
  cylinder, 388
  exponential convergence, 329, 356, 386
  hyperbolicity, 387
  parabolic, 388, 389
  Rauzy–Veech, 257
  shadowing property, 330
  trail, 231
Restricted domain, 264
Riemann surface lamination, 364
Rigidity
  conjecture, 251
  counterexamples, 273
  quasisymmetric, 173
  smooth, 249, 250
Rotation
  irrational, 4
  rational, 4
Rotation number, 26
  bounded type, 101, 198
  Diophantine, 101
  even-type, 236
  function, 89

S

Saturation, 386
Schwarzian derivative, 106
  chain rule, 107
  kernel, 107
  minimum principle, 110
Seven-point argument, 130
Signature, 126, 251
Skew product, 232, 233
Small denominators, 77, 81
Spectrum (of a Cantor set), 59
Standard family
  of commuting pairs, 297
Surgery
  saddle-node, 274

T

Teichmüller distance, 363
Theorem
  Arnold's conjugacy, 78, 86
  Avila–Kocsard, 67
  Birkhoff's ergodic, 36, 216
  Carleson's, 185
  Denjoy's, 45
  Douady–Yoccoz, 64
  Gottschalk–Hedlund, 74
  Hahn–Banach, 66, 72
  Herman–Yoccoz, 99
  Katznelson–Ornstein, 99
  Khanin–Teplinsky, 99
  Markeley's, 58
  McDuff's, 59
  measurable Riemann mapping, 280, 282
  monotone convergence, 190
  Morrey's, 282
  Navas–Triestino, 67
  Schauder–Tychonoff, 70
  Siegel's, 77
  Stone–Weierstrass, 15
  straightening, 362
  Trujillo's, 219
  Weyl's equidistribution, 14
  Yoccoz's, 129, 250
  Yoccoz's global conjugacy, 100
  Yoccoz's local conjugacy, 98
  Zakeri's, 127
Topology
  geometric, 385
  Hausdorff, 385
  weak*, 70, 72
Transfer isomorphism, 390
Transformation
  fractional linear, 7
  Möbius, 7
Translation number, 26
Twisting, 386
  uniform, 386

U

Unit circle, 3

V

Variation
  bounded, 46, 69
  total, 46

Y

Yoccoz's inequality, 167, 177, 196, 262

Z

Zygmund class, 49

SPRINGER NATURE

GPSR Compliance

The European Union's (EU) General Product Safety Regulation (GPSR) is a set of rules that requires consumer products to be safe and our obligations to ensure this.

If you have any concerns about our products, you can contact us on ProductSafety@springernature.com

In case Publisher is established outside the EU, the EU authorized representative is:

Springer Nature Customer Service Center GmbH
Europaplatz 3
69115 Heidelberg, Germany

The manufacturer's authorised representative in the EU is Springer Nature Customer Service Centre GmbH, Europaplatz 3, 69115 Heidelberg, Germany. If you have any concerns regarding our products, please contact ProductSafety@springernature.com

Printed and bound by CPI Group (UK) Ltd, Croydon, CR0 4YY
28/07/2026
02179207-0005